Derivatives

1. $\dfrac{d(au)}{dx} = a\dfrac{du}{dx}$

2. $\dfrac{d(u + v - w)}{dx} = \dfrac{du}{dx} + \dfrac{dv}{dx} - \dfrac{dw}{dx}$

3. $\dfrac{d(uv)}{dx} = u\dfrac{dv}{dx} + v\dfrac{du}{dx}$

4. $\dfrac{d(u/v)}{dx} = \dfrac{v(du/dx) - u(dv/dx)}{v^2}$

5. $\dfrac{d(u^n)}{dx} = nu^{n-1}\dfrac{du}{dx}$

6. $\dfrac{d(u^v)}{dx} = vu^{v-1}\dfrac{du}{dx} + u^v(\ln u)\dfrac{dv}{dx}$

7. $\dfrac{d(e^u)}{dx} = e^u\dfrac{du}{dx}$

8. $\dfrac{d(e^{au})}{dx} = ae^{au}\dfrac{du}{dx}$

9. $\dfrac{da^u}{dx} = a^u(\ln a)\dfrac{du}{dx}$

10. $\dfrac{d(\ln u)}{dx} = \dfrac{1}{u}\dfrac{du}{dx}$

11. $\dfrac{d(\log_a u)}{dx} = \dfrac{1}{u(\ln a)}\dfrac{du}{dx}$

12. $\dfrac{d\sin u}{dx} = \cos u\dfrac{du}{dx}$

13. $\dfrac{d\cos u}{dx} = -\sin u\dfrac{du}{dx}$

14. $\dfrac{d\tan u}{dx} = \sec^2 u\dfrac{du}{dx}$

15. $\dfrac{d\cot u}{dx} = -\csc^2 u\dfrac{du}{dx}$

16. $\dfrac{d\sec u}{dx} = \tan u \sec u\dfrac{du}{dx}$

17. $\dfrac{d\csc u}{dx} = -(\cot u)(\csc u)\dfrac{du}{dx}$

18. $\dfrac{d\sin^{-1}u}{dx} = \dfrac{1}{\sqrt{1 - u^2}}\dfrac{du}{dx}$

19. $\dfrac{d\cos^{-1}u}{dx} = \dfrac{-1}{\sqrt{1 - u^2}}\dfrac{du}{dx}$

20. $\dfrac{d\tan^{-1}u}{dx} = \dfrac{1}{1 + u^2}\dfrac{du}{dx}$

21. $\dfrac{d\cot^{-1}u}{dx} =$

22. $\dfrac{d\sec}{dx} =$

23. $\dfrac{d\csc^{-1}u}{dx} = \dfrac{}{u\sqrt{u^2 - 1}}\dfrac{du}{dx}$

24. $\dfrac{d\sinh u}{dx} = \cosh u\dfrac{du}{dx}$

25. $\dfrac{d\cosh u}{dx} = \sinh u\dfrac{du}{dx}$

26. $\dfrac{d\tanh u}{dx} = \operatorname{sech}^2 u\dfrac{du}{dx}$

27. $\dfrac{d\coth u}{dx} = -(\operatorname{csch}^2 u)\dfrac{du}{dx}$

28. $\dfrac{d\operatorname{sech} u}{dx} = -(\operatorname{sech} u)(\tanh u)\dfrac{du}{dx}$

29. $\dfrac{d\operatorname{csch} u}{dx} = -(\operatorname{csch} u)(\coth u)\dfrac{du}{dx}$

30. $\dfrac{d\sinh^{-1}u}{dx} = \dfrac{1}{\sqrt{1 + u^2}}\dfrac{du}{dx}$

31. $\dfrac{d\cosh^{-1}u}{dx} = \dfrac{1}{\sqrt{u^2 - 1}}\dfrac{du}{dx}$

32. $\dfrac{d\tanh^{-1}u}{dx} = \dfrac{1}{1 - u^2}\dfrac{du}{dx}$

33. $\dfrac{d\coth^{-1}u}{dx} = \dfrac{1}{1 - u^2}\dfrac{du}{dx}$

34. $\dfrac{d\operatorname{sech}^{-1}u}{dx} = \dfrac{-1}{u\sqrt{1 - u^2}}\dfrac{du}{dx}$

35. $\dfrac{d\operatorname{csch}^{-1}u}{dx} = \dfrac{-1}{|u|\sqrt{1 + u^2}}\dfrac{du}{dx}$

Continued on overleaf.

A Brief Table of Integrals

(An arbitrary constant may be added to each integral.)

1. $\int x^n \, dx = \dfrac{1}{n+1} x^{n+1} \ (n \neq -1)$

2. $\int \dfrac{1}{x} \, dx = \ln|x|$

3. $\int e^x \, dx = e^x$

4. $\int a^x \, dx = \dfrac{a^x}{\ln a}$

5. $\int \sin x \, dx = -\cos x$

6. $\int \cos x \, dx = \sin x$

7. $\int \tan x \, dx = -\ln|\cos x|$

8. $\int \cot x \, dx = \ln|\sin x|$

9. $\int \sec x \, dx = \ln|\sec x + \tan x|$
   $$= \ln\left|\tan\left(\dfrac{1}{2}x + \dfrac{1}{4}\pi\right)\right|$$

10. $\int \csc x \, dx = \ln|\csc x - \cot x|$
    $$= \ln\left|\tan \dfrac{1}{2}x\right|$$

11. $\int \sin^{-1} \dfrac{x}{a} \, dx = x \sin^{-1} \dfrac{x}{a} + \sqrt{a^2 - x^2} \quad (a > 0)$

12. $\int \cos^{-1} \dfrac{x}{a} \, dx = x \cos^{-1} \dfrac{x}{a} - \sqrt{a^2 - x^2} \quad (a > 0)$

13. $\int \tan^{-1} \dfrac{x}{a} \, dx = x \tan^{-1} \dfrac{x}{a} - \dfrac{a}{2} \ln(a^2 + x^2) \quad (a > 0)$

14. $\int \sin^2 mx \, dx = \dfrac{1}{2m}(mx - \sin mx \cos mx)$

15. $\int \cos^2 mx \, dx = \dfrac{1}{2m}(mx + \sin mx \cos mx)$

16. $\int \sec^2 x \, dx = \tan x$

17. $\int \csc^2 x \, dx = -\cot x$

18. $\int \sin^n x \, dx = -\dfrac{\sin^{n-1}x \cos x}{n} + \dfrac{n-1}{n} \int \sin^{n-2}x \, dx$

19. $\int \cos^n x \, dx = \dfrac{\cos^{n-1}x \sin x}{n} + \dfrac{n-1}{n} \int \cos^{n-2}x \, dx$

20. $\int \tan^n x \, dx = \dfrac{\tan^{n-1}x}{n-1} - \int \tan^{n-2}x \, dx \quad (n \neq 1)$

21. $\int \cot^n x \, dx = -\dfrac{\cot^{n-1}x}{n-1} - \int \cot^{n-2}x \, dx \quad (n \neq 1)$

22. $\int \sec^n x \, dx = \dfrac{\tan x \sec^{n-2}x}{n-1} + \dfrac{n-2}{n-1} \int \sec^{n-2}x \, dx \quad (n \neq 1)$

23. $\int \csc^n x \, dx = -\dfrac{\cot x \csc^{n-2}x}{n-1} + \dfrac{n-2}{n-1} \int \csc^{n-2}x \, dx \quad (n \neq 1)$

24. $\int \sinh x \, dx = \cosh x$

25. $\int \cosh x \, dx = \sinh x$

26. $\int \tanh x \, dx = \ln|\cosh x|$

27. $\int \coth x \, dx = \ln|\sinh x|$

28. $\int \text{sech}\, x \, dx = \tan^{-1}(\sinh x)$

This table is continued on the endpapers at the back.

Undergraduate Texts in Mathematics

Apostol: Introduction to Analytic
Number Theory.

Armstrong: Basic Topology.

Bak/Newman: Complex Analysis.

Banchoff/Wermer: Linear Algebra
Through Geometry.

Childs: A Concrete Introduction to
Higher Algebra.

Chung: Elementary Probability Theory
with Stochastic Processes.

Croom: Basic Concepts of Algebraic
Topology.

Curtis: Linear Algebra:
An Introductory Approach.

Dixmier: General Topology.

Driver: Why Math?

Ebbinghaus/Flum/Thomas
Mathematical Logic.

Fischer: Intermediate Real Analysis.

Fleming: Functions of Several Variables.
Second edition.

Foulds: Optimization Techniques: An
Introduction.

Foulds: Combination Optimization for
Undergraduates.

Franklin: Methods of Mathematical
Economics.

Halmos: Finite-Dimensional Vector
Spaces. Second edition.

Halmos: Naive Set Theory.

Iooss/Joseph: Elementary Stability and
Bifurcation Theory.

Jänich: Topology.

Kemeny/Snell: Finite Markov Chains.

Klambauer: Aspects of Calculus.

Lang: Undergraduate Analysis.

Lang: A First Course in Calculus. Fifth
Edition.

Lang: Calculus of One Variable. Fifth Edition.

Lang: Introduction to Linear Algebra. Second
Edition.

Lax/Burstein/Lax: Calculus with
Applications and Computing, Volume 1.
Corrected Second Printing.

LeCuyer: College Mathematics with APL.

Lidl/Pilz: Applied Abstract Algebra.

Macki/Strauss: Introduction to Optimal
Control Theory.

Malitz: Introduction to Mathematical
Logic.

Marsden/Weinstein: Calculus I, II, III.
Second edition.

Jerrold Marsden
Alan Weinstein

CALCULUS I

Second Edition

With 528 Figures

Springer-Verlag New York Berlin Heidelberg Tokyo

Jerrold Marsden
Department of Mathematics
University of California
Berkeley, California 94720
U.S.A.

Alan Weinstein
Department of Mathematics
University of California
Berkeley, California 94720
U.S.A.

Editorial Board

F. W. Gehring
Department of Mathematics
University of Michigan
Ann Arbor, Michigan 48109
U.S.A

P. R. Halmos
Department of Mathematics
Indiana University
Bloomington, Indiana 47405
U.S.A.

AMS Subject Classification: 26-01

Cover photograph by Nancy Williams Marsden.

Library of Congress Cataloging in Publication Data
Marsden, Jerrold E.
      Calculus I.
      (Undergraduate texts in mathematics)
      Includes index.
      1. Calculus. I. Weinstein, Alan.
II. Marsden, Jerrold E.   Calculus. III. Title.
IV. Title: Calculus one.   V. Series.
QA303.M3372   1984   515   84-5478

Typeset by Computype, Inc., St. Paul, Minnesota.
Printed and bound by Halliday Lithograph, West Hanover, Massachusetts.
Printed in the United States of America.

9 8 7 6 5 4 3 2 (Corrected Second Printing, 1986)

ISBN 0-387-90974-5 Springer-Verlag New York Berlin Heidelberg Tokyo
ISBN 3-540-90974-5 Springer-Verlag Berlin Heidelberg New York Tokyo

To Nancy and Margo

Preface

The goal of this text is to help students learn to use calculus intelligently for solving a wide variety of mathematical and physical problems.

This book is an outgrowth of our teaching of calculus at Berkeley, and the present edition incorporates many improvements based on our use of the first edition. We list below some of the key features of the book.

Examples and Exercises

The exercise sets have been carefully constructed to be of maximum use to the students. With few exceptions we adhere to the following policies.

- The section *exercises are graded* into three consecutive groups:

   (a) The first exercises are routine, modelled almost exactly on the examples; these are intended to give students confidence.
   (b) Next come exercises that are still based directly on the examples and text but which may have variations of wording or which combine different ideas; these are intended to train students to think for themselves.
   (c) The last exercises in each set are difficult. These are marked with a star (★) and some will challenge even the best students. Difficult does not necessarily mean theoretical; often a starred problem is an interesting application that requires insight into what calculus is really about.

- The *exercises come in groups* of two and often four similar ones.
- *Answers* to odd-numbered exercises are available in the back of the book, and every other odd exercise (that is, Exercise 1, 5, 9, 13, . . .) has a complete solution in the student guide. Answers to even-numbered exercises are not available to the student.

Placement of Topics

Teachers of calculus have their own pet arrangement of topics and teaching devices. After trying various permutations, we have arrived at the present arrangement. Some highlights are the following.

- *Integration* occurs early in Chapter 4; *antidifferentiation* and the $\int$ notation with motivation already appear in Chapter 2.

- *Trigonometric functions* appear in the first semester in Chapter 5.
- The *chain rule* occurs early in Chapter 2. We have chosen to use rate-of-change problems, square roots, and algebraic functions in conjunction with the chain rule. Some instructors prefer to introduce $\sin x$ and $\cos x$ early to use with the chain rule, but this has the penalty of fragmenting the study of the trigonometric functions. We find the present arrangement to be smoother and easier for the students.
- *Limits* are presented in Chapter 1 along with the derivative. However, while we do not try to hide the difficulties, technicalities involving epsilonics are deferred until Chapter 11. (Better or curious students can read this concurrently with Chapter 2.) Our view is that it is very important to teach students to differentiate, integrate, and solve calculus problems as quickly as possible, without getting delayed by the intricacies of limits. After some calculus is learned, the details about limits are best appreciated in the context of l'Hôpital's rule and infinite series.
- *Differential equations* are presented in Chapter 8 and again in Sections 12.7, 12.8, and 18.3. Blending differential equations with calculus allows for more interesting applications early and meets the needs of physics and engineering.

Prerequisites and Preliminaries

A historical introduction to calculus is designed to orient students before the technical material begins.

Prerequisite material from algebra, trigonometry, and analytic geometry appears in Chapters R, 5, and 14. These topics are treated completely: in fact, analytic geometry and trigonometry are treated in enough detail to serve as a first introduction to the subjects. However, high school algebra is only lightly reviewed, and knowledge of some plane geometry, such as the study of similar triangles, is assumed.

Several *orientation quizzes* with answers and a *review section* (Chapter R) contribute to bridging the gap between previous training and this book. Students are advised to assess themselves and to take a pre-calculus course if they lack the necessary background.

Chapter and Section Structure

The book is intended for a three-semester sequence with six chapters covered per semester. (Four semesters are required if pre-calculus material is included.)

The length of chapter sections is guided by the following typical course plan: If six chapters are covered per semester (this typically means four or five student contact hours per week) then approximately two sections must be covered each week. Of course this schedule must be adjusted to students' background and individual course requirements, but it gives an idea of the pace of the text.

Proofs and Rigor

Proofs are given for the most important theorems, with the customary omission of proofs of the intermediate value theorem and other consequences of the completeness axiom. Our treatment of integration enables us to give particularly simple proofs of some of the main results in that area, such as the fundamental theorem of calculus. We de-emphasize the theory of limits, leaving a detailed study to Chapter 11, after students have mastered the fundamentals of calculus—differentiation and integration. Our book *Calculus Unlimited* (Benjamin/Cummings) contains all the proofs omitted in this text and additional ideas suitable for supplementary topics for good students. Other references for the theory are Spivak's *Calculus* (Benjamin/Cummings & Publish or Perish), Ross' *Elementary Analysis: The Theory of Calculus* (Springer) and Marsden's *Elementary Classical Analysis* (Freeman).

Calculators

Calculator applications are used for motivation (such as for functions and composition on pages 40 and 112) and to illustrate the numerical content of calculus (see, for instance, p. 142). Special calculator discussions tell how to use a calculator and recognize its advantages and shortcomings.

Applications

Calculus students should not be treated as if they are already the engineers, physicists, biologists, mathematicians, physicians, or business executives they may be preparing to become. Nevertheless calculus is a subject intimately tied to the physical world, and we feel that it is misleading to teach it any other way. Simple examples related to distance and velocity are used throughout the text. Somewhat more special applications occur in examples and exercises, some of which may be skipped at the instructor's discretion. Additional connections between calculus and applications occur in various section supplements throughout the text. For example, the use of calculus in the determination of the length of a day occurs at the end of Chapters 5, 9, and 14.

Visualization

The ability to visualize basic graphs and to interpret them mentally is very important in calculus and in subsequent mathematics courses. We have tried to help students gain facility in forming and using visual images by including plenty of carefully chosen artwork. This facility should also be encouraged in the solving of exercises.

Computer-Generated Graphics

Computer-generated graphics are becoming increasingly important as a tool for the study of calculus. High-resolution plotters were used to plot the graphs of curves and surfaces which arose in the study of Taylor polynomial approximation, maxima and minima for several variables, and three-dimensional surface geometry. Many of the computer drawn figures were kindly supplied by Jerry Kazdan.

Supplements

Student Guide Contains

- Goals and guides for the student
- Solutions to every other odd-numbered exercise
- Sample exams

Instructor's Guide Contains

- Suggestions for the instructor, section by section
- Sample exams
- Supplementary answers

Misprints

Misprints are a plague to authors (and readers) of mathematical textbooks. We have made a special effort to weed them out, and we will be grateful to the readers who help us eliminate any that remain.

Acknowledgments

We thank our students, readers, numerous reviewers and assistants for their help with the first and current edition. For this edition we are especially grateful to Ray Sachs for his aid in matching the text to student needs, to Fred Soon and Fred Daniels for their unfailing support, and to Connie Calica for her accurate typing. Several people who helped us with the first edition deserve our continued thanks. These include Roger Apodaca, Grant Gustafson, Mike Hoffman, Dana Kwong, Teresa Ling, Tudor Ratiu, and Tony Tromba.

Jerry Marsden
Alan Weinstein

Berkeley, California

How to Use this Book: A Note to the Student

Begin by orienting yourself. Get a rough feel for what we are trying to accomplish in calculus by rapidly reading the Introduction and the Preface and by looking at some of the chapter headings.

Next, make a preliminary assessment of your own preparation for calculus by taking the quizzes on pages 13 and 14. If you need to, study Chapter R in detail and begin reviewing trigonometry (Section 5.1) as soon as possible.

You can learn a little bit about calculus by reading this book, but you can learn to use calculus only by practicing it yourself. You should do many more exercises than are assigned to you as homework. The answers at the back of the book and solutions in the student guide will help you monitor your own progress. There are a lot of examples with complete solutions to help you with the exercises. The end of each example is marked with the symbol ▲.

Remember that even an experienced mathematician often cannot "see" the entire solution to a problem at once; in many cases it helps to begin systematically, and then the solution will fall into place.

Instructors vary in their expectations of students as far as the degree to which answers should be simplified and the extent to which the theory should be mastered. In the book we have arranged the theory so that only the proofs of the most important theorems are given in the text; the ends of proofs are marked with the symbol ■. Often, technical points are treated in the starred exercises.

In order to prepare for examinations, try reworking the examples in the text and the sample examinations in the Student Guide without looking at the solutions. Be sure that you can do all of the assigned homework problems.

When writing solutions to homework or exam problems, you should use the English language liberally and correctly. A page of disconnected formulas with no explanatory words is incomprehensible.

We have written the book with your needs in mind. Please inform us of shortcomings you have found so we can correct them for future students. We wish you luck in the course and hope that you find the study of calculus stimulating, enjoyable, and useful.

Jerry Marsden
Alan Weinstein

Contents

Contents of Volume II

Contents of Volume III

Introduction

Calculus has earned a reputation for being an essential tool in the sciences. Our aim in this introduction is to give the reader an idea of what calculus is all about and why it is useful.

Calculus has two main divisions, called differential calculus and integral calculus. We shall give a sample application of each of these divisions, followed by a discussion of the history and theory of calculus.

Differential Calculus

The graph in Fig. I.1 shows the variation of the temperature y (in degrees Centigrade) with the time x (in hours from midnight) on an October day in New Orleans.

Figure I.1. Temperature in °C as a function of time.

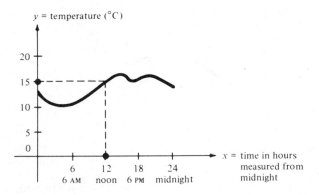

Each point on the graph indicates the temperature at a particular time. For example, at $x = 12$ (noon), the temperature was 15°C. The fact that there is exactly one y for each x means that y is a *function of x*.

The graph as a whole can reveal information more readily than a table. For example, we can see at a glance that, from about 5 A.M to 2 P.M., the temperature was rising, and that at the end of this period the maximum temperature for the day was reached. At 2 P.M. the air cooled (perhaps due to a brief shower), although the temperature rose again later in the afternoon. We also see that the lowest temperature occurred at about 5 A.M.

We know that the sun is highest at noon, but the highest temperature did not occur until 2 hours later. How, then, is the high position of the sun at noon reflected in the shape of the graph? The answer lies in the concept of *rate of change*, which is the central idea of differential calculus.

At any given moment of time, we can consider the rate at which temperature is changing with respect to time. What is this rate? If the graph of temperature against time were a segment of a straight line, as it is in Fig. I.2, the answer would be easy. If we compare the temperature measurement at

Figure I.2. The ratio $(y_2 - y_1)/(x_2 - x_1)$ is the ratio of change of temperature with respect to time.

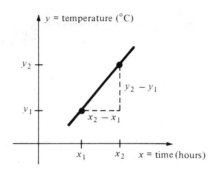

times x_1 and x_2, the ratio $(y_2 - y_1)/(x_2 - x_1)$ of change in temperature to change in time, measured in degrees per hour, is the rate of change. It is a basic property of straight lines that this ratio, called the *slope* of the line, does not depend upon which two points are used to form the ratio.

Returning to Fig. I.1, we may ask for the rate of change of temperature with respect to time at noon. We cannot just use a ratio $(y_2 - y_1)/(x_2 - x_1)$; since the graph is no longer a straight line, the answer would depend on which points on the graph we chose. One solution to our problem is to draw the line l which best fits the graph at the point $(x, y) = (12, 15)$, and to take the slope of this line (see Fig. I.3). The line l is called the *tangent line* to the temperature

Figure I.3. The rate of change of temperature with respect to time when $x = 12$ is the slope of the line l.

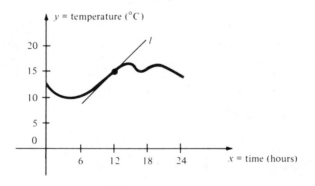

curve at $(12, 15)$; its slope can be measured with a ruler to be about $1°C$ per hour. By drawing tangent lines to the curve at other points, the reader will find that for no other point is the slope of the tangent line as great as $1°C$ per hour. Thus, the high position of the sun at noon is reflected by the fact that the rate of change of temperature with respect to time was greatest then.

The example just given shows the importance of rates of change and tangent lines, but it leaves open the question of just what the tangent line *is*. Our definition of the tangent line as the one which "best fits" the curve leaves much to be desired, since it appears to depend on personal judgment. Giving a mathematically precise definition of the tangent line to the graph of a function in the xy plane is the first step in the development of *differential calculus*. The slope of the tangent line, which represents the rate of change of y with respect to x, is called the *derivative* of the function. The process of determining the derivative is called *differentiation*.

The principal tool of differential calculus is a series of rules which lead to a formula for the rate of change of y with respect to x, given a formula for y in terms of x. (For instance, if $y = x^2 + 3x$, the derivative at x turns out to be $2x + 3$.) These rules were discovered by Isaac Newton (1642–1727) in England and, independently, by Gottfried Leibniz (1642–1716), a German working in France. Newton and Leibniz had many precursors. The ancient Greeks, notably Archimedes of Syracuse (287–212 B.C.), knew how to construct the tangent lines to parabolas, hyperbolas, and certain spirals. They were, in effect, computing derivatives. After a long period with little progress, development of Archimedes' ideas revived around 1600. By the middle of the seventeenth century, mathematicians could differentiate powers (i.e., the functions $y = x, x^2, x^3$, and so on) and some other functions, but a *general* method, which could be used by anyone with a little training, was first developed by Newton and Leibniz in the 1670's. Thanks to their work, it is no longer difficult or time-consuming to differentiate functions.

Integral Calculus and the Fundamental Theorem

The second fundamental operation of calculus is called *integration*. To illustrate this operation, we consider another question about Fig. I.1: What was the *average* temperature on this day?

We know that the average of a list of numbers is found by adding the entries in the list and then dividing by the number of entries. In the problem at hand, though, we do not have a finite list of numbers, but rather a continuous graph.

As we did with rates of change, let us look at a simpler example. Suppose that the temperature changed by jumps every two hours, as in Fig. I.4. Then we could simply add the 12 temperature readings and divide by 12 to get the average.

We can interpret this averaging process graphically in the following way. Let $y_1, \ldots, y_{12}$ be the 12 temperature readings, so that their average is $y_{ave} = \frac{1}{12}(y_1 + \cdots + y_{12})$. The region under the graph, shaded in Fig. I.5, is composed of 12 rectangles. The area of the ith rectangle is (base) $\times$ (height) $= 2y_i$, so the total area is $A = 2y_1 + 2y_2 + \cdots + 2y_{12} = 2(y_1 + \cdots + y_{12})$. Comparing this with the formula for the average, we find that $y_{ave} = A/24$. In other words, the average temperature is equal to the area under the graph,

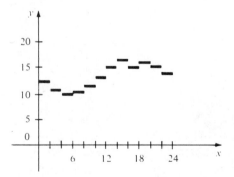

Figure I.4. If the temperature changes by jumps, the average is easy to find.

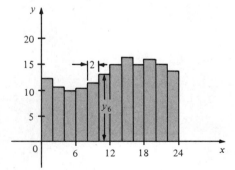

Figure I.5. The area of the ith rectangle is $2y_i$.

divided by the length of the time interval. Now we can guess how to define the average temperature for Fig. I.1. It is simply the area of the region under the graph (shaded in Fig. I.6) divided by 24.

Figure I.6. The average temperature is 1/24 times the shaded area.

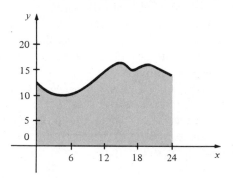

The area under the graph of a function on an interval is called the *integral* of the function over the interval. Finding integrals, or *integrating*, is the subject of the *integral calculus*.

Progress in integration was parallel to that in differentiation, and eventually the two problems became linked. The ancient Greeks knew the area of simple geometric figures bounded by lines, circles and parabolas. By the middle of the seventeenth century, areas under the graph of x, x^2, x^3, and other functions could be calculated. Mathematicians at that time realized that the slope and area problems were related. Newton and Leibniz formulated this relationship precisely in the form of the *fundamental theorem of calculus*, which states that integration and differentiation are inverse operations. To suggest the idea behind this theorem, we observe that if a list of numbers b_1, $b_2, \ldots, b_n$ is given, and the differences $d_1 = b_2 - b_1$, $d_2 = b_3 - b_2, \ldots$, $d_{n-1} = b_n - b_{n-1}$ are taken (this corresponds to differentiation), then we can recover the original list from the d_i's and the initial entry b_1 by adding (this corresponds to integration): $b_2 = b_1 + d_1$, $b_3 = b_1 + d_1 + d_2, \ldots$, and finally $b_n = b_1 + d_1 + d_2 + \cdots + d_{n-1}$.

The fundamental theorem of calculus, together with the rules of differentiation, brings the solution of many integration problems within reach of anyone who has learned the differential calculus.

The importance and applicability of calculus lies in the fact that a wide

Figure I.7. Quantities related by the operations of calculus. (The independent variable is in brackets.)

Distance traveled along a road	Differentiation / Integration	Velocity	[Time]
Velocity	Differentiation / Integration	Acceleration	[Time]
Cost of living	Differentiation / Integration	Inflation rate	[Time]
Total cost of some goods	Differentiation / Integration	Marginal cost	[Quantity]
Height above sea level on a trail	Differentiation / Integration	Steepness	[Distance]
Mass of a rod	Differentiation / Integration	Linear density	[Length]
Height of a tree	Differentiation / Integration	Growth rate	[Time]

variety of quantities are related by the operations of differentiation and integration. Some examples are listed in Fig. I.7.

The primary aim of this book is to help you learn *how* to carry out the operations of differentiation and integration and *when* to use them in the solution of many types of problems.

The Theory of Calculus

We shall describe three approaches to the theory of calculus. It will be simpler, as well as more faithful to history, if we begin with integration.

The simplest function to integrate is a constant $y = k$. Its integral over the interval $[a, b]$ is simply the area $k(b - a)$ of the rectangle under its graph (see Fig. I.8). Next in simplicity are the functions whose graphs are composed of several horizontal straight lines, as in Fig. I.9. Such functions are called step functions. The integral of such a function is the sum of the areas of the rectangles under its graph, which is easy to compute.

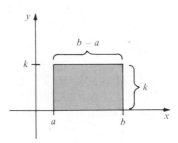

Figure I.8. The integral of the constant function $y = k$ over the interval $[a, b]$, is just the area $k(b - a)$ of this rectangle.

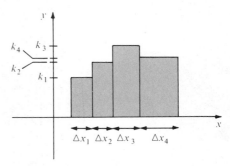

Figure I.9. The integral over $[a, b]$ of this step function is $k_1\Delta x_1 + k_2\Delta x_2 + k_3\Delta x_3 + k_4\Delta x_4$, where k_i is the value of y on the ith interval, and Δx_i is the length of that interval.

There are three ways to go from the simple problem of integrating step functions to the interesting problem of integrating more general functions, like $y = x^2$ or the function in Fig. I.1. These three ways are the following.

1. **The method of exhaustion.** This method was invented by Eudoxus of Cnidus (408–355 B.C.) and was exploited by Archimedes of Syracuse (287–212 B.C.) to calculate the areas of circles, parabolic segments, and other figures. In terms of functions, the basic idea is to *compare* the function to be integrated with step functions. In Fig. I.10, we show the graph of $y = x^2$ on $[0, 1]$, and step functions whose graphs lie below and above it. Since a figure inside another figure has a smaller area, we may conclude that the integral of $y = x^2$

Figure I.10. The integral of $y = x^2$ on the interval $[0, 1]$ lies between the integrals of the two functions.

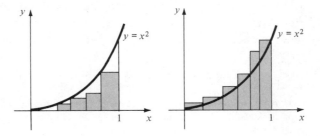

on [0, 1] lies between the integrals of these two step functions. In this way, we can get lower and upper estimates for the integral. By choosing step functions with shorter and shorter "steps," it is reasonable to expect that we can *exhaust* the area between the rectangles and the curve and, thereby, calculate the area to any accuracy desired. By reasoning with *arbitrarily* small steps, we can in some cases determine the exact area—that is just what Archimedes did.

2. **The method of limits.** This method was fundamental in the seventeenth-century development of calculus and is the one which is most important today. Instead of comparing the function to be integrated with step functions, we *approximate* it by step functions, as in Fig. I.11. If, as we allow the steps to get shorter and shorter, the approximation gets better and better, we say that the integral of the given function is the *limit* of these approximations.

Figure I.11. The integral of this step function is an approximation for the integral of x^2.

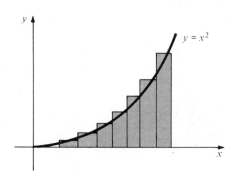

3. **The method of infinitesimals.** This method, too, was invented by Archimedes, but he kept it for his personal use since it did not meet the standards of rigor demanded at that time. (Archimedes' use of infinitesimals was not discovered until 1906. It was found as a *palimpsest*, a parchment which had been washed and reused for some religious writing.)[1] The infinitesimal method was also used in the seventeenth century, especially by Leibniz. The idea behind this method is to consider any function as being a step function whose graph has infinitely many steps, each of them infinitely small, or *infinitesimal*, in length. It is impossible to represent this idea faithfully by a drawing, but Fig. I.12 suggests what is going on.

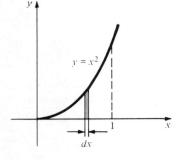

Figure I.12. The integral of x^2 on [0, 1] may be thought of as the sum of the areas of infinitely many rectangles of infinitesimal width dx.

Each of these three methods—exhaustion, limits, and infinitesimals—has its advantages and disadvantages. The method of exhaustion is the easiest to comprehend and to make rigorous, but it is usually cumbersome in applications. Limits are much more efficient for calculation, but their theory is considerably harder to understand; indeed, it was not until the middle of the nineteenth century with the work of Augustin-Louis Cauchy (1789–1857) and Karl Weierstrass (1815–1897), among others, that limits were given a firm mathematical foundation. Infinitesimals lead most quickly to answers to many problems, but the idea of an "infinitely small" quantity is hard to comprehend fully,[2] and the method can lead to wrong answers if it is not used carefully. The mathematical foundations of the method of infinitesimals were not

---

[1] See S. H. Gould, *The method of Archimedes*, American Mathematical Monthly **62**(1955), 473–476.

[2] An early critic of infinitesimals was Bishop George Berkeley, who referred to them as "ghosts of departed quantities" in his anticalculus book, *The Analyst* (1734). The city in which this calculus book has been written is named after him.

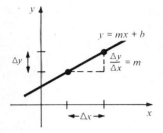

Figure I.13. If $y = mx + b$, the rate of change of y with respect to x is constant and equal to m.

established until the twentieth century with the work of the logician, Abraham Robinson (1918–1974).[3]

The three methods used to define the integral can be applied to differentiation as well. In this case, we replace the piecewise constant functions by the linear functions $y = mx + b$. For a function of this form, a change of Δx in x produces a change $\Delta y = m\Delta x$ in y, so the rate of change, given by the ratio $\Delta y / \Delta x$, is equal to m, independent of x and of Δx (see Fig. I.13).

1. **The method of exhaustion.** To find the rate of change of a general function, we may compare the function with linear functions by seeing how straight lines with various slopes cross the graph at a given point. In Fig. I.14, we show the graph of $y = x^2$, together with lines which are more and less steep at the point $x = 1$, $y = 1$. By bringing our comparison lines closer and closer together, we can calculate the rate of change to any accuracy desired; if the algebra is simple enough, we can even calculate the rate of change exactly.

The historical origin of this method can be found in the following definition of tangency used by the ancient Greeks: "the tangent line touches the curve, and in the space between the line and curve, no other straight line can be interposed."[4]

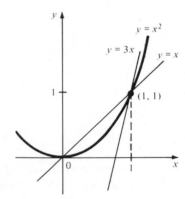

Figure I.14. The rate of change of $y = x^2$ at $x = 1$ lies between 1 and 3.

2. **The method of limits.** To approximate the tangent line to a curve we draw the *secant* line through two nearby points. As the two points become closer and closer, the slope of the secant approaches a limiting value which is the rate of change of the function (see Fig. I.15).

Figure I.15. The rate of change of a function is the limit of the slopes of secant lines drawn through two points on the graph.

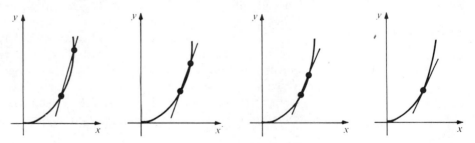

[3] A calculus textbook based upon this work is H. J. Keisler, *Elementary Calculus*, Prindle, Weber, and Schmidt, Boston (1976).

[4] See C. Boyer, *The History of the Calculus and Its Conceptual Development*, Dover, New York, p. 57. The method of exhaustion is not normally used in calculus courses for differentiation, and this book is no exception. However, it could be used and it is intellectually satisfying to do so; see *Calculus Unlimited*, Benjamin/Cummings (1980) by J. Marsden and A. Weinstein.

This approach to rates of change derives from the work of Pierre de Fermat[5] (1601–1665), whose interest in tangents arose from the idea, due originally to Kepler, that the slope of the tangent line should be zero at a maximum or minimum point (Fig. I.16).

Figure I.16. The slope of the tangent line is zero at a maximum or minimum point.

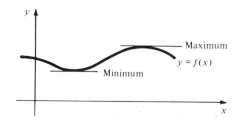

3. **The method of infinitesimals.** In this method, we simply think of the tangent line to a curve as a secant line drawn through two infinitesimally close points on the curve, as suggested by Fig. I.17. This idea seems to go back to Galileo[6] (1564–1642) and his student Cavalieri (1598–1647), who defined instantaneous velocity as the ratio of an infinitely small distance to an infinitely short time.

Figure I.17. The tangent line may be thought of as the secant line through a pair of infinitesimally near points.

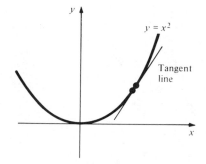

As with integration, infinitesimals lead most quickly to answers (but not always the right ones), and the method of exhaustion is conceptually simplest. Because of is computational power, the method of limits has become the most widely used approach to differential calculus. It is this method which we shall use in this book.

The Power of Calculus (The Calculus of Power)

To end this introduction, we shall give an example of a practical problem which calculus can help us to solve.

The sun, which is the ultimate source of nearly all of the earth's energy, has always been an object of fascination. The relation between the sun's position and the seasons was predicted by early agricultural societies, some of which developed quite sophisticated astronomical techniques. Today, as the earth's resources of fossil fuels dwindle, the sun has new importance as a *direct* source of energy. To use this energy efficiently, it is useful to know just how

---

[5] Fermat is also famous for his work in number theory. Fermat's last theorem: "If n is an integer greater than 2, there are no positive integers x, y, and z such that $x^n + y^n = z^n$," remains unproven today. Fermat claimed to have proved it, but his proof has not been found, and most mathematicians now doubt that it could have been correct.

[6] Newton's acknowledgment, "If I have seen further than others, it is because I have stood on the shoulders of giants," probably refers chiefly to Galileo, who died the year Newton was born. (A similar quotation from Lucan (39–65 A.D.) was cited by Robert Burton in the early 1600's—"Pygmies see further than the giants on whose shoulders they stand.")

much solar radiation is available at various locations at different times of the year.

From basic astronomy we know that the earth revolves about the sun while rotating about an axis inclined at 23.5° to the plane of its orbit (see Fig. I.18). Even assuming idealized conditions, such as a perfectly spherical earth

Figure I.18. The earth revolving about the sun.

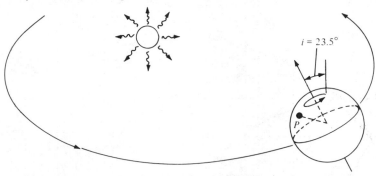

revolving in a circle about the sun, it is not a simple matter to predict the length of the day or the exact time of sunset at a given latitude on the earth on a given day of the year.

In 1857, an American scientist named L. W. Meech published in the *Smithsonian Contributions to Knowledge* (Volume 9, Article II) a paper entitled "On the relative intensity of the heat and light of the sun upon different latitudes of the earth." Meech was interested in determining the extent to which the variation of temperature on the surface of the earth could be correlated with the variations of the amount of sunlight impinging on different latitudes at different times. One of Meech's ultimate goals was to predict whether or not there was an open sea near the north pole—a region then unexplored. He used the integral calculus to sum the total amount of sunlight arriving at a given latitude on a given day of the year, and then he summed this quantity over the entire year. Meech found that the amount of sunlight reaching the atmosphere above polar regions was surprisingly large during the summer due to the long days (see Fig. I.19). The differential calculus is used to predict the shape of graphs like those in Fig. I.19 by calculating the slopes of their tangent lines.

Meech realized that, since the sunlight reaching the polar regions arrives at such a low angle, much of it is absorbed by the atmosphere, so one cannot conclude the existence of "a brief tropical summer with teeming forms of vegetable and animal life in the centre of the frozen zone." Thus, Meech's calculations fell short of permitting a firm conclusion as to the existence or not of an open sea at the North Pole, but his work has recently taken on new importance. Graphs like Fig. I.19 on the next page have appeared in books devoted to meteorology, geology, ecology (with regard to the biological energy balance), and solar energy engineering.

Even if one takes into account the absorption of energy by the atmosphere, on a summer day the middle latitudes still receive more energy at the earth's surface than does the equator. In fact, the hottest places on earth are not at the equator but in bands north and south of the equator. (This is enhanced by climate: the low-middle latitudes are much freer of clouds than the equatorial zone.)[7]

---

[7] According to the *Guinness Book of World Records*, the world's highest temperatures (near 136°F) have occurred at Ouargla, Algeria (latitude 32°N), Death Valley, California (latitude 36°N), and Al'Aziziyah, Libya (latitude 32°N). Locations in Chile, Southern Africa, and Australia approach these records.

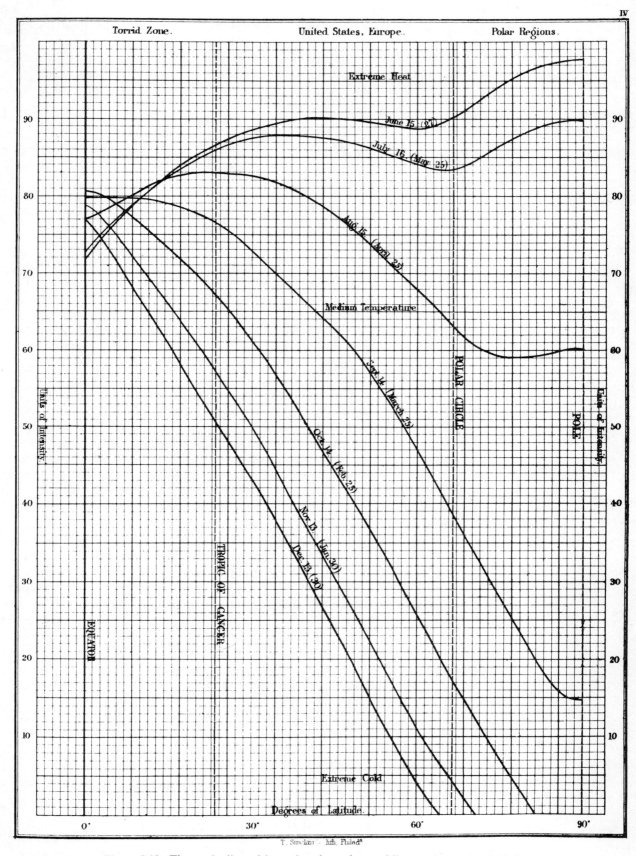

Figure I.19. The sun's diurnal intensity along the meridian, at intervals of 30 days.

As we carry out our study of calculus in this book, we will from time to time in supplementary sections reproduce parts of Meech's calculations (slightly simplified) to show how the material being learned may be applied to a substantial problem. By the time you have finished this book, you should be able to read Meech's article yourself.

Orientation Quizzes

Guidelines

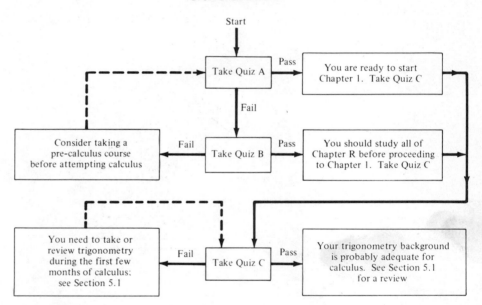

Quiz A

(Passing score is 8/10. Answers are on p. A.1.)

1. What is the slope of the line $3y + 4x = 2$?
2. For which values of x is $3x + 2 > 0$?
3. For which values of x is $3x^2 - 2x - 1 > 0$?
4. Solve for x: $2x^2 + 8x - 11 = 0$.
5. Sketch the graph of $f(x) = x^2 - x - 2$.
6. Let $g(x) = (3x^2 + 2x - 8)/(2x^2 - x)$. Compute $g(2)$ and state the domain of g.
7. For what values of x is $3x + 2 > 2x - 8$?
8. Where does the graph of $f(x) = 3x - 2$ intersect the graph of $g(x) = x^2$?
9. Sketch the curve $x^2 + y^2 - 2x - 4y + 1 = 0$.
10. Find the distance between the point $(1, 1)$ and the intersection point of the lines $y = -2x + 1$ and $y = 4x - 5$.

Quiz B

(Passing score is 8/10. Answers are on p. A.1.)

1. $\frac{3}{2} + \frac{5}{4} =$ _____ .
2. Factor: $x^2 + 3x =$ _____ .
3. $(-6)(-3) + 8(-1) =$ _____ .
4. If a bag of sand weighs 80 kilograms, 5% of the bag weighs _____ .
5. $[(3x - 1)/2x] - \frac{1}{2} =$ _____ . (Bring to a common denominator.)
6. $x^3 \cdot x^4 =$ _____ .
7. Arrange these numbers in ascending order (smallest to largest): 8, -6, $\frac{1}{2}$, 0, -4.
8. Solve for x: $6x + 2 = -x - 1$.
9. If $x = 3$ and $y = 9$, then $\sqrt{y}/3x =$ _____ .
10. $(x^2 - 16)/(x - 4) =$ _____ . (Simplify.)

Quiz C

(Passing score is 7/10. Answers are on p. A.1.)

1. Find the coordinates of the point P:

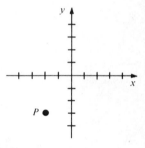

2. Find x:

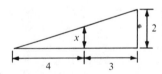

3. Find θ:

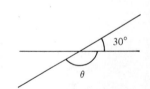

4. Find $\sin\theta$:

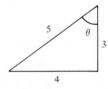

5. Find y:

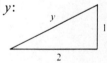

6. What is the circumference of a circle whose radius is 4 centimeters?
7. A rectangle has area 10 square meters and one side of length 5 meters. What is its perimeter?
8. The volume of a cylindrical can with radius 3 centimeters and height 2 centimeters is _____ .
9. $\cos 60° =$ _____ .
10. $2\sin^2 x + 2\cos^2 x =$ _____ .

Review of Fundamentals

Functions are to calculus as numbers are to algebra.

Success in the study of calculus depends upon a solid understanding of algebra and analytic geometry. In this chapter, we review topics from these subjects which are particularly important for calculus.

R.1 Basic Algebra: Real Numbers and Inequalities

The real numbers are ordered like the points on a line.

The most important facts about the real numbers concern algebraic operations (addition, multiplication, subtraction, and division) and order (greater than and less than). In this section, we review some of these facts.

The positive whole numbers $1, 2, 3, 4, \ldots$ ($\ldots$ means "and so on") that arise from the counting process are called the *natural numbers*. The arithmetic operations of addition and multiplication can be performed within the natural numbers, but the "inverse" operations of subtraction and division lead to the introduction of zero ($3 - 3 = 0$), negative numbers ($2 - 6 = -4$), and fractions ($3 \div 5 = \frac{3}{5}$). The whole numbers, positive, zero, and negative, are called *integers*. All numbers which can be put in the form m/n, where m and n are integers, are called *rational numbers*.

Example 1 Determine whether or not the following numbers are natural numbers, integers or rational numbers.

(a) 0   (b) $3 - \dfrac{4+5}{2}$ (c) $7 - 6$ (d) $\dfrac{4+5}{-3}$

Solution (a) 0 is not a natural number (they are only the numbers $1, 2, 3, \ldots$), but it is an integer, and so it is a rational number—every integer m is rational since it can be written as $m/1$.

(b) $3 - (4+5)/2 = 3 - 9/2 = -3/2$ is not a natural number or an integer, but it is rational.

(c) $7 - 6 = 1$ is a natural number, an integer, and a rational number.

(d) $(4+5)/(-3) = -3$ is not a natural number, but is an integer and a rational number. ▲

The ancient Greeks already knew that lines in simple geometric figures could have lengths which did not correspond to ratios of whole numbers. For instance, the length $\sqrt{2}$ of the diagonal of a square with sides of unit length cannot be expressed in the form m/n, with m and n integers. (The same turns out to be true of π, the circumference of a circle with unit diameter.) Numbers which are not ratios of integers are called *irrational numbers*. These, together with the rational numbers, comprise the *real numbers*.

The usual arithmetic operations of addition, multiplication, subtraction, and division (except by zero) may be performed on real numbers, and these operations satisfy the usual algebraic rules. You should be familiar with these rules; some examples are "if equals are added to equals, the results are equal," "$a + b = b + a$," and "if $ab = ac$ and $a \neq 0$, we can divide both sides by a to conclude that $b = c$."

For example, to solve the equation $3x + 2 = 8$ for x, subtract 2 from both sides of the equation:

$$3x + 2 - 2 = 8 - 2,$$
$$3x = 6.$$

Dividing both sides by 3 gives $x = 2$.

Two fundamental identities from algebra that will be useful for us are

$$(a + b)(a - b) = a^2 - b^2 \quad \text{and} \quad (a + b)^2 = a^2 + 2ab + b^2.$$

Example 2   Simplify: $(a + b)(a - b) + b^2$.

Solution   Since $(a + b)(a - b) = a^2 - b^2$, we have $(a + b)(a - b) + b^2 = a^2 - b^2 + b^2 = a^2$. ▲

Example 3   Expand: $(a + b)^3$.

Solution   We have $(a + b)^2 = a^2 + 2ab + b^2$. Therefore

$$\begin{aligned}(a + b)^3 &= (a + b)^2(a + b) \\ &= (a^2 + 2ab + b^2)(a + b) \\ &= (a^2 + 2ab + b^2)a + (a^2 + 2ab + b^2)b \\ &= a^3 + 2a^2b + ab^2 + a^2b + 2ab^2 + b^3 \\ &= a^3 + 3a^2b + 3ab^2 + b^3. \; ▲\end{aligned}$$

Another important algebraic operation is *factoring*. We try to reverse the process of expanding: $(x + r)(x + s) = x^2 + (r + s)x + rs$.

Example 4   Factor: $2x^2 - 4x - 6$.

Solution   We notice first that $2x^2 + 4x - 6 = 2(x^2 + 2x - 3)$. Using the fact that the only integer factors of -3 are ± 1 and ± 3, we find by trial and error that $x^2 + 2x - 3 = (x + 3)(x - 1)$, so we have $2x^2 + 4x - 6 = 2(x + 3)(x - 1)$. ▲

The *quadratic formula* is used to solve for x in equations of the form $ax^2 + bx + c = 0$ when the left-hand side cannot be readily factored. The method of *completing the square*, by which the quadratic formula may be derived, is often more important than the formula itself.

Example 5   Solve the equation $x^2 - 5x + 3 = 0$ by completing the square.

Solution   We transform the equation by adding and subtracting $\left(\frac{5}{2}\right)^2$ on the left-hand side:

$$x^2 - 5x + \left(\tfrac{5}{2}\right)^2 - \left(\tfrac{5}{2}\right)^2 + 3 = 0,$$

$$\left(x - \tfrac{5}{2}\right)^2 - \tfrac{13}{4} = 0,$$

$$\left(x - \tfrac{5}{2}\right)^2 = \tfrac{13}{4},$$

$$x - \frac{5}{2} = \pm \frac{\sqrt{13}}{2},$$

$$x = \frac{5}{2} \pm \frac{\sqrt{13}}{2}. \;\blacktriangle$$

Another method for completing the square is to write

$$x^2 - 5x + 3 = (x + p)^2 + q.$$

and then expand to

$$x^2 - 5x + 3 = x^2 + 2px + p^2 + q.$$

Equating coefficients, we see that $p = -\frac{5}{2}$ and $q = 3 - p^2 = 3 - \frac{25}{4} = -\frac{13}{4}$, so $x^2 - 5x + 3 = \left(x - \frac{5}{2}\right)^2 - \frac{13}{4}$. This can be used to solve $x^2 - 5x + 3 = 0$ as above.

Completing the Square

To complete the square in the expression $ax^2 + bx + c$, factor out a and then add and subtract $(b/2a)^2$:

$$ax^2 + bx + c = a\left[\left(x + \frac{b}{2a}\right)^2 + \left(\frac{c}{a} - \frac{b^2}{4a^2}\right)\right].$$

When the method of completing the square is applied to the *general* quadratic equation $ax^2 + bx + c = 0$, one obtains the following general formula for the solution of the equation. (See Exercise 53).

Quadratic Formula

To solve $ax^2 + bx + c = 0$, where $a \neq 0$, compute

$$x = \frac{-b \pm \sqrt{b^2 - 4ac}}{2a}.$$

If $b^2 - 4ac > 0$, there are two solutions.
If $b^2 - 4ac = 0$, there is one solution.
If $b^2 - 4ac < 0$, there are no solutions.
(The expression $b^2 - 4ac$ is called the *discriminant*.)

In case $b^2 - 4ac < 0$, there is no real number $\sqrt{b^2 - 4ac}$, because the square of every real number is greater than or equal to zero. (Square roots of negative numbers can be found if we extend the real-number system to encompass the so-called *imaginary numbers*.)[1] Thus the symbol $\sqrt{r}$ represents a real number only when $r \geqslant 0$, in which case we always take $\sqrt{r}$ to mean the *non-negative* number whose square is r.

Example 6   Solve for x: (a) $4x^2 = 2x + 5$ and (b) $2x^2 + 4x - 6 = 0$.

Solution   (a) Subtracting $2x + 5$ from both sides of the equation gives $4x^2 - 2x - 5 = 0$, which is in the form $ax^2 + bx + c = 0$ with $a = 4$, $b = -2$, and $c = -5$. The quadratic formula gives the two roots

$$x = \frac{-(-2) \pm \sqrt{(-2)^2 - 4(4)(-5)}}{2(4)}$$

$$= \frac{2 \pm \sqrt{4 + 80}}{8} = \frac{2 \pm \sqrt{84}}{8} = \frac{1}{4} \pm \frac{\sqrt{21 \cdot 4}}{8} = \frac{1}{4} \pm \frac{\sqrt{21}}{4}.$$

(b) An alternative to the quadratic formula is to factor. From Example 4, $2x^2 + 4x - 6 = 2(x + 3)(x - 1)$. Thus the two roots are $x = -3$ and $x = 1$. The reader may check that the quadratic formula gives the same roots. ▲

Example 7   Solve for x: $x^2 - 5x + 20 = 0$.

Solution   We use the quadratic formula:

$$x = \frac{5 \pm \sqrt{25 - 4 \cdot 20}}{2}.$$

The discriminant is negative, so there are no real solutions. ▲

The real numbers have a relation of order: if two real numbers are unequal, one of them is less than the other. We may represent the real numbers as points on a line, with larger numbers to the right, as shown in Fig. R.1.1. If the number a is less than b, we write $a < b$. In this case, we also say that b is greater than a and write $b > a$.

Figure R.1.1. The real number line.

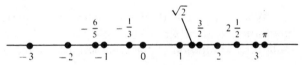

Given any two numbers, a and b, exactly one of the following three possibilities holds:

(1) $a < b$,
(2) $a = b$,
(3) $a > b$.

Combinations of these possibilities have special names and notations.

If (1) or (2) holds, we write $a \leqslant b$ and say that "a is less than or equal to b."

If (2) or (3) holds, we write $a \geqslant b$ and say that "a is greater than or equal to b."

If (1) or (3) holds, we write $a \neq b$ and say that "a is unequal to b."

For example, $3 \leqslant 3$ is true, $(-2)^2 \leqslant 0$ is false (since $(-2)^2 = 4 > 0$) and $-\pi < -\frac{1}{2}\pi$ is true; note that $-\pi$ and $-\frac{1}{2}\pi$ both lie to the left of zero on the

---

[1] Imaginary numbers are discussed in Section 12.6.

number line and since $-\frac{1}{2}\pi$ is only half as far from zero as $-\pi$, it lies to the right of $-\pi$.

If x is any real number, we know that $x^2 \geq 0$. If $x \neq 0$, we can make the stronger statement that $x^2 > 0$.

Example 8   Write the proper inequality sign between each of the following pairs of numbers:

(a) 0.0000025 and $-100{,}000$ (b) $\frac{3}{4}$ and $\frac{6}{7}$ (c) $\sqrt{12}$ and 4

Solution   (a) $0.0000025 > -100{,}000$ since a positive number is always to the right of a negative number.

(b) $\frac{3}{4} < \frac{6}{7}$ since $\frac{3}{4} = \frac{21}{28}$ and $\frac{6}{7} = \frac{24}{28}$.

(c) $\sqrt{12} < 4$ since $12 < 4^2$. ▲

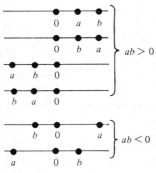

Figure R.1.2. Possible positions of a and b when they have the same sign ($ab > 0$) or opposite signs ($ab < 0$).

We can summarize the most important properties of inequalities as follows.

1. If $a < b$ and $b < c$, then $a < c$.
2. If $a < b$, then $a + c < b + c$ for any c, and $ac < bc$ if $c > 0$, while $ac > bc$ if $c < 0$. (Multiplication by a negative number reverses the sign of inequality. For instance, $3 < 4$, and multiplication by -2 gives $-6 > -8$.)
3. $ab > 0$ when a and b have the same sign; $ab < 0$ when a and b have opposite signs. (See Fig. R.1.2.)
4. If a and b are any two numbers, then $a < b$ when $a - b < 0$ and $a > b$ when $a - b > 0$.

Example 9   Transform $a + (b - c) > b - a$ to an inequality with a alone on one side.

Solution   We transform by reversible steps:

$$a + b - c > b - a,$$

$$2a + b - c > b \qquad \text{(add } a \text{ to both sides),}$$

$$2a - c > 0 \qquad \text{(add } -b \text{ to both sides),}$$

$$2a > c \qquad \text{(add } c \text{ to both sides),}$$

$$a > \tfrac{1}{2}c \qquad \text{(multiply both sides by } \tfrac{1}{2}\text{).} \ ▲$$

Example 10   (a) Find all numbers x for which $x^2 < 9$.
(b) Find all numbers x such that $x^2 - 2x - 3 > 0$.

Solution   (a) We transform the inequality as follows (all steps are reversible):

$$x^2 < 9,$$

$$x^2 - 9 < 0, \qquad \text{(add } -9 \text{ to both sides),}$$

$$(x + 3)(x - 3) < 0 \qquad \text{(factor).}$$

Since the product $(x + 3)(x - 3)$ is negative, the factors $x + 3$ and $x - 3$ must have opposite signs. Thus, either $x + 3 > 0$ and $x - 3 < 0$, so that $x > -3$ and $x < 3$ (that is, $-3 < x < 3$); or $x + 3 < 0$ and $x - 3 > 0$, in which case $x < -3$ and $x > 3$, which is impossible. We conclude that $x^2 < 9$ if and only if $-3 < x < 3$.

(b) The inequality $x^2 - 2x - 3 > 0$ is the same as $(x - 3)(x + 1) > 0$. That is, $x - 3$ and $x + 1$ have the same sign. There are two cases to consider:

Case 1: $x - 3$ and $x + 1$ are both positive; that is, $x - 3 > 0$ and $x + 1 > 0$; that is, $x > 3$ and $x > -1$, which is the same as $x > 3$ (since any number greater than 3 is certainly greater than -1).

Case 2: $x - 3 < 0$ and $x + 1 < 0$; that is, $x < 3$ and $x < -1$, which is the same as $x < -1$.

Thus $x^2 - 2x - 3 > 0$ whenever $x > 3$ or $x < -1$. These numbers x are illustrated in Fig. R.1.3. (The open dot indicates that this point is not included in the shaded region—if it were included, we would have used a solid dot.) ▲

Figure R.1.3. Solution of the inequality $x^2 - 2x - 3 > 0$.

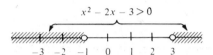

$x^2 - 2x - 3 > 0$

Exercises for Section R.1

In Exercises 1–4, determine whether or not each given number is a natural number, an integer, or a rational number.

1. $\frac{8}{6} - \frac{9}{4}$
2. $(-1) \div (-1)$
3. $\left(\frac{1}{\sqrt{2}} + \frac{1}{\sqrt{3}} \right)\left(\frac{1}{\sqrt{2}} - \frac{1}{\sqrt{3}} \right)$
4. $\pi - \frac{1}{2}$

Simplify the expressions in Exercises 5–8.

5. $(a - 3)(b + c) - (ac + 2b)$,
6. $(b^2 - a)^2 - a(2b^2 - a)$
7. $a^2c + (a - b)(b - c)(a + b)$
8. $(3a + 2)^2 - (4a + b)(2a - 1)$

Expand the expressions in Exercises 9–12.

9. $(a - b)^3$
10. $(3a + b^2 + c)^2$
11. $(b + c)^4$
12. $(2c - b)^2(2c + b)^2$

Factor the expressions in Exercises 13–20.

13. $x^2 + 5x + 6$
14. $x^2 - 5x + 6$
15. $x^2 - 5x - 6$
16. $x^2 + 5x - 6$
17. $3x^2 - 6x - 24$
18. $-5x^2 + 15x - 10$
19. $x^2 - 1$
20. $4x^2 - 9$

Solve for x in Exercises 21–24.

21. $2(3x - 7) - (4x - 10) = 0$
22. $3(3 + 2x) + (2x - 1) = 8$
23. $(2x + 1)^2 + (9 - 4x^2) + (x - 5) = 10$
24. $8(x + 1)^2 - 8x + 10 = 0$

25. Verify that $x^3 - 1 = (x - 1)(x^2 + x + 1)$.
26. Factor $x^3 + 1$ into linear and quadratic factors.
27. Factor $x^3 + x^2 - 2x$ into linear factors.
28. Factor $x^4 - 2x^2 + 1$ into linear factors. [*Hint:* First consider x^2 as the variable.]

Solve the equations in Exercises 29 and 30 in three ways: (a) by factoring; (b) by completing the square; (c) by using the quadratic formula.

29. $x^2 + 5x + 4 = 0$
30. $4x^2 - 12x + 9 = 0$

Solve for x in Exercises 31–36.

31. $x^2 + \frac{1}{2}x - \frac{1}{2} = 0$
32. $4x^2 - 18x + 20 = 0$
33. $-x^2 + 5x + 0.3 = 0$
34. $5x^2 + 2x - 1 = 0$
35. $x^2 - 5x + 7 = 0$
36. $0.1x^2 - 1.3x + 0.7 = 0$

Solve for x in Exercises 37–42.

37. $x^2 + 4 = 3x^2 - x$
38. $4x = 3x^2 + 7$
39. $2x + x^2 = 9 + x^2$
40. $(5 - x)(2 - x) = 1$
41. $2x^2 - 2\sqrt{7}\,x + \frac{7}{2} = 0$
42. $x^2 + 9x = 0$

43. Put the following list of numbers in ascending order. (Try to do it without finding decimal equivalents for the numbers.)

$$-\tfrac{5}{3}, \quad -\sqrt{2}, \quad -\tfrac{7}{5}, \quad \tfrac{22}{7}, \quad 3, \quad \tfrac{23}{8}, \quad 0, \quad \tfrac{9}{5}, \quad -\sqrt{3}$$

44. Put the following numbers in ascending order. (Do not use a calculator):

$$-\sqrt{90}, \quad 8, \quad -\tfrac{81}{8}, \quad \tfrac{81}{10}, \quad \sqrt{90}, \quad 9, \quad -9, \quad -8$$

Simplify the inequalities in Exercises 45–50.

45. $(a - b) + c > 2c - b$
46. $(a + c^2) + c(a - c) \geqslant ac + 1$
47. $ab - (a - 2b)b < b^2 + c$
48. $2(a + ac) - 4ac > 2a - c$
49. $b(b + 2) > (b + 1)(b + 2)$
50. $(a - b)^2 > 3 - 2ab + a^2$

51. Find all numbers x such that: (a) $4x - 13 < 3$, (b) $2(7 - x) \geqslant x + 1$, (c) $5(x - 3) - 2x + 6 > 0$. Sketch your solutions on a number line.

52. Find all numbers x such that: (a) $2(x^2 - x) > 0$, (b) $3x^2 + 2x - 1 \geqslant 0$, (c) $x^2 - 5x + 6 < 0$. Sketch your solutions on a number line.

★53. (a) Prove the quadratic formula by the method of completing the square. (b) Show that the equation $ax^2 + bx + c = 0$, where $a \neq 0$, has two equal roots if and only if $b^2 = 4ac$.

R.2 Intervals and Absolute Values

The number x belongs to the interval $[a - r, a + r]$ when $|x - a| \leqslant r$.

In this section, two important notations are discussed. The first is that of intervals on the real-number line, and the second is the absolute value $|x|$, which is the distance from the origin (zero) to x.

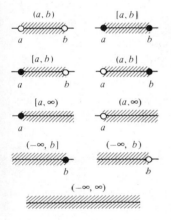

Figure R.2.1. The nine types of intervals.

We begin by listing the notations used for different kinds of intervals.

(a, b) means all x such that $a < x < b$ (open interval).
$[a, b]$ means all x such that $a \leqslant x \leqslant b$ (closed interval).
$[a, b)$ means all x such that $a \leqslant x < b$ (half-open interval).
$(a, b]$ means all x such that $a < x \leqslant b$ (half-open interval).
$[a, \infty)$ means all x such that $a \leqslant x$ (half-open interval).
(a, ∞) means all x such that $a < x$ (open interval).
$(-\infty, b]$ means all x such that $x \leqslant b$ (half-open interval).
$(-\infty, b)$ means all x such that $x < b$ (open interval).
$(-\infty, \infty)$ means all real numbers (open interval).

These collections of real numbers are illustrated in Fig. R.2.1. A black dot indicates that the corresponding endpoint is included in the interval; a white circle indicates that the endpoint is not included in the interval. Notice that a closed interval contains both its endpoints, a half-open interval contains one endpoint, and an open interval contains none.

Warning The symbol ∞ ("infinity") does not denote a real number. It is merely a placeholder to indicate that an interval extends without limit.

In the formation of intervals, we can allow $a = b$. Thus the interval $[a, a]$ consists of the number a alone (if $a \leqslant x \leqslant a$, then $x = a$), while (a, a), $(a, a]$, and $[a, a)$ contain no numbers at all.

Many collections of real numbers are not intervals. For example, the integers, $\ldots, -3, -2, -1, 0, 1, 2, 3, \ldots$, form a collection of real numbers which cannot be designated as a single interval. The same goes for the rational numbers, as well as the collection of all x for which $x^2 - 2x - 3 > 0$. (See Fig. R.1.3.) A collection of real numbers is also called a *set* of real numbers. Intervals are examples of sets of real numbers, but not every set is an interval.

We will often use capital letters to denote sets of numbers. If A is a set and x is a number, we write $x \in A$ and say that "x is an element of A" if x belongs to the collection A. For example, if we write $x \in [a, b]$ (read "x is an element of $[a, b]$"), we mean that x is a member of the collection $[a, b]$; that is, $a \leqslant x \leqslant b$. Similar notation is used for the other types of intervals.

Example 1   True or false: (a) $3 \in [1, 8]$, (b) $-1 \in (-\infty, 2)$, (c) $1 \in [0, 1)$, (d) $8 \in (-8, \infty)$, (e) $3 - 5 \in \mathbb{Z}$, where $\mathbb{Z}$ denotes the set of integers.

Solution   (a) True, because $1 \leqslant 3 \leqslant 8$ is true;
(b) true, because $-1 < 2$ is true;
(c) false, because $0 \leqslant 1 < 1$ is false ($1 < 1$ is false);
(d) true, because $-8 < 8$ is true.
(e) true, because $3 - 5 = -2$, which is an integer. ▲

Example 2   Prove: If $a < b$, then $(a + b)/2 \in (a, b)$.

Solution   We must show that $a < (a + b)/2 < b$. For the first inequality:

$$a < b,$$
$$2a < a + b \qquad \text{(adding } a\text{)},$$
$$a < \frac{a + b}{2} \qquad \text{(multiplying by } \tfrac{1}{2}\text{)}.$$

The proof that $(a + b)/2 < b$ is done similarly; add b to $(a < b)$ and divide by 2. Thus $a < (a + b)/2 < b$; i.e., the average of two numbers lies between them. ▲

Example 3   Let A be the set consisting of those x for which $x^2 - 2x - 3 > 0$. Describe A in terms of intervals.

Solution   From Example 10(b), Section R.1, A consists of those x for which $x > 3$ or $x < -1$. In terms of intervals, A consists of $(3, \infty)$ and $(-\infty, -1)$, as in Fig. R.1.3. ▲

If a real number x is considered as a point on the number line, the distance between this point and zero is called the *absolute value* of x. If x is positive or zero, the absolute value of x is equal to x itself. If x is negative, the absolute value of x is equal to the positive number $-x$ (see Fig. R.2.2). The absolute value of x is denoted by $|x|$. For instance, $|8| = 8$, $|-7| = 7$, $|-10^8| = 10^8$.

Figure R.2.2. The absolute value measures the distance to the origin.

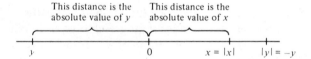

This distance is the absolute value of y This distance is the absolute value of x

y 0 $x = |x|$ $|y| = -y$

Absolute Value

The absolute value $|x|$ of a real number x is equal to x if $x \geqslant 0$ and $-x$ if $x < 0$. To compute $|x|$, change the sign of x, if necessary, to make a non-negative number.

Example 4   Find (a) $|-6|$, (b) $|8 \cdot (-3)|$, (c) $|(-2) \cdot (-5)|$, (d) all x such that $|x| = 2$.

Solution   (a) $|-6| = 6$.
(b) $|8 \cdot (-3)| = |-24| = 24$.
(c) $|(-2)(-5)| = |10| = 10$.
(d) If $|x| = 2$ and $x \geqslant 0$, we must have $x = 2$. If $|x| = 2$ and $x < 0$, we must have $-x = 2$; that is, $x = -2$. Thus $|x| = 2$ if and only if $x = \pm 2$. ▲

For any real number x, $|x| \geqslant 0$, and $|x| = 0$ exactly when $x = 0$. If b is a positive number, there are two numbers having b as their absolute value: b and $-b$. Geometrically, if $x < 0$, $|x|$ is the "mirror image" point which is obtained from x by flipping the line over, keeping zero fixed.

If x_1 and x_2 are any two real numbers, the distance between x_1 and x_2 is $x_1 - x_2$ if $x_1 > x_2$ and $x_2 - x_1$ if $x_1 < x_2$. (See Fig. R.2.3 and note that the position of zero in this figure is unimportant.) Since $x_1 \geqslant x_2$ if and only if $x_1 - x_2 \geqslant 0$, and $x_2 - x_1 = -(x_1 - x_2)$, we have the result shown in the next box.

Figure R.2.3. The distance between x_1 and x_2 is $|x_1 - x_2|$.

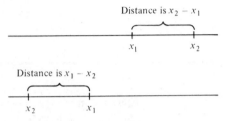

Distance Formula on the Line

If x_1 and x_2 are points on the number line, the distance between x_1 and x_2 is equal to $|x_1 - x_2|$.

Example 5 Describe as an interval the set of real numbers x for which $|x - 8| \leqslant 3$.

Solution $|x - 8| \leqslant 3$ means that $x - 8 \leqslant 3$ in case $x - 8 \geqslant 0$ and $-(x - 8) \leqslant 3$ in case $x - 8 < 0$. In the first case, we have $x \leqslant 11$ and $x \geqslant 8$. In the second case, we have $x \geqslant 5$ and $x < 8$. Thus $|x - 8| \leqslant 3$ if and only if $x \in [5, 11]$. ▲

Example 6 Describe the interval $(4, 9)$ by a single inequality involving absolute values.

Solution Let m be the midpoint of the interval $(4, 9)$; that is, $m = \frac{1}{2}(4 + 9) = \frac{13}{2}$. A number x belongs to $(4, 9)$ if and only if the distance from x to m is less than the distance from 9 to m, which is $|9 - \frac{13}{2}| = \frac{5}{2}$. (Note that the distance from 4 to m is $|4 - \frac{13}{2}| = |-\frac{5}{2}| = \frac{5}{2}$ as well.) So we have $x \in (4, 9)$ if and only if $|x - \frac{13}{2}| < \frac{5}{2}$. (See Fig. R.2.4.) ▲

The most important algebraic properties of absolute values are listed below.

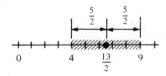

Figure R.2.4. The interval $(4, 9)$ may be described by the inequality $|x - \frac{13}{2}| < \frac{5}{2}$.

Properties of Absolute Values

If x and y are any real numbers:

1. $|x + y| \leqslant |x| + |y|$
2. $|xy| = |x| \, |y|$
3. $|x| = \sqrt{x^2}$

Example 7 Show by example that $|x + y|$ is not always equal to $|x| + |y|$.

Solution Let $x = 3$ and $y = -5$. Then $|x + y| = |3 - 5| = 2$, while $|x| + |y| = 3 + 5 = 8$. (Many other numbers will work as well. In fact, $|x + y|$ will be less than $|x| + |y|$ whenever x and y have opposite signs.) ▲

The preceding example illustrates the general relation between $|x + y|$ and $|x| + |y|$: they are equal if x and y have the same sign, $|x + y| < |x| + |y|$ if x and y have opposite signs.

Example 8   Prove that $|x| = \sqrt{x^2}$.

Solution   For any number x, we have $(-x)^2 = x^2$, so $|x|^2 = x^2$ whatever the sign of x. Thus $|x|$ is a number such that $|x| \geqslant 0$ and $|x|^2 = x^2$, so it is the square root of x^2. ▲

Exercises for Section R.2

1. True or false: (a) $-7 \in [-8, 1]$; (b) $5 \in (\frac{11}{2}, 6]$; (c) $4 \in (-4, 6]$; (d) $4 \in (4, 6)$; (e) $\frac{3}{2} + \frac{7}{2} \in \mathbb{Z}$?

2. Which numbers in the list $-54, -9, -\frac{2}{3}, 0, \frac{5}{16}, \frac{1}{2}$, $8, 32, 100$ belong to which of the following intervals?

    (a) $[-10, 1)$ (b) $(-\infty, 44)$

    (c) $(75, 500)$ (d) $(-20, -\frac{2}{3})$

    (e) $(-9, \frac{5}{16}]$

3. Prove: If $a < b$ then $a < (2a + b)/3$.
4. Prove: If $a < b$ then $(a - 4b)/3 < -b$.

Describe the solutions of the inequalities in Exercises 5–12 in terms of intervals.

5. $x + 4 \geqslant 7$
6. $2x + 5 \leqslant -x + 1$
7. $x > 4x - 6$
8. $5 - x > 4 - 2x$
9. $x^2 + 2x - 3 > 0$
10. $2x^2 - 6 \leqslant 0$
11. $x^2 - x \geqslant 0$
12. $(2x + 1)(x - 5) \leqslant 0$

Find the absolute values in Exercises 13–20.

13. $|3 - 5|$
14. $|3 + 5|$
15. $|-3 - 5|$
16. $|-3 + 5|$
17. $|3 \cdot 5|$
18. $|(-3)(-5)|$
19. $|(-3) \cdot 5|$
20. $|3 \cdot (-5)|$

21. Find all x such that $|x| = 8$.
22. Find all x such that $|-x| = 9$.

Describe in terms of absolute values the set of x satisfying the inequalities in Exercises 23–26.

23. $x^2 + 5x > 0$
24. $x^2 - 2x < 0$
25. $x^2 - x - 2 > 0$
26. $x^2 + 5x + 7 < 0$

Express each of the inequalities in Exercises 27–32 in the form "x belongs to the interval ...".

27. $3 < x \leqslant 4$
28. $x > 5$
29. $|x| < 5$
30. $|x - 3| \leqslant 6$
31. $|3x + 1| < 2$
32. $x^2 - 3x + 2 \geqslant 0$

Express each of the statements in Exercises 33–38 in terms of an inequality involving absolute values:

33. $x \in (-3, 3)$
34. $-x \in (-4, 4)$
35. $x \in (-6, 6)$
36. $x \in (2, 6)$
37. $x \in [-8, 12]$
38. $|x| \in (0, 1)$

39. Show by example that $|x + y + z|$ is not always equal to $|x| + |y| + |z|$.
40. Show by example that $|x - y|$ need not equal $||x| - |y||$.
41. Prove that $x = \sqrt[3]{x^3}$.
42. Prove that $|x| = \sqrt[4]{x^4}$.
43. Is the formula $|x - y| \leqslant |x| - |y|$ always true?
44. Using the formula $|xy| = |x||y|$, find a formula for $|a/b|$. [*Hint:* Let $x = b$ and $y = a/b$.]

R.3 Laws of Exponents

Fractional and integer exponents obey similar laws.

The expression b^n, where b is a real number called the *base* and n is a natural number called the *exponent*, is defined as the product of b with itself n times:

$$b^n = b \cdot b \cdot \cdots \cdot b \qquad (n \text{ times}).$$

This operation of raising a number to a power, or *exponentiation*, has the following properties, called *laws of exponents*:

Laws of Exponents: Integer Powers

1. $b^n b^m = b^{n+m}$
2. $(b^n)^m = b^{nm}$
3. $(bc)^n = b^n c^n$

These laws can all be understood and remembered using common sense. For example, $b^{m+n} = b^m b^n$ because

$$\underbrace{b \cdot b \cdot b \cdot \cdots \cdot b}_{m+n \text{ times}} = \underbrace{\underbrace{(b \cdot b \cdot \cdots \cdot b)}_{m \text{ times}} \underbrace{(b \cdot b \cdot \cdots \cdot b)}_{n \text{ times}}}_{m+n \text{ times in all}}.$$

Likewise

$$(b^n)^m = \underbrace{(b^n)(b^n) \cdots (b^n)}_{m \text{ times}}$$

$$= \underbrace{(b \cdots b)(b \ldots b)(b \ldots b)}_{nm \text{ times}}$$

$$= b^{nm}.$$

and

$$(bc)^n = \underbrace{bcbc \ldots bc}_{n \text{ times}} = \underbrace{(b \ldots b)}_{n \text{ times}} \underbrace{(c \ldots c)}_{n \text{ times}}$$

$$= b^n c^n.$$

Example 1   Simplify

(a) $2^{10} \cdot 5^{10}$ (b) $\dfrac{(3 \cdot 2)^{10} + 3^9}{3^9}$

Solution   (a) $2^{10} \cdot 5^{10} = (2 \cdot 5)^{10} = 10^{10}$.

(b) $\dfrac{(3 \cdot 2)^{10} + 3^9}{3^9} = \dfrac{3^{10} \cdot 2^{10} + 3^9}{3^9} = \dfrac{3^9 \cdot 3 \cdot 2^{10} + 3^9}{3^9} = 3 \cdot 2^{10} + 1.$ ▲

The first of the three laws of exponents is particularly important; it is the basis for extending the operation of exponentiation to allow negative exponents. If b^0 were defined, we ought to have $b^0 b^n = b^{0+n} = b^n$. If $b \neq 0$, then $b^n \neq 0$, and the equation $b^0 b^n = b^n$ implies that b^0 must be 1. We take this as the *definition* of b^0, noting that 0^0 is not defined (see Exercise 31).

If n is a natural number, then b^{-n} is defined in order to make $b^{-n}b^n = b^{-n+n} = b^0 = 1$; that is, $b^{-n} = 1/b^n$.

Negative Powers: Definition and Laws of Exponents

If b is a real number and n is a positive integer, we define

$$b^{-n} = \frac{1}{b^n} \, .$$

The laws of exponents given in the preceding box remain valid for integers n, m; positive, negative, or zero.

For example, let us show that $b^{n+m} = b^n b^m$ is valid if $n = -q$ is negative and m is positive, with $m > q$. Then $b^{n+m} = b^{m-q} = b \cdot \cdots \cdot b$ ($m - q$ times). Also,

$$b^n b^m = b^{-q} b^m = \frac{b^m}{b^q} = \frac{\overbrace{b \cdots bb}^{m \text{ times}}}{\underbrace{b \ldots b}_{q \text{ times}}} = \overbrace{b \ldots b}^{m - q \text{ times}} \, .$$

Thus $b^{n+m} = b^n b^m$. The other cases and laws are verified similarly.

Example 2   Simplify

(a) $\dfrac{(2 \cdot 3)^{-2} \cdot 4}{(1/3)^2}$.

(b) $[(8/3)^2 - (3/8)^3][(8/3)^{-2} + (3/8)^{-3}]$.

Solution   (a) $\dfrac{(2 \cdot 3)^{-2} 4}{(1/3)^2} = \dfrac{4 \cdot 3^2}{2^2 \cdot 3^2} = 1.$

(b) Multiplying out, the given expression becomes

$$\left(\frac{8}{3}\right)^2 \left(\frac{8}{3}\right)^{-2} + \left(\frac{8}{3}\right)^2 \left(\frac{3}{8}\right)^{-3} - \left(\frac{3}{8}\right)^3 \left(\frac{8}{3}\right)^{-2} - \left(\frac{3}{8}\right)^3 \left(\frac{3}{8}\right)^{-3}$$

$$= 1 + \left(\frac{8}{3}\right)^2 \left(\frac{8}{3}\right)^3 - \left(\frac{3}{8}\right)^3 \left(\frac{3}{8}\right)^2 - 1$$

$$= \left(\frac{8}{3}\right)^5 - \left(\frac{3}{8}\right)^5. \quad \blacktriangle$$

To define $b^{1/n}$, we require $b^{1/n} \cdot b^{1/n} \cdot \cdots \cdot b^{1/n} = b^{1/n + \cdots + 1/n} = b^1 = b$ to hold—that is, $(b^{1/n})^n$ ought to be equal to b. Thus we declare $b^{1/n}$ to be $\sqrt[n]{b}$, that positive number whose nth power is b; i.e., $b^{1/n}$ is defined by the equation $(b^{1/n})^n = b$. (If n is odd, then $\sqrt[n]{b}$ may be defined even if b is negative, but we will reserve the notation $b^{1/n}$ for the case $b > 0$.)

Finally, if $r = m/n$ is a rational number, we define $b^r = b^{m/n} = (b^m)^{1/n}$. We leave it to you to verify that the result is independent of the way in which r is expressed as a quotient of positive integers; for instance, $(b^4)^{1/6} = (b^6)^{1/9}$ (see Exercise 32).

Having defined b^r for $b > 0$ and r rational, one can go back and prove the laws of exponents for this general case. These laws are useful for calculations with rational exponents.

Let us first check that $(bc)^{1/n} = b^{1/n}c^{1/n}$. Now $(bc)^{1/n}$ is that number whose nth power is bc; but $(b^{1/n}c^{1/n})^n = (b^{1/n})^n(c^{1/n})^n = bc$, by Property 3 for integer powers and the fact that $(b^{1/n})^n = b$. Thus $(b^{1/n}c^{1/n})^n = bc$, which means that $b^{1/n}c^{1/n} = (bc)^{1/n}$.

Using this, we can check that $b^{p+q} = b^p b^q$ as follows. Let $p = m/n$ and $q = k/l$. Then

$$b^{p+q} = b^{m/n+k/l} = b^{(ml+kn)/nl} = \left(b^{ml+kn}\right)^{1/nl}$$

$$= \left(b^{ml}b^{kn}\right)^{1/nl} \qquad \text{(by Property 1 for integer powers)}$$

$$= \left(b^{ml}\right)^{1/nl}\left(b^{kn}\right)^{1/nl} \qquad \left(\text{by the law } (bc)^{1/n} = b^{1/n}c^{1/n} \text{ just proved}\right)$$

$$= b^{ml/nl}b^{kn/nl} \qquad \text{(by the definition of } b^{m/n})$$

$$= b^p b^q.$$

The other properties are checked in a similar way (Exercises 33 and 34).

Rational Powers

Rational powers are defined by:

$$b^n = b \cdot \cdots \cdot b \ (n \text{ times}); \ b^0 = 1$$
$$b^{-n} = 1/b^n$$
$$b^{1/n} = \sqrt[n]{b} \ \text{if } b > 0 \text{ and } n \text{ is a natural number}$$
$$b^{m/n} = (b^m)^{1/n}$$

If $b, c > 0$ and p, q are rational, then:

1. $b^{p+q} = b^p b^q$
2. $b^{pq} = (b^p)^q$
3. $(bc)^p = b^p c^p$
4. $b^p < b^q$ if $b > 1$ and $p < q$; $b^p > b^q$ if $b < 1$ and $p < q$.

Example 3 Find $8^{-2/3}$ and $9^{3/2}$.

Solution $8^{-2/3} = 1/8^{2/3} = 1/(\sqrt[3]{8}\,)^2 = 1/2^2 = 1/4.$ $\qquad 9^{3/2} = (\sqrt{9}\,)^3 = 3^3 = 27.$ ▲

Example 4 Simplify $[x^{2/3}(x^{-3/2})]^{8/3}$ and $(x^{2/3})^{5/2}/x^{1/4}$.

Solution $(x^{2/3}x^{-3/2})^{8/3} = (x^{2/3-3/2})^{8/3} = (x^{-5/6})^{8/3} = x^{-20/9} = 1/\sqrt[9]{x^{20}}$.

$\qquad \left(x^{2/3}\right)^{5/2}/x^{1/4} = x^{2/3 \cdot 5/2 - 1/4} = x^{5/3 - 1/4} = x^{17/12}.$ ▲

Example 5 We defined $b^{m/n}$ as $(b^m)^{1/n}$. Show that $b^{m/n} = (b^{1/n})^m$ as well.

Solution We must show that $(b^{1/n})^m$ is the nth root of b^m. But $[(b^{1/n})^m]^n = (b^{1/n})^{mn} = (b^{1/n})^{nm} = [(b^{1/n})^n]^m = b^m$; this calculation used only the laws of *integer* exponents and the fact that $(b^{1/n})^n = b$. ▲

Example 6 Remove the square roots in the denominator:

$$\frac{1}{\sqrt{x-a}+\sqrt{x-b}}.$$

Solution There is a useful trick called *rationalizing*. We multiply top and bottom by $\sqrt{x-a}-\sqrt{x-b}$, giving

$$\frac{\sqrt{x-a}-\sqrt{x-b}}{(\sqrt{x-a}+\sqrt{x-b})(\sqrt{x-a}-\sqrt{x-b})}=\frac{\sqrt{x-a}-\sqrt{x-b}}{(\sqrt{x-a})^2-(\sqrt{x-b})^2}$$

$$=\frac{\sqrt{x-a}-\sqrt{x-b}}{b-a}.\ \blacktriangle$$

Example 7 Assume that the cost of food doubles every 6 years. By what factor has it increased after

(a) 12 years?   (b) 18 years?   (c) 3 years?   (d) 20 years?

Solution (a) Since the cost doubles in 6 years, in 12 years it increases by a factor of $2 \cdot 2 = 4$.

(b) In $18 = 6 + 6 + 6$ years it increases by a factor $2 \cdot 2 \cdot 2 = 8$.

(c) In 3 years, let it increase by a factor k. Then in 6 years we get $k \cdot k = 2$, so $k = \sqrt{2} = 2^{3/6} \approx 1.4142$.

(d) In 20 years the factor is $2^{20/6} = 2^{10/3} = \sqrt[3]{2^{10}} \approx 10.0794.$ $\blacktriangle$

Exercises for Section R.3

Simplify the expressions in Exercises 1–20.

1. $3^2 \cdot \left(\dfrac{1}{3}\right)^2$

2. $8^3 \cdot \left(\dfrac{1}{4}\right)^2$

3. $\dfrac{(4 \cdot 3)^{10} + 4^9}{8^4}$

4. $\dfrac{(2 \cdot 3)^{16} + 3^{15}}{3^{15}}$

5. $\dfrac{(4 \cdot 3)^{-6} \cdot 8}{9^3}$

6. $\dfrac{(27)^{-1}}{3^{-3}}$

7. $\left(\dfrac{8^{-2}}{4^{-4}}\right)^4$

8. $\left[\left(\dfrac{1}{2}\right)^{-1} + \left(\dfrac{1}{3}\right)^{-3}\left(\dfrac{1}{2}\right)^{-1} - \left(\dfrac{1}{3}\right)^{-2}\right]$

9. $9^{1/2}$

10. $16^{1/4}$

11. $(1/9)^{-1/2}$

12. $(1/16)^{-1/4}$

13. $2^{5/3}/4^{7/3}$

14. $3^{-8/11}(1/9)^{-4/11}$

15. $12^{2/3} \cdot 18^{2/3}$

16. $20^{7/2} \cdot 5^{-7/2}$

17. $(x^{3/2} + x^{5/2})x^{-3/2}$

18. $(x^{3/4})^{8/3}$

19. $x^{5/2}(x^{-3/2} + 2x^{1/2} + 3x^{7/2})$

20. $y^{1/2}(1/y + 2\sqrt{y} + y^{-1/3})$

Using the laws of rational exponents, verify the root formulas in Exercises 21 and 22.

21. $\sqrt[a]{\sqrt[b]{x}} = \sqrt[ab]{x}$

22. $\sqrt[ac]{x^{ab}} = \sqrt[c]{x^b}$

Simplify the expressions in Exercises 23 and 24 by writing with rational exponents.

23. $\left[\dfrac{\sqrt[4]{ab^3}}{\sqrt{b}}\right]^6$

24. $\sqrt[3]{\dfrac{\sqrt{a^3b^9}}{\sqrt[4]{a^6b^6}}}$

25. The price of housing doubles every 10 years. By what factor does it increase after 20 years? 30 years? 50 years?

📱[1]26. Money in a certain bank account grows by a factor of 1.1 every year. If an initial deposit of $100 is made, how much money will be in the account after 10 years?

Factor the expressions in Exercises 27–30 using fractional exponents. For example: $x + 2\sqrt{2xy} + 2y = (x^{1/2} + (2y)^{1/2})^2$.

27. $x - \sqrt{xy} - 2y$

28. $\sqrt[3]{xy^2} + \sqrt[3]{yx^2} + x + y$

29. $x - 2\sqrt{x} - 8$

30. $x + 2\sqrt{3x} + 3$

★31. Since $0^x = 0$ for any positive rational x, 0^0 ought to be zero. On the other hand, $b^0 = 1$ for any $b > 0$, so 0^0 ought to be 1. Are *both* choices consistent with the laws of exponents?

★32. Suppose that $b > 0$ and that $p = m/n = m'/n'$. Show, using the definition of rational powers, that $b^{m/n} = b^{m'/n'}$; that is, b^p is unambiguously defined. [*Hint:* Raise both $b^{m/n}$ and $b^{m'/n'}$ to the power nn'.]

★33. Prove Rules 2 and 3 for rational powers.

★34. Let $b > 1$ and p and q be rational numbers with $p < q$. Prove that $b^p < b^q$. Deduce the corresponding result for $b < 1$ by using $b^p = (1/b)^{-p}$.

R.4 Straight Lines

The graph $y = ax + b$ is a straight line in the xy plane.

In this section, we review some basic analytic geometry. We will develop the point-slope form of the equation of a straight line which will be essential for calculus.

One begins the algebraic representation of the plane by drawing two perpendicular lines, called the x and y axes, and the placing the real numbers on each of these lines, as shown in Fig. R.4.1. Any point P in the plane can

Figure R.4.1. The x and y axes in the plane.

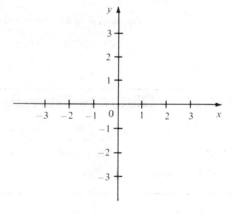

now be described by the pair (a, b) of real numbers obtained by dropping perpendiculars to the x and y axes, as shown in Fig. R.4.2. The numbers which describe the point P are called the *coordinates* of P: the first coordinate listed is called the x coordinate; the second is the y coordinate. We can use any letters we wish for the coordinates, including x and y themselves.

Often the point with coordinates (a, b) is simply called "the point (a, b)." Drawing a point (a, b) on a graph is called *plotting* the point; some points are plotted in Fig. R.4.3. Note that the point $(0, 0)$ is located at the intersection of the coordinate axes; it is called the *origin* of the coordinate system.

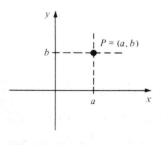

Figure R.4.2. The point P has coordinates (a, b).

[1] 📱 This symbol denotes exercises or discussions that may require use of a hand-held calculator.

Figure R.4.3. Examples of plotted points.

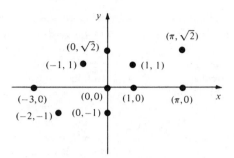

Example 1   Let $a = 3$ and $b = 2$. Plot the points (a, b), (b, a), $(-a, b)$, $(a, -b)$, and $(-a, -b)$.

Solution   The points to be plotted are $(3, 2)$, $(2, 3)$, $(-3, 2)$, $(3, -2)$, and $(-3, -2)$; they are shown in Fig. R.4.4. ▲

The theorem of Pythagoras leads to a simple formula for the distance between two points (see Fig. R.4.5):

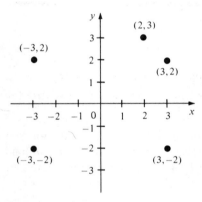

Figure R.4.4. More plotted points.

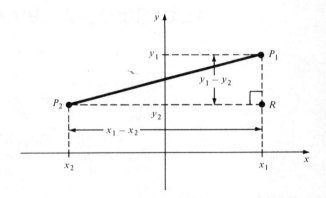

Figure R.4.5. By the Pythagorean theorem,
$$|P_1 P_2| = \sqrt{|P_2 R|^2 + |P_1 R|^2} = \sqrt{(x_1 - x_2)^2 + (y_1 - y_2)^2} .$$

Distance Formula

If P_1 has coordinates (x_1, y_1) and P_2 has coordinates (x_2, y_2), the distance from P_1 to P_2 is $\sqrt{(x_1 - x_2)^2 + (y_1 - y_2)^2}$. The distance between P_1 and P_2 is denoted $|P_1 P_2|$.

Example 2   Find the distance from $(6, -10)$ to $(2, -1)$.

Solution   The distance is

$$\sqrt{(6 - 2)^2 + \left[-10 - (-1) \right]^2} = \sqrt{4^2 + (-9)^2}$$
$$= \sqrt{16 + 81} = \sqrt{97} \approx 9.85. ▲$$

If we have two points on the x axis, $(x_1, 0)$ and $(x_2, 0)$, the distance between them is $\sqrt{(x_1 - x_2)^2 + (0 - 0)^2} = \sqrt{(x_1 - x_2)^2} = |x_1 - x_2|$. Thus *the distance formula in the plane includes the distance formula on the line* as a special case.

Draw a line[2] l in the plane and pick two distinct points P_1 and P_2 on l. Let P_1 have coordinates (x_1, y_1) and P_2 have coordinates (x_2, y_2). The ratio $(y_2 - y_1)/(x_2 - x_1)$ (assuming that $x_2 \neq x_1$) is called the *slope* of the line l and is often denoted by the letter m. See Fig. R.4.6.

Figure R.4.6. The slope of this line is $(y_2 - y_1)/(x_2 - x_1)$.

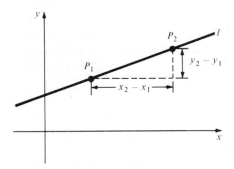

Slope Formula

If (x_1, y_1) and (x_2, y_2) lie on the line l, the slope of l is

$$m = \frac{y_2 - y_1}{x_2 - x_1}.$$

An important feature of the slope m is that it does not depend upon which two points we pick, so long as they lie on the line l. To verify this, we observe (see Fig. R.4.7) that the right triangles $P_1 P_2 R$ and $P_1' P_2' R'$ are similar, since corresponding angles are equal, so $P_2 R / P_1 R = P_2' R' / P_1' R'$. In other words, the slope calculated using P_1 and P_2 is the same as the slope calculated using P_1' and P_2'. The slopes of some lines through the origin are shown in Fig. R.4.8.

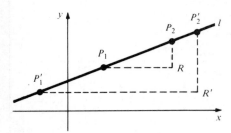

Figure R.4.7. The slope does not depend on which two points on l are used.

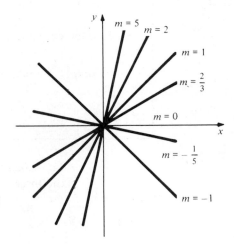

Figure R.4.8. Slopes of some lines through the origin.

Example 3   What is the slope of the line which passes through the points $(0, 1)$ and $(1, 0)$?

Solution   By the slope formula, with $x_1 = 0$, $y_1 = 1$, $x_2 = 1$, and $y_2 = 0$, the slope is $(0 - 1)/(1 - 0) = -1$. ▲

---

[2] In this book when we use the term *line* with no other qualification, we shall mean a *straight line*. Rather than referring to "curved lines," we will use the term *curve*.

Warning   A line which is parallel to the y axis does not have a slope. In fact, any two points on such a line have the same x coordinates, so when we form the ratio $(y_2 - y_1)/(x_2 - x_1)$, the denominator becomes zero, which makes the expression meaningless. A vertical line has the equation $x = x_1$; y can take any value.

To find the equation satisfied by the coordinates of the points on a line, we consider a line l with slope m and which passes through the point (x_1, y_1). If (x, y) is any *other* point on l, the slope formula gives

$$\frac{y - y_1}{x - x_1} = m.$$

That is,

$$y = y_1 + m(x - x_1).$$

This is called the *point-slope form* of the equation of l; a general point (x, y) lies on l exactly when the equation holds.

If, for the point in the point-slope form of the equation, we take the point $(0, b)$ where l intersects the y axis (the number b is called the y *intercept* of l), we have $x_1 = 0$ and $y_1 = b$ and obtain the *slope-intercept form* $y = mx + b$.

If we are given two points (x_1, y_1) and (x_2, y_2) on a line, we know that the slope is $(y_2 - y_1)/(x_2 - x_1)$. Substituting this term for m in the point-slope form of the equation gives the *point-point form*:

$$y = y_1 + \left(\frac{y_2 - y_1}{x_2 - x_1} \right)(x - x_1).$$

Straight Lines

Name	Data needed	Formula
point-slope	one point (x_1, y_1) on the line and the slope m	$y = y_1 + m(x - x_1)$
slope-intercept	the slope m of the line and the y-intercept b	$y = mx + b$
point-point	two points (x_1, y_1) and (x_2, y_2) on the line	$y = y_1 + \left(\dfrac{y_2 - y_1}{x_2 - x_1} \right)(x - x_1)$

For calculus, the point-slope form will turn out to be the most important of the three forms of the equation of a line, illustrated in Fig. R.4.9(a).

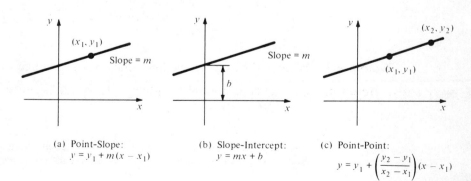

Figure R.4.9. Three forms for the equation of a line.

(a) Point-Slope:
$y = y_1 + m(x - x_1)$

(b) Slope-Intercept:
$y = mx + b$

(c) Point-Point:
$y = y_1 + \left(\dfrac{y_2 - y_1}{x_2 - x_1} \right)(x - x_1)$

Example 4   Find the equation of the line through $(1, 1)$ with slope 5. Put the equation into slope-intercept form.

Solution   Using the point-slope form, with $x_1 = 1$, $y_1 = 1$, and $m = 5$, we get $y = 1 + 5(x - 1)$. This simplifies to $y = 5x - 4$, which is the slope-intercept form. ▲

Example 5   Let l be the line through the points $(3, 2)$ and $(4, -1)$. Find the point where this line intersects the x axis.

Solution   The equation of the line, in point-point form, with $x_1 = 3$, $y_1 = 2$, $x_2 = 4$, and $y_2 = -1$, is

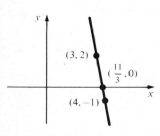

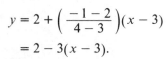

$$y = 2 + \left(\frac{-1 - 2}{4 - 3} \right)(x - 3)$$
$$= 2 - 3(x - 3).$$

The line intersects the x axis at the point where $y = 0$, that is, where
$$0 = 2 - 3(x - 3).$$

Solving this equation for x, we get $x = \frac{11}{3}$, so the point of intersection is $(\frac{11}{3}, 0)$. (See Fig. R.4.10.) ▲

Figure R.4.10. Finding where the line through $(3, 2)$ and $(4, -1)$ meets the x axis.

Example 6   Find the slope and y intercept of the line $3y + 8x + 5 = 0$.

Solution   The following equations are equivalent:
$$3y + 8x + 5 = 0,$$
$$3y = -8x - 5,$$
$$y = -\tfrac{8}{3}x - \tfrac{5}{3}.$$

The last equation is in slope-intercept form, with slope $-\frac{8}{3}$ and y intercept $-\frac{5}{3}$. ▲

Using the method of Example 6, one can show that any equation of the form $Ax + By + C = 0$ describes a straight line, as long as A and B are not both zero. If $B \neq 0$, the slope of the line is $-A/B$; if $B = 0$, the line is vertical; and if $A = 0$, it is horizontal.

Finally, we recall without proof the fact that lines with slopes m_1 and m_2 are perpendicular if and only if $m_1 m_2 = -1$. In other words, *the slopes of perpendicular lines are negative reciprocals of each other*.

Example 7   Find the equation of the line through $(0, 0)$ which is perpendicular to the line $3y - 2x + 8 = 0$.

Solution   The given equation has the form $Ax + By + C = 0$, with $A = -2$, $B = 3$, and $C = 8$; the slope of the line it describes is $-A/B = \frac{2}{3} = m_1$. The slope m_2 of the perpendicular line must satisfy $m_1 m_2 = -1$, so $m_2 = -\frac{3}{2}$. The line through the origin with this slope has the equation $y = -\frac{3}{2}x$. ▲

Exercises for Section R.4

1. Plot the points $(0, 0)$, $(1, 1)$, $(-1, -1)$, $(2, 8)$, $(-2, -8)$, $(3, 27)$, and $(-3, -27)$.
2. Plot the points $(-1, 2)$, $(-1, -2)$, $(1, -2)$, and $(1, 2)$.

3. Plot the points (x, x^2) for $x = -2, -\frac{3}{2}, -1, -\frac{1}{2}, 0, \frac{1}{2}, 1, \frac{3}{2}, 2$.
4. Plot the points $(x, x^4 - x^2)$ for $x = -2, -\frac{3}{2}, -1, \frac{1}{2}, 0, \frac{1}{2}, 1, \frac{3}{2}, 2$.

Find the distance between each of the pairs of points in Exercises 5–10.

5. $(1, 1), (1, -1)$
6. $(-1, 1), (-1, -1)$
7. $(-3, 9), (2, -8)$
8. $(0, 0), (-3, 27)$
9. $(43721, 56841), (3, 56841)$
10. $(839, 8400), (840, 8399)$

Find the distance or a formula for the distance between each pair of points in Exercises 11–16.

11. $(2, 1), (3, 2)$
12. $(a, 2), (3 + a, 6)$
13. $(x, y), (3x, y + 10)$
14. $(a, 0), (a + b, b)$
15. $(a, a), (-a, -a)$
16. $(a, b), (10a, 10b)$

Find the slope of the line through the points in Exercises 17–20.

17. $(1, 3), (2.6)$
18. $(0, 1), (2, -4)$
19. $(-1, 6), (1, -1)$
20. $(0, 0), (-1, -1)$

In Exercises 21–24 find the equation of the line through the point P with slope m, and sketch a graph of the line.

21. $P = (2, 3)$, $m = 2$
22. $P = (-2, 6)$, $m = -\frac{1}{2}$
23. $P = (-1, 7)$, $m = 0$
24. $P = (7, -1)$, $m = 1$

Find the equation of the line through the pairs of points in Exercises 25–28.

25. $(5, 7), (-1, 4)$
26. $(1, 1), (3, 2)$
27. $(1, 4), (3, 4)$
28. $(1, 4), (1, 6)$

Find the slope and y intercept of each of the lines in Exercises 29–36.

29. $x + 2y + 4 = 0$
30. $\frac{1}{2}x - 3y + \frac{1}{3} = 0$
31. $4y = 17$
32. $2x + y = 0$
33. $13 - 4x = 7(x + y)$
34. $x - y = 14(x + 2y)$
35. $y = 17$
36. $x = 60$

37. (a) Find the slope of the line $4x + 5y - 9 = 0$.
    (b) Find the equation of the line through $(1, 1)$ which is perpendicular to the line in part (a).
38. (a) Find the slope of the line $2x - 8y - 10 = 0$.
    (b) Find the equation of the line through $(1, 0)$ which is perpendicular to the line in (a).

Find the equation of the line with the given data in Exercises 39–42.

39. Slope $= 5$; y intercept $= 14$
40. y intercept $= 6$; passes through $(7, 8)$
41. Passes through $(4, 2)$ and $(2, 4)$
42. Passes through $(-1, -1)$; slope $= -10$

43. Find the coordinates of the point which is a distance 3 from the x axis and a distance 5 from $(1, 2)$.
★44. (a) Find the coordinates of a point whose distance from $(0, 0)$ is $2\sqrt{2}$ and whose distance from $(4, 4)$ is $2\sqrt{2}$.
    (b) If $\lambda > 2\sqrt{2}$ show both algebraically and geometrically that there are exactly two points whose distance from $(0, 0)$ is λ and whose distance from $(4, 4)$ is also λ.

R.5 Circles and Parabolas

$(x - a)^2 + (y - b)^2 = r^2$ is a circle and $y = a(x - p)^2 + q$ is a parabola.

We now consider two more geometric figures which can be described by simple algebraic formulas: the circle and the parabola.

The circle C with radius $r > 0$ and center at (a, b) consists of those points (x, y) for which the distance from (x, y) to (a, b) is equal to r. (See Fig. R.5.1.)

The distance formula shows that $\sqrt{(x - a)^2 + (y - b)^2} = r$ or, equivalently, $(x - a)^2 + (y - b)^2 = r^2$. If the center of the circle is at the origin, this equation takes the simpler form $x^2 + y^2 = r^2$.

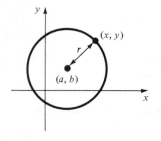

Figure R.5.1. The point (x, y) is a typical point on the circle with radius r and center (a, b).

Example 1   Find the equation of the circle with center $(1,0)$ and radius 5.

Solution   Here $a = 1$, $b = 0$, and $r = 5$, so $(x - a)^2 + (y - b)^2 = r^2$ becomes the equation $(x - 1)^2 + y^2 = 25$ or $x^2 - 2x + y^2 = 24$. ▲

Example 2   Find the equation of the circle whose center is $(2, 1)$ and which passes through the point $(5, 6)$.

Solution   The equation must be of the form $(x - 2)^2 + (y - 1)^2 = r^2$; the problem is to determine r^2. Since the point $(5, 6)$ lies on the circle, it must satisfy the equation. That is,

$$r^2 = (5 - 2)^2 + (6 - 1)^2 = 3^2 + 5^2 = 34,$$

so the correct equation is $(x - 2)^2 + (y - 1)^2 = 34$. ▲

Example 3   Show that the graph of $x^2 + y^2 - 6x - 16y + 8 = 0$ is a circle. Find its center.

Solution   Complete the squares:

$$0 = x^2 + y^2 - 6x - 16y + 8 = x^2 - 6x + y^2 - 16y + 8$$
$$= (x^2 - 6x + 9) + (y^2 - 16y + 64) - 9 - 64 + 8$$
$$= (x - 3)^2 + (y - 8)^2 - 65.$$

Thus the equation becomes $(x - 3)^2 + (y - 8)^2 = 65$, whose graph is a circle with center $(3, 8)$ and radius $\sqrt{65} \approx 8.06$. ▲

Consider next the equation $y = x^2$. If we plot a number of points whose coordinates satisfy this equation, by choosing values for x and computing y, we find that these points may be joined by a smooth curve as in Fig. R.5.2. This curve is called a *parabola*. It is also possible to give a purely geometric definition of a parabola and derive the equation from geometry as was done for the line and circle. In fact, we will do so in Section 14.1.

If x is replaced by $-x$, the value of y is unchanged, so the graph is symmetric about the y axis. Similarly, we can plot $y = 3x^2$, $y = 10x^2$, $y = -\frac{1}{2}x^2$, $y = -8x^2$, and so on. (See Fig. R.5.3.) These graphs are also parabolas. The general parabola of this type has the equation $y = ax^2$, where a

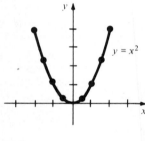

x	2	1.5	1	0.5	0
y	4	2.25	1	0.25	0

Figure R.5.2. The parabola $y = x^2$.

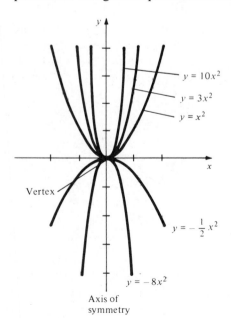

$y = 10x^2$

$y = 3x^2$

$y = x^2$

Vertex

$y = -\frac{1}{2}x^2$

$y = -8x^2$

Axis of symmetry

Figure R.5.3. Parabolas $y = ax^2$ for various values of a.

is a nonzero constant; these parabolas all have their *vertex* at the origin. If $a > 0$ the parabola opens upwards, and if $a < 0$ it opens downwards.

Example 4   Let C be the parabola with vertex at the origin and passing through the point $(2, 8)$. Find the point on C whose x coordinate is 10.

Solution   The equation of C is of the form $y = ax^2$. To find a, we use the fact that $(2, 8)$ lies on C. Thus $8 = a \cdot 2^2 = 4a$, so $a = 2$ and the equation is $y = 2x^2$. If the x coordinate of a point on C is 10, the y coordinate is $2 \cdot 10^2 = 200$, so the point is $(10, 200)$. ▲

A special focusing property of parabolas is of practical interest: a parallel beam of light rays (as from a star) impinging upon a parabola in the direction of its axis of symmetry will focus at a single point as shown in Fig. R.5.4. The property follows from the law that the angle of incidence equals the angle of reflection, together with some geometry or calculus. (See Review Exercises 86 and 87 at the end of Chapter 1.)

Figure R.5.4. The focusing property of a parabolic reflector.

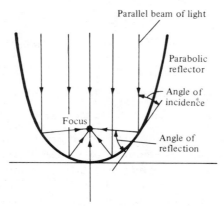

Just as we considered circles with center at an arbitrary point (a, b), we can consider parabolas with vertex at any point (p, q). The equation of such a figure is $y = a(x - p)^2 + q$. We have started with $y = ax^2$, then replaced x by $x - p$ and y by $y - q$ to get

$$y - q = a(x - p)^2, \quad \text{i.e.,} \quad y = a(x - p)^2 + q.$$

This process is illustrated in Fig. R.5.5. Notice that if (x, y) lies on the (shifted) parabola, then the corresponding point on the original parabola is $(x - p, y - q)$, which must therefore satisfy the equation of the original parabola; i.e., $y - q$ must equal $a(x - p)^2$.

Figure R.5.5. The equation $y = q + a(x - p)^2$ is the parabola $y = ax^2$ shifted from $(0, 0)$ to (p, q).

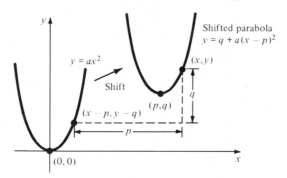

Given an equation of the form $y = ax^2 + bx + c$, we can complete the square on the right-hand side to put it in the form $y = a(x - p)^2 + q$. Thus the graph of any equation $y = ax^2 + bx + c$ is a parabola.

Example 5   Graph $y = -2x^2 + 4x + 1$.

Solution   Completing the square gives

$$y = -2x^2 + 4x + 1 = -2\left(x^2 - 2x - \tfrac{1}{2}\right)$$
$$= -2\left(x^2 - 2x + 1 - 1 - \tfrac{1}{2}\right) = -2\left(x^2 - 2x + 1 - \tfrac{3}{2}\right)$$
$$= -2(x - 1)^2 + 3.$$

The vertex is thus at $(1, 3)$ and the parabola opens downward like $y = -2x^2$ (see Fig. R.5.6). ▲

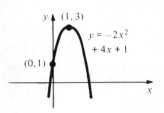

Figure R.5.6. A parabola with vertex at $(1, 3)$.

Equations of Circles and Parabolas

The equation of the circle with radius r and center at (a, b) is

$$(x - a)^2 + (y - b)^2 = r^2.$$

The equation of a parabola with vertex at (p, q) is

$$y = a(x - p)^2 + q.$$

Analytic geometry provides an algebraic technique for finding the points where two geometric figures intersect. If each figure is given by an equation in x and y, we solve for those pairs (x, y) which satisfy both equations.

Two lines will have either zero, one, or infinitely many intersection points; there are none if the two lines are parallel and different, one if the lines have different slopes, and infinitely many if the two lines are the same. For a line and a circle or parabola, there may be zero, one, or two intersection points. (See Fig. R.5.7.)

Figure R.5.7. Intersections of some geometric figures.

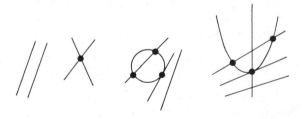

Example 6   Where do the lines $x + 3y + 8 = 0$ and $y = 3x + 4$ intersect?

Solution   To find the intersection point, we solve the simultaneous equations

$$x + 3y + 8 = 0,$$
$$-3x + y - 4 = 0.$$

Multiply the first equation by 3 and add to the second to get the equation $0 + 10y + 20 = 0$, or $y = -2$. Substituting $y = -2$ into the first equation gives $x - 6 + 8 = 0$, or $x = -2$. The intersection point is $(-2, -2)$. ▲

Example 7   Where does the line $x + y = 1$ meet the parabola $y = 2x^2 + 4x + 1$?

Solution   We look for pairs (x, y) which satisfy both equations. We may substitute

$2x^2 + 4x + 1$ for y in the equation of the line to obtain

$$x + 2x^2 + 4x + 1 = 1,$$
$$2x^2 + 5x = 0,$$
$$x(2x + 5) = 0,$$

so $x = 0$ or $-\frac{5}{2}$. We may use either equation to find the corresponding values of y. The linear equation $x + y = 1$ is simpler; it gives $y = 1 - x$, so $y = 1$ when $x = 0$ and $y = \frac{7}{2}$ when $x = -\frac{5}{2}$. Thus the points of intersection are $(0, 1)$ and $(-\frac{5}{2}, \frac{7}{2})$. (See Fig. R.5.8.) ▲

Figure R.5.8. Intersections of the line $x + y = 1$ and the parabola $y = 2x^2 + 4x + 1$ occur at $(0, 1)$ and $(-\frac{5}{2}, \frac{7}{2})$.

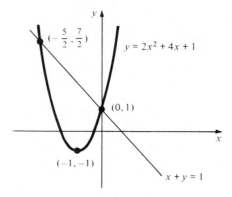

Example 8   (a) Where does the line $y = 3x + 4$ intersect the parabola $y = 8x^2$?
(b) For which values of x is $8x^2 < 3x + 4$?

Solution   (a) We solve these equations simultaneously:

$$-3x + y - 4 = 0,$$
$$y = 8x^2.$$

Substituting the second equation into the first gives $-3x + 8x^2 - 4 = 0$, so $8x^2 - 3x - 4 = 0$.

By the quadratic formula,

$$x = \frac{3 \pm \sqrt{9 + 4 \cdot 8 \cdot 4}}{16} = \frac{3 \pm \sqrt{137}}{16} \approx \frac{3 \pm 11.705}{16}$$
$$\approx 0.919 \quad \text{and} \quad -0.544.$$

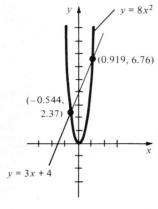

When $x \approx 0.919$, $y \approx 8(0.919)^2 \approx 6.76$. Similarly, when $x \approx -0.544$, $y \approx 2.37$, so the two points of intersection are approximately $(0.919, 6.76)$ and $(-0.544, 2.37)$. (See Fig. R.5.9.) As a check, you may substitute these pairs into the equation $-3x + y - 4 = 0$.

If the final quadratic equation had just one root—a double root—there would have been just one point of intersection; if no (real) roots, then no points of intersection.

(b) The inequality is satisfied where the parabola lies below the line, that is, for x in the interval between the x-values of the intersection points. Thus we have $8x^2 < 3x + 4$ whenever

$$x \in \left(\frac{3 - \sqrt{137}}{16}, \frac{3 + \sqrt{137}}{16} \right) \approx (-0.544, 0.919). \quad ▲$$

Figure R.5.9. The line $y = 3x + 4$ intersects the parabola $y = 8x^2$ at two points.

> ## Intersection Points
>
> To find the intersection points of two figures, find pairs (x, y) which simultaneously satisfy the equations describing the two figures.

Exercises for Section R.5

In Exercises 1–4, find the equation of the circle with center at P and radius r. Sketch.

1. $P = (1, 1)$; $r = 3$
2. $P = (-1, 7)$; $r = 5$
3. $P = (0, 5)$; $r = 5$
4. $P = (1, 1)$; $r = 1$

5. Find the equation of the circle whose center is at $(-1, 4)$ and which passes through the point $(0, 1)$. Sketch.
6. Find the equation of the circle with center $(0, 3)$ and which passes through the point $(1, 1)$. Sketch.

Find the center and radius of each of the circles in Exercises 7–10. Sketch.

7. $x^2 + y^2 - 2x + y - \frac{3}{4} = 0$
8. $2x^2 + 2y^2 + 8x + 4y + 3 = 0$
9. $-x^2 - y^2 + 8x - 4y - 11 = 0$
10. $3x^2 + 3y^2 - 6x + 36 - 8 = 0$

Find the equation of the parabola whose vertex is at V and which passes through the point P in Exercises 11–14.

11. $V = (1, 2)$; $P = (0, 1)$
12. $V = (0, 1)$; $P = (1, 2)$
13. $V = (5, 5)$; $P = (0, 0)$
14. $V = (2, 1)$; $P = (1, 4)$

Sketch the graph of each of the parabolas in Exercises 15–18, marking the vertex in each case.

15. $y = x^2 - 4x + 7$
16. $y = -x^2 + 4x - 1$
17. $y = -2x^2 + 8x - 5$
18. $y = 3x^2 + 6x + 2$

Graph each of the equations in Exercises 19–24.

19. $y = -3x^2$
20. $y = -3x^2 + 4$
21. $y = -6x^2 + 8$
22. $y = -3(x + 4)^2 + 4$
23. $y = 4x^2 + 4x + 1$
24. $y = 2(x + 1)^2 - x^2$

Find the points where the pairs of figures described by the equations in Exercises 25–30 intersect. Sketch a graph.

25. $y = -2x + 7$ and $y = 5x + 1$
26. $y = \frac{1}{3}x - 4$ and $y = 2x^2$
27. $y = 5x^2$ and $y = -6x + 7$
28. $x^2 + y^2 - 2y - 3 = 0$ and $y = 3x + 1$
29. $y = 3x^2$ and $y - x + 1 = 0$
30. $y = x^2$ and $x^2 + (y - 1)^2 = 1$

31. What are the possible numbers of points of intersection between two circles? Make a drawing similar to Fig. R.5.7.
32. What are the possible numbers of points of intersection between a circle and a parabola? Make a drawing similar to Fig. R.5.7.

Find the points of intersection between the graphs of each of the pairs of equations in Exercises 33–36. Sketch your answers.

33. $y = 4x^2$ and $x^2 + 2y + y^2 - 3 = 0$
34. $x^2 + 2x + y^2 = 0$ and $x^2 - 2x + y^2 = 0$
35. $y = x^2 + 4x + 5$ and $y = x^2 - 1$
36. $x^2 + (y - 1)^2 = 1$ and $y = -x^2 + 1$

In Exercises 37–40, determine for which x the inequality is true and explain your answer geometrically.

37. $9x^2 < x + 1$
38. $x^2 - 1 < x$
39. $-4x^2 > 2x - 1$
40. $x^2 - 7x + 6 < 0$

R.6 Functions and Graphs

A curve which intersects each vertical line at most once is the graph of a function.

In arithmetic and algebra, we operate with *numbers* (and letters which represent them). The mathematical objects of central interest in calculus are functions. In this section, we review some basic material concerning functions in preparation for their appearance in calculus.

A *function* f on the real-number line is a rule which associates to each real number x a uniquely specified real number written $f(x)$ and pronounced "f of x." Very often, $f(x)$ is given by a formula (such as $f(x) = x^3 + 3x + 2$)

which tells us how to compute $f(x)$ when x is given. The process of calculating $f(x)$ is called *evaluating* f at x. We call x the *independent variable* or the *argument* of f.

Example 1   If $f(x) = x^3 + 3x + 2$, what is $f(-2)$, $f(2.9)$, $f(q)$?

Solution   Substituting -2 for x in the formula defining f, we have $f(-2) = (-2)^3 + 3(-2) + 2 = -8 - 6 + 2 = -12$. Similarly, $f(2.9) = (2.9)^3 + 3(2.9) + 2 = 24.389 + 8.7 + 2 = 35.089$. Finally, substituting q for x in the formula for f gives $f(q) = q^3 + 3q + 2$. ▲

A function need not be given by a single formula, nor does it have to be denoted by the letter f. For instance, we can define a function H as follows:

$$H(x) = 0 \quad \text{if} \quad x < 0 \quad \text{and} \quad H(x) = 1 \quad \text{if} \quad x \geqslant 0.$$

We have a uniquely specified value $H(x)$ for any given x, so H is a function. For example, $H(3) = 1$ since $3 \geqslant 0$; $H(-5) = 0$ since $-5 < 0$; $H(0) = 1$ since $0 \geqslant 0$. Finally, since $|x| \geqslant 0$ no matter what the value of x, we may write $H(|x|) = 1$.

⌨ [3]

Calculator Discussion

We can think of a function as a *machine* or a *program* in a calculator or computer which yields the output $f(x)$ when we feed in the number x. (See Fig. R.6.1.)

Figure R.6.1. A function on a calculator.

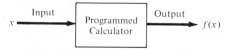

Many pocket calculators have functions built into them. Take, for example, the key labeled x^2. Enter a number x, say, 3.814. Now press the x^2 key and read: 14.546596. The x^2 key represents a function, the "squaring function." Whatever number x is fed in, pressing this key causes the calculator to give x^2 as an output.

We remark that the functions computed by calculators are often only *approximately* equal to the idealized mathematical functions indicated on the keys. For instance, entering 2.000003 and pressing the x^2 key gives the result 4.000012, while squaring 2.000003 by hand gives 4.000012000009. For a more extreme example, let $f(x) = [x^2 - 4.000012] \cdot 10^{13} + 2$. What is $f(2.000003)$? Carrying out the operations by hand gives

$$f(2.000003) = \left[(2.000003)^2 - 4.000012\right] \cdot 10^{13} + 2$$
$$= \left[4.000012000009 - 4.000012\right] \cdot 10^{13} + 2$$
$$= 0.000000000009 \cdot 10^{13} + 2$$
$$= 90 + 2 = 92.$$

If we used the calculator to square 2.000003, we would obtain $f(2.000003) = 0 \cdot 10^{13} + 2 = 2$, which is nowhere near the correct answer. ▲

Some very simple functions turn out to be quite useful. For instance,

$$f(x) = x$$

defines a perfectly respectable function called the *identity function* ("identity" because if we feed in x we get back the identical number x). Similarly,

---

[3] ⌨ This symbol denotes exercises or discussions that may require use of a hand-held calculator.

$$f(x) = 0$$

is the *zero function* and

$$f(x) = 1$$

is the *constant function* whose value is always 1, no matter what x is fed in.

Some formulas are not defined for all x. For instance, $1/x$ is defined only if $x \neq 0$. With a slightly more general definition, we can still consider $f(x) = 1/x$ as a function.

Definition of a Function

Let D be a set of real numbers. A function f with domain D is a rule which assigns a unique real number $f(x)$ to each number x in D.

If we specify a function by a formula like $f(x) = (x - 2)/(x - 3)$, its domain may be assumed to consist of all x for which the formula is defined (in this case all $x \neq 3$), unless another domain is explicitly mentioned. If we wish, for example, to consider the squaring function applied only to positive numbers, we would write: "Let f be defined by $f(x) = x^2$ for $x > 0$."

Example 2   (a) What is the domain of $f(x) = 3x/(x^2 - 2x - 3)$? (b) Evaluate $f(1.6)$.

Solution   (a) The domain of f consists of all x for which the denominator is not zero. But $x^2 - 2x - 3 = (x - 3)(x + 1)$ is zero just at $x = 3$ and $x = -1$. Thus, the domain consists of all real numbers except 3 and -1.

(b) $\quad f(1.6) = 3(1.6)/\left[(1.6)^2 - 2(1.6) - 3\right]$

$\qquad\qquad = 4.8/\left[2.56 - 3.2 - 3\right]$

$\qquad\qquad = 4.8/(-3.64) \approx -1.32.$ ▲

To visualize a function, we can draw its *graph*.

Definition of the Graph

Let f be a function with domain D. The set of all points (x, y) in the plane with x in D and $y = f(x)$ is called the graph of f.

Example 3   (a) Let $f(x) = 3x + 2$. Evaluate $f(-1)$, $f(0)$, $f(1)$, and $f(2.3)$.
(b) Draw the graph of f.

Solution   (a) $f(-1) = 3(-1) + 2 = -1$; $f(0) = 3 \cdot 0 + 2 = 2$; $f(1) = 3 \cdot 1 + 2 = 5$; $f(2.3) = 3(2.3) + 2 = 8.9$.

(b) The graph is the set of all (x, y) such that $y = 3x + 2$. This is then just the straight line $y = 3x + 2$. It has y intercept 2 and slope 3, so we can plot it directly (Fig. R.6.2). ▲

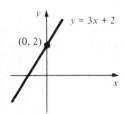

Figure R.6.2. The graph of $f(x) = 3x + 2$ is a line.

Example 4 Draw the graph of $f(x) = 3x^2$.

Solution The graph of $f(x) = 3x^2$ is just the parabola $y = 3x^2$, drawn in Fig. R.6.3 (see Section R.5). ▲

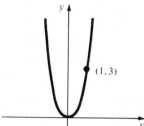

Figure R.6.3. The graph of $f(x) = 3x^2$ is a parabola.

Example 5 Let g be the *absolute value function* defined by $g(x) = |x|$. (The domain consists of all real numbers.) Draw the graph of g.

Solution We begin by choosing various values of x in the domain, computing $g(x)$, and plotting the points $(x, g(x))$. Connecting these points results in the graph shown in Fig. R.6.4. Another approach is to use the definition

$$g(x) = |x| = \begin{cases} x & \text{if } x \geqslant 0, \\ -x & \text{if } x < 0. \end{cases}$$

We observe that the part of the graph of g for $x \geqslant 0$ is a line through $(0,0)$ with slope 1, while the part for $x < 0$ is a line through $(0,0)$ with slope -1. It follows that the graph of g is as drawn in Fig. R.6.4. ▲

Figure R.6.4. Some points on the graph of $y = |x|$.

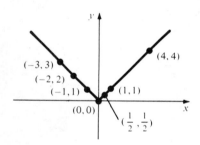

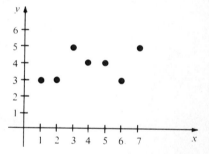

Figure R.6.5. The graph of the number of letters in "one, two, three, four, five, six, seven."

If the domain of a function consists of finitely many points, then the graph consists of isolated dots; there is no line to be filled in. For instance, Fig. R.6.5 shows the graph of a function l whose domain is $\{1, 2, 3, 4, 5, 6, 7\}$ and for which $l(x)$ is the number of letters in the English name for x.

Example 6 Draw the graph $y = \sqrt{x}$.

Solution The domain consists of all $x \geqslant 0$. The graph passes through $(0, 0)$ and may be obtained by plotting a number of points. Alternatively, we can note that for $x \geqslant 0$ and $y \geqslant 0$, $y = \sqrt{x}$ is the same as $y^2 = x$, which is a parabola with the roles of x and y reversed; see Fig. R.6.6. ▲

In plotting a complicated function such as $f(x) = 0.3x^4 - 0.2x^2 - 0.1$, we must be sure to take enough values of x, for we might otherwise miss some important details.

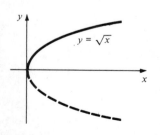

Figure R.6.6. $y = \sqrt{x}$ is half a parabola on its side.

📱 **Example 7**   Plot the graph of $f(x) = 0.3x^4 - 0.2x^2 - 0.1$ using

(a) $x = -2, -1, 0, 1, 2,$
(b) x between -2 and 2 at intervals of 0.1.

Solution   (a) Choosing $x = -2, -1, 0, 1, 2$ gives the points $(-2, 3.9), (-1, 0), (0, -0.1),$ $(1, 0), (2, 3.9)$ on the graph. (See Fig. R.6.7.) Should we draw a smooth curve through these points? How can we be sure there are no other little bumps in the graph?

(b) To answer this question, we can do some serious calculating:[4] let us plot points on the graph of $f(x) = 0.3x^4 - 0.2x^2 - 0.1$ for values of x at intervals of 0.1 between -2 and 2. If we notice that $f(x)$ is unchanged if x is replaced by $-x$, we can cut the work in half. It is only necessary to calculate $f(x)$ for $x \geqslant 0$, since the values for negative x are the same, and so the graph of f is symmetric about the y axis. The results of this calculation are tabulated and plotted in Fig. R.6.8.

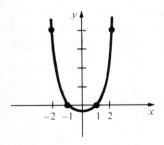

Figure R.6.7. Correct appearance of graph?

x	$f(x)$	x	$f(x)$	x	$f(x)$	x	$f(x)$
0	-0.10000	0.5	-0.13125	1.0	0.00000	1.5	0.96875
0.1	-0.10197	0.6	-0.13312	1.1	0.09723	1.6	1.35408
0.2	-0.10752	0.7	-0.12597	1.2	0.23408	1.7	1.82763
0.3	-0.11557	0.8	-0.10512	1.3	0.41883	1.8	2.40128
0.4	-0.12432	0.9	-0.06517	1.4	0.66048	1.9	3.08763
						2.0	3.90000

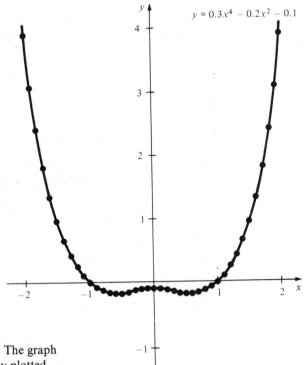

Figure R.6.8. The graph more carefully plotted.

Thus we see that indeed our original guess (Fig. R.6.7) was wrong and that the more refined calculation gives Fig. R.6.8. How can we be sure not to have missed still more bumps and wiggles? By plotting many points we can

[4] Some hints for carrying out these calculations on your calculator are contained in the Student Guide.

make good guesses but can never know for sure. The calculus we will develop in Chapters 1 to 3 of this book can tell us exactly how many wiggles the graph of a function can have and so will greatly facilitate plotting. ▲

Graphs of functions can have various shapes, but not every set of points in the plane is the graph of a function. Consider, for example, the circle $x^2 + y^2 = 5$ (Fig. R.6.9). If this circle were the graph of a function f, what would $f(2)$ be? Since $(2, 1)$ lies on the circle, we must have $f(2) = 1$. Since $(2, -1)$ also lies on the circle, $f(2)$ should also be equal to -1. But our definition of a function requires that $f(2)$ should have a definite value. Our only escape from this apparent contradiction is the conclusion that *the circle is not the graph of any function*. However, the upper semicircle alone is the graph of $y = \sqrt{5 - x^2}$ and the lower semicircle is the graph of $y = -\sqrt{5 - x^2}$, each with domain $[-\sqrt{5}, \sqrt{5}]$. Thus, while the circle is not a graph, it can be broken into two graphs.

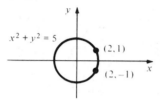

Figure R.6.9. The circle is not the graph of a function.

Example 8 Which straight lines in the plane are graphs of functions?

Solution If a line is not vertical, it has the form $y = mx + b$, so it is the graph of the function $f(x) = mx + b$. (If $m = 0$, the function is a constant function.) A vertical line is not the graph of a function—if the line is $x = a$, then $f(a)$ is not determined since y can take on any value. ▲

There is a test for determining whether a set of points in the plane is the graph of a function. If the number x_0 belongs to the domain of a function f, the vertical line $x = x_0$ intersects the graph of f at the point $(x_0, f(x_0))$ and at no other point. If x_0 does not belong to the domain, (x_0, y) is not on the graph for any value of y, so the vertical line does not intersect the graph at all. Thus we have the following criterion:

Recognizing Graphs of Functions

A set of points in the plane is the graph of a function if and only if every vertical line intersects the set in at most one point.

The domain of the function is the set of x_0 such that the vertical line $x = x_0$ meets the graph.

If C is a set of points satisfying this criterion, we can reconstruct the function f of which C is the graph. For each value x_0 of x, look for a point where the line $x = x_0$ meets C. The y coordinate of this point is $f(x_0)$. If there is no such point, x_0 is not in the domain of f.

Example 9 For each of the sets in Fig. R.6.10:

(i) Tell whether it is the graph of a function.
(ii) If the answer to part (i) is yes, tell whether $x = 3$ is in the domain of the function.
(iii) If the answer to part (ii) is yes, evaluate the function at $x = 3$.

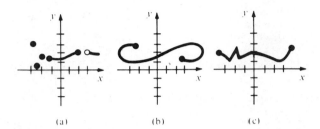

Figure R.6.10. Which curves are graphs of functions?

(a)   (b)   (c)

Solution   (a) (i) yes; (ii) no; this is indicated by the white dot.
(b) (i) no; for example, the line $x = 3$ cuts the curve in two points.
(c) (i) yes; (ii) yes; (iii) 1. ▲

Example 10   Which of the curves $y^2 = x$ and $y^3 = x$ is the graph of a function?

Solution   We begin with $y^2 = x$. Note that each value of y determines a unique value of x; we plot a few points in Fig. R.6.11.

We see immediately that the vertical line $x = 4$ meets $y^2 = x$ in two points, so $y^2 = x$ cannot be the graph of a function. (The curve $y^2 = x$ is a parabola whose axis of symmetry is horizontal.) Now look at $y^3 = x$. We begin by plotting a few points (Fig. R.6.12).

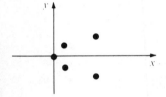

Figure R.6.11. Five points satisfying $y^2 = x$.

Figure R.6.12. Five points on $y^3 = x$.

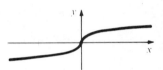

Figure R.6.13. The curve $y^3 = x$.

These points could all lie on the graph of a function. In fact, the full curve $y^3 = x$ appears as in Fig. R.6.13. We see by inspection that the curve intersects each vertical line exactly once, so there is a function f whose graph is the given curve. Since $f(x)$ is a number whose cube is x, f is called the *cube root* function. ▲

Exercises for Section R.6

Evaluate each of the functions in Exercises 1–6 at $x = -1$ and $x = 1$:

1. $f(x) = 5x^2 - 2x$
2. $f(x) = -x^2 + 3x - 5$
3. $f(x) = x^3 - 2x^2 + 1$
4. $f(x) = 4x^2 + x - 2$
5. $f(x) = -x^3 + x^2 - x + 1$
6. $f(x) = (x - 1)^2 + (x + 1)^2 + 2$

Find the domain of each of the functions in Exercises 7–12 and evaluate each function at $x = 10$.

7. $f(x) = \dfrac{x^2}{x - 1}$

8. $f(x) = \dfrac{x^2}{x^2 + 2x - 1}$

9. $f(x) = 5x\sqrt{1 - x^2}$

10. $f(x) = \dfrac{x^2 - 1}{\sqrt{x - 4}}$

11. $f(x) = \dfrac{5x + 2}{x^2 - x - 6}$

12. $f(x) = \dfrac{x}{(x^2 - 2)^2}$

Draw the graphs of the indicated functions in Exercises 13–20.

13. f in Exercise 1
14. f in Exercise 2
15. f in Exercise 4
16. f in Exercise 6
17. $f(x) = (x - 1)^2 + 3$
18. $f(x) = x^2 - 9$
19. $f(x) = 3x + 10$
20. $f(x) = x^2 + 4x + 2$

Draw the graphs of the functions in Exercises 21–24.

21. $f(x) = |x - 1|$
22. $f(x) = 3|x - 2|$
23. $f(x) = \sqrt{x - 1}$

24. $f(x) = 2\sqrt{x - 2}$

▥ Plot 10 points on the graphs of the functions in Exercises 25 and 26.

25. f in Exercise 7
26. f in Exercise 10

Plot the graphs of the functions in Exercises 27 and 28 using the given sets of values for x.

27. $f(x) = 3x^3 - x^2 + 1$:
   (a) at $x = -2, -1, 0, 1, 2$;
   ▥ (b) at 0.2 intervals in $[-2, 2]$.
28. $f(x) = 2x^4 - x^3$:
   (a) at $x = -2, -1, 0, 1, 2$;
   ▥ (b) at 0.1 intervals in $[-2, 2]$.
29. Which of the curves in Fig. R.6.14 are graphs of functions?

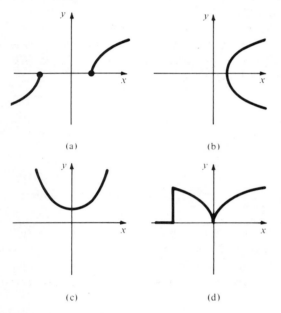

(a)   (b)   (c)   (d)

Figure R.6.14. Which curves are graphs of functions?

30. Match the following formulas with the sets in Fig. R.6.15:
   (a) $x + y \geqslant 1$
   (b) $x - y \leqslant 1$
   (c) $y = (x - 1)^2$
   (d) $y = \left(\dfrac{x + |x|}{2}\right)^2$
   (e) Not the graph of a function
   (f) None of the above

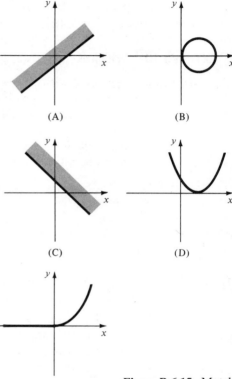

(A)   (B)   (C)   (D)   (E)

Figure R.6.15. Match the set with the formula.

Tell whether the curve defined by each of the equations in Exercises 31–36 is the graph of a function. If the answer is yes, find the domain of the function.

31. $xy = 1$
32. $x^2 - y^2 = 1$
33. $y = \sqrt{x^2 - 1}$
34. $x + y^2 = 3$
35. $y + x^2 = 3$
36. $x^2 - y^3 = 1$

Review Exercises for Chapter R

In Exercises 1–8, solve for x.
   1. $3x + 2 = 0$
   2. $x^2 + 2x + 1 = 0$
   3. $x(x - 1) = 1$
   4. $(x - 1)^2 - 9 = 0$
   5. $(x + 1)^2 - (x - 1)^2 = 2$
   6. $x^2 - 8(x - 2) + 10 = 0$
   7. $(x - 1)^3 = 8$
   8. $x^3 + 3x^2 - 3x = 9$

In Exercises 9–16, find all x satisfying the given inequality.
   9. $8x + 2 > 0$
   10. $10x - 6 < 5$
   11. $x^2 - 5 < 0$
   12. $x^2 - 5x + 6 < 0$
   13. $x^2 - (x - 1)^2 > 2$
   14. $x^3 > 1$
   15. $3x^2 - 7 < 0$
   16. $x^2 - 2x + 1 > 0$

Describe in terms of intervals the set of x satisfying the conditions in Exercises 17–24.

17. $x^2 < 1$
18. $|x - 1| < 2$
19. $x^2 - 2x < 0$
20. $8x - 3 \geqslant 0$
21. $|x - 1|^2 \geqslant 2$
22. $x^2 - 7x + 12 > 0$
23. $x^2 - 11x + 30 < 0$
24. $|x - 3| \leqslant 2$

Find the x satisfying the conditions of Exercises 25–28.

25. $x < 10$ and $x^3 \in (-8, 27)$
26. $-4 \leqslant x < 3$ and $x^2 \geqslant 2$
27. $x \in [5, 9)$ and $20 \leqslant x^2 < 36$
28. $x \in (-4, \frac{3}{4}]$ and $x \in [-\sqrt{2}, 3)$

Find the x satisfying the conditions of Exercises 29–32, and express your answer in terms of intervals.

29. $2(7 - x) \geqslant 1$ or $3x - 22 > 0$.
30. $3x^2 - 7 < 0$ and $x \leqslant 1$.
31. $(2x - 5)(x - 3) > 0$ or $x \in (-5, 1]$, but *not* both.
32. $(x - \frac{7}{5})(3x - 10) \leqslant 0$ and $x^2(5x - 15) > 0$.

Simplify the expressions in Exercises 33–36.

33. $|-8| + 5$
34. $|(a + 1)^2 - 2a - 1|$
35. $|10 - (6 + 12)|$
36. $|-18| + |-2|$

Simplify the expressions in Exercises 37–44.

37. $\sqrt{2} \cdot 2^{-1/2}$
38. $\dfrac{8}{4^{-2}}$
39. $\dfrac{(2 - 3^{1/2})(2 + 3^{1/2})}{2^{-2}}$
40. $\dfrac{(\sqrt{2} - \sqrt{3})(\sqrt{2} + \sqrt{3})\sqrt{2}}{2^{1/2}3^{-2}}$
41. $\dfrac{x^{1/4}\sqrt{x}\, y}{x^{1/2}y^{3/4}}$
42. $(x + y)^{1/4}(\sqrt{x} + 2\sqrt[4]{x}\sqrt[4]{y} + \sqrt{y})^{-1/4}$
43. $\dfrac{1}{\sqrt{x - 1} + \sqrt{x - 8}}$
44. $\dfrac{\sqrt{x}}{\sqrt{x + 2} - \sqrt{x + 5}}$

Find the distance between the pairs of points in Exercises 45–48.

45. $(-1, 1)$ and $(2, 0)$
46. $(2, 0)$ and $(10, 1)$
47. $(5, 5)$ and $(10, 10)$
48. $(100, 100)$ and $(-100, -100)$

Find the equation of the line passing through each of the pairs of points in Exercises 49–52.

49. $(\frac{1}{2}, -1), (7, 3)$
50. $(-1, 6), (1, -6)$
51. $(-2, 3), (1, 0)$
52. $(5, 5), (-3, -2)$

Find the equation of the line passing through the given point P with slope m in Exercises 53–56.

53. $P = (\frac{3}{4}, 13)$, $m = -3$
54. $P = (15, -1)$, $m = 9$
55. $P = (-2, 10)$, $m = \frac{7}{2}$
56. $P = (-9, -5)$, $m = -1$

Find the equations of the straight lines with the given data in Exercises 57–60.

57. Passes through $(1, 1)$ and is perpendicular to the line $5y + 8x = 3$.
58. Passes through $(2, 3)$ and is parallel to the line $y + 7x = 1$.
59. Passes through $(2, 4)$ and is horizontal.
60. Passes through $(-2, -4)$ and is vertical.

Find the equation of the circle with center P and radius r in Exercises 61–64.

61. $P = (12, 5)$, $r = 8$
62. $P = (-9, 3\frac{1}{2})$, $r = 3\frac{1}{2}$
63. $P = (-1, 7)$, $r = 3$
64. $P = (-1, -1)$, $r = 1$

Graph the parabolas in Exercises 65–68.

65. $y = 3x^2 + 4x + 2$
66. $y = 3x^2 - 4x + 2$
67. $y = -2x^2 + 1$
68. $y = -x^2 + 2x$

Find the points where the pairs of graphs described in Exercises 69–72 intersect.

69. $x^2 + y^2 = 4$ and $y = x$
70. $x^2 = y$ and $x = y^2$
71. $y = 3x + 4$ and $y = 3(x + 2)$
72. $2x + 4y = 6$ and $y = x^2 + 1$

Sketch the graphs of the functions in Exercises 73–76.

73. $f(x) = 3|x|$
74. $f(x) = |x| - x$
75. $f(x) = 5x^3 + 1$
76. $f(x) = 1 - x^3$

▦ Plot the graphs of the functions in Exercises 77–80 by evaluating f as indicated.

77. $f(x) = \frac{1}{10}(x^3 - x)$:
    (a) at $x = -2, 0, 2$;
    (b) at intervals of 0.5 in $[-2, 2]$.
78. $f(x) = -x^4 + 3x^2$
    (a) at $x = -2, -1, 0, 1, 2$;
    (b) at intervals of 0.1 in $[-2, 2]$.
79. $f(x) = (x - \frac{1}{2})^{2/3}$
    (a) at $x = -2, -1, 0, 1, 2$;
    (b) at intervals of 0.1 in $[-2, 2]$.
80. $f(x) = x^3 + 1/x^2$
    (a) at $x = -2, -1, -\frac{1}{2}, \frac{1}{2}, 1, 2$;
    (b) at intervals of 0.1 in $[-1, 1]$, $x \neq 0$.

81. Which of the sets in Figure R.R.1 are graphs of functions?

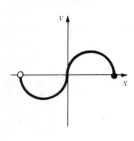

(a)

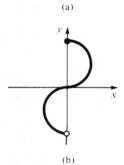

(b)

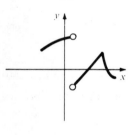

(c)

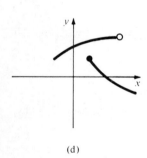

(d)

Figure R.R.1. Which sets are graphs of functions?

82. Match the graphs with the formulas.

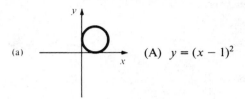

(a)   (A) $y = (x - 1)^2$

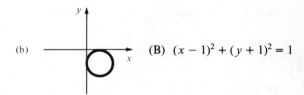

(b)   (B) $(x - 1)^2 + (y + 1)^2 = 1$

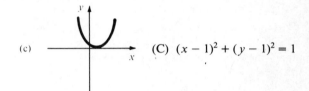

(c)   (C) $(x - 1)^2 + (y - 1)^2 = 1$

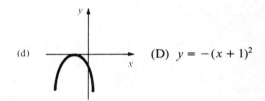

(d)   (D) $y = -(x + 1)^2$

Figure R.R.2. Matching.

★83. (a) If k and l are positive, show that
$$\frac{1}{1 + 1/(k + l)} > \frac{1}{1 + 1/l} \,.$$

(b) Using the result in (b), show that, if k and l are positive numbers, then $f(k + l) > f(l)$, where $f(x) = x/(x + 1)$.

★84. (a) Prove that a circle and a parabola can intersect in at most four points.

(b) Give examples to show that 0, 1, 2, 3, or 4 intersection points are possible.

Chapter 1

Derivatives and Limits

Differentiation is one of the two fundamental operations of calculus.

Differential calculus describes and analyzes change. The position of a moving object, the population of a city or a bacterial colony, the height of the sun in the sky, and the price of cheese all change with time. Altitude can change with position along a road; the pressure inside a balloon changes with temperature. To measure the rate of change in all these situations, we introduce in this chapter the operation of differentiation.

1.1 Introduction to the Derivative

Velocities and slopes are both derivatives.

This section introduces the basic idea of the derivative by studying two problems. The first is the problem of finding the velocity of a moving object, and the second is the problem of finding the slope of the line tangent to a graph.

To analyze velocity, imagine a bus which moves due east on a straight highway. Let x designate the time in seconds that has passed since we first observed the bus. (Using "x" for time rather than the more common "t" will make it easier to compare velocities with slopes.) Suppose that after x seconds the bus has gone a distance y meters to the east (Fig. 1.1.1). Since the distance y depends on the time x, we have a distance function $y = f(x)$. For example, if

Figure 1.1.1. What is the velocity of the bus in terms of its position?

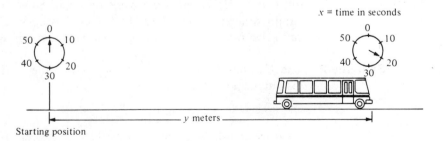

Starting position

$f(x)$ happens to be $f(x) = 2x^2$ for $0 \leqslant x \leqslant 5$, then the bus has gone $2 \cdot (3)^2$ $= 18$ meters after 3 seconds and $2 \cdot (5)^2 = 50$ meters after 5 seconds.

The velocity of the bus at any given moment, measured in meters per second, is a definite physical quantity; it can be measured by a speedometer on the bus or by a stationary radar device. Since this velocity refers to a single instant, it is called the *instantaneous velocity*. Given a distance function such as $y = f(x) = 2x^2$, how can we calculate the instantaneous velocity at a specific time x_0, such as $x_0 = 3$ seconds? To answer this question, we will relate the instantaneous velocity to the average velocity during short time intervals.

Suppose that the distance covered is measured at time x_0 and again at a later time x; these distances are $y_0 = f(x_0)$ and $y = f(x)$. Let $\Delta x = x - x_0$ designate the time elapsed between our two measurements.[1] Then the extra distance covered is $y - y_0$, which we designate by $\Delta y = y - y_0$. The *average velocity* during the time interval Δx is defined simply as the distance travelled divided by the elapsed time; that is, average velocity $= \Delta y/\Delta x = [f(x) - f(x_0)]/\Delta x$. Since $x = x_0 + \Delta x$, we can also write

$$\text{average velocity} = \frac{f(x_0 + \Delta x) - f(x_0)}{\Delta x}.$$

Example 1 A bus travels $2x^2$ meters in x seconds. Find Δx, Δy and the average velocity during the time interval Δx for the following situations: (a) $x_0 = 3$, $x = 4$; (b) $x_0 = 3$, $x = 3.1$; (c) $x_0 = 3$, $x = 3.01$.

Solution (a) $\Delta x = x - x_0 = 4 - 3 = 1$ second, $\Delta y = f(x_0 + \Delta x) - f(x_0) = f(4) - f(3)$ $= 2 \cdot 4^2 - 2 \cdot 3^2 = 14$ meters, average velocity $= \Delta y/\Delta x = 14$ meters per second; (b) $\Delta x = 0.1$, $\Delta y = 1.22$, average velocity $= 12.2$; (c) $\Delta x = 0.01$, $\Delta y = .1202$, average velocity $= 12.02$ meters per second. ▲

If we specify the accuracy to which we want to determine the instantaneous velocity, we can expect to get this accuracy by calculating the average velocity $\Delta y/\Delta x$ for Δx sufficiently small. As the desired accuracy increases, Δx may need to be made even smaller; the *exact* velocity may then be described as the number v which $\Delta y/\Delta x$ approximates as Δx becomes very small. For instance, in Example 1, you might guess that the instantaneous velocity at $x_0 = 3$ seconds is $v = 12$ meters per second; this guess is correct, as we will see shortly.

Our description of v as the number which $\Delta y/\Delta x$ approximates for Δx very small is a bit vague, because of the ambiguity in what is meant by "approximates" and "very small." Indeed, these ideas were the subject of controversy during the early development of calculus around 1700. It was thought that Δx ultimately becomes "infinitesimal," and for centuries people argued about what, if anything, "infinitesimal" might mean. Using the notion of "limit," a topic taken up in the next section, one can resolve these difficulties. However, if we work on an intuitive basis with such notions as "approximates," "gets close to," "small," "very small," "nearly zero," etc., we can solve problems and arrive at answers that will be fully justified later.

Example 2 The bus has gone $f(x) = 2x^2$ meters at time x (in seconds). Calculate its instantaneous velocity at $x_0 = 3$.

---

[1] Δ is the capital Greek letter "delta," which corresponds to the Roman D and stands for difference. The combination "Δx", read "delta-x", is not the product of Δ and x but rather a single quantity: the difference between two values of x.

Solution   We choose Δx arbitrarily and calculate the average velocity for a time interval Δx starting at time $x_0 = 3$:

$$\frac{\Delta y}{\Delta x} = \frac{f(3 + \Delta x) - f(3)}{\Delta x} = \frac{2(3 + \Delta x)^2 - 2 \cdot 3^2}{\Delta x}$$

$$= \frac{2(9 + 6\Delta x + (\Delta x)^2) - 2 \cdot 9}{\Delta x} = \frac{18 + 12\Delta x + 2(\Delta x)^2 - 18}{\Delta x}$$

$$= \frac{12\Delta x + 2(\Delta x)^2}{\Delta x} = 12 + 2\Delta x.$$

If we let Δx become very small in this last expression, $2\Delta x$ becomes small as well, and so $\Delta y/\Delta x = 12 + 2\Delta x$ approximates 12. Thus the required instantaneous velocity at $x_0 = 3$ is 12 meters per second. Note how nicely the 18's cancelled. This allowed us to divide through by Δx and avoid ending up with a zero in the denominator. ▲

Warning   In calculating what $\Delta y/\Delta x$ approximates for Δx nearly zero, it usually does no good to set $\Delta x = 0$ directly, for then we merely get $0/0$, which gives us no information.

The following more general procedure is suggested by Example 2.

Instantaneous Velocity

To calculate the instantaneous velocity at x_0 when the position at time x is $y = f(x)$:

1. Form the average velocity over the interval from x_0 to $x_0 + \Delta x$:

$$\frac{\Delta y}{\Delta x} = \frac{f(x_0 + \Delta x) - f(x_0)}{\Delta x}.$$

2. Simplify your expression for $\Delta y/\Delta x$ as much as possible, cancelling Δx from numerator and denominator wherever you can.
3. Find the number v that is approximated by $\Delta y/\Delta x$ for Δx small.

Example 3   The position of a bus at time x is $y = 3x^2 + 8x$ for $x \geq 0$. (a) Find the instantaneous velocity at an arbitrary positive time x_0. (b) At what time is the instantaneous velocity 11 meters per second?

Solution   (a) The calculation is similar to that of Example 2, except that x_0 no longer has the specific value $x_0 = 3$. The average velocity for a time interval Δx starting at x_0 is

$$\frac{\Delta y}{\Delta x} = \frac{f(x_0 + \Delta x) - f(x_0)}{\Delta x}$$

where $f(x) = 3x^2 + 8x$. Thus

$$\frac{\Delta y}{\Delta x} = \frac{\left[3(x_0 + \Delta x)^2 + 8(x_0 + \Delta x)\right] - (3x_0^2 + 8x_0)}{\Delta x}$$

$$= \frac{6x_0\Delta x + 3(\Delta x)^2 + 8\Delta x}{\Delta x} = 6x_0 + 8 + 3\Delta x.$$

As Δx gets close zero, the term $3\Delta x$ gets close to zero as well, so $\Delta y/\Delta x$ gets close to (that is, approximates) $6x_0 + 8$. Thus our instantaneous velocity is $v = 6x_0 + 8$ meters per second at the positive time x_0.

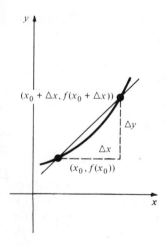

Figure 1.1.2. $\Delta y/\Delta x$ is the slope of the secant line.

(b) We set the velocity equal to 11: $6x_0 + 8 = 11$. Solving for x_0 gives $x_0 = \frac{1}{2}$ second. ▲

The second problem we study is a geometric one—to find the slope of the line tangent to the graph of a given function. We shall see that this problem is closely related to the problem of finding instantaneous velocities.

To solve the slope problem for the function $y = f(x)$, we begin by drawing the straight line which passes through the points $(x_0, f(x_0))$ and $(x_0 + \Delta x, f(x_0 + \Delta x))$, where Δx is a positive number; see Fig. 1.1.2. This straight line is called a *secant line*, and $\Delta y/\Delta x = [f(x_0 + \Delta x) - f(x_0)]/\Delta x$ is its slope.

As Δx becomes small, x_0 being fixed, it appears that the secant line comes close to the tangent line, so that the slope $\Delta y/\Delta x$ of the secant line comes close to the slope of the tangent line. See Fig. 1.1.3.

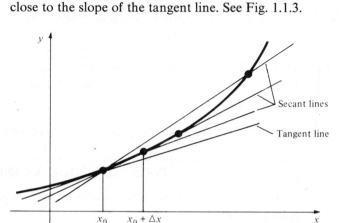

Figure 1.1.3. The secant line comes close to the tangent line as the second point moves close to x_0.

Slope of the Tangent Line

Given a function $y = f(x)$, the slope m of the line tangent to its graph at (x_0, y_0) is calculated as follows:

1. Form the slope of the secant line:

$$\frac{\Delta y}{\Delta x} = \frac{f(x_0 + \Delta x) - f(x_0)}{\Delta x}.$$

2. Simplify the expression for $\Delta y/\Delta x$, cancelling Δx if possible.
3. Find the number m that is approximated by $\Delta y/\Delta x$ for Δx small.

Example 4   Calculate the slope of the tangent line to the graph of $f(x) = x^2 + 1$ at $x_0 = -1$. Indicate your result on a sketch.

Solution   We form the slope of the secant line:

$$\frac{\Delta y}{\Delta x} = \frac{f(x_0 + \Delta x) - f(x_0)}{\Delta x}$$

$$= \frac{\left[(-1 + \Delta x)^2 + 1\right] - \left[(-1)^2 + 1\right]}{\Delta x} = \frac{-2\Delta x + (\Delta x)^2}{\Delta x}$$

$$= -2 + \Delta x.$$

For Δx small, this approximates -2, so the required slope is -2. Figure 1.1.4

Figure 1.1.4. The tangent line to $y = x^2 + 1$ at $x_0 = -1$ has slope -2.

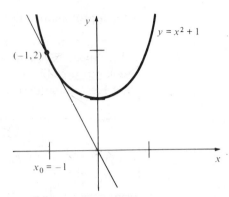

shows the graph of the parabola $y = x^2 + 1$. We have sketched the tangent line through the point $(-1, 2)$. ▲

We define the *slope* of the graph of the function f at $(x_0, f(x_0))$ to be the slope of the tangent line there.

Up to this point, we have drawn all the pictures with Δx positive. However, the manipulations in Examples 2, 3, and 4 are valid if Δx has any sign, as long as $\Delta x \neq 0$. From now on we will allow Δx to be either positive or negative.

Comparing the two previous boxes, we see that the procedures for calculating instantaneous velocities and for calculating slopes are actually identical; for example, the velocity calculation of Example 2 also tells us the slope m of $y = 2x^2$ at $(3, 18)$, namely $m = 12$. We will later find that the same procedure applies to many other situations. It is thus convenient and economical to introduce terms which apply to all the different situations: instead of calling $\Delta y / \Delta x$ an average velocity or the slope of a secant, we call it a *difference quotient*; we call the final number obtained a *derivative* rather than an instantaneous velocity or a slope. We use the notation $f'(x_0)$ to designate the derivative of f at x_0.

The Derivative

To calculate the derivative $f'(x_0)$ of a function $y = f(x)$ at x_0:

1. Form the difference quotient
$$\frac{\Delta y}{\Delta x} = \frac{f(x_0 + \Delta x) - f(x_0)}{\Delta x}.$$

2. Simplify $\Delta y / \Delta x$, cancelling Δx if possible.
3. The derivative is the number $f'(x_0)$ that $\Delta y / \Delta x$ approximates for Δx small.

This operation of finding a derivative is called *differentiation*.

The reader should be aware that the precise version of Step 3 involves the notion of a limit, which is discussed in the next section.

Example 5 Suppose that m is a constant. Differentiate $f(x) = mx + 2$ at $x_0 = 10$.

Solution Here the function is linear, so the derivative should be equal to the slope: $f'(10) = m$. To see this algebraically, calculate
$$\frac{\Delta y}{\Delta x} = \frac{[m(10 + \Delta x) + 2] - (m \cdot 10 + 2)}{\Delta x} = \frac{m \Delta x}{\Delta x} = m.$$

This approximates (in fact equals) m for Δx small, so $f'(10) = m$. ▲

Glancing back over our examples, we notice that all the functions have been either linear or quadratic. By treating a general quadratic function, we can check our previous results and point the way to the goal of developing general rules for finding derivatives.

Quadratic Function Rule

Let $f(x) = ax^2 + bx + c$, where a, b, and c are constants, and let x_0 be any real number. Then $f'(x_0) = 2ax_0 + b$.

To justify the quadratic function rule, we form the difference quotient

$$\frac{\Delta y}{\Delta x} = \frac{f(x_0 + \Delta x) - f(x_0)}{\Delta x}$$

$$= \frac{a(x_0 + \Delta x)^2 + b(x_0 + \Delta x) + c - ax_0^2 - bx_0 - c}{\Delta x}$$

$$= \frac{ax_0^2 + 2ax_0\Delta x + a(\Delta x)^2 + bx_0 + b\Delta x + c - ax_0^2 - bx_0 - c}{\Delta x}$$

$$= \frac{2ax_0\Delta x + a(\Delta x)^2 + b\Delta x}{\Delta x}$$

$$= 2ax_0 + b + a\Delta x.$$

As Δx approaches zero, $a\Delta x$ approaches zero, too, so $\Delta y/\Delta x$ approximates $2ax_0 + b$. Therefore $2ax_0 + b$ is the derivative of $ax^2 + bx + c$ at $x = x_0$.

Example 6   Find the derivative of $f(x) = 3x^2 + 8x$ at (a) $x_0 = -2$ and (b) $x_0 = \frac{1}{2}$.

Solution   (a) Applying the quadratic function rule with $a = 3$, $b = 8$, $c = 0$, and $x_0 = -2$, we find $f'(-2) = 2(3)(-2) + 8 = -4$.

(b) Taking $a = 3$, $b = 8$, $c = 0$ and $x_0 = \frac{1}{2}$, we get $f'(\frac{1}{2}) = 2 \cdot 3 \cdot (\frac{1}{2}) + 8 = 11$, which agrees with our answer in Example 3(b). ▲

If we set $a = 0$ in the quadratic function rule, we find that the derivative of any linear function $bx + c$ is the constant b, independent of x_0: the slope of a linear function is constant. For a general quadratic function, though, the derivative $f'(x_0)$ does depend upon the point x_0 at which the derivative is taken. In fact, we can consider f' as a *new function*; writing the letter x instead of x_0, we have $f'(x) = 2ax + b$. We can rephrase the quadratic function rule with x_0 replaced by x as in the following box, which also summarizes the special cases $a = 0$ and $a = 0 = b$.

Differentiating the Simplest Functions

The derivative of the quadratic function $f(x) = ax^2 + bx + c$ is the linear function $f'(x) = 2ax + b$.

The derivative of the linear function $f(x) = bx + c$ is the constant function $f'(x) = b$.

The derivative of the constant function $f(x) = c$ is the zero function $f'(x) = 0$.

The next example illustrates the use of thinking of the derivative as a function.

Example 7   There is one point on the graph of the parabola $y = f(x) = x^2 - 4x + 5$ where the slope is zero, so that the tangent line is horizontal (Fig. 1.1.5). Find that point using: (a) derivatives; and (b) algebra.

Solution

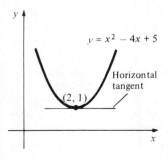

(a) By the quadratic function rule with $a = 1$, $b = -4$, $c = 5$, the derivative function is $f'(x) = 2x - 4$. For zero slope we have $0 = f'(x) = 2x - 4$, i.e., $x = 2$. Then $y = 1$, so our point is $(2, 1)$. This point is called the *vertex* of the parabola.

(b) Completing the square gives $f(x) = x^2 - 4x + 4 + 1 = (x - 2)^2 + 1$. Now $(x - 2)^2$ is zero for $x = 2$ and positive otherwise, so the parabola has its lowest point at $x = 2$. It is plausible from the figure, and true, that this low point is the point where the slope is zero. ▲

Figure 1.1.5. The vertex of the parabola is the point where its slope is zero.

We conclude this section with some examples of standard terms and notations. When we are dealing with functions given by specific formulas, we often omit the function names. Thus in Example 7(a) we can say "the derivative of $x^2 - 4x + 5$ is $2x - 4$." Another point is that we can use letters different from x, y, and f. For example, the area A of a circle depends on its radius r; we can write $A = g(r) = \pi r^2$. The quadratic function rule with $a = \pi$, $b = 0 = c$, with f replaced by g and with x replaced by r, tells us that $g'(r) = 2\pi r$. Thus for a circle the derivative of the area function is the circumference function—a fact whose geometric interpretation will be discussed in Section 2.1. Similarly, the time is often denoted by t in velocity problems.

Example 8   A stunt woman is on a moving passenger train. Her distance function is $3t^2 + t$. On the adjacent track is a long moving freight train. The distance function for the center of this freight train is $t^2 + 7t$. She must jump to the freight train. What time is best?

Solution   The safest time to jump is when the stunt woman has the same velocity as the freight train (see Fig. 1.1.6). Her instantaneous velocity v is the derivative of

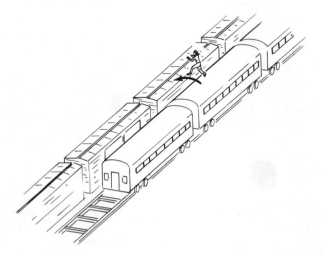

Figure 1.1.6. The stunt woman should jump when she has the same velocity as the freight train.

$3t^2 + t$. By the quadratic function rule, $v = 6t + 1$; similarly the instantaneous velocity of the freight train is $2t + 7$. The velocities are equal when $2t + 7 = 6t + 1$, i.e., $t = \frac{3}{2}$. That is the safest time. ▲

In this section, we have discussed the derivative, one of the two most basic concepts of calculus. We showed how to find derivatives in some cases and indicated a few of their applications. Before we can usefully discuss other applications of derivatives, we need to develop efficient techniques for calculating them. The next section begins that task.

Exercises for Section 1.1

In Exercises 1–4, y represents the distance a bus has travelled after x seconds. Find Δy and the average velocity during the time interval Δx for the following situations.

(a) $x_0 = 2$, $\Delta x = 0.5$ (b) $x_0 = 2$, $\Delta x = 0.01$
(c) $x_0 = 4$, $\Delta x = 0.1$ (d) $x_0 = 4$, $\Delta x = 0.01$

1. $y = x^2 + 3x$ 2. $y = 3x^2 + x$
3. $y = x^2 + 10x$ 4. $y = 2x$

In Exercises 5–8, $f(x)$ is the number of meters a bus has gone at a time x (in seconds). Find the instantaneous velocity at the given time x_0.

5. $x^2 + 3x$; $x_0 = 2$ 6. $x^2 + 3x$; $x_0 = 4$
7. $3x^2 + x$; $x_0 = 2$ 8. $3x^2 + x$; $x_0 = 4$

In Exercises 9–12, y is the position (measured in meters) of a bus at time x (in seconds). (a) Find the instantaneous velocity at an arbitrary (positive) time x_0. (b) At what time is the instantaneous velocity 10 meters per second?

9. $y = x^2 + 3x$ 10. $y = 3x^2 + x$
11. $y = x^2 + 10x$ 12. $y = 2x$

In Exercises 13–16, use the $\Delta y / \Delta x$ method of Example 4 to find the slope of the tangent line to the graph of the given function at the given point. Sketch.

13. $y = x^2$; $x_0 = 1$
14. $y = -x^2$; $x_0 = 2$
15. $y = 5x^2 - 3x + 1$; $x_0 = 0$
16. $y = x + 1 - x^2$; $x_0 = 2$

In Exercises 17–20 use the $\Delta y / \Delta x$ method of Example 5 to compute the derivative of $f(x)$ at x_0; a is a constant in each case.

17. $f(x) = ax + 2$; $x_0 = 0$
18. $f(x) = 2x + a$; $x_0 = 0$
19. $f(x) = ax^2$; $x_0 = 1$
20. $f(x) = 8x^2 + a$; $x_0 = 2$

In Exercises 21–24, use the quadratic function rule to find the derivative of the given function at the indicated point.

21. $f(x) = x^2 + x - 1$; $x_0 = 1$
22. $f(x) = x^2 - x$; $x_0 = 2$
23. $f(x) = 3x^2 + x - 2$; $x_0 = -2$
24. $f(x) = -3x^2 - x + 1$; $x_0 = -1$

In Exercises 25–28, find the vertex of the given parabola using (a) derivatives and (b) algebra.

25. $y = x^2 - 16x + 2$
26. $y = x^2 + 8x + 2$
27. $y = -2x^2 - 8x - 1$
28. $y = -2x^2 - 3x + 5$

Differentiate the functions in Exercises 29–36 using the quadratic function rule.

29. $f(x) = x^2 + 3x - 1$ 30. $f(x) = -3x + 4$
31. $f(x) = (x - 1)(x + 1)$ 32. $f(x) = (9 - x)(1 - x)$
33. $g(t) = -4t^2 + 3t + 6$ 34. $g(r) = \pi r^2 + 3$
35. $g(s) = 1 - s^2$ 36. $h(t) = 3t^2 - 5t + 9$

37. Inspector Clumseaux is on a moving passenger train. His distance function is $2t^2 + 3t$. On the adjacent track is a long moving freight train; the distance function for the center of the freight train is $3t^2 + t$. What is the best time for him to jump to the freight train?

38. Two trains, A and B, are moving on adjacent tracks with positions given by the functions $A(t) = t^2 + t + 5$ and $B(t) = 3t + 4$. What is the best time for a hobo on train B to make a moving transfer to train A?

39. An apple falls from a tall tree toward the earth. After t seconds, it has fallen $4.9t^2$ meters. What is the velocity of the apple when $t = 3$?

40. A rock thrown down from a bridge has fallen $4t + 4.9t^2$ meters after t seconds. Find its velocity at $t = 3$.

41. $f(x) = x^2 - 2$; find $f'(3)$
42. $f(x) = -13x^2 - 9x + 5$; find $f'(1)$
43. $f(x) = 1$; find $f'(7)$
44. $g(s) = 0$; find $g'(3)$
45. $k(y) = (y + 4)(y - 7)$; find $k'(-1)$
46. $x(f) = 1 - f^2$; find $x'(0)$
47. $f(x) = -x + 2$; find $f'(3.752764)$
48. $g(a) = 10a - 8$; find $g'(3.1415)$

In Exercises 49–54, find the derivative of each of the given functions by finding the value approximated by $\Delta y / \Delta x$ for Δx small:

49. $4x^2 + 3x + 2$ 50. $(x - 3)(x + 1)$
51. $1 - x^2$ 52. $-x^2$
53. $-2x^2 + 5x$ 54. $1 - x$

55. Let $f(x) = 2x^2 + 3x + 1$. (a) For which values of x is $f'(x)$ negative, positive, and zero? (b) Identify these points on a graph of f.

56. Show that two quadratic functions which have the same derivative must differ by a constant.

57. Let $A(x)$ be the area of a square of side length x. Show that $A'(x)$ is half the perimeter of the square.

58. Let $A(r)$ be the area of a circle of radius r. Show that $A'(r)$ is the circumference.

59. Where does the line tangent to the graph of $y = x^2$ at $x_0 = 2$ intersect the x axis?

60. Where does the line tangent to the graph of $y = 2x^2 - 8x + 1$ at $x_0 = 1$ intersect the y axis?

61. Find the equation of the line tangent to the graph of $f(x) = 3x^2 + 4x + 2$ at the point where $x_0 = 1$. Sketch.

62. Find the tangent line to the parabola $y = x^2 - 3x + 1$ when $x_0 = 2$. Sketch.

★63. Find the lines through the point $(4, 7)$ which are tangent to the graph of $y = x^2$. Sketch. (*Hint:* Find and solve an equation for the x coordinate of the point of tangency.)

★64. Given a point $(\bar{x}, \bar{y})$, find a general rule for determining how many lines through the point are tangent to the parabola $y = x^2$.

★65. Let R be any point on the parabola $y = x^2$. Draw the horizontal line through R and draw the perpendicular to the tangent line at R. Show that the distance between the points where these lines cross the y axis is equal to $\frac{1}{2}$, regardless of the value of x. (Assume, however, that $x \neq 0$.)

★66. If $f(x) = ax^2 + bx + c = a(x - r)(x - s)$ (r and s are the roots of f), show that the values of $f'(x)$ at r and s are negatives of one another. Explain this by appeal to the symmetry of the graph.

★67. Using your knowledge of circles, sketch the graph of $f(x) = \sqrt{4 - x^2}$. Use this to guess the values of $f'(0)$ and $f'(\sqrt{2}\,)$.

★68. A trained flea crawls along the parabola $y = x^2$ in such a way that its x coordinate at time t is $2t + 1$. The sun is shining from the east (positive x axis) so that a shadow of the flea is projected on a wall built along the y axis. What is the velocity of this shadow when $t = 3$?

★69. A ball is thrown upward at $t = 0$; its height in meters until it strikes the ground is $24.5t - 4.9t^2$ when the time is t seconds. Find:
   (a) The velocity at $t = 0, 1, 2, 3, 4, 5$.
   (b) The time when the ball is at its highest point.
   (c) The time when the velocity is zero.
   (d) The time when the ball strikes the ground.

★70. A toolbox falls from a building, its height y in feet from the ground after t seconds being given by $y = 100 - 16t^2$.
   (a) Find the *impact time* t^*, i.e., the positive time for which $y = 0$.
   (b) Find the *impact velocity*, i.e., the velocity at t^*.
   (c) The *momentum p* is defined by $p = Wv/32$, where W is the weight in pounds, and v is the velocity in feet per second. Find the impact momentum for a 20-lb toolbox.

1.2 Limits

The limit of a function $f(t)$ at a point $x = x_0$ is the value which $f(x)$ approximates for x close to x_0.

In this section, we introduce limits and study their properties. In the following sections, we will use limits to clarify statements such as "$\Delta y/\Delta x$ approximates $f'(x_0)$ for Δx small," and to systematize the computing of derivatives. Some technical points in the theory of limits have been deferred to Chapter 11, where limits are needed again for other purposes. Readers who wish to see more of the theory now can read Section 11.1 together with the present section.

We illustrate the idea of a limit by looking at the function

$$f(x) = \frac{2x^2 - 7x + 3}{x - 3}$$

which is defined for all real numbers except 3. Computing values of $f(x)$ for some values of x near 3, we obtain the following tables:

x	3.5	3.1	3.01	3.0001	3.000001
$f(x)$	6	5.2	5.02	5.0002	5.000002

x	2.5	2.9	2.99	2.9999	2.999999
$f(x)$	4	4.8	4.98	4.9998	4.999998

It appears that, as x gets closer and closer to $x_0 = 3$, $f(x)$ gets closer and closer to 5, i.e., $f(x)$ approximates 5 for x close to 3. As in our discussion of the derivative, it does no good to set $x = 3$, because $f(3)$ is not defined. In the special case we are considering, there is another way to see that $f(x)$ approximates 5:

$$\frac{2x^2 - 7x + 3}{x - 3} = \frac{(2x - 1)(x - 3)}{(x - 3)} = 2x - 1.$$

The cancellation of $(x - 3)$ is valid for $x \neq 3$. Now for x close to 3, $2x - 1$ approximates $2 \cdot 3 - 1 = 5$. Note that after cancelling $x - 3$, the function becomes defined at $x_0 = 3$.

In general, suppose that we have a function $f(x)$ and are interested in its behavior near some value x_0. Assume that $f(x)$ is defined for all x near x_0, but not necessarily at $x = x_0$ itself. If the value $f(x)$ of f approximates a number l as x gets close to a number x_0, we say that "l is the limit of $f(x)$ as x

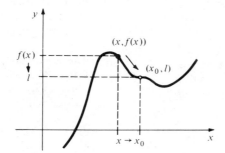

Figure 1.2.1. The notion of limit: as x approaches x_0, $f(x)$ gets near to l.

approaches x_0" or "$f(x)$ approaches l as x approaches x_0." See Fig. 1.2.1. Two usual notations for this are

$$f(x) \to l \quad \text{as} \quad x \to x_0$$

or

$$\lim_{x \to x_0} f(x) = l.$$

For example, the discussion above suggests that

$$\frac{2x^2 - 7x + 3}{x - 3} \to 5 \quad \text{as} \quad x \to 3;$$

that is

$$\lim_{x \to 3} \frac{2x^2 - 7x + 3}{x - 3} = 5.$$

Example 1 Using numerical computations, guess the value of $\lim_{x \to 4}[1/(4x - 2)]$.

Solution We make a table using a calculator and round off to three significant figures:

x	4.1	4.01	4.001	3.9	3.99	3.999
$1/(4x - 2)$	0.0694	0.0712	0.0714	0.0735	0.0716	0.0714

It appears that the limit is a number which, when rounded to three decimal places, is 0.071. In addition, we may notice that as $x \to 4$, the expression $4x - 2$ in the denominator of our fraction approaches 14. The decimal expansion of $\frac{1}{14}$ is $0.071428 \ldots$, so we may guess that

$$\lim_{x \to 4} \frac{1}{4x - 2} = \frac{1}{14}. \quad \blacktriangle$$

We summarize the idea of limit in the following display.

The Notion of Limit

If the value of $f(x)$ approximates the number l for x close to x_0, then we say that f *approaches the limit l as x approaches x_0*, and we write

$$f(x) \to l \quad \text{as} \quad x \to x_0, \quad \text{or} \quad \lim_{x \to x_0} f(x) = l.$$

The following points should be noted.

1. The quantity $\lim_{x \to x_0} f(x)$ depends upon the values of $f(x)$ for x near x_0, but not for x equal to x_0. Indeed, even if $f(x_0)$ is defined, it can be changed arbitrarily without affecting the value of the limit.
2. As x gets nearer and nearer to x_0, the values of $f(x)$ might not approach any fixed number. In this case, we say that $f(x)$ has no limit as $x \to x_0$, or that $\lim_{x \to x_0} f(x)$ does not exist.
3. In determining $\lim_{x \to x_0} f(x)$, we must consider values of x on both sides of x_0.
4. Just as in our discussion of the derivative, one can still legitimately complain that the definition of limit given in the preceding display is too vague. Readers who wish to see an air-tight definition should now read the first few pages of Section 11.1. (Section 11.1 is needed for other theoretical points in Chapter 11 and for proofs, but not for what follows here.)

Example 2   Reading the graph in Fig. 1.2.2, find $\lim_{t \to b} g(t)$ if it exists, for $b = 1, 2, 3, 4$, and 5.

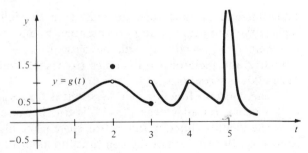

Figure 1.2.2. Find the limits of g at the indicated points. A small circle means that the indicated point does not belong to the graph.

Solution   Notice first of all that we have introduced new letters; $\lim_{t \to b} g(t)$ means the value approached by $g(t)$ as t approaches b.

$b = 1$: $\lim_{t \to 1} g(t) = 0.5$. In this case, $g(b)$ is defined and happens to be equal to the limit.

$b = 2$: $\lim_{t \to 2} g(t) = 1$. In this case, $g(b)$ is defined and equals 1.5, which is not the same as the limit.

$b = 3$: $\lim_{t \to 3} g(t)$ does not exist. For t near 3, $g(t)$ has values near 0.5 (for $t < 3$) and near 1 (for $t > 3$). There is no *single* number approached by $g(t)$ as t approaches 3.

$b = 4$: $\lim_{t \to 4} g(t) = 1$. In this case, $g(b)$ is not defined.

$b = 5$: $\lim_{t \to 5} g(t)$ does not exist. As t approaches 5, $g(t)$ grows larger and larger and does not approach any limit. ▲

The computation of limits is aided by certain properties, which we list in the following display. We will make no attempt to prove them until Chapter 11. Instead, we will present some remarks and graphs which suggest that they are reasonable.

Basic Properties of Limits

Assume that $\lim_{x \to x_0} f(x)$ and $\lim_{x \to x_0} g(x)$ exist:

Sum rule:

$$\lim_{x \to x_0} \left[f(x) + g(x) \right] = \lim_{x \to x_0} f(x) + \lim_{x \to x_0} g(x)$$

Product rule:

$$\lim_{x \to x_0} \left[f(x)g(x) \right] = \left[\lim_{x \to x_0} f(x) \right]\left[\lim_{x \to x_0} g(x) \right]$$

Reciprocal rule:

$$\lim_{x \to x_0} \left[\frac{1}{f(x)} \right] = \frac{1}{\lim_{x \to x_0} f(x)} \qquad \text{if} \quad \lim_{x \to x_0} f(x) \neq 0$$

Constant function rule:

$$\lim_{x \to x_0} c = c$$

Identity function rule:

$$\lim_{x \to x_0} x = x_0$$

Replacement rule: If the functions f and g have the same values for all x near x_0, but not necessarily including $x = x_0$, then

$$\lim_{x \to x_0} f(x) = \lim_{x \to x_0} g(x).$$

The sum and product rules are based on the following observation: If we replace the numbers y_1 and y_2 by numbers z_1 and z_2 which are close to y_1 and y_2, then $z_1 + z_2$ and $z_1 z_2$ will be close to $y_1 + y_2$ and $y_1 y_2$, respectively. Similarly, the reciprocal rule comes from such common sense statements as "$1/14.001$ is close to $1/14$."

The constant function rule says that if $f(x)$ is identically equal to c, then $f(x)$ is near c for all x near x_0. This is true because c is near c.

The identity function rule is true since it merely says that x is near x_0 if x is near x_0. Illustrations of the constant function rule and the identity function rule are presented in Fig. 1.2.3.

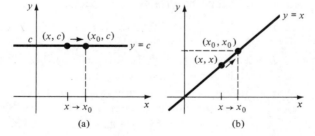

Figure 1.2.3. In (a) $\lim_{x \to x_0} c = c$ and in (b) $\lim_{x \to x_0} x = x_0$.

Finally, the replacement rule follows from the fact that $\lim_{x \to x_0} f(x)$ depends only on the values of $f(x)$ for x near x_0, and not at x_0 nor on values of x far away from x_0. The situation is illustrated in Fig. 1.2.4.

Example 3 Use the basic properties of limits: (a) to find $\lim_{x \to 3}(x^2 + 2x + 5)$; (b) to show $\lim_{x \to 3}[(2x^2 - 7x + 3)/(x - 3)] = 5$ as we guessed in the introductory calculation at the start of this section, and (c) to find $\lim_{u \to 2}[(8u^2 + 2)/(u - 1)]$.

Figure 1.2.4. If the graphs of f and g are identical near x_0, except possibly at the single point where $x = x_0$, then $\lim_{x \to x_0} f(x) = \lim_{x \to x_0} g(x)$.

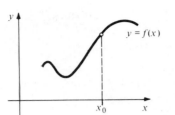 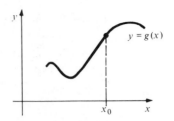

Solution

(a) Common sense suggests that the answer should be $3^2 + 2 \cdot 3 + 5 = 20$. In fact this is correct.

By the product and identity function rules,

$$\lim_{x \to 3} x^2 = \lim_{x \to 3} (x \cdot x) = \left(\lim_{x \to 3} x \right)\left(\lim_{x \to 3} x \right) = 3 \cdot 3 = 9.$$

By the product, constant function, and identity function rules,

$$\lim_{x \to 3} 2x = \left(\lim_{x \to 3} 2 \right)\left(\lim_{x \to 3} x \right) = 2 \cdot 3 = 6.$$

By the sum rule,

$$\lim_{x \to 3} (x^2 + 2x) = \lim_{x \to 3} x^2 + \lim_{x \to 3} 2x = 9 + 6 = 15.$$

Finally, by the sum and constant function rules,

$$\lim_{x \to 3} (x^2 + 2x + 5) = \lim_{x \to 3} (x^2 + 2x) + \lim_{x \to 3} 5 = 15 + 5 = 20.$$

(b) We cannot use common sense or the quotient rule, since

$$\lim_{x \to 3} (x - 3) = \lim_{x \to 3} x - \lim_{x \to 3} 3 = 3 - 3 = 0.$$

Since substituting $x = 3$ into the numerator yields zero, $x - 3$ must be a factor; in fact, $2x^2 - 7x + 3 = (2x - 1)(x - 3)$, and we have

$$\frac{2x^2 - 7x + 3}{x - 3} = \frac{(2x - 1)(x - 3)}{x - 3}.$$

For $x \neq 3$, we can divide numerator and denominator by $x - 3$ to obtain $2x - 1$. Now we apply the replacement rule, with

$$f(x) = \frac{2x^2 - 7x + 3}{x - 3} \quad \text{and} \quad g(x) = 2x - 1$$

since these two functions agree for $x \neq 3$. Therefore

$$\lim_{x \to 3} \frac{2x^2 - 7x + 3}{x - 3} = \lim_{x \to 3} (2x - 1) = 2 \left(\lim_{x \to 3} x \right) - 1 = 2 \cdot 3 - 1 = 5.$$

(c) Here the letter "u" is used in place of "x," but we do not need to change our procedures. By the sum, identity, and constant function rules, we get $\lim_{u \to 2} (u - 1) = \lim_{u \to 2} u - \lim_{u \to 2} 1 = 2 - 1 = 1$. Similarly,

$$\lim_{u \to 2} (8u^2 + 2)$$

$$= \lim_{u \to 2} 8u^2 + \lim_{u \to 2} 2 \qquad \text{(sum rule)}$$

$$= \left(\lim_{u \to 2} 8 \right)\left(\lim_{u \to 2} u^2 \right) + 2 \qquad \text{(product and constant function rules)}$$

$$= 8\left(\lim_{u \to 2} u \right)\left(\lim_{u \to 2} u \right) + 2 \qquad \text{(product and constant function rules)}$$

$$= 8 \cdot 2 \cdot 2 + 2 = 34 \qquad \text{(identity function rule)}.$$

Thus, by the product and reciprocal rules,

$$\lim_{u \to 2} \frac{8u^2 + 2}{u - 1} = \lim_{u \to 2} \left[(8u^2 + 2) \cdot \frac{1}{(u - 1)} \right]$$

$$= \lim_{u \to 2} (8u^2 + 2) \cdot \frac{1}{\displaystyle\lim_{u \to 2} (u - 1)} = 34 \cdot \frac{1}{1} = 34.$$

This agrees with the common sense rule obtained by substituting $u = 2$. ▲

As you gain experience with limits, you can eliminate some of the steps used in the solution of Example 3. Moreover, you can use some further rules which can be derived from the basic properties.

Derived Properties of Limits

Assume that the limits on the right-hand sides below exist. Then we have:

Extended sum rule:

$$\lim_{x \to x_0} \left[f_1(x) + \cdots + f_n(x) \right] = \lim_{x \to x_0} f_1(x) + \cdots + \lim_{x \to x_0} f_n(x)$$

Extended product rule:

$$\lim_{x \to x_0} \left[f_1(x) \cdots f_n(x) \right] = \lim_{x \to x_0} f_1(x) \cdots \lim_{x \to x_0} f_n(x)$$

Constant multiple rule:

$$\lim_{x \to x_0} cf(x) = c \lim_{x \to x_0} f(x)$$

Quotient rule:

$$\lim_{x \to x_0} \frac{f(x)}{g(x)} = \frac{\displaystyle\lim_{x \to x_0} f(x)}{\displaystyle\lim_{x \to x_0} g(x)} \qquad \text{if} \quad \lim_{x \to x_0} g(x) \neq 0$$

Power rule:

$$\lim_{x \to x_0} x^n = x_0^n$$

$$(n = 0, \pm 1, \pm 2, \pm 3, \ldots \text{ and } x_0 \neq 0 \text{ if } n \text{ is not positive}).$$

We outline how these derived properties can be obtained from the basic properties. To prove the extended sum rule with three summands from the basic properties of limits, we must work out $\lim_{x \to x_0}(f_1(x) + f_2(x) + f_3(x))$ when $\lim_{x \to x_0} f_i(x)$ is known to exist. The idea is to use the basic sum rule for two summands. In fact $f_1(x) + f_2(x) + f_3(x) = f_1(x) + g(x)$, where $g(x) = f_2(x) + f_3(x)$. Note that $\lim_{x \to x_0} g(x) = \lim_{x \to x_0} f_2(x) + \lim_{x \to x_0} f_3(x)$ by the basic sum rule. Moreover $\lim_{x \to x_0}(f_1(x) + g(x)) = \lim_{x \to x_0} f_1(x) + \lim_{x \to x_0} g(x)$ by the same rule. Putting these results together, we have

$$\lim_{x \to x_0} \left[f_1(x) + f_2(x) + f_3(x) \right] = \lim_{x \to x_0} \left[f_1(x) + g(x) \right]$$

$$= \lim_{x \to x_0} f_1(x) + \lim_{x \to x_0} g(x)$$

$$= \lim_{x \to x_0} f_1(x) + \lim_{x \to x_0} f_2(x) + \lim_{x \to x_0} f_3(x),$$

as we set out to show. The extended sum rule with more than three terms is now plausible; it can be proved by induction (see Exercise 65). The extended product rule can be proved by very similar arguments. To get the constant multiple rule, we may start with the basic product rule $\lim_{x \to x_0}[f(x)g(x)] = [\lim_{x \to x_0} f(x)][\lim_{x \to x_0} g(x)]$. Let $g(x)$ be the constant function $g(x) = c$; the constant function rule gives $\lim_{x \to x_0}[cf(x)] = [\lim_{x \to x_0} c][\lim_{x \to x_0} f(x)] = c \lim_{x \to x_0} f(x)$, as we wanted to show. Similarly, the quotient rule follows from the basic product rule and the reciprocal rule by writing $f/g = f \cdot 1/g$. The power rule follows from the extended product rule with $f_1(x) = x, \ldots, f_n(x) = x$ and the identity function rule. The next example illustrates the use of the derived properties.

Example 4 Find $\lim_{x \to 1} \dfrac{x^3 - 3x^2 + 14x}{x^6 + x^3 + 2}$.

Solution Common sense correctly suggests that the answer is $(1^3 - 3 \cdot 1^2 + 14 \cdot 1)/(1^6 + 1^3 + 2) = 3$. To get this answer systematically, we shall write $f(x) = x^3 - 3x^2 + 14x$, $g(x) = x^6 + x^3 + 2$, and use the quotient rule. First of all, $\lim_{x \to 1} x^6 = 1^6 = 1$ and $\lim_{x \to 1} x^3 = 1$ by the power rule; $\lim_{x \to 1} 2 = 2$ by the constant function rule; since all three limits exist, $\lim_{x \to 1} g(x) = 1 + 1 + 2 = 4$ by the extended sum rule. Similarly, $\lim_{x \to 1} f(x) = 12$. Since $\lim_{x \to 1} g(x) \neq 0$, the quotient rule applies and so $\lim_{x \to 1}[f(x)/g(x)] = \frac{12}{4} = 3$, as we anticipated. ▲

Clearly the common sense method of just setting $x = 1$ is far simpler when it works. A general term to describe those situations where it does work is "continuity."

Definition of Continuity

A function $f(x)$ is said to be continuous at $x = x_0$ if $\lim_{x \to x_0} f(x) = f(x_0)$.

Thus if $f(x)$ is continuous at x_0, two things are true: (1) $\lim_{x \to x_0} f(x)$ exists and (2) this limit can be calculated by merely setting $x = x_0$ in $f(x)$, much as in Example 4. The geometric meaning of continuity will be analyzed extensively in Section 3.1.

We now discuss certain functions which are continuous at many or all values of x_0. Instead of the specific function $(x^3 - 3x^2 + 14x)/(x^6 + x^3 + 2)$, we consider more generally a ratio $r(x) = f(x)/g(x)$ of two polynomials. Such a ratio is called a *rational function*, just as a ratio of integers is called a rational number. Note that a polynomial $f(x)$ is itself a rational function—we can simply choose the denominator $g(x)$ in the ratio $r(x)$ to be $g(x) = 1$. Suppose that we are interested in the rational function $r(x) = f(x)/g(x)$ for values of x near x_0. Moreover, suppose that $g(x_0) \neq 0$ so that $r(x_0)$ is defined; for instance, in Example 4 we had $g(x_0) = 4 \neq 0$ at $x_0 = 1$. Using the limit rules in almost exactly the same way as we did in Example 4 leads to the conclusion that the common sense approach works for the rational function $r(x)$. We summarize in the following box.

Continuity of Rational Functions

If $f(x)$ is a polynomial or a ratio of polynomials and $f(x_0)$ is defined, then

$$\lim_{x \to x_0} f(x) = f(x_0).$$

As an example of the use of the continuity of rational functions, note that to calculate $\lim_{x \to 4}[1/(4x - 2)]$, we can now just set $x = 4$ to get $\frac{1}{14}$, as we guessed in Example 1 above. Indeed, students seduced by the simplicity of this rule often believe that a *limit* is nothing more than a *value*. The next example should help you avoid this trap.

Example 5 Find

(a) $\displaystyle\lim_{x \to 2} \frac{x^2 + x - 6}{x^2 + 2x - 8}$

and

(b) $\displaystyle\lim_{\Delta x \to 0} \frac{(\Delta x)^2 + 2\Delta x}{(\Delta x)^2 + \Delta x}$,

where Δx is a variable.

Solution (a) The denominator vanishes when $x = 2$, so we cannot use the continuity of rational functions as yet. Instead we factor. When the denominator is not zero we have

$$\frac{x^2 + x - 6}{x^2 + 2x - 8} = \frac{(x + 3)(x - 2)}{(x + 4)(x - 2)} = \frac{x + 3}{x + 4} .$$

Thus

$$\lim_{x \to 2} \frac{x^2 + x - 6}{x^2 + 2x - 8} = \lim_{x \to 2} \frac{x + 3}{x + 4} \quad \text{(by the replacement rule)}$$

$$= \frac{2 + 3}{2 + 4} = \frac{5}{6} \quad \text{(by the continuity of rational functions).}$$

(b) The denominator vanishes when $\Delta x = 0$, so again we use the replacement rule:

$$\lim_{\Delta x \to 0} \frac{(\Delta x)^2 + 2\Delta x}{(\Delta x)^2 + \Delta x} = \lim_{\Delta x \to 0} \frac{\Delta x + 2}{\Delta x + 1} \quad \text{(replacement rule)}$$

$$= 2 \quad \text{(continuity of rational functions).} \ \blacktriangle$$

There are many limits that cannot be dealt with by the laws of limits we have so far. For example, we claim that if x_0 is positive, then

$$\lim_{x \to x_0} \sqrt{x} = \sqrt{x_0} ,$$

i.e., the function $f(x) = \sqrt{x}$ is continuous at x_0. To make this result plausible, *assume* that $\lim_{x \to x_0} \sqrt{x} = l$ exists. Then by the product rule,

$$l^2 = \left(\lim_{x \to x_0} \sqrt{x} \right)\left(\lim_{x \to x_0} \sqrt{x} \right) = \lim_{x \to x_0} x = x_0 .$$

Now l must be positive since $\sqrt{x} > 0$ for all x which are positive, and all x which are close enough to x_0 are positive. Hence, $l = \sqrt{x_0}$. This limit is consistent with the appearance of the graph of $y = \sqrt{x}$. (See Fig. 1.2.5.)

In Section 11.1, we give a careful proof of the continuity of $\sqrt{x}$.

Figure 1.2.5. The graph of $y = \sqrt{x}$ suggests that $\lim_{x \to x_0}\sqrt{x} = \sqrt{x_0}$.

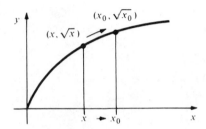

Example 6   Find

$$\lim_{x \to 3} \frac{8x^2}{1 + \sqrt{x}} .$$

Solution   By using the properties of limits and the continuity of $\sqrt{x}$, we get $\lim_{x \to 3}(1 + \sqrt{x}) = \lim_{x \to 3}1 + \lim_{x \to 3}\sqrt{x} = 1 + \sqrt{x} \neq 0$. Thus

$$\lim_{x \to 3} \frac{8x^2}{1 + \sqrt{x}} = \frac{\lim_{x \to 3} 8x^2}{\lim_{x \to 3}\left(1 + \sqrt{x}\right)} = \frac{8 \cdot 3^2}{1 + \sqrt{3}} = \frac{72}{1 + \sqrt{3}} . \; \blacktriangle$$

Sometimes limits can fail to exist even when a function is given by a simple formula; the following is a case in point.

Example 7   Does $\lim_{x \to 0}(|x|/x)$ exist?

Solution   The function in question has the value 1 for $x > 0$ and -1 for $x < 0$. For $x = 0$, it is undefined. (See Fig. 1.2.6.) There is no number l which is

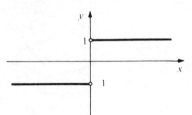

Figure 1.2.6. The graph of the function $|x|/x$.

approximated by $|x|/x$ as $x \to 0$, since $|x|/x$ is sometimes 1 and sometimes -1, according to the sign of x. We conclude that $\lim_{x \to 0}(|x|/x)$ does not exist. $\blacktriangle$

It is possible to define a notion of *one-sided limit* so that a function like $|x|/x$ has limits from the left and right (see Section 11.1 for details). Since the one-sided limits are different, the limit *per se* does not exist. The reader might wonder if any function of interest in applications actually shows a jump similar to that in Fig. 1.2.6. The answer is "yes." For example, suppose that a ball is dropped and, at $t = 0$, bounces off a hard floor. Its velocity will change very rapidly from negative (that is, downward) to positive (that is, upward). It is often convenient to idealize this situation by saying that the velocity function jumps from a negative to a positive value exactly at $t = 0$, much as in Fig. 1.2.6.

We conclude this section with some limits involving $\pm \infty$. We shall be quite informal and emphasize examples, again leaving a more careful discussion to Chapter 11. First, it is often useful to consider limits of the form $\lim_{x \to \infty} f(x)$. This symbol refers to the value approached by $f(x)$ as x becomes arbitrarily large. Likewise, $\lim_{x \to -\infty} f(x)$ is the value approached by $f(x)$ as x gets large in the negative sense. Limits as $x \to \pm \infty$ obey similar rules to those with $x \to x_0$.

Example 8 Find

    (a) $\displaystyle \lim_{x \to \infty} \frac{1}{x}$;

    (b) $\displaystyle \lim_{x \to \infty} \frac{2x + 1}{3x + 1}$;

and

    (c) $\displaystyle \lim_{x \to -\infty} \frac{5x^2 - 3x + 2}{x^2 + 1}$.

Solution As x gets very large, $1/x$ gets very small. Thus

    (a) $\displaystyle \lim_{x \to \infty} \frac{1}{x} = 0$.

We shall do (b) and (c) by writing the given expression in terms of $1/x$.

    (b) $\displaystyle \lim_{x \to \infty} \frac{2x + 1}{3x + 1} = \lim_{x \to \infty} \frac{2 + 1/x}{3 + 1/x} = \frac{2 + 0}{3 + 0} = \frac{2}{3}$,

    (c) $\displaystyle \lim_{x \to -\infty} \frac{5x^2 - 3x + 2}{x^2 + 1} = \lim_{x \to -\infty} \frac{5 - 3/x + 2/x^2}{1 + 1/x^2}$

$$= \frac{5 - 0 + 0}{1 + 0} = 5. \ \blacktriangle$$

Example 9 Find $\lim_{x \to \infty} f(x)$ and $\lim_{x \to -\infty} f(x)$ for the function f in Figure 1.2.7.

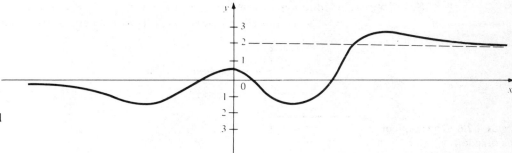

Figure 1.2.7. Find
$\lim_{x \to \infty} f(x)$ and
$\lim_{x \to -\infty} f(x)$.

Solution Assuming that the ends of the graph continue as they appear to be going, we conclude that $\lim_{x \to \infty} f(x) = 2$ and $\lim_{x \to -\infty} f(x) = 0$. $\blacktriangle$

Another kind of limit occurs when the value of $f(x)$ becomes arbitrarily large and positive as x approaches x_0. We then write $\lim_{x \to x_0} f(x) = \infty$. In this case $\lim_{x \to x_0} f(x)$ does not, strictly speaking, exist (infinity is not a real number). Similarly $\lim_{x \to x_0} f(x) = -\infty$ is read "the limit of $f(x)$ as x approaches x_0 is minus infinity," which means that while $\lim_{x \to x_0} f(x)$ does not exist, as x approaches x_0 from either side $f(x)$ becomes arbitrarily large in the negative sense.

Example 10 Find

    (a) $\displaystyle \lim_{x \to 2} \frac{-3x}{x^2 - 4x + 4}$

and

    (b) $\displaystyle \lim_{x \to 0} \frac{3x + 2}{x}$.

Solution (a) The denominator vanishes when $x = 2$, so the quotient rule does not apply. We may factor the denominator to get $-3x/(x^2 - 4x + 4) = -3x/(x - 2)^2$.

For x near 2, the numerator is near -6, while the denominator is small and positive, so the quotient is large and negative. Thus,

$$\lim_{x \to 2} \frac{-3x}{(x-2)^2} = -\infty.$$

(See Fig. 1.2.8(a).)

(b) We write $(3x+2)/x = 3 + 2/x$. When x is near 0, $2/x$ is *either* large and positive *or* large and negative, according to the sign of x (Fig. 1.2.8(b)). Hence $\lim_{x \to 0}[(3x+2)/x]$ does not have any value, finite or infinite. (To get $+\infty$ or $-\infty$, one-sided limits must be used.) ▲

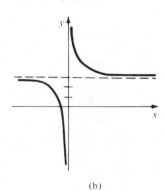

Figure 1.2.8. In (a) the limit is $-\infty$ and in (b) it does not exist.

(a)

$$\lim_{x \to 2} \frac{-3x}{x^2 - 4x + 4} = -\infty$$

(b)

$y = \dfrac{3x+2}{x}$ is infinitely schizophrenic near $x = 0$

Exercises for Section 1.2

1. Guess $\lim_{x \to 1}[(x^3 - 3x^2 + 5x - 3)/(x - 1)]$ by doing numerical calculations. Verify your guess by using the properties of limits.
2. Find $\lim_{x \to -1}[2x/(4x^2 + 5)]$, first by numerical calculation and guesswork, then by the basic properties of limits, and finally by the continuity of rational functions.

Refer to Fig. 1.2.9 for Exercises 3 and 4.

3. Find $\lim_{x \to -3} f(x)$ and $\lim_{x \to 3} f(x)$ if they exist.
4. Find $\lim_{x \to -1} f(x)$ and $\lim_{x \to 1} f(x)$ if they exist.

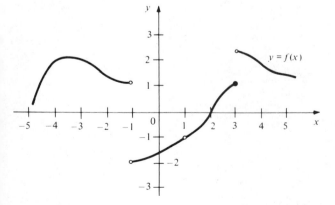

Figure 1.2.9. Find the limits at $x = -3$, -1, 1, and 3 if they exist. A small circle means that the indicated point does not belong to the graph.

Use the basic properties of limits to find the limits in Exercises 5–8.

5. $\lim_{x \to 3} (17 + x)$

6. $\lim_{x \to 3} x^2$

7. $\lim_{u \to -1} \dfrac{u+1}{u-1}$

8. $\lim_{s \to 2} \dfrac{s^2 - 1}{s}$

Use the basic and derived properties of limits to find the limits in Exercises 9–12.

9. $\lim_{x \to 3} \dfrac{x^2 - 9}{x^2 + 3}$

10. $\lim_{x \to 2} \dfrac{(x^2 + 3x - 10)}{(x + 2)}$

11. $\lim_{x \to 1} \dfrac{x^{10} + 8x^3 - 7x^2 - 2}{x + 1}$

12. $\lim_{x \to 2} \dfrac{(x^2 + 3x - 9)}{x + 2}$

Use the continuity of rational functions and the replacement rule, if necessary, to evaluate the limits in Exercises 13–22.

13. $\lim_{u \to \sqrt{3}} \dfrac{u - \sqrt{3}}{u^2 - 3}$

14. $\lim_{t \to \sqrt{5}} \dfrac{t - \sqrt{5}}{t^2 - 5}$

15. $\lim_{x \to 2} \dfrac{x - 2}{x - 2}$

16. $\lim_{x \to 3} \dfrac{x^2 - 3}{x^2 - 3}$

17. $\lim_{x \to 3} \dfrac{x^2 - 4x + 3}{x^2 - 2x - 3}$

18. $\lim_{x \to -5} \dfrac{x^2 + x - 20}{x^2 + 6x + 5}$

19. $\lim_{\Delta x \to 0} \dfrac{(\Delta x)^2 + 3(\Delta x)}{\Delta x}$

20. $\lim_{\Delta x \to 0} \dfrac{(\Delta x)^3 + 2(\Delta x)^2}{\Delta x}$

21. $\lim_{\Delta x \to 0} \dfrac{3(\Delta x)^2 + 2(\Delta x)}{\Delta x}$

22. $\lim_{\Delta x \to 0} \dfrac{(\Delta x)^3 + 2(\Delta x)^2 + 7(\Delta x)}{\Delta x}$

Find the limits in Exercises 23–26 using the continuity of $\sqrt{x}$.

23. $\lim_{x \to 4} \dfrac{2x}{1 - \sqrt{x}}$

24. $\lim_{x \to 9} \dfrac{2x^2 - x}{\sqrt{x}}$

25. $\lim_{x \to 3}(1 - \sqrt{x})(2 + \sqrt{x})$ 26. $\lim_{x \to 2}(x^2 + 2x)\sqrt{x}$

Find the limits in Exercises 27–30 if they exist. Justify your answer.

27. $\lim_{x \to 0}\left(\dfrac{x}{|x|} + 1 \right)$

28. $\lim_{x \to 0}\left(x^2 + \dfrac{x}{|x|} \right)$

29. $\lim_{x \to 1} \dfrac{|x - 1|}{x - 1}$

30. $\lim_{x \to 2} \dfrac{|x - 2|}{x - 2}$

Find the limits in Exercises 31–36 as $x \to \pm\infty$.

31. $\lim_{x \to \infty} \dfrac{x - 1}{2x + 1}$

32. $\lim_{x \to \infty} \dfrac{2x^2 + 1}{3x^2 + 2}$

33. $\lim_{x \to -\infty} \dfrac{2x - 1}{3x + 1}$

34. $\lim_{x \to -\infty} \dfrac{3x^3 + 2x^2 + 1}{4x^3 - x^2 + x + 2}$

35. $\lim_{x \to \infty} \dfrac{|x|}{x}$

36. $\lim_{x \to -\infty} \dfrac{x}{|x|}$

37. For the function in Fig. 1.2.10, find $\lim_{x \to a}f(x)$ for $a = 0, 1, 2, 3, 4$ if it exists. In each case, tell whether $\lim_{x \to a}f(x) = f(a)$.

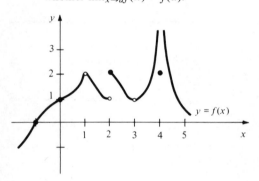

Figure 1.2.10. Find the limits at $0, 1, 2, 3, 4$.

38. Find $\lim_{x \to a}f(x)$, where $a = -2, 0$, and 1 for f sketched in Fig. 1.2.11.

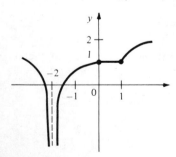

Figure 1.2.11. Find $\lim_{x \to a}f(x)$ at the indicated points.

Refer to Fig. 1.2.12 for Exercises 39 and 40 (assume that the functions keep going as they appear to).

39. Find $\lim_{x \to \infty} f(x)$ and $\lim_{x \to -\infty} f(x)$.

40. Find $\lim_{x \to \infty} g(x)$ and $\lim_{x \to -\infty} g(x)$.

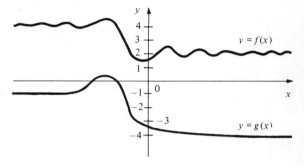

Figure 1.2.12. Find the limits at $\pm\infty$.

Find the limits in Exercises 41–44. If the limit is $\pm\infty$, give that as your answer.

41. $\lim_{x \to 0}\left(-\dfrac{x}{x^3} \right)$

42. $\lim_{y \to 3} \dfrac{y - 4}{y^2 - 6y + 9}$

43. $\lim_{x \to \sqrt{5}} \dfrac{2}{x^2 - 5}$

44. $\lim_{x \to 0} \dfrac{x^2 + 5x}{x^2}$

Find the limits in Exercises 45–58 if they exist.

45. $\lim_{u \to 0} \dfrac{u^3 + 2u^2 + u}{u}$

46. $\lim_{x \to \infty} \dfrac{x^3 + 2}{3x^3 + x}$

47. $\lim_{x \to 2} \dfrac{2x}{(x - 2)^2}$

48. $\lim_{x \to -2} \dfrac{x + 2}{|x + 2|}$

49. $\lim_{x \to 2} \dfrac{x^2 - 5x + 6}{x^2 - 6x + 8}$

50. $\lim_{x \to 4} \dfrac{x^2 - 5x + 6}{x^2 - 6x + 8}$

51. $\lim_{t \to 4} \dfrac{t^2 + 2\sqrt{t}}{|t|}$

52. $\lim_{s \to \infty} \dfrac{s^2 - 2s + 1}{2s^2 + 3s + 2}$

53. $\lim_{\Delta x \to 0} \dfrac{(5 + \Delta x)^3 - 5^3}{\Delta x}$

54. $\lim_{\Delta x \to 0} \dfrac{(\Delta x)^4 + 2(\Delta x)^3 + 2\Delta x}{\Delta x}$

55. $\lim_{x \to 1} \dfrac{3(x^3 - 1)}{x - 1}$

56. $\lim_{q \to 3} \sqrt{|q - 3|}$

57. $\lim_{s \to 3} \dfrac{3s^2 - 2s - 21}{(s - 3)^2}$

58. $\lim_{x \to -\infty} \dfrac{\sqrt{x}}{x^2 + 1}$

59. How should $f(x) = (x^5 - 1)/(x - 1)$ be defined at $x = 1$ in order that $\lim_{x \to 1} f(x) = f(1)$?

60. How should $g(t) = (t^2 + 4t)/(t^2 - 4t)$ be defined at $t = 0$ to make $\lim_{t \to 0} g(t) = g(0)$?

★61. A block of ice melts in a room held at 75°F. Let $f(t)$ be the base area of the block and $g(t)$ the height of the block, measured with a ruler at time t.

    (a) Assume that the block of ice melts completely at time T. What values would you assign to $f(T)$ and $g(T)$?

    (b) Give physical reasons why $\lim_{t \to T} f(t) = f(T)$ and $\lim_{t \to T} g(t) = g(T)$ need not both hold. What are the limits?

    (c) The limiting volume of the ice block at time T is zero. Write this statement as a limit formula.

    (d) Using (b) and (c), illustrate the product rule for limits.

★62. A thermometer is stationed at x centimeters from a candle flame. Let $f(x)$ be the Celsius scale reading on the thermometer. Assume that the glass in the thermometer will crack upon contact with the flame.

    (a) Explain physically why $f(0)$ doesn't make any sense.

    (b) Describe in terms of the thermometer scale the meaning of $\lim_{x \to 0+} f(x)$ (i.e., the limit of $f(x)$ as x approaches zero through positive values).

    (c) Draw a realistic graph of $f(x)$ for a scale with maximum value 200°C. (Assume that the flame temperature is 400°C.)

    (d) Repeat (c) for a maximum scale value of 500°C.

★63. Suppose that $f(x) \neq 0$ for all $x \neq x_0$ and that $\lim_{x \to x_0} f(x) = \infty$. Can you conclude that $\lim_{x \to x_0} [1/f(x)] = 0$? Explain.

★64. Draw a figure, similar to Figs. 1.2.3 and 1.2.4, which illustrates the sum rule in our box on basic properties of limits.

★65. (a) Prove the extended sum rule in the box on derived properties of limits for the case $n = 4$ by using the basic sum rule and using the extended sum rule for the case $n = 3$ proved in the text.

    (b) Assume that the extended sum rule holds when $n = 16$; prove from your assumption that it holds when $n = 17$.

    (c) Assume that the extended sum rule holds for some given integer $n \geqslant 2$; prove that it holds for the integer $n + 1$.

    (d) According to the principle of induction, if a statement is true for $n + 1$ whenever it is true for n, and is true for some specific integer, m, then the statement is also true for $m + 1, m + 2, m + 3, \ldots$, i.e., it is true for all integers larger than m. Use induction and the basic sum rule to prove the extended sum rule.

★66. Prove the extended product rule for limits by induction (see Exercise 65) and the basic properties of limits.

1.3 The Derivative as a Limit and the Leibniz Notation

The derivative is the limit of a difference quotient.

We are now ready to tie together the discussion of the derivative in Section 1.1 with the discussion of limits in Section 1.2.

Let $f(x)$ be a function such as the one graphed in Fig. 1.3.1. Recall the following items from Section 1.1: If $(x_0, f(x_0))$ and $(x_0 + \Delta x, f(x_0 + \Delta x))$ are two points on the graph, we write $\Delta y = f(x_0 + \Delta x) - f(x_0)$ and call $\Delta y / \Delta x$ the

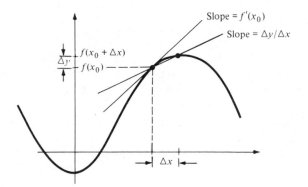

Figure 1.3.1. The limit of $\Delta y / \Delta x$ as $\Delta x \to 0$ is $f'(x_0)$.

difference quotient. This difference quotient is the slope of a secant line, as shown in the figure; moreover, if $f(x)$ is distance as a function of time, then $\Delta y/\Delta x$ is an average velocity. If Δx is small, then $\Delta y/\Delta x$ approximates the derivative $f'(x_0)$. Using these ideas, we were led to conclude that $f'(x_0)$ is the slope of the tangent line; moreover if $f(x)$ represents distance as a function of time, $f'(x_0)$ is the instantaneous velocity at time x_0. We can now make our discussion of $f'(x_0)$ more precise using the language of limits.

Suppose that the domain of a function $f(x)$ contains an open interval about a given number x_0. (For example, we might have $x_0 = 3$, and $f(x)$ might be defined for all x which obey $1 < x < 4$.) Consider the difference quotient

$$\frac{\Delta y}{\Delta x} = \frac{f(x_0 + \Delta x) - f(x_0)}{\Delta x}$$

as a function of the variable Δx. The domain of the difference quotient then consists of those Δx, positive or negative, which are near enough to zero so that $f(x_0 + \Delta x)$ is defined. Since Δx appears in the denominator, $\Delta x = 0$ is not in the domain of the difference quotient. (For instance, in the example just mentioned with $x_0 = 3$ and $1 < x < 4$, $\Delta y/\Delta x$ would be defined for $-2 < \Delta x < 0$ and $0 < \Delta x < 1$.) As the examples in Section 1.1 indicated, we should look at the limit of $\Delta y/\Delta x$ as $\Delta x \to 0$. This leads to the following definition of the derivative in terms of limits.

Formal Definition of the Derivative

Let $f(x)$ be a function whose domain contains an open interval about x_0. We say that f is *differentiable at* x_0 when the following limit exists:

$$f'(x_0) = \lim_{\Delta x \to 0} \frac{f(x_0 + \Delta x) - f(x_0)}{\Delta x} \; ;$$

$f'(x_0)$ is then called the *derivative* of $f(x)$ at x_0.

Example 1   Suppose that $f(x) = x^2$. Then $f'(3) = 6$ by the quadratic function rule with $a = 1$, $b = 0 = c$ and $x_0 = 3$. Justify that $f'(3) = 6$ directly from the formal definition of the derivative and the rules for limits.

Solution   We write the difference quotient and simplify:

$$\frac{\Delta y}{\Delta x} = \frac{f(x_0 + \Delta x) - f(x_0)}{\Delta x} = \frac{(3 + \Delta x)^2 - 3^2}{\Delta x} = \frac{6 \Delta x + (\Delta x)^2}{\Delta x} .$$

The independent variable is now Δx, but, of course, we can still use the rules for limits given in the previous section. By the replacement rule, we can cancel:

$$\lim_{\Delta x \to 0} \frac{6 \Delta x + (\Delta x)^2}{\Delta x} = \lim_{\Delta x \to 0} (6 + \Delta x),$$

provided the latter limit exists. However, $6 + \Delta x$ is a polynomial in the variable Δx and is defined at $\Delta x = 0$, so by the continuity of rational functions, $\lim_{\Delta x \to 0}(6 + \Delta x) = 6 + 0 = 6$. ▲

Example 2   Use the formal definition of the derivative and the rules for limits to differentiate x^3.

Solution   Letting $f(x) = x^3$, we have

$$f'(x_0) = \lim_{\Delta x \to 0} \frac{f(x_0 + \Delta x) - f(x_0)}{\Delta x} = \lim_{\Delta x \to 0} \frac{(x_0 + \Delta x)^3 - x_0^3}{\Delta x}$$

$$= \lim_{\Delta x \to 0} \frac{x_0^3 + 3x_0^2 \Delta x + 3x_0 (\Delta x)^2 + (\Delta x)^3 - x_0^3}{\Delta x} \quad \text{(expanding the cube)}$$

$$= \lim_{\Delta x \to 0} \frac{3x_0^2 \Delta x + 3x_0 (\Delta x)^2 + (\Delta x)^3}{\Delta x}$$

$$= \lim_{\Delta x \to 0} \left(3x_0^2 + 3x_0 \Delta x + (\Delta x)^2 \right) \quad \text{(by the replacement rule)}$$

$$= 3x_0^2$$

(using the continuity of rational functions and setting $\Delta x = 0$).

The derivative of x^3 at x_0 is therefore $3x_0^2$. ▲

As the next example shows, we can write x instead of x_0 when differentiating by the limit method, as long as we remember that x is to be held constant when we let $\Delta x \to 0$.

Example 3   If $f(x) = 1/x$, find $f'(x)$ for $x \neq 0$.

Solution   The difference quotient is

$$\frac{\Delta y}{\Delta x} = \frac{1/(x + \Delta x) - 1/x}{\Delta x} = \frac{x - (x + \Delta x)}{x(x + \Delta x) \Delta x} = - \frac{\Delta x}{x(x + \Delta x) \Delta x}.$$

Here x is being held constant at some nonzero value, and $\Delta y/\Delta x$ is considered as a function of Δx. Note that Δx is in the domain of the difference quotient provided that $\Delta x \neq 0$ and $\Delta x \neq -x$.

For $\Delta x \neq 0$, $\Delta y/\Delta x$ equals $-1/x(x + \Delta x)$, so, by the replacement rule,

$$\lim_{\Delta x \to 0} \frac{\Delta y}{\Delta x} = \lim_{\Delta x \to 0} \left(- \frac{1}{x(x + \Delta x)} \right)$$

$$= - \frac{1}{x^2} \quad \text{(by the continuity of rational functions)}.$$

Thus, $f'(x) = -1/x^2$. ▲

If we look back over the examples we have done, we may see a pattern. The derivative of x^3 is $3x^2$ by Example 2. The derivative of x^2 is given by the quadratic function rule as $2x^1 = 2x$. The derivative of $x = x^1$ is $1 \cdot x^0 = 1$, and the derivative of $1/x = x^{-1}$ is $(-1)x^{-2}$ by Example 3. In each case, when we differentiate x^n, we get nx^{n-1}. This general rule makes it unnecessary to memorize individual cases. In the next section, we will prove the rule for n a positive integer, and eventually we will prove it for all numbers n. For now, let us see how to prove the rule for $x^{1/2} = \sqrt{x}$. We should get $\frac{1}{2} x^{(1/2) - 1} = \frac{1}{2} x^{-1/2} = 1/2\sqrt{x}$.

Example 4   Differentiate $\sqrt{x}$ $(x > 0)$.

Solution   The difference quotient is

$$\frac{\Delta y}{\Delta x} = \frac{\sqrt{x + \Delta x} - \sqrt{x}}{\Delta x}.$$

In order to cancel Δx, we perform a trick: rationalize by multiplying numerator and denominator by $\sqrt{x + \Delta x} + \sqrt{x}$:

$$\frac{\Delta y}{\Delta x} = \frac{(\sqrt{x + \Delta x} - \sqrt{x})(\sqrt{x + \Delta x} + \sqrt{x})}{(\Delta x)(\sqrt{x + \Delta x} + \sqrt{x})}$$

$$= \frac{(x + \Delta x) - x}{\Delta x(\sqrt{x + \Delta x} + \sqrt{x})} = \frac{1}{\sqrt{x + \Delta x} + \sqrt{x}}.$$

Notice that this trick enabled us to cancel Δx in the numerator and denominator.

Now recall from the previous section that $\lim_{x \to x_0} \sqrt{x} = \sqrt{x_0}$. Thus, $\lim_{\Delta x \to 0} \sqrt{x + \Delta x} = \sqrt{x}$. Hence, by the quotient rule for limits,

$$\lim_{\Delta x \to 0} \frac{\Delta y}{\Delta x} = \frac{1}{\lim_{\Delta x \to 0} (\sqrt{x + \Delta x} + \sqrt{x})}$$

$$= \frac{1}{\lim_{\Delta x \to 0} \sqrt{x + \Delta x} + \lim_{\Delta x \to 0} \sqrt{x}} \qquad \text{(sum rule)}$$

$$= \frac{1}{\sqrt{x} + \sqrt{x}} = \frac{1}{2\sqrt{x}} \qquad \text{(continuity of } \sqrt{x} \text{).}$$

Thus, the derivative is indeed $1/2\sqrt{x}$. ▲

Next, let us establish a general relationship between differentiability and continuity.

Theorem: Differentiability Implies Continuity

If $f'(x_0)$ exists, then f is continuous at x_0; that is, $\lim_{x \to x_0} f(x) = f(x_0)$.

Proof We first note that $\lim_{x \to x_0} f(x) = f(x_0)$ is the same as $\lim_{x \to x_0}(f(x) - f(x_0)) = 0$ (by the sum rule and then the constant function rule applied to the constant $f(x_0)$). With $\Delta x = x - x_0$, and $\Delta y = f(x_0 + \Delta x) - f(x_0)$, this is, in turn, the same as $\lim_{\Delta x \to 0} \Delta y = 0$. Now we use again the trick of multiplying numerator and denominator by an appropriate factor:

$$\lim_{\Delta x \to 0} \Delta y = \lim_{\Delta x \to 0} \left(\frac{\Delta y}{\Delta x} \cdot \Delta x \right) \qquad \text{(replacement rule)}$$

$$= \left(\lim_{\Delta x \to 0} \frac{\Delta y}{\Delta x} \right)\left(\lim_{\Delta x \to 0} \Delta x \right) \qquad \text{(product rule)}$$

$$= f'(x_0) \cdot 0 \qquad \left(\text{since } \lim_{\Delta x \to 0} \Delta x = 0 \right)$$

$$= 0.$$

This proves our claim. ∎

The converse theorem is not true; the following is a counterexample.

Example 5 Show that $f(x) = |x|$ has no derivative at $x_0 = 0$, yet is continuous. (See Section R.2 for a review of absolute values.)

Solution   The difference quotient at $x_0 = 0$ is $(|0 + \Delta x| - |0|)/\Delta x = |\Delta x|/\Delta x$, which is 1 for $\Delta x > 0$ and -1 for $\Delta x < 0$. As we saw in Example 7 of Section 1.2, the function $|\Delta x|/\Delta x$ has no limit at $\Delta x = 0$, so the derivative of $|x|$ at $x_0 = 0$ does not exist.

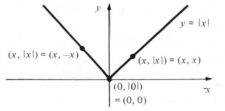

Figure 1.3.2. As $x \to 0$ from either direction, $|x| \to 0$, so $f(x) = |x|$ is continuous at 0.

On the other hand, as $x \to 0$, $f(x) = |x| \to 0$ as well (see Fig. 1.3.2), so $\lim_{x \to 0}|x| = |0|$; $|x|$ is continuous at 0. ▲

We have seen that the derivative $f'(x_0)$ of $y = f(x)$ at x_0 is approximated by the difference quotient $\Delta y/\Delta x$, where $\Delta x = x - x_0$.

In the view of Gottfried Wilhelm von Leibniz (1646–1716), one of the founders of calculus, one could think of Δx as becoming "infinitesimal." The resulting quantity he denoted as dx, the letters d and Δ being the Roman and Greek equivalents of one another. When Δx became the infinitesimal dx, Δy simultaneously became the infinitesimal dy and the ratio $\Delta y/\Delta x$ became dy/dx, which was no longer an approximation to the derivative but exactly equal to it. The notation dy/dx has proved to be extremely convenient—not as a ratio of infinitesimal quantities but as a *synonym* for $f'(x)$.[2]

Leibniz Notation

If $y = f(x)$, the derivative $f'(x)$ may be written

$$\frac{dy}{dx}, \quad dy/dx, \quad \frac{df(x)}{dx}, \quad (d/dx)\,f(x), \quad \text{or} \quad d(f(x))/dx.$$

This is just a notation and does not represent division. If we wish to denote the value $f'(x_0)$ of f' at a specific point x_0, we may write

$$\frac{dy}{dx}\bigg|_{x_0} \quad \text{and} \quad \frac{df(x)}{dx}\bigg|_{x_0}.$$

dy/dx is read "the derivative of y with respect to x" or "dy by dx."

Of course, we can use this notation if the variables are named other than x and y. For instance, the area A of a square of side l is $A = l^2$ so we can write $dA/dl = 2l$.

In the f' notation, if $f(x) = 3x^2 + 2x$, then $f'(x) = 6x + 2$. Using the Leibniz notation we may write:

$$\text{if} \quad y = 3x^2 + 2x, \quad \text{then} \quad \frac{dy}{dx} = 6x + 2.$$

---

[2] Modern developments in mathematics have made it possible to give rigorous definitions of dx and dy. The earlier objections to infinitesimals as quantities which were supposed to be smaller than any real number but still nonzero have been circumvented through the work of the logician Abraham Robinson (1918–1974). A calculus textbook based upon this approach is H. J. Keisler, *Elementary Calculus*, Prindle, Weber, and Schmidt, Boston (1976).

We can also use the even more compact notation

$$\frac{d(3x^2 + 2x)}{dx} = 6x + 2 \quad \text{or} \quad \frac{d}{dx}(3x^2 + 2x) = 6x + 2.$$

Here the d/dx may be thought of as a symbol for the *operation* of differentiation. It takes the place of the prime (′) in the functional notation.

Example 6   (a) Find the slope m of the graph $y = \sqrt{x}$ at $x = 4$. (b) Find the velocity v of a bus whose distance function is t^3.

Solution   (a) The slope is a derivative. The derivative of $\sqrt{x}$ is $dy/dx = d(\sqrt{x})/dx = 1/(2\sqrt{x})$ by Example 4. Evaluating at $x = 4$ gives $m = 1/(2\sqrt{4}) = \frac{1}{4}$.
(b) $v = (d/dt)(t^3) = 3t^2$. ▲

Supplement to Section 1.3
Filling a Pond

We conclude this section with a harder and perhaps more interesting application that previews some important topics to be considered in detail later: rates of change (Section 2.1) and integration (Chapter 4).

Suppose that a mountain brook swells from a trickle to a torrent each year as the snows melt. At the time t (days after midnight on March 31), the flow rate is known to be $3t^2$ thousand liters per day. We wish to build a large pond which holds the runoff for the entire month of April. How big must the pond be?

The main difficulty here is that a flow rate of, say $3 \cdot (5)^2$ at midnight of April 5 does not tell us directly how much water will be in the pond on April 5, but merely how fast water will be pouring in at that moment. Let's see if we can somehow handle that difficulty.

Designate the unknown amount of water in the pond at time t by $A = f(t)$. During a short time interval Δt starting at t, the amount of water entering the pond will be at least $3t^2 \Delta t$ and no more than $3(t + \Delta t)^2 \Delta t$. Thus, $\Delta A = A(t + \Delta t) - A(t)$ is slightly larger than $3t^2 \Delta t$. For Δt very small, we can presumably take $\Delta A \approx 3t^2 \Delta t$, i.e., $\Delta A / \Delta t \approx 3t^2$. However, for Δt very small, $\Delta A / \Delta t$ approximates the derivative dA/dt. Thus our problem becomes the following. Find the "amount" function $f(t)$, given that the derivative obeys $f'(t) = 3t^2$.

Now, turning Example 2 around, we know one function which obeys $f'(t) = 3t^2$, namely $f(t) = t^3$. This solution is reasonable in the sense that $f(0) = 0$, i.e., the pond is empty at midnight of March 31. Could there also be a different amount function that works? Not really. If a capacity of t^3 thousand liters is exactly right to accommodate all the influx up to time t, no other capacity will be exactly right. We thus have our answer: at midnight on April 30, $A = f(30) = (30)^3$; our pond must hold 27,000 thousand liters.

Exercises for Section 1.3

Use the formal definition of the derivative and the rules for limits to find the derivatives of the functions in Exercises 1–12.

1. $f(x) = x^2 + x$
2. $f(x) = 2x^2 - 3x$
3. $f(x) = 5x^3$
4. $f(x) = 2x^3$
5. $f(x) = \dfrac{3}{x}$
6. $f(x) = \dfrac{10}{x}$
7. $f(x) = x^2 + \dfrac{3}{x}$
8. $f(x) = x^3 - \dfrac{2}{x}$
9. $f(x) = 2\sqrt{x}$
10. $f(x) = 8\sqrt{x}$
11. $f(x) = 2x^2 - \sqrt{x} + \dfrac{1}{x}$
12. $f(x) = x^3 + 2x - \dfrac{1}{\sqrt{x}}$

Show that the functions in Exercises 13 and 14 have no derivative at x_0, yet are continuous.

13. $f(x) = 1 + |x|$; $x_0 = 0$
14. $g(x) = |x + 1|$; $x_0 = -1$.

Find dy/dx in the Exercises 15–18.

15. $y = x^2 - x$
16. $y = x - 5x^2$
17. $y = 3x^3 + x$
18. $y = x^2 - x^3$

In Exercises 19 and 20, find the slope of the line tangent to the given graph at the given point.

19. $y = 8\sqrt{x}$; $x_0 = 9$
20. $y = 2x^2 - \sqrt{x} + 1/x$; $x_0 = 1$

In Exercises 21 and 22, $f(t)$ is the position of a car on a straight road at time t. Find its velocity at the given time.

21. $f(t) = 5t^3$; $t = 1$
22. $f(t) = t^2 - t^3$; $t = \frac{1}{2}$

In Exercises 23–26, evaluate the derivatives.

23. $\dfrac{d(3/t)}{dt}$ at $t = 1$

24. $\dfrac{d}{dt}\left[t^3 + 2t - \dfrac{1}{\sqrt{t}} \right]$ at $t = 2$

25. $\dfrac{d}{dx}(3x^3 + x)\Big|_{x=1}$

26. $\dfrac{d}{dx}\left(\dfrac{10}{x} \right)\Big|_{x=2}$

27. Using the limit method, find the derivative of $2x^3 + x^2 - 3$ at $x_0 = 1$.

28. (a) Expand $(a + b)^4$. (b) Use the limit method to differentiate x^4.

Use limits to find the derivatives of the functions in Exercises 29–32.

29. $f(x) = 1/x^2$
30. $\sqrt[3]{x}$
31. $f(x) = (x^2 + x)/2x$
32. $f(x) = x/(1 + x^2)$

★33. Find an example of a function which is continuous everywhere and which is differentiable everywhere except at *two* points.

★34. (a) Show by the quadratic function rule that if $f(x) = ax^2 + bx + c$, $g(x) = dx^2 + ex + f$, and $h(x) = f(x) + g(x)$, then $h'(x) = f'(x) + g'(x)$; i.e., $(d/dx)[f(x) + g(x)] = (d/dx) f(x) + (d/dx) g(x)$.

(b) Show from the rules for limits that if $f(x)$ and $g(x)$ are differentiable functions, then

$$\frac{d}{dx}[f(x) + g(x)] = \frac{d}{dx} f(x) + \frac{d}{dx} g(x)$$

and

$$\frac{d}{dx}[f(x) - g(x)] = \frac{d}{dx} f(x) - \frac{d}{dx} g(x).$$

(c) Argue geometrically, using graphs and slopes, that a function $C(x)$ for which $C'(x) = 0$ must be a constant function.

(d) Combining (b) and (c), show that if $f'(x) = g'(x)$, then there is some constant C such that $f(x) = g(x) + C$. Illustrate your result graphically.

(e) In (d) show that if $f(0) = 0 = g(0)$, then $f(x) = g(x)$ for all x.

(f) Use (e) to argue that in the pond example discussed in the Supplement, $A(t) = t^3$ is the only appropriate solution of $A'(t) = 3t^2$.

★35. (a) Do some calculator experiments to guess $\lim_{x \to 0}(\sin x / x)$ and $\lim_{x \to 0}[(1 - \cos x)/x]$, where the angle x is measured in radians.

(b) *Given* the facts that $\lim_{x \to 0}(\sin x / x) = 1$ and $\lim_{x \to 0}[(1 - \cos x)/x] = 0$, use trigonometric identities to show:

$$\frac{d(\sin x)}{dx} = \cos x,$$

$$\frac{d(\cos x)}{dx} = -\sin x.$$

★36. Suppose that the mountain brook in the Supplement has a flow rate of $t^2/12 + 2t$ thousand liters per day t days after midnight on March 31. What is the runoff for the first 15 days of April? The entire month?

1.4 Differentiating Polynomials

Polynomials can be differentiated using the power rule, the sum rule, and the constant multiple rule.

In Section 1.3, we learned how to compute derivatives of some simple functions using limits. Now we shall use the limit method to find a general rule for differentiating polynomials like $f(x) = 3x^5 - 8x^4 + 4x + 2$. To do this systematically, we shall break apart a polynomial using two basic operations.

First, we recognize that a polynomial is a sum of monomials: for example, $f(x) = 3x^5 - 8x^4 + 4x + 2$ is the sum of $3x^5$, $-8x^4$, $4x$ and 2. Second, a monomial is a product of a constant and a power of x. For example, $3x^5$ is the product of 3 and x^5.

Let us work backward, starting with powers of x. Thus our first goal is to differentiate x^n, where n is a positive integer. We have already seen that for $n = 1, 2,$ or 3 (as well as $n = -1$ or $\frac{1}{2}$), the derivative of x^n is nx^{n-1}.

We can establish this rule for any positive integer n by using limits. Let $f(x) = x^n$. To compute $f'(x)$, we must find the limit

$$\lim_{\Delta x \to 0} \frac{f(x + \Delta x) - f(x)}{\Delta x}.$$

Now $f(x) = (x + \Delta x)^n = (x + \Delta x)(x + \Delta x) \dots (x + \Delta x)$, n times. To expand a product like this, we select one term from each factor, multiply these n terms, and then add all such products. For example,

$$(x + \Delta x)(x + \Delta x) = x^2 + x\,\Delta x + (\Delta x)x + (\Delta x)^2$$
$$= x^2 + 2x\,\Delta x + (\Delta x)^2,$$
$$(x + \Delta x)(x + \Delta x)(x + \Delta x) = x^3 + x^2\Delta x + x(\Delta x)x + (\Delta x)x^2 + (\Delta x)^2 x$$
$$+ x(\Delta x)^2 + (\Delta x)x(\Delta x) + (\Delta x)^3$$
$$= x^3 + 3x^2\,\Delta x + 3x(\Delta x)^2 + (\Delta x)^3.$$

For $(x + \Delta x)^n$, notice that the coefficient of Δx will be nx^{n-1} since there will be exactly n terms which contain $n - 1$ factors of x and one of Δx. Thus

$$(x + \Delta x)^n = x^n + nx^{n-1}\Delta x + \left(\text{terms involving } (\Delta x)^2, (\Delta x)^3, \dots, (\Delta x)^n\right).$$

If you are familiar with the binomial theorem, you will know the remaining terms; however, their exact form is not needed here. For $\Delta x \neq 0$, dividing out Δx now gives

$$\frac{f(x + \Delta x) - f(x)}{\Delta x} = \frac{(x + \Delta x)^n - x^n}{\Delta x}$$
$$= \frac{nx^{n-1}\Delta x + \left(\text{terms involving } (\Delta x)^2, \dots, (\Delta x)^n\right)}{\Delta x}$$
$$= nx^{n-1} + \left(\text{terms involving } (\Delta x), \dots, (\Delta x)^{n-1}\right).$$

The terms involving $\Delta x, \dots, (\Delta x)^{n-1}$ add up to a polynomial in Δx, so the limit as $\Delta x \to 0$ is obtained by setting $\Delta x = 0$ and by using the continuity of rational functions (Section 1.2). Therefore,

$$f'(x) = \lim_{\Delta x \to 0} \frac{f(x + \Delta x) - f(x)}{\Delta x} = nx^{n-1}.$$

Power Rule

To differentiate a power x^n, bring down the exponent as a factor and then reduce the exponent by 1.

If $f(x) = x^n$, then $f'(x) = nx^{n-1}$; that is

$$\frac{d}{dx}(x^n) = nx^{n-1}, \qquad n = 1, 2, 3, \dots.$$

Example 1   Compute the derivatives of x^8, x^{12}, and x^{99}.

Solution $(d/dx)x^8 = 8x^7$, $(d/dx)x^{12} = 12x^{11}$, and $(d/dx)x^{99} = 99x^{98}$. ▲

Next, we consider the *constant multiple rule*, stated in the following box.

Constant Multiple Rule

To differentiate the product of a number k with $f(x)$, multiply the number k by the derivative $f'(x)$:

$$(kf)'(x) = kf'(x),$$

$$\frac{d}{dx}(kf(x)) = k\frac{d}{dx}f(x).$$

Proof of the Constant Multiple Rule   Let $h(x) = kf(x)$. By the definition of the derivative and the basic properties of limits, we get

$$h'(x) = \lim_{\Delta x \to 0} \frac{h(x + \Delta x) - h(x)}{\Delta x}$$

$$= \lim_{\Delta x \to 0} \frac{kf(x + \Delta x) - kf(x)}{\Delta x} = \lim_{\Delta x \to 0} k\left(\frac{f(x + \Delta x) - f(x)}{\Delta x}\right)$$

$$= \left(\lim_{\Delta x \to 0} k\right) \lim_{\Delta x \to 0}\left[\frac{f(x + \Delta x) - f(x)}{\Delta x}\right] = kf'(x). \blacksquare$$

Example 2   Differentiate

(a) $-3x^7$ (b) $5\sqrt{x}$ (c) $\dfrac{8}{x}$ and   (d) $-6ax^2$.

Solution   (a) By the constant multiple and power rules,

$$\frac{d}{dx}(-3x^7) = (-3)\frac{d}{dx}x^7 = (-3)(7)x^6 = -21x^6.$$

(b) From Example 4, Section 1.3, $(d/dx)\sqrt{x} = 1/2\sqrt{x}$. Thus, by the constant multiple rule,

$$\frac{d}{dx}(5\sqrt{x}) = 5\frac{d}{dx}\sqrt{x} = \frac{5}{2\sqrt{x}}.$$

(c) By Example 3, Section 1.3, $(d/dx)(1/x) = -1/x^2$. Thus

$$\frac{d}{dx}\left(\frac{8}{x}\right) = \frac{-8}{x^2}.$$

(d) Although it is not explicitly stated, we assume that a is constant (letters from the beginning of the alphabet are often used for constants). Thus, by the constant multiple rule

$$\frac{d}{dx}(-6ax^2) = -6a\frac{d}{dx}x^2 = -12ax. \blacktriangle$$

The final basic technique we need is the *sum rule*.

If f and g are two functions, the sum $f + g$ is defined by the formula $(f + g)(x) = f(x) + g(x)$.

Example 3   Let $f(x) = 3x^2 + 5x + 9$ and $g(x) = 2x^2 + 5x$. Use the quadratic function rule to verify that $(f + g)' = f' + g'$.

Solution   By the quadratic function rule, $f'(x) = 6x + 5$ and $g'(x) = 4x + 5$, thus $f'(x) + g'(x) = 10x + 10$. On the other hand, $f(x) + g(x) = 5x^2 + 10x + 9$, so $(f + g)'(x) = 10x + 10 = f'(x) + g'(x). \blacktriangle$

Sum Rule

To differentiate a sum, take the sum of the derivatives:

$$(f + g)' = f' + g'$$

or

$$\frac{d}{dx}\left[f(x) + g(x) \right] = \frac{d}{dx}\left[f(x) \right] + \frac{d}{dx}\left[g(x) \right]$$

or

$$\frac{d}{dx}(u + v) = \frac{du}{dx} + \frac{dv}{dx}.$$

To be convinced that a mathematical statement such as the sum rule is true, one should ideally do three things:

1. Check some simple examples directly.
2. Have a mathematical justification (proof).
3. Have a simple physical model, application, or diagram that makes the result plausible.

In Example 3 we checked the sum rule in a simple case. In the next paragraph we give a mathematical justification for the sum rule. In the Supplement at the end of the section, we give a simple physical model.

Proof of the Sum Rule

By the definition of the derivative as a limit, $(f + g)'(x_0)$ is equal to

$$\lim_{\Delta x \to 0} \frac{(f + g)(x_0 + \Delta x) - (f + g)(x_0)}{\Delta x}$$

(if this limit exists). We can rewrite the limit as

$$\lim_{\Delta x \to 0} \frac{f(x_0 + \Delta x) + g(x_0 + \Delta x) - f(x_0) - g(x_0)}{\Delta x}$$

$$= \lim_{\Delta x \to 0} \left[\frac{f(x_0 + \Delta x) - f(x_0)}{\Delta x} + \frac{g(x_0 + \Delta x) - g(x_0)}{\Delta x} \right].$$

By the sum rule for *limits*, this is

$$\lim_{\Delta x \to 0} \frac{f(x_0 + \Delta x) - f(x_0)}{\Delta x} + \lim_{\Delta x \to 0} \frac{g(x_0 + \Delta x) - g(x_0)}{\Delta x}.$$

If f and g are differentiable at x_0, these two limits are just $f'(x_0)$ and $g'(x_0)$. Thus $f + g$ is differentiable at x_0, and $(f + g)'(x_0) = f'(x_0) + g'(x_0)$. ∎

The sum rule extends to several summands. For example, to find a formula for the derivative of $f(x) + g(x) + h(x)$, we apply the sum rule twice:

$$\frac{d}{dx}\left[f(x) + g(x) + h(x) \right] = \frac{d}{dx}\left[f(x) + (g(x) + h(x)) \right]$$

$$= \frac{d}{dx} f(x) + \frac{d}{dx}(g(x) + h(x))$$

$$= \frac{d}{dx} f(x) + \frac{d}{dx} g(x) + \frac{d}{dx} h(x).$$

Example 4   Find the formula for the derivative $8f(x) - 10g(x)$.

Solution   We use the sum and constant multiple rules:

$$\frac{d}{dx}\big[8f(x) - 10g(x)\big] = \frac{d}{dx}\big[8f(x)\big] + \frac{d}{dx}\big[-10g(x)\big]$$

$$= 8\frac{d}{dx}f(x) - 10\frac{d}{dx}g(x). \; \blacktriangle$$

Now we can differentiate any polynomial.

Example 5   Differentiate (a) $(x^{95} + x^{23} + 2x^2 + 4x + 1)$; (b) $4x^9 - 6x^5 + 3x$.

Solution   (a) $\frac{d}{dx}(x^{95} + x^{23} + 2x^2 + 4x + 1)$

$$= \frac{d}{dx}(x^{95}) + \frac{d}{dx}(x^{23}) + \frac{d}{dx}(2x^2) + \frac{d}{dx}(4x) + \frac{d}{dx}(1)$$

(sum rule)

$$= 95x^{94} + 23x^{22} + 4x + 4$$

(power rule and constant multiple rule).

(b) $\frac{d}{dx}(4x^9 - 6x^5 + 3x) = 36x^8 - 30x^4 + 3. \; \blacktriangle$

Here, for reference, is a general rule, but you need not memorize it, since you can readily do any example by using the sum, power, and constant multiple rules.

Derivative of a Polynomial

If $f(x) = c_n x^n + \cdots + c_2 x^2 + c_1 x + c_0$, then

$$f'(x) = nc_n x^{n-1} + (n-1)c_{n-1}x^{n-2} + \cdots + 2c_2 x + c_1.$$

Example 6   Find the derivative of $x^3 + 5x^2 - 9x + 2$.

Solution $\frac{d}{dx}(x^3 + 5x^2 - 9x + 2) = \frac{d}{dx}x^3 + \frac{d}{dx}(5x^2) - \frac{d}{dx}(9x) + \frac{d}{dx}2$

$$= 3x^2 + 10x - 9. \; \blacktriangle$$

Example 7   (a) Compute $f'(s)$ if $f(s) = (s^2 + 3)(s^3 + 2s + 1)$.

(b) Find $\frac{d}{dx}(10x^3 - 8/x + 5\sqrt{x})$.

Solution   (a) First we expand the product

$$(s^2 + 3)(s^3 + 2s + 1) = (s^5 + 2s^3 + s^2) + (3s^3 + 6s + 3)$$

$$= s^5 + 5s^3 + s^2 + 6s + 3.$$

Now we differentiate this polynomial:

$$f'(s) = 5s^4 + 15s^2 + 2s + 6.$$

(b) $\frac{d}{dx}\left(10x^3 - \frac{8}{x} + 5\sqrt{x}\right) = \frac{d}{dx}(10x^3) + \frac{d}{dx}\left(\frac{-8}{x}\right) + \frac{d}{dx}(5\sqrt{x})$

$$= 10\frac{d}{dx}(x^3) - 8\frac{d}{dx}(x^{-1}) + 5\frac{d}{dx}(x^{1/2})$$

$$= 30x^2 + \frac{8}{x^2} + \frac{5}{2}x^{-1/2} = 30x^2 + \frac{8}{x^2} + \frac{5}{2\sqrt{x}}. \; \blacktriangle$$

The differentiation rules we have learned can be applied to the problems of finding slopes and velocities.

Example 8   (a) Find the slope of the tangent line to the graph of

$$y = x^4 - 2x^3 + 1 \qquad \text{at} \quad x = 1.$$

(b) A train has position $x = 3t^2 + 2 - \sqrt{t}$ at time t. Find the velocity of the train at $t = 2$.

Solution   (a) The slope is the derivative at $x = 1$. The derivative is

$$\frac{dy}{dx} = \frac{d}{dx}(x^4 - 2x^3 + 1) = 4x^3 - 6x^2.$$

At $x = 1$, this is $4 \cdot 1^3 - 6 \cdot 1^2 = -2$, the required slope.

(b) The velocity is

$$\frac{dx}{dt} = \frac{d}{dt}\left(3t^2 + 2 - \sqrt{t}\right) = 6t - \frac{1}{2\sqrt{t}}.$$

At $t = 2$, we get

$$\frac{dx}{dt}\bigg|_2 = 6 \cdot 2 - \frac{1}{2\sqrt{2}} = 12 - \frac{1}{2\sqrt{2}} = \frac{24\sqrt{2} - 1}{2\sqrt{2}} = \frac{48 - \sqrt{2}}{4}. \quad \blacktriangle$$

Supplement to Section 1.4
A Physical Model for the Sum Rule

Imagine a train, on a straight track, whose distance at time x from a fixed reference point on the ground is $f(x)$. There is a runner on the train whose distance from a reference point on the train is $g(x)$. Then the distance of the runner from the fixed reference point on the ground is $f(x) + g(x)$. (See Fig. 1.4.1.) Suppose that, at a certain time x_0, the runner is going at 20 kilometers

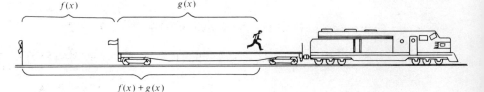

Figure 1.4.1. The sum rule illustrated in terms of velocities.

per hour with respect to the train while the train is going at 140 kilometers per hour—that is, $f'(x_0) = 140$ and $g'(x_0) = 20$. What is the velocity of the runner as seen from an observer on the ground? It is the sum of 140 and 20—that is, 160 kilometers per hour. Considered as the sum of two velocities, the number 160 is $f'(x_0) + g'(x_0)$; considered as the velocity of the runner with respect to the ground, the number 160 is $(f + g)'(x_0)$. Thus we have $f'(x_0) + g'(x_0)$ $= (f + g)'(x_0)$.[3]

---

[3] The fact that one does not add velocities this way in the theory of special relativity does not violate the sum rule. In classical mechanics, velocities are derivatives, but in relativity, velocities are not simply derivatives, so the formula for their combination is more complicated.

Exercises for Section 1.4

Differentiate the functions in Exercises 1–12.

1. x^{10}
2. x^{14}
3. x^{33}
4. x^5
5. $-5x^4$
6. $-53x^{20}$
7. $3x^{10}$
8. $8x^{100}$
9. $3\sqrt{x}$
10. $2/x$
11. $-8\sqrt{x}$
12. $-6/x$

In Exercises 13–16, verify the sum rule for the given pair of functions.

13. $f(x) = 3x^2 + 6$, $g(x) = x + 7$
14. $f(x) = 8x + 9$, $g(x) = x^2 - 1$
15. $f(x) = x^2 + x + 1$, $g(x) = -1$
16. $f(x) = 2x^2 - 3x + 6$, $g(x) = -x^2 + 8x - 9$

17. Find a formula for the derivative of $f(x) - 2g(x)$.
18. Find a formula for the derivative of $3f(x) + 2g(x)$.

Differentiate the functions in Exercises 19–22.

19. $x^5 + 8x$
20. $5x^3$
21. $t^5 + 6t^2 + 8t + 2$
22. $s^{10} + 8s^9 + 5s^8 + 2$

Differentiate the functions in Exercises 23–34.

23. $f(x) = x^4 - 7x^2 - 3x + 1$
24. $h(x) = 3x^{11} + 8x^5 - 9x^3 - x$
25. $g(s) = s^{13} + 12s^8 - \frac{3}{8}s^7 + s^4 + \frac{4}{3}s^3$
26. $f(y) = -y^3 - 8y^2 - 14y - \frac{1}{3}$
27. $f(x) = x^4 - 3x^3 + 2x^2$
28. $f(t) = t^4 + 4t^3$
29. $g(h) = 8h^{10} + h^9 - 56.5h^2$
30. $h(\gamma) = \pi\gamma^{10} + \frac{21}{7}\gamma^9 - \sqrt{2}\,\gamma^2$
31. $p(x) = (x^2 + 1)^3$
32. $r(t) = (t^4 + 2t^2)^2$
33. $f(t) = (t^3 - 17t + 9)(3t^5 - t^2 - 1)$
34. $h(x) = (x^4 - 1)(x^2 + x + 2)$

35. Find $f'(r)$ if $f(r) = -5r^6 + 5r^4 - 13r^2 + 15$.
36. Find $g'(s)$ if $g(s) = s^7 + 13s^6 - 18s^3 + \frac{3}{2}s^2$.
37. Find $h'(t)$ if $h(t) = (t^4 + 9)(t^3 - t)$.
38. Differentiate $x^5 + 2x^4 + 7$.
39. Differentiate $(u^4 + 5)(u^3 + 7u^2 + 19)$.
40. Differentiate $(3t^5 + 9t^3 + 5t)(t + 1)$.

Differentiate the functions in Exercises 41–46.

41. $f(x) = x^2 - \sqrt{x}$
42. $f(x) = 3x^3 + \dfrac{1}{x}$
43. $f(x) = x^3 - 2x + 2\sqrt{x}$

44. $f(x) = x^5 - \sqrt{x} + \dfrac{1}{x}$
45. $f(x) = (1 - \sqrt{x})(1 + \sqrt{x})$
46. $f(x) = (1 + \sqrt{x})\sqrt{x}$
47. A particle moves on a line with position $f(t) = 16t^2 + (0.03)t^4$ at time t. Find the velocity at $t = 8$.
48. Suppose that the position x of a car at time t is $(t - 2)^3$.
    (a) What is the velocity at $t = -1, 0, 1$?
    (b) Show that the average velocity over every interval of time is positive.
    (c) There is a stop sign at $x = 0$. A police officer gives the driver a ticket because there was no period of time during which the car was stopped. The driver argues that, since his velocity was zero at $t = 2$, he obeyed the stop sign. Who is right?
49. Find the slope of the tangent line to the graph of $x^4 - x^2 + 3x$ at $x = 1$.
50. Find the slope of the line tangent to the graph of $f(x) = x^8 + 2x^2 + 1$ at $(1, 4)$.

For each of the functions in Exercises 51–54, find a function whose derivative is $f(x)$. (Do not find $f'(x)$.)

51. $f(x) = x^2$
52. $f(x) = x^2 + 2x + 3$
53. $f(x) = x^n$ (n any positive integer)
54. $f(x) = (x + 3)(x^2 + 1)$

55. Verify the constant multiple rule for general quadratic functions, i.e., show that $(kf)'(x) = kf'(x)$ if $f(x) = ax^2 + bx + c$.
56. Verify the sum rule for general cubic functions using the formula for the derivative of a polynomial.
57. Let $V(r)$ be the volume V of a sphere as a function of the radius r. Show that $V'(r)$ is the surface area.
58. Let $V(l)$ be the volume of a cube as a function of l, where $2l$ is length of one of its edges. Show that $V'(l)$ is the surface area.
★59. Explain the constant multiple rule in terms of a change of units in distance from miles to kilometers.
★60. Show that if two polynomials have the same derivative, they must differ by a constant.

1.5 Products and Quotients

To differentiate a product, differentiate each factor in turn and sum the results.

We have given general rules for the derivative of a sum and a constant multiple. We now turn to products and quotients.

The product fg and quotient f/g of two functions are defined by $(fg)(x) = f(x)g(x)$ and $(f/g)(x) = f(x)/g(x)$, the latter being defined only when $g(x) \neq 0$. The formulas for $(fg)'$ and $(f/g)'$ are more complicated than those for $(f + g)'$ and $(kf)'$, but they are just as straightforward to apply. Before developing the correct formulas, let us convince ourselves that $(fg)'$ is *not* $f'g'$.

Example 1   Let $f(x) = x^2$ and $g(x) = x^3$. Is $(fg)'$ equal to $f'g'$?

Solution   Notice that the product function is obtained simply by multiplying the formulas for f and g: $(fg)(x) = (x^2)(x^3) = x^5$. Thus, $(fg)'(x) = 5x^4$. On the other hand, $f'(x) = 2x$ and $g'(x) = 3x^2$, so $(f'g')(x) = f'(x)g'(x) = 6x^3$. Since $5x^4$ and $6x^3$ are not the same function, $(fg)'$ is *not* equal to $f'g'$. ▲

Example 1 shows that the derivative of the product of two functions is not the product of their derivatives. We state the correct rule for products now and discuss below why it is true.

Product Rule

To differentiate a product $f(x)g(x)$, differentiate each factor and multiply it by the other one, then add the two products:

$$(fg)'(x) = f(x)g'(x) + f'(x)g(x)$$

or

$$\frac{d}{dx}(uv) = u\frac{dv}{dx} + \frac{du}{dx}v.$$

Example 2   (a) Verify the product rule for f and g in Example 1.
(b) Verify the product rule for $f(x) = x^m$ and $g(x) = x^n$, where m and n are natural numbers.

Solution   (a) We know that $(fg)'(x) = 5x^4$. On the other hand, $f(x)g'(x) + f'(x)g(x) = (x^2)(3x^2) + (2x)(x^3) = 5x^4$, so the product rule gives the right answer.
(b) By the power rule in Section 1.4, $f'(x) = mx^{m-1}$ and $g'(x) = nx^{n-1}$, so that $(fg)'(x) = f(x)g'(x) + f'(x)g(x) = x^m(nx^{n-1}) + (mx^{m-1})x^n = (n + m)x^{m+n-1}$. On the other hand, $(fg)(x) = x^m x^n = x^{m+n}$, so again by the power rule $(fg)'(x) = (m + n)x^{m+n-1}$, which checks. ▲

The form of the product rule may be a surprise to you. Why should that strange combination of f, g, and their derivatives be the derivative of fg? The following mathematical justification should convince you that the product rule is correct.

Proof of the Product Rule

To find $(fg)'(x_0)$, we take the limit

$$\lim_{\Delta x \to 0} \frac{(fg)(x_0 + \Delta x) - (fg)(x_0)}{\Delta x}$$

$$= \lim_{\Delta x \to 0} \frac{f(x_0 + \Delta x)g(x_0 + \Delta x) - f(x_0)g(x_0)}{\Delta x}. \tag{1}$$

Simplifying this expression is not as straightforward as for the sum rule. We may make use of a geometric device: think of $f(x)$ and $g(x)$ as the lengths of the sides of a rectangle; then $f(x)g(x)$ is its area. The rectangles for $x = x_0$ and $x = x_0 + \Delta x$ are shown in Fig. 1.5.1. The area of the large rectangle is

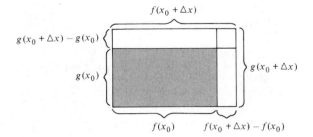

Figure 1.5.1 The geometry behind the proof of the product rule.

$f(x_0 + \Delta x)g(x_0 + \Delta x)$; that of the darker rectangle is $f(x_0)g(x_0)$. The difference $f(x_0 + \Delta x)g(x_0 + \Delta x) - f(x_0)g(x_0)$ is the area of the lighter region, which can be decomposed into three rectangles having areas

$$[f(x_0 + \Delta x) - f(x_0)]g(x_0),$$
$$[f(x_0 + \Delta x) - f(x_0)][g(x_0 + \Delta x) - g(x)],$$

and

$$f(x_0)[g(x_0 + \Delta x) - g(x)].$$

Thus we have the identity:

$$f(x_0 + \Delta x)g(x_0 + \Delta x) - f(x_0)g(x_0)$$
$$= [f(x_0 + \Delta x) - f(x_0)]g(x_0) + f(x_0)[g(x_0 + \Delta x) - g(x_0)]$$
$$+ [f(x_0 + \Delta x) - f(x_0)][g(x_0 + \Delta x) - g(x_0)]. \tag{2}$$

(If you do not like geometric arguments, you can verify this identity algebraically.)

Substituting (2) into (1), we obtain

$$\lim_{\Delta x \to 0} \left\{ \frac{f(x_0 + \Delta x) - f(x_0)}{\Delta x} g(x_0) + f(x_0) \frac{g(x_0 + \Delta x) - g(x_0)}{\Delta x} \right.$$
$$\left. + \frac{[f(x_0 + \Delta x) - f(x_0)][g(x_0 + \Delta x) - g(x_0)]}{\Delta x} \right\}. \tag{3}$$

By the sum and constant multiple rules for limits, (3) equals

$$\left[\lim_{\Delta x \to 0} \frac{f(x_0 + \Delta x) - f(x_0)}{\Delta x} \right] g(x_0) + f(x_0) \left[\lim_{\Delta x \to 0} \frac{g(x_0 + \Delta x) - g(x_0)}{\Delta x} \right]$$
$$+ \lim_{\Delta x \to 0} \frac{[f(x_0 + \Delta x) - f(x_0)][g(x_0 + \Delta x) - g(x_0)]}{\Delta x}. \tag{4}$$

We recognize the first two limits in (4) as $f'(x_0)$ and $g'(x_0)$, so the first two terms give $f'(x_0)g(x_0) + f(x_0)g'(x_0)$—precisely the product rule. To show that the third limit, represented geometrically by the small rectangle in the upper right-hand corner of Fig. 1.5.1, is zero, we use continuity of g (see Section 1.3). The product rule for *limits* yields

$$\lim_{\Delta x \to 0} \left[\frac{f(x_0 + \Delta x) - f(x_0)}{\Delta x} \right] \cdot \lim_{\Delta x \to 0} \left[g(x_0 + \Delta x) - g(x_0) \right]$$

$$= f'(x_0) \cdot 0 = 0. \ \blacksquare$$

Example 3 Using the product rule, differentiate $(x^2 + 2x - 1)(x^3 - 4x^2)$. Check your answer by multiplying out first.

Solution

$$\frac{d}{dx} \left[(x^2 + 2x - 1)(x^3 - 4x^2) \right]$$

$$= \frac{d(x^2 + 2x - 1)}{dx} (x^3 - 4x^2) + (x^2 + 2x - 1) \frac{d(x^3 - 4x^2)}{dx}$$

$$= (2x + 2)(x^3 - 4x^2) + (x^2 + 2x - 1)(3x^2 - 8x)$$

$$= (2x^4 - 6x^3 - 8x^2) + (3x^4 - 2x^3 - 19x^2 + 8x)$$

$$= 5x^4 - 8x^3 - 27x^2 + 8x.$$

Multiplying out first,

$$(x^2 + 2x - 1)(x^3 - 4x^2) = x^5 - 4x^4 + 2x^4 - 8x^3 - x^3 + 4x^2$$

$$= x^5 - 2x^4 - 9x^3 + 4x^2.$$

The derivative of this is $5x^4 - 8x^3 - 27x^2 + 8x$, so our answer checks. ▲

Example 4 Differentiate $x^{3/2}$ by writing $x^{3/2} = x \cdot \sqrt{x}$ and using the product rule.

Solution We know that $(d/dx)x = 1$ and $(d/dx)\sqrt{x} = 1/(2\sqrt{x})$. Thus, the product rule gives

$$\frac{d}{dx} (x^{3/2}) = \frac{d}{dx} (x \cdot \sqrt{x}) = \left(\frac{d}{dx} x \right) \sqrt{x} + x \frac{d}{dx} \sqrt{x}$$

$$= \sqrt{x} + x \cdot \frac{1}{2\sqrt{x}} = \sqrt{x} + \frac{1}{2} \sqrt{x} = \frac{3}{2} \sqrt{x}.$$

This result may be written $(d/dx) x^{3/2} = \frac{3}{2} x^{1/2}$, which is another instance of the rule $(d/dx)(x^n) = nx^{n-1}$ for noninteger n. ▲

We now come to quotients. Let $h(x) = f(x)/g(x)$, where f and g are differentiable at x_0, and suppose $g(x_0) \neq 0$ so that the quotient is defined at x_0. If we *assume* the existence of $h'(x_0)$, it is easy to compute its value from the product rule.

Since $h(x) = f(x)/g(x)$, we have $f(x) = g(x)h(x)$. Apply the product rule to obtain

$$f'(x_0) = g'(x_0)h(x_0) + g(x_0)h'(x_0).$$

Solving for $h'(x_0)$, we get

$$h'(x_0) = \frac{f'(x_0) - g'(x_0)h(x_0)}{g(x_0)} = \frac{f'(x_0) - g'(x_0)\left[f(x_0)/g(x_0) \right]}{g(x_0)}$$

$$= \frac{f'(x_0)g(x_0) - f(x_0)g'(x_0)}{\left[g(x_0)\right]^2}.$$

This is the quotient rule.

Quotient Rule

To differentiate a quotient $f(x)/g(x)$ (where $g(x) \neq 0$), take the derivative of the numerator times the denominator, subtract the numerator times the derivative of the denominator, and divide the result by the square of the denominator:

$$\left(\frac{f}{g}\right)'(x) = \frac{f'(x)g(x) - f(x)g'(x)}{\left[g(x)\right]^2} \quad \text{or}$$

$$\frac{d}{dx}\left(\frac{u}{v}\right) = \frac{(du/dx)v - u(dv/dx)}{v^2}.$$

When you use the quotient rule, it is important to remember which term in the numerator comes first. (In the product rule, both terms occur with a plus sign, so the order does not matter.) One memory aid is the following: Write your guess for the right formula and set $g = 1$ and $g' = 0$. Your formula should reduce to f'. If it comes out as $-f'$ instead, you have the terms in the wrong order.

Example 5 Differentiate $\dfrac{x^2}{x^3 + 5}$.

Solution By the quotient rule, with $f(x) = x^2$ and $g(x) = x^3 + 5$,

$$\frac{d}{dx}\left(\frac{x^2}{x^3 + 5}\right) = \frac{2x(x^3 + 5) - x^2(3x^2)}{(x^3 + 5)^2}$$

$$= \frac{x}{(x^3 + 5)^2}(2x^3 + 10 - 3x^3) = \frac{x(-x^3 + 10)}{(x^3 + 5)^2}. \quad \blacktriangle$$

Example 6 Find the derivative of (a) $h(x) = (2x + 1)/(x^2 - 2)$ and (b) $\sqrt{x}/(1 + 3x^2)$.

Solution (a) By the quotient rule with $f(x) = 2x + 1$ and $g(x) = x^2 - 2$,

$$h'(x) = \frac{2(x^2 - 2) - (2x + 1)2x}{(x^2 - 2)^2} = \frac{2x^2 - 4 - 4x^2 - 2x}{(x^2 - 2)^2}$$

$$= -\frac{2x^2 + 2x + 4}{(x^2 - 2)^2}.$$

(b) $\dfrac{d}{dx}\left(\dfrac{\sqrt{x}}{1 + 3x^2}\right) = \dfrac{(1 + 3x^2) \cdot (1/2\sqrt{x}) - \sqrt{x} \cdot 6x}{(1 + 3x^2)^2} = \dfrac{1 - 9x^2}{2\sqrt{x}(1 + 3x^2)^2}. \quad \blacktriangle$

In the argument given for the quotient rule, we assumed that $h'(x_0)$ exists; however, we can prove the quotient rule more carefully by the method of limits.

Proof of the Quotient Rule

The derivative of $h(x) = f(x)/g(x)$ at x_0 is given by the following limit:

$$h'(x_0) = \lim_{\Delta x \to 0} \frac{h(x_0 + \Delta x) - h(x_0)}{\Delta x}$$

$$= \lim_{\Delta x \to 0} \frac{f(x_0 + \Delta x)/g(x_0 + \Delta x) - f(x_0)/g(x_0)}{\Delta x}$$

$$= \lim_{\Delta x \to 0} \frac{f(x_0 + \Delta x)g(x_0) - f(x_0)g(x_0 + \Delta x)}{g(x_0)g(x_0 + \Delta x)\Delta x}.$$

A look at the calculations in the limit derivation of the product rule suggests that we add $-f(x_0)g(x_0) + f(x_0)g(x_0) = 0$ to the numerator. We get

$$h'(x_0) = \lim_{\Delta x \to 0} \frac{f(x_0 + \Delta x)g(x_0) - f(x_0)g(x_0) + f(x_0)g(x_0) - f(x_0)g(x_0 + \Delta x)}{g(x_0)g(x_0 + \Delta x)\Delta x}$$

$$= \lim_{\Delta x \to 0} \left\{ \frac{1}{g(x_0)g(x_0 + \Delta x)} \left[\frac{f(x_0 + \Delta x) - f(x_0)}{\Delta x} g(x_0) \right. \right.$$

$$\left. \left. -f(x_0) \frac{g(x_0 + \Delta x) - g(x_0)}{\Delta x} \right] \right\}$$

$$= \frac{1}{\lim_{\Delta x \to 0} g(x_0 + \Delta x)} \frac{1}{g(x_0)} \left[f'(x_0)g(x_0) - f(x_0)g'(x_0) \right]. \qquad (5)$$

Since g is differentiable at x_0, it is continuous there (see Section 1.3), and so

$$\lim_{\Delta x \to 0} g(x_0 + \Delta x) = g(x_0). \qquad (6)$$

Substituting (6) into (5) gives the quotient rule. ∎

Certain special cases of the quotient rule are particularly useful. If $f(x) = 1$, then $h(x) = 1/g(x)$ and we get the reciprocal rule:

Reciprocal Rule

To differentiate the reciprocal $1/g(x)$ of a function (where $g(x) \neq 0$), take the negative of the derivative of the function and divide by the square of the function:

$$\left(\frac{1}{g} \right)'(x) = \frac{-g'(x)}{[g(x)]^2} \qquad \text{or} \qquad \frac{d}{dx}\left(\frac{1}{u} \right) = -\frac{1}{u^2} \frac{du}{dx}.$$

Example 7 Differentiate (a) $1/(x^3 + 3x^2)$ and (b) $1/(\sqrt{x} + 2)$.

Solution (a) $\dfrac{d}{dx}\left(\dfrac{1}{x^3 + 3x^2} \right) = -\dfrac{1}{(x^3 + 3x^2)^2} \dfrac{d}{dx}(x^3 + 3x^2)$

$$= -\frac{3x^2 + 6x}{(x^3 + 3x^2)^2}.$$

(b) $\dfrac{d}{dx} \dfrac{1}{(\sqrt{x}+2)} = -\dfrac{1}{(\sqrt{x}+2)^2} \dfrac{d}{dx}(\sqrt{x}+2)$

$$= -\dfrac{1}{2\sqrt{x}\,(\sqrt{x}+2)^2} \cdot \blacktriangle$$

Combining the reciprocal rule with the power rule from Section 1.4 enables us to differentiate negative powers.[4] If k is a positive integer, then

$$\dfrac{d}{dx}(x^{-k}) = \dfrac{d}{dx}\left(\dfrac{1}{x^k}\right) = -\dfrac{1}{(x^k)^2}\dfrac{d}{dx}(x^k)$$

$$= -\dfrac{1}{x^{2k}}(kx^{k-1}) = -kx^{-k-1}.$$

Writing n for the negative integer $-k$, we have $(d/dx)(x^n) = nx^{n-1}$, just as for positive n. Recalling that $(d/dx)(x^0) = (d/dx)(1) = 0$, we have established the following general rule.

Integer Power Rule

If n is any (positive, negative or zero) integer, $(d/dx)x^n = nx^{n-1}$. (When $n < 0$, x must be unequal to zero.)

Example 8   Differentiate $1/x^6$

Solution $(d/dx)(1/x^6) = (d/dx)(x^{-6}) = -6x^{-6-1} = -6x^{-7}.$ $\blacktriangle$

We conclude this section with a summary of the differentiation rules obtained so far. Some of these rules are special cases of the others. For instance, the linear and quadratic function rules are special cases of the polynomial rule, and the reciprocal rule is the quotient rule for $f(x) = 1$. Remember that the basic idea for differentiating a complicated function is to break it into its component parts and combine the derivatives of the parts according to the rules.

Example 9   Differentiate (a) $3x^4 + \dfrac{2}{x} - \dfrac{5}{x^3}$ and (b) $\dfrac{1}{(x^2+3)(x^2+4)}$.

Solution   (a) By the sum, power, and constant multiple rules,

$$\dfrac{d}{dx}\left(3x^4 + \dfrac{2}{x} - \dfrac{5}{x^3}\right) = 3\dfrac{d}{dx}(x^4) + 2\dfrac{d}{dx}(x^{-1}) - 5\dfrac{d}{dx}(x^{-3})$$

$$= 3 \cdot 4x^3 + 2(-1)x^{-2} - 5(-3)x^{-4}$$

$$= 12x^3 - \dfrac{2}{x^2} + \dfrac{15}{x^4}.$$

(b) Let $f(x) = (x^2+3)(x^2+4)$. By the product rule, $f'(x) = 2x(x^2+4) + (x^2+3)2x = 4x^3 + 14x$. By the reciprocal rule, the derivative of $1/f(x)$ is

$$-\dfrac{f'(x)}{[f(x)]^2} = -\dfrac{4x^3 + 14x}{(x^2+3)^2(x^2+4)^2} \cdot \blacktriangle$$

---

[4] Students requiring a review of negative exponents should read Section R.3.

Example 10   We derived the reciprocal rule from the quotient rule. By writing $f(x)/g(x) = f(x) \cdot [1/g(x)]$, show that the quotient rule also follows from the product rule and the reciprocal rule.

Solution

$$\left(\frac{f}{g}\right)'(x) = \left(f \cdot \frac{1}{g}\right)'(x)$$

$$= f'(x) \cdot \frac{1}{g(x)} + f(x) \cdot \left(\frac{1}{g}\right)'(x)$$

$$= \frac{f'(x)}{g(x)} + f(x) \cdot \frac{-g'(x)}{[g(x)]^2}$$

$$= \frac{f'(x)g(x) - f(x)g'(x)}{[g(x)]^2}.$$

(This calculation gives another way to reconstruct the quotient rule if you forget it—assuming, of course, that you remembered the reciprocal rule.) ▲

Differentiation Rules

	The derivative of	is	In Leibniz notation
Linear function	$bx + c$	b	$\frac{d}{dx}(bx + c) = b$
Quadratic function	$ax^2 + bx + c$	$2ax + b$	$\frac{d}{dx}(ax^2 + bx + c) = 2ax + b$
Sum	$f(x) + g(x)$	$f'(x) + g'(x)$	$\frac{d}{dx}(u + v) = \frac{du}{dx} + \frac{dv}{dx}$
Constant multiple	$kf(x)$	$kf'(x)$	$\frac{d}{dx}(ku) = k\frac{du}{dx}$
Power	x^n {n any integer}	nx^{n-1}	$\frac{d}{dx}(x^n) = nx^{n-1}$
Polynomial	$c_n x^n + \cdots + c_2 x^2$ $+ c_1 x + c_0$	$nc_n x^{n-1} + \cdots$ $+ 2c_2 x + c_1$	$\frac{d}{dx}(c_n x^n + \cdots + c_2 x^2 + c_1 x + c_0)$ $= nc_n x^{n-1} + \cdots + 2c_2 x + c_1$
Product	$f(x)g(x)$	$f'(x)g(x) + f(x)g'(x)$	$\frac{d}{dx}(uv) = \frac{du}{dx}v + u\frac{dv}{dx}$
Quotient	$f(x)/g(x)$ {$g(x) \neq 0$}	$\dfrac{f'(x)g(x) - f(x)g'(x)}{[g(x)]^2}$	$\frac{d}{dx}\left(\frac{u}{v}\right) = \dfrac{(du/dx)v - u(dv/dx)}{v^2}$
Reciprocal	$1/g(x)$ {$g(x) \neq 0$}	$-g'(x)/[g(x)]^2$	$\frac{d}{dx}\left(\frac{1}{v}\right) = -\frac{1}{v^2}\frac{dv}{dx}$

Exercises for Section 1.5

Compute the derivatives of the functions in Exercises 1–12 by using the product rule. Verify your answer by multiplying out and differentiating the resulting polynomials.

1. $(x^2 + 2)(x + 8)$
2. $(x + 1)(x - 1)$
3. $(x^4 + x)(x^3 - 2)$
4. $(x^2 + 3x)(2x - 1)$
5. $(x^2 + 2x + 1)(x - 1)$
6. $(x^3 + 3x^2 + 3x + 1)(x - 1)$
7. $(x^2 + 2x + 2)(x^2 + 3x)$
8. $(x^2 + 4x + 8)(x^2 + 2x - 1)$
9. $(x - 1)(x^2 + x + 1)$
10. $(x - 2)(x^2 + 2x + 1)$
11. $(x - 1)(x^3 + x^2 + x + 1)$
12. $(x^3 + 2)(x^2 + 2x + 1)$

In Exercises 13–16, differentiate the given function by writing it as indicated and using the product rule.

13. $x^{5/2} = x^2 \cdot \sqrt{x}$ 14. $x = \sqrt{x} \cdot \sqrt{x}$
15. $x^{7/2} = x^3 \cdot \sqrt{x}$ 16. $x^2 = \sqrt{x} \cdot x^{3/2}$

Differentiate the functions in Exercises 17–30.

17. $\dfrac{x - 2}{x^2 + 3}$ 18. $\dfrac{x^3 - 3x + 5}{x^4 - 1}$

19. $\dfrac{x^7 - x^2}{x^3 + 1}$ 20. $\dfrac{5x^3 + x - 10}{3x^4 + 2}$

21. $\dfrac{x^2 + 2}{x^2 - 2}$ 22. $\dfrac{x}{1 - x^2}$

23. $\dfrac{1}{x^2} + \dfrac{x}{x^2 + 1}$ 24. $\dfrac{1}{t^9}$

25. $\dfrac{r^2 + 2}{r^8}$ 26. $\dfrac{x}{2} + \dfrac{3}{x + 1}$

27. $\left(\dfrac{s}{1 - s}\right)^2$ 28. $\dfrac{(x^3 - 1)^2}{x^3 + 1}$

29. $\dfrac{(x^2 + 1)^2 + 1}{(x^2 + 1)^2 - 1}$ 30. $\dfrac{4}{(x^2 - 1)(x + 7)^2}$

Find the indicated derivatives in Exercises 31–38.

31. $\dfrac{d}{dx}\left(\dfrac{1}{x^4}\right)$

32. $\dfrac{d}{dx}\left(\dfrac{1}{x^5 + 5x^2}\right)$

33. $\dfrac{d}{dx}\left(\dfrac{1}{(x + 1)^2}\right)$

34. $\dfrac{d}{dx}\left(\dfrac{1}{(x^2 + 9)^2}\right)$

35. $\dfrac{d}{ds}(s^3 + 1)(s^5 + 1)$

36. $\dfrac{d}{du}(u^4 + 2u)(u^3 + 2u)$

37. $\dfrac{d}{dy}\left[4(y + 1)^2 - 2(y + 1) - \dfrac{1}{y + 1}\right]$

38. $\dfrac{d}{dz}\left[\dfrac{(z - 3)^2 + 2}{(z - 3)^2 + 3}\right]$

Differentiate the functions in Exercises 39–46.

39. $\dfrac{\sqrt{x} - 1}{\sqrt{x} + 1}$ 40. $\dfrac{1}{(\sqrt{x} + 1)^2}$

41. $(3\sqrt{x} + 1)x^2$ 42. $\dfrac{1}{x + \sqrt{x}}$

43. $\dfrac{x}{1 + \sqrt{x}}$ 44. $\dfrac{1 - \sqrt{x}}{x^2 + 2}$

45. $\dfrac{\sqrt{2}}{1 + 3\sqrt{x}}$ 46. $\dfrac{8x^{3/2} + \sqrt{x}}{1 - \sqrt{x}}$

47. Use the reciprocal rule twice to differentiate $1/[1/g(x)]$ and show that the result is $g'(x)$.
48. Differentiate x^m/x^n by the quotient rule and compare your answer with the derivative of x^{m-n} obtained by the power rule.
49. Find the slope of the line tangent to the graph of $f(x) = 1/\sqrt{x}$ at $x = 2$.
50. Find the slope of the line tangent to the graph of $f(x) = (2x + 1)/(3x + 1)$ at $x = 1$.

Let $f(x) = 4x^5 - 13x$ and $g(x) = x^3 + 2x - 1$. Find the derivatives of the functions in Exercises 51–56.

51. $f(x)g(x)$
52. $[f(x) + x^3 - 7x][g(x)]$
53. $xf(x) + g(x)$
54. $\dfrac{f(x)}{g(x)} + (x^3 - 3x) - 7$
55. $\dfrac{g(x)}{f(x)}$
56. $\dfrac{f(x)}{f(x) + g(x) - 4x^5 - x^3 + 10x + 1}$

★57. Let $P(x)$ be a quadratic polynomial. Show that $(d/dx)(1/P(x))$ is zero for at most one value of x in its domain. Find an example of $P(x)$ for which $(d/dx)(1/P(x))$ is never zero on its domain.

★58. Calculate the following limits by expressing each one as the derivative of some function:

(a) $\displaystyle\lim_{x \to 1} \dfrac{x^8 - x^7 + 3x^2 - 3}{x - 1}$,

(b) $\displaystyle\lim_{x \to 2} \dfrac{1/x^3 - 1/2^3}{x - 2}$,

(c) $\displaystyle\lim_{x \to -1} \dfrac{x^2 + x}{(x + 2)(x + 1)}$.

1.6 The Linear Approximation and Tangent Lines

A good approximation to $f(x_0 + \Delta x)$ is $f(x_0) + f'(x_0)\Delta x$.

In Section 1.1, we saw that the derivative $f'(x_0)$ is the slope of the tangent line to the graph $y = f(x)$. This section explores the relationship between the graph of f and its tangent line a little further.

Recall from Section R.4 that the equation of the straight line through (x_0, y_0) with slope m is

$$y = y_0 + m(x - x_0).$$

In particular we get the following formula for the tangent line to $y = f(x)$ (see Figure 1.6.1).

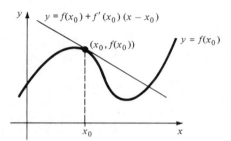

Figure 1.6.1. The tangent line to the graph $y = f(x)$ at $(x_0, f(x_0))$.

Equation of the Tangent Line

The equation of the line tangent to $y = f(x)$ at $(x_0, f(x_0))$ is

$$y = f(x_0) + f'(x_0)(x - x_0).$$

Example 1 (a) Find the equation of the line tangent to the graph $y = \sqrt{x} + 1/(2(x + 1))$ at $x = 1$.

(b) Find the equation of the line tangent to the graph of the function $f(x) = (2x + 1)/(3x + 1)$ at $x = 1$.

Solution (a) Here $x_0 = 1$ and $f(x) = \sqrt{x} + 1/2(x + 1)$. We compute

$$f'(x) = \frac{1}{2\sqrt{x}} - \frac{1}{2(x + 1)^2},$$

so $f'(1) = \frac{1}{2} - \frac{1}{8} = \frac{3}{8}$.

Since $f(1) = 1 + \frac{1}{4} = \frac{5}{4}$, the tangent line has equation $y = \frac{5}{4} + \frac{3}{8}(x - 1)$, i.e., $8y = 3x + 7$.

(b) By the quotient rule, $f'(x)$ equals $[2(3x + 1) - (2x + 1)3]/(3x + 1)^2 = -1/(3x + 1)^2$. The equation of the tangent line is

$$y = f(1) + f'(1)(x - 1) = \tfrac{3}{4} - \tfrac{1}{16}(x - 1)$$

or $y = -\frac{1}{16}x + \frac{13}{16}$. ▲

Example 2   Where does the line tangent to $y = \sqrt{x}$ at $x = 2$ cross the x axis?

Solution   Here $dy/dx = 1/2\sqrt{x}$, which equals $1/2\sqrt{2}$ at $x = 2$. Since $y = \sqrt{2}$ at $x = 2$, the equation of the tangent line is

$$y = \sqrt{2} + \frac{1}{2\sqrt{2}}\,(x - 2).$$

This line crosses the x axis when $y = 0$ or $0 = \sqrt{2} + (1/2\sqrt{2})(x - 2)$. Solving for x, we get $x = -2$. Thus the tangent line crosses the x axis at $x = -2$. ▲

We have used the idea of limit to pass from difference quotients to derivatives. We can also go in the other direction: given $f(x_0)$ and $f'(x_0)$, we can use the derivative to get an approximate value for $f(x)$ when x is near x_0.

According to the definition of the derivative, the difference quotient $\Delta y/\Delta x = [f(x_0 + \Delta x) - f(x_0)]/\Delta x$ is close to $f'(x_0)$ when Δx is small. That is, the *d*ifference

$$\frac{\Delta y}{\Delta x} - f'(x_0) = \frac{f(x_0 + \Delta x) - f(x_0)}{\Delta x} - f'(x_0) = d$$

is small when Δx is small. Multiplying the preceding equation by Δx and rearranging gives

$$f(x_0 + \Delta x) = f(x_0) + f'(x_0)\,\Delta x + d\,\Delta x. \tag{1}$$

Suppose now that we know $f(x_0)$ and $f'(x_0)$ and that we wish to evaluate f at the nearby point $x = x_0 + \Delta x$. Formula (1) expresses $f(x)$ as a sum of three terms, the third of which becomes small—even compared to Δx—as $\Delta x \to 0$. By dropping this term, we obtain the approximation

$$f(x) \approx f(x_0) + f'(x_0)\,\Delta x. \tag{2}$$

In terms of $x = x_0 + \Delta x$, we have

$$f(x) \approx f(x_0) + f'(x_0)(x - x_0). \tag{3}$$

The right-hand side of (3) is a linear function of x, called the *linear approximation* to f at x_0. Notice that its graph is just the tangent line to the graph of f at $(x_0, f(x_0))$. (See Fig. 1.6.2.)

Figure 1.6.2. As Δx approaches zero, the difference between $f(x)$ and the approximation $f(x_0) + f'(x_0)(x - x_0)$ becomes arbitrarily small compared to $\Delta x = x - x_0$.

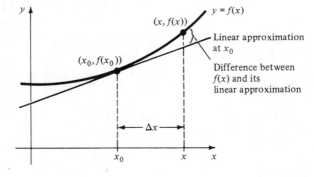

The linear approximation is also called the *first-order* approximation. Second-order and higher-order approximations are introduced in Section 12.5.

Example 3 (a) Show that the linear approximation to $(x_0 + \Delta x)^2$ is $x_0^2 + 2x_0 \Delta x$. (b) Calculate an approximate value for $(1.03)^2$. Compare with the actual value. Do the same for $(1.0003)^2$ and $(1.0000003)^2$.

Solution (a) Let $f(x) = x^2$, so $f'(x) = 2x$. Thus the linear approximation to $f(x_0 + \Delta x)$ is $f(x_0) + f'(x_0) \Delta x = x_0^2 + 2x_0 \Delta x$.
(b) Let $x_0 = 1$ and $\Delta x = 0.03$; from (a), the approximate value is $1 + 2\Delta x = 1.06$. The exact value is 1.0609. If $\Delta x = 0.0003$, the approximate value is 1.0006 (very easy to compute), while the exact value is 1.00060009 (slightly harder to compute). If $\Delta x = 0.0000003$, the approximate value is 1.0000006, while the exact value is 1.00000060000009. Notice that the error decreases even faster than Δx. ▲

The Linear Approximation

For x near x_0, $f(x_0) + f'(x_0)(x - x_0)$ is a good approximation for $f(x)$.

$$f(x_0 + \Delta x) \approx f(x_0) + f'(x_0) \Delta x \quad \text{or} \quad \Delta y \approx f'(x_0) \Delta x$$

The error becomes arbitrarily small, compared with Δx, as $\Delta x \to 0$.

▦ **Example 4** Calculate an approximate value for the following quantities using the linear approximation around $x_0 = 9$. Compare with the values on your calculator.

(a) $\sqrt{9.02}$ (b) $\sqrt{10}$ (c) $\sqrt{8.82}$ (d) $\sqrt{8}$

Solution Let $f(x) = \sqrt{x}$ and recall that $f'(x) = 1/2\sqrt{x}$. Thus the linear approximation is
$$f(x_0 + \Delta x) \approx f(x_0) + f'(x_0) \Delta x,$$
i.e.,
$$\sqrt{x_0 + \Delta x} \approx \sqrt{x_0} + \frac{1}{2\sqrt{x_0}} \Delta x.$$

(a) Let $x_0 = 9$ and $\Delta x = 0.02$, so $x_0 + \Delta x = 9.02$. Thus
$$\sqrt{9.02} \approx \sqrt{9} + \frac{1}{2\sqrt{9}} 0.02 = 3 + \frac{0.02}{6} = 3.0033 \ldots .$$

On our calculator we get 3.0033315.
(b) Let $x_0 = 9$ and $\Delta x = 1$; then
$$\sqrt{10} \approx \sqrt{9} + \frac{1}{2\sqrt{9}} 1 = 3 + \frac{1}{6} = 3.166 \ldots .$$

On our calculator we get 3.1622777.
(c) Let $x_0 = 9$ and $\Delta x = -0.18$; then
$$\sqrt{8.82} \approx \sqrt{9} + \frac{1}{2\sqrt{9}} (-0.18)$$
$$= 3 + \frac{1}{6}(-0.18) = 3 - 0.03 = 2.97.$$

On our calculator we get 2.9698485.

(d) Let $x_0 = 9$ and $\Delta x = -1$; then

$$\sqrt{8} = \sqrt{9} - \frac{1}{2\sqrt{9}} = 3 - \frac{1}{6} = 2.833\ldots.$$

On our calculator, we get 2.8284271.

Notice that the linear approximation gives the best answers in (a) and (c), where Δx is smallest. ▲

Example 5   Calculate the linear approximation to the area of a square whose side is 2.01. Draw a geometric figure, obtained from a square of side 2, whose area is exactly that given by the linear approximation.

Solution $A = f(r) = r^2$. The linear approximation near $r_0 = 2$ is given by $f(r_0) + f'(r_0)$ $(r - r_0) = r_0^2 + 2r_0(r - r_0) = 4 + 4(r - r_0)$. When $r - r_0 = 0.01$, this is 4.04.

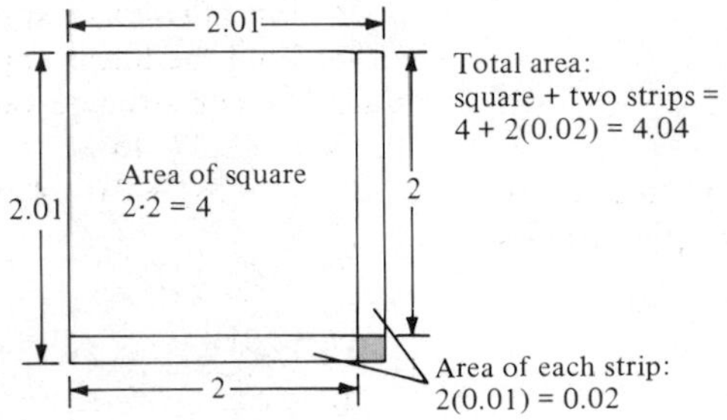

Figure 1.6.3. The linear approximation to the change in area with respect to a side has error equal to the shaded area. (Diagram not to scale.)

The required figure is shown in Fig. 1.6.3. It differs from the square of side 2.01 only by the small shaded square in the corner, whose area is $(0.01)^2 = 0.0001$. ▲

▦ **Example 6**   Calculate an approximate value for

$$\frac{2}{\sqrt{0.99} + (0.99)^2}$$

and compare with the numerical value on your calculator.

Solution   We let $f(x) = 2/(\sqrt{x} + x^2)$ and note that we are asked to calculate $f(1 - 0.01)$. By the linear approximation,

$$f(1 - 0.01) \approx f(1) - f'(1)(0.01).$$

Note that $f(1) = 1$. We calculate $f'(x)$ by the quotient rule:

$$f'(x) = -\frac{2(1/2\sqrt{x} + 2x)}{(\sqrt{x} + x^2)^2}$$

$$= -\frac{1 + 4x\sqrt{x}}{\sqrt{x}(\sqrt{x} + x^2)^2}.$$

At $x = 1$, $f'(1) = -\frac{5}{4}$, so

$$f(1 - 0.01) \approx 1 + \tfrac{5}{4}(0.01) = 1.0125.$$

On our calculator we find $f(0.99) = 1.0126134$, in rather good agreement. ▲

Exercises for Section 1.6

In Exercises 1–4, find the equation of the line tangent to the graph of the given function at the indicated point and sketch:

1. $y = 1 - x^2$; $x_0 = 1$
2. $y = x^2 - x$; $x_0 = 0$
3. $y = x^2 - 2x + 1$; $x_0 = 2$
4. $y = 3x^2 + 1 - x$; $x_0 = 5$

Find the equation of the tangent line to the graph of $f(x)$ at $(x_0, f(x_0))$ in Exercises 5–8.

5. $f(x) = (x^2 - 7)\dfrac{3x}{x + 2}$; $x_0 = 0$

6. $f(x) = \dfrac{1}{(x^2 + 4)^2}$; $x_0 = 0$

7. $f(x) = \left[\dfrac{1}{x} - 2x\right](x^2 + 2)$; $x_0 = \frac{1}{2}$

8. $f(x) = \dfrac{x^2}{1 + x^2}$; $x_0 = 1$

In Exercises 9–12, find where the tangent line to the graph of the given function at the given point crosses the x axis.

9. $y = \dfrac{x}{x + 1}$; $x_0 = 1$

. 10. $y = \dfrac{\sqrt{x}}{x + 1}$; $x_0 = 2$

11. $y = \dfrac{2}{1 + \sqrt{x}}$; $x_0 = 4$

12. $y = x(\sqrt{x} + 1)$; $x_0 = 1$

▦ Calculate an approximate value for each of the squares in Exercises 13–16 and compare with the exact value:

13. $(2.02)^2$
14. $(199)^2$
15. $(4.999)^2$
16. $(-1.002)^2$

▦ In Exercises 17–20, calculate an approximate value for the square root using the linear approximation at $x_0 = 16$. Compare with the value on your calculator.

17. $\sqrt{16.016}$
18. $\sqrt{17}$
19. $\sqrt{15.92}$
20. $\sqrt{15}$

Using the linear approximation, find an approximate value for the quantities in Exercises 21–24.

21. $(2.94)^4$
22. $(1.03)^4$
23. $(3.99)^3$
24. $(101)^8$

25. The radius of a circle is increased from 3 to 3.04. Using the linear approximation, what do you find to be the increase in the area of the circle?
26. The radius r of the base of a right circular cylinder of fixed height h is changed from 4 to 3.96. Using the linear approximation, approximate the change in volume V.
27. A sphere is increased in radius from 5 to 5.01. Using the linear approximation, estimate the increase in surface area (the surface area of a sphere of radius r is $4\pi r^2$).
28. Redo Exercise 27 replacing surface area by volume (the volume inside a sphere of radius r is $\frac{4}{3}\pi r^3$).

Calculate approximate values in Exercises 29–32.

29. $(x^2 + 3)(x + 2)$ if $x = 3.023$

30. $\dfrac{x^2}{x^3 + 2}$ if $x = 2.004$

31. $\dfrac{1}{(2.01)^2 + (2.01)^3}$

32. $\dfrac{1}{(1.99)^2 + (1.99)^4}$

33. Find the equation of the line tangent to the graph of $f(x) = x^8 + 2x^2 + 1$ at $(1, 4)$.
34. Find the equation of the tangent line to the graph of $x^4 - x^2 + 3x$ at $x = 1$.
35. Find the linear approximation for $1/0.98$.
36. Find the linear approximation for $1/1.98$.

Calculate approximate values for each of the quantities in Exercises 37–40.

37. $s^4 - 5s^3 + 3s - 4$; $s = 0.9997$

38. $\dfrac{x^4}{x^5 - 2x^2 - 1}$; $x = 2.0041$

39. $(2.01)^{20}$

40. $\dfrac{1}{(1.99)^2}$

41. Let $h(t) = -4t^2 + 7t + \frac{3}{4}$. Use the linear approximation to approximate values for $h(3.001)$, $h(1.97)$, and $h(4.03)$.
42. Let $f(x) = 3x^2 - 4x + 7$. Using the linear approximation, find approximate values for $f(2.02)$, $f(1.98)$, and $f(2.004)$. Compute the actual values without using a calculator and compare with the approximations. Compare the amount of time you spend in computing the approximations with the time spent in obtaining the actual values.
43. Let $g(x) = -4x^2 + 8x + 13$. Find $g'(3)$. Show that the linear approximation to $g(3 + \Delta x)$ always gives an answer which is too large, regardless of whether Δx is positive or negative. Interpret your answer geometrically by drawing a graph of g and its tangent line when $x_0 = 3$.
44. Let $f(x) = 3x^2 - 4x + 7$. Show that the linear approximation to $f(2 + \Delta x)$ always gives an answer which is too small, regardless of whether Δx is positive or negative. Interpret your answer geometrically by drawing a graph of f and its tangent line at $x_0 = 2$.
★45. Let $f(x) = x^4$.
   (a) Find the linear approximation to $f(x)$ near $x = 2$.
   (b) Is the linear approximation larger or smaller than the actual value of the function?
   ▦ (c) Find the largest interval containing $x = 2$ such that the linear approximattion is accurate within 10% when x is in the interval.
★46. (a) Give numerical examples to show that linear approximations to $f(x) = x^3$ may be either too large or too small.

(b) Illustrate your examples by sketching a graph of $y = x^3$, using calculated values of the function.

47. Show that a good approximation to $1/(1 + x)$ when x is small is $1 - x$.

★48. If you travel 1 mile in $60 + x$ seconds, show that a good approximation to your average speed, for x small, is $60 - x$ miles per hour. (This works quite well on roads which have mileposts.) Find the error in this approximation if $x = 1, -1, 5, -5, 10, -10$.

📖 49. Return to Exercise 47. Show experimentally that a better approximation to $1/(1 + x)$ is $1 - x + x^2$. Use this result to refine the speedometer checking rule in Exercise 48.

★50. Devise a speedometer checking rule for metric units which works for speeds in the vicinity of about 90 or 100 kilometers per hour.

Review Exercises for Chapter 1

Differentiate the functions in Exercises 1–20.

1. $f(x) = x^2 - 1$
2. $f(x) = 3x^2 + 2x - 10$
3. $f(x) = x^3 + 1$
4. $f(x) = x^4 - 8$
5. $f(x) = 2x - 1$
6. $f(x) = 8x + 1$
7. $f(s) = s^2 + 2s$
8. $f(r) = r^4 + 10r + 2$
9. $f(x) = -10x^5 + 8x^3$
10. $f(x) = 12x^3 + 2x^2 + 2x - 8$
11. $f(x) = (x^2 - 1)(x^2 + 1)$
12. $f(x) = (x^3 + 2x + 3)(x^2 + 2)$
13. $f(x) = 3x^3 - 2\sqrt{x}$
14. $f(x) = x^4 + 9\sqrt{x}$
15. $f(x) = x^{50} + \dfrac{1}{x}$
16. $f(x) = x^9 - \dfrac{8}{x}$
17. $f(x) = \dfrac{x^2 + 1}{x^2 - 1}$
18. $f(x) = \dfrac{\sqrt{x} + 2}{x^2 - 1}$
19. $f(x) = \dfrac{1}{\sqrt{x}\,(x^2 + 2)}$
20. $f(x) = \dfrac{\sqrt{x}}{(x^2 + 2)^2}$

Find the derivatives indicated in Exercises 21–30.

21. $\dfrac{d}{ds}(s + 1)^2(\sqrt{s} + 2)$

22. $\dfrac{d}{du}\dfrac{u^2 + 2 + 3\sqrt{u}}{\sqrt{u}}$

23. $\dfrac{d}{dr}\dfrac{\pi r^2}{1 + \sqrt{r}}$

24. $\dfrac{d}{dv}\dfrac{\sqrt{3}\,v + 1}{\sqrt{v} + 2}$

25. $\dfrac{d}{dt}(3t^2 + 2)^{-1}$

26. $\dfrac{d}{dx}\dfrac{3}{x^3 + 2x + 1}$

27. $\dfrac{d}{dp}\dfrac{\sqrt{2}\,p^2}{p^2 + 1}$

28. $\dfrac{d}{dq}(q + 2)^{-3}$

29. $\dfrac{d}{dx}\dfrac{1}{\sqrt{x}\,(\sqrt{x} - 1)}$

30. $\dfrac{d}{dx}\dfrac{x^2 + 1}{x^2 + x + 1}$

Find the limits in Exercises 31–46.

31. $\displaystyle\lim_{x \to 1}(x^2 + 1)$

32. $\displaystyle\lim_{x \to 1}\dfrac{x^3 + 1}{x + 1}$

33. $\displaystyle\lim_{x \to 1}\dfrac{x^3 - 1}{x - 1}$

34. $\displaystyle\lim_{x \to 1}\dfrac{x^5 - 1}{x - 1}$

35. $\displaystyle\lim_{h \to 0}\dfrac{(h - 2)^6 - 64}{h}$

36. $\displaystyle\lim_{h \to 0}\dfrac{(h - 2)^6 + 64}{h}$

37. $\displaystyle\lim_{x \to 3}\dfrac{3x^2 + 2x}{x}$

38. $\displaystyle\lim_{x \to 0}\dfrac{3x^2 + 2x}{x}$

39. $\displaystyle\lim_{x \to 3}\dfrac{x^2 - 9}{x - 3}$

40. $\displaystyle\lim_{s \to 0}\dfrac{(s + 3)^9 - 3^9}{s}$

41. $\displaystyle\lim_{\Delta x \to 0}\dfrac{f(x + \Delta x) - f(x)}{\Delta x}$
where $f(x) = x^4 + 3x^2 + 2$

42. $\displaystyle\lim_{\Delta x \to 0}\dfrac{f(x + \Delta x) - f(x)}{\Delta x}$
where $f(x) = 3x^8 - 8x^7 + 10$.

43. $\displaystyle\lim_{x \to \infty}\dfrac{3\sqrt{x} + 2}{5\sqrt{x} + 1}$

44. $\displaystyle\lim_{x \to \infty}\dfrac{5x^2 + 4}{3x^2 + 9}$

45. $\displaystyle\lim_{x \to \infty}\dfrac{5x^2 + 4}{5x^3 + 9}$

46. $\displaystyle\lim_{x \to \infty}\dfrac{5x^3 + 4}{5x^2 + 9}$

47. For the function in Fig. 1.R.1, find $\lim_{x \to x_0} f(x)$ for $x_0 = -3, -2, -1, 0, 1, 2, 3$. If the limit is not defined or does not exist, say so.

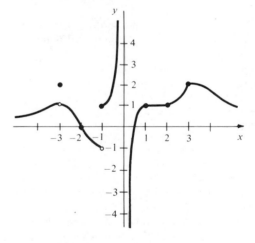

Figure 1.R.1. Find $\lim_{x \to x_0} f(x)$ at the indicated points.

48. Do as in Exercise 47 for the function in Fig. 1.R.2.

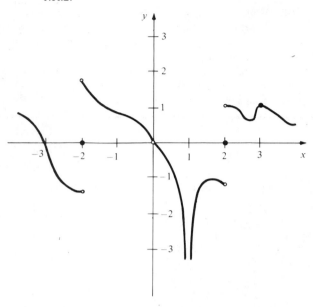

Figure 1.R.2. Find the limit at the indicated points.

49. Use the limit method directly to find $f'(1)$, where $f(x) = 3x^3 + 8x$.
50. Use the limit method directly to compute $(d/dx)(x^3 - 8x^2)$ at $x = -1$.
51. Use the limit method directly to compute $(d/dx)(x - \sqrt{x})$.
52. Use the limit method directly to compute the derivative of $f(x) = x^3 - \frac{1}{3}$.

Find the slope of the tangent line to the graphs of the functions in Exercises 53–58 at the indicated points.

53. $y = x^3 - 8x^2$; $x_0 = 1$
54. $y = x^4 + 2x$; $x_0 = -1$
55. $y = (x^2 + 1)(x^3 - 1)$; $x_0 = 0$
56. $y = x^4 + 10x + 2$; $x_0 = 2$
57. $y = 3x^4 - 10x^9$; $x_0 = 0$
58. $y = 3x + 1$; $x_0 = 5$

59. Two long trains, A and B, are moving on adjacent tracks with positions given by the functions $A(t) = t^3 + 2t$ and $B(t) = 7t^2/2 + 8$. What are the best times for a hobo on train B to make a moving transfer to train A?
60. A backpack is thrown down from a cliff at $t = 0$. It has fallen $2t + 4.9t^2$ meters after t seconds. Find its velocity at $t = 3$.
61. A bus moving along a straight road has moved $f(t) = (t^2 + \sqrt{t})/(1 + \sqrt{t})$ meters after time t (in seconds). What is its velocity at $t = 1$?
62. A car has position $x = (\sqrt{t} - 1)/(2\sqrt{t} + 1)$ at time t. What is its velocity at $t = 4$?

Calculate approximate values for the quantities in Exercises 63–70 using the linear approximation.

63. $(1.009)^8$
64. $(-1.008)^4 - 3(-1.008)^3 + 2$
65. $\sqrt{4.0001}$
66. $\sqrt{8.97}$
67. $f(2.003)$ where $f(x) = \dfrac{3x^3 - 10x^2 + 8x + 2}{x}$.
68. $g(1.0005)$ where $g(x) = x^4 - 10x^3$.
69. $h(2.95)$, where $h(s) = 4s^3 - s^4$.
70. $\dfrac{1 + (0.99)^2}{1 + (0.99)^3}$.

Find the equation of the line tangent to the graph of the function at the indicated point in Exercises 71–74.

71. $f(x) = x^3 - 6x$; $(0, 0)$
72. $f(x) = \dfrac{x^4 - 1}{6x^2 + 1}$; $(1, 0)$
73. $f(x) = \dfrac{x^3 - 7}{x^3 + 11}$; $(2, \frac{1}{19})$
74. $f(x) = \dfrac{x^5 - 6x^4 + 2x^3 - x}{x^2 + 1}$; $(1, -2)$
75. A sphere is increased in radius from 2 meters to 2.01 meters. Find the increase in volume using the linear approximation. Compare with the exact value.
76. A rope is stretched around the earth's equator. If it is to be raised 10 feet off the ground, approximately how much longer must it be? (The earth is 7,927 miles in diameter.)

In Exercises 77–80, let $f(x) = 2x^2 - 5x + 2$, $g(x) = \frac{3}{4}x^2 + 2x$ and $h(x) = -3x^2 + x + 3$.

77. Find the derivative of $f(x) + g(x)$ at $x = 1$.
78. Find the derivative of $3f(x) - 2h(x)$ at $x = 0$.
79. Find the equation of the tangent line to the graph of $f(x)$ at $x = 1$.
80. Find the equation of the tangent line to the graph of $g(x)$ at $x = -1$.

81. Let B be a rectangular box with a square end of side length r. Suppose B is three times as long as it is wide. Let V be the volume of B. Compute dV/dr. What fraction of the surface area of B is your answer?
82. Calculate $\lim_{x \to \infty}(x - \sqrt{x^2 - a^2})$ and interpret your answer geometrically by drawing a right triangle with hypotenuse of length x and short leg of length a.
83. Suppose that $z = 2y^2 + 3y$ and $y = 5x + 1$.
    (a) Find dz/dy and dy/dx.
    (b) Express z in terms of x and find dz/dx.
    (c) Compare dz/dx with $(dz/dy) \cdot (dy/dx)$. (Write everything in terms of x.)
    (d) Solve for x in terms of y and find dx/dy.
    (e) Compare dx/dy with dy/dx.

84. Differentiate both sides of the equation

$$\frac{f(x)}{g(x)} = \frac{1}{g(x)/f(x)}$$

and show that you get the same result on each side.

★85. Find the equation of a line through the origin, with positive slope, which is tangent to the parabola $y = x^2 - 2x + 2$.

★86. Prove that the parabola $y = x^2$ has the optical focusing property mentioned in Section R.5. (This problem requires trigonometry; consult Section 5.1 for a review.) *Hint*: Refer to Fig. 1.R.3 and carry out the following program:
(a) Express $\tan\phi$ and $\tan\theta$ in terms of x.
(b) Prove that $90° - \theta = \theta - \phi$ by using the trigonometric identities:

$$\tan 2\theta = \frac{2\tan\theta}{1 - \tan^2\theta}$$

and

$$\tan(\phi + 90°) = -\frac{1}{\tan\phi}.$$

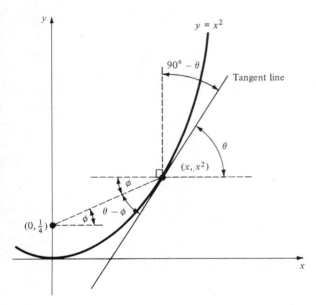

Figure 1.R.3. The geometry needed to prove that the parabola has the optical focusing property.

★87. Prove that the parabola $y = ax^2$ has the optical focusing property. (You should start by figuring out where the focal point will be.)

★88. The following is a useful technique for drawing the tangent line at a point P on a curve on paper (not given by a formula). Hold a mirror perpen-

dicular to the paper and rotate it until the graph and its reflection together form a differentiable curve through P. Draw a line l along the edge of the mirror. Then the line through P perpendicular to l is the tangent line. (See Fig. 1.R.4.) Justify this procedure.

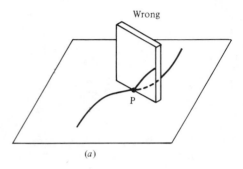

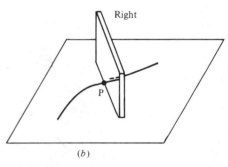

Figure 1.R.4. How to draw a tangent line with a mirror.

★89. The polynomial $a_n x^n + a_{n-1} x^{n-1} + \cdots + a_0$ is said to have *degree* n if $a_n \neq 0$. For example: $\deg(x^3 - 2x + 3) = 3$, $\deg(x^4 + 5) = 4$, $\deg(0x^2 + 3x + 1) = 1$. The degree of the rational function $f(x)/g(x)$, where $f(x)$ and $g(x)$ are polynomials, is defined to be the degree of f minus the degree of g.
(a) Prove that, if $f(x)$ and $g(x)$ are polynomials, then $\deg f(x)g(x) = \deg f(x) + \deg g(x)$.
(b) Prove the result in part (a) when $f(x)$ and $g(x)$ are rational functions.
(c) Prove that, if $f(x)$ is a rational function with nonzero degree, then $\deg f'(x) = \deg f(x) - 1$. What if $\deg f(x) = 0$?

★90. Show that $f(x) = x$ and $g(x) = 1/(1 - x)$ obey the "false product rule" $(fg)'(x) \overset{?}{=} f'(x)g'(x)$.

★91. (a) Prove that if f/g is a rational function (i.e., a quotient of polynomials) with derivative zero, then f/g is a constant.
(b) Conclude that if the rational functions F and G are both antiderivatives for a function h, then F and G differ by a constant.

Chapter 2

Rates of Change
and the
Chain Rule

The rate at which one variable is changing with respect to another can be computed using differential calculus.

In Chapter 1, we learned how to differentiate algebraic functions and, thereby, to find velocities and slopes. In this chapter, we will learn some applications involving rates of change. We will also develop a new rule of differential calculus called the chain rule. This rule is important for our study of related rates in this chapter and will be indispensable when we come to use trigonometric and exponential functions.

2.1 Rates of Change and the Second Derivative

If $y = f(x)$, then $f'(x)$ is the rate of change of y with respect to x.

The derivative concept applies to more than just velocities and slopes. To explain these other applications of the derivative, we shall begin with the situation where two quantities are related linearly.

Suppose that two quantities x and y are related in such a way that a change Δx in x always produces a change Δy in y which is proportional to Δx; that is, the ratio $\Delta y / \Delta x$ equals a constant, m. We say that y changes *proportionally* or *linearly* with x.

For instance, consider a hanging spring to which objects may be attached. Let x be the weight of the object in grams, and let y be the resulting length of the spring in centimeters. It is an experimental fact called Hooke's law that (for values of Δx which are not too large) a change Δx in the weight of the object produces a proportional change Δy in the length of the spring. (See Fig. 2.1.1.)

If we graph y against x, we get a segment of a straight line with slope

$$m = \frac{\Delta y}{\Delta x}$$

as shown in Fig. 2.1.2. The equation of the line is $y = mx + b$, and the

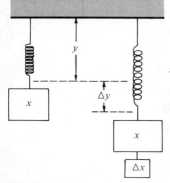

Figure 2.1.1. Hooke's law states that the change in length Δy is proportional to the change in weight Δx.

function $f(x) = mx + b$ is a *linear function*. The slope m of a straight line represents *the rate of change of y with respect to x*. (The quantity b is the length of the spring when the weight is removed.)

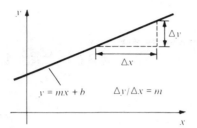

Figure 2.1.2. y changes proportionally to x when $\Delta y / \Delta x$ is constant.

Example 1   Suppose that y changes proportionally with x, and the rate of change is 3. If $y = 2$ when $x = 0$, find the equation relating y to x.

Solution   The rate of change is the slope: $m = 3$. The equation of a straight line with this slope is $y = 3x + b$, where b is to be determined. Since $y = 2$ when $x = 0$, b must be 2; hence $y = 3x + 2$. ▲

Linear or Proportional Change

The variable y changes proportionally with x when y is related to x by a linear function: $y = mx + b$, where $\Delta y / \Delta x = m$. The number m is the rate of change of y with respect to x.

Example 2   Let S denote the supply of hogs in Chicago, measured in thousands, and let P denote the price of pork in cents per pound. Suppose that, for S between 0 and 100, P changes linearly with S. On April 1, $S = 50$ and $P = 163$; on April 3, a rise in S of 10 leads to a decline in P to 161. What happens if S falls to 30?

Solution   Watch the words *of* and *to*! The rise in S *of* 10 means that $\Delta S = 10$; the decline of P *to* 161 means that $\Delta P = 161 - 163 = -2$. Thus the rate of change is $-\frac{2}{10} = -\frac{1}{5}$. (The minus sign indicates that the direction of price change is opposite to the direction of supply change.) We have $P = -\frac{1}{5} S + b$ for some b. Since $P = 163$ when $S = 50$, we have $163 = -\frac{1}{5} \cdot 50 + b$, or $b = 163 + 10 = 173$, so $P = -\frac{1}{5} S + 173$. When $S = 30$, this gives $P = -6 + 173 = 167$. At this point, then, pork will cost \$1.67 a pound. ▲

If the dependence of $y = f(x)$ on x is not linear, we can still introduce the notion of the *average rate of change* of y with respect to x, just as we introduced the average velocity in Section 1.1. Namely, the difference quotient

$$\frac{\Delta y}{\Delta x} = \frac{f(x_0 + \Delta x) - f(x_0)}{\Delta x}$$

is called *the average rate of change of y with respect to x on the interval between x_0 and $x_0 + \Delta x$*. For functions f which are not linear, this average rate of change depends on the interval chosen. If we fix x_0 and let Δx approach 0, the limit of the average rate of change is the derivative $f'(x_0)$, which we refer to as the *rate of change of y with respect to x at the point x_0*. This may be referred to as an *instantaneous* rate of change, especially when the independent variable represents time.

Example 3   An oil slick has area $y = 30x^3 + 100x$ square meters x minutes after a tanker explosion. Find the average rate of change in area with respect to time during the period from $x = 2$ to $x = 3$ and from $x = 2$ to $x = 2.1$. What is the instantaneous rate of change of area with respect to time at $x = 2$?

Solution   The average rate of change from $x = 2$ to $x = 3$ is

$$\frac{(30 \cdot 3^3 + 100 \cdot 3 - 30 \cdot 2^3 - 100 \cdot 2) \text{ square meters}}{1 \text{ minute}} = 670 \frac{\text{square meters}}{\text{minute}} .$$

From $x = 2$ to $x = 2.1$, the average rate is calculated in a similar way to be

$$\frac{47.83}{0.1} = 478.3 \frac{\text{square meters}}{\text{minute}} .$$

Finally, the instantaneous rate of change is found by evaluating the derivative $90x^2 + 100$ at $x = 2$ to obtain 460. Since the instantaneous rate of change is a limit of average rates, it is measured in the same units, so the oil slick is growing at a rate of 460 square meters per minute after 2 minutes. ▲

Rates of Change

If two quantities x and y are related by $y = f(x)$, the derivative $f'(x_0)$ represents the rate of change of y with respect to x at the point x_0. It is measured in (units of y)/(units of x).

A positive rate of change is sometimes called a *rate of increase*.

Example 4   A circle with radius r millimeters has area $A = \pi r^2$ square millimeters. Find the rate of increase of area with respect to radius at $r_0 = 5$. Interpret your answer geometrically.

Solution   Here $A = f(r) = \pi r^2$. Since π is a constant, the derivative $f'(r)$ is $2\pi r$, and $f'(5) = 10\pi$. Notice that the rate of change is measured in units of (square millimeters)/millimeters, which are just millimeters. The value $2\pi r$ of the rate of change can be interpreted as the circumference of the circle (Fig. 2.1.3). ▲

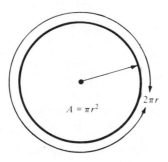

Figure 2.1.3. The rate of change of A with respect to r is $2\pi r$, the circumference of the circle.

In the next two examples, a negative rate of change indicates that one quantity decreases when another increases. Since $\Delta y = f(x_0 + \Delta x) - f(x_0)$, it follows that Δy is negative when $f(x_0 + \Delta x) < f(x_0)$. Thus, if $\Delta y / \Delta x$ is negative, an increase in x produces a decrease in y. This leads to our stated interpretation of negative rates of change. If a rate of change is negative, its absolute value is sometimes called a *rate of decrease*.

Example 5   Suppose that the price of pork P depends on the supply S by the formula $P = 160 - 3S + (0.01)S^2$. Find the rate of change of P with respect to S when $S = 50$. (See Example 2 for units.)

Solution   The rate of change is the derivative of $f(S) = 0.01S^2 - 3S + 160$ with respect to $x = S$ at $x_0 = 50$. The derivative is $f'(S) = 0.02S - 3$. When $S = 50$, we get $f'(50) = 1 - 3 = -2$. Thus the price is *decreasing* by 2 cents per pound per thousand hogs when $S = 50$ thousand. ▲

Example 6   A reservoir contains $10^8 - 10^4 t - 80t^2 - 10t^3 + 5t^5$ liters of water at time t, where t is the time in hours from when the gates are opened. How many liters per hour are leaving the reservoir after one hour?

Solution   The rate of change of the amount of water in the reservoir is the derivative of the polynomial $10^8 - 10^4 t - 80t^2 - 10t^3 + 5t^5$, namely $-10^4 - 160t - 30t^2 + 25t^4$. At $t = 1$, this equals $-10^4 - 160 - 30 + 25 = -10{,}165$ liters per hour. This is negative, so 10,165 liters per hour are *leaving* the reservoir after one hour. ▲

We now reconsider the velocity and acceleration of a particle moving on a straight line. Suppose for the moment that the line is vertical and designate one direction as "+" and the other as "−". We shall usually choose the upward direction as "+," but consistently using the other sign would give equivalent results. We also choose some point as the origin, designated by $x = 0$, as well as a unit of length, such as meters. Thus, if our designated origin represents the level of the Golden Gate Bridge, $x = 100$ would designate a location 100 meters above the bridge along our vertical straight line, and $x = -10$ would indicate a location 10 meters beneath the bridge (Fig. 2.1.4).

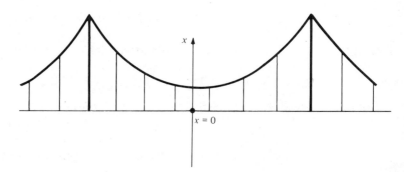

Figure 2.1.4. A coordinate system with "+" upwards and $x = 0$ at bridge level.

Suppose that, at time t, a particle has location $x = f(t)$ along our line. We call $[f(t + \Delta t) - f(t)]/\Delta t$ the average velocity and $dx/dt = f'(t)$ the instantaneous velocity; this can either be positive, indicating upward motion, or can be negative, indicating downward motion.

Example 7   Suppose that $x = 0$ represents the level of the Golden Gate Bridge and that $x = f(t) = 8 + 6t - 5t^2$ represents the position of a stone at time t in seconds.

(a) Is the stone above the bridge, at the bridge, or below the bridge at $t = 0$? How about at $t = 2$?

(b) Suppose that the average velocity during the interval from t_0 to $t_0 + \Delta t$ is negative; what can be said about the height at time $t_0 + \Delta t$?

(c) What is the instantaneous velocity at $t = 1$?

Solution   (a) At $t = 0$, $x = 8$, so the stone is 8 meters above the level of the bridge. At $t = 2$, $x = 8 + 6 \cdot 2 - 5 \cdot 4 = 0$, so the stone is at the level of the bridge.

(b) It is less than that at time t_0.

(c) We compute $dx/dt = 6 - 10 t$, which at $t = 1$ is -4. Thus, the instantaneous velocity is 4 meters per second downward. ▲

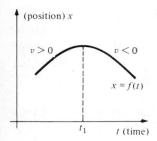

(position) x

$v > 0$ $v < 0$

$x = f(t)$

t_1 t (time)

Figure 2.1.5. The velocity is positive until $t = t_1$ and negative afterward.

Of course, these interpretations of positive and negative velocity also apply to horizontal motion, like that which we discussed at the beginning of Chapter 1. In particular, we may let x denote the position along a road with larger values of x corresponding to points further east, say. Then $dx/dt > 0$ indicates that motion is eastward; $dx/dt < 0$ indicates westward motion. The magnitude, or absolute value, of the velocity is called the *speed*.

If we graph $x = f(t)$ against t, the slope of the graph indicates whether the velocity is positive or negative. (See Fig. 2.1.5.)

Note that the instantaneous velocity $v = dx/dt = f'(t)$ is usually itself changing with time. The rate of change of v with respect to time is called *acceleration*; it may be computed by differentiating $v = f'(t)$ once again.

Example 8   Suppose that $x = f(t) = \frac{1}{4}t^2 - t + 2$ denotes the position of a bus at time t.

(a) Find the velocity as a function of time; plot its graph.
(b) Find and plot the speed as a function of time.
(c) Find the acceleration.

Solution   (a) The velocity is $v = dx/dt = \frac{1}{2}t - 1$ (see Fig. 2.1.6(a)).

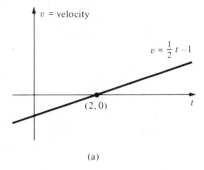

v = velocity

$v = \frac{1}{2}t - 1$

$(2, 0)$

t

(a)

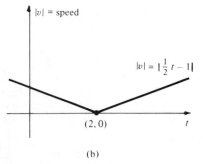

$|v|$ = speed

$|v| = |\frac{1}{2}t - 1|$

$(2, 0)$

t

(b)

Figure 2.1.6. (a) Velocity and (b) speed as functions of time for $x = \frac{1}{4}t^2 - t + 2$. The acceleration dv/dt is constant and positive.

(b) The speed $|v| = |\frac{1}{2}t - 1|$ [see Fig. 2.1.6(b)].
(c) The acceleration is $dv/dt = \frac{1}{2}$. ▲

In this example, the acceleration happens to be constant and positive, indicating that the *velocity* is increasing at a constant rate. Note, though, that the *speed* decreases and then increases; it decreases when the velocity and acceleration have opposite signs and increases when the signs are the same. This may be illustrated by an example. If your car is moving backwards (negative velocity) but you have a positive acceleration, your speed decreases until your car reverses direction, moves forward (positive velocity), and the speed increases.

Since acceleration is the derivative of the velocity and velocity is already a derivative, we have an example of the general concept of the *second derivative*, i.e., the derivative of the derivative. If $y = f(x)$, the second derivative is denoted $f''(x)$ and is defined to be the derivative of $f'(x)$. In Leibniz notation we write d^2y/dx^2 for the second derivative of $y = f(x)$. Note that d^2y/dx^2 is *not* the square of dy/dx, but rather represents the result of the operation d/dx performed twice.

If an object has position x (in meters) which is a function $x = f(t)$ of time t (in seconds), the acceleration is thus denoted by $f''(t)$ or d^2x/dt^2. It is measured in meters per second per second, i.e., meters per second^2 or feet per second^2.

Second Derivatives

To compute the second derivative $f''(x)$:

1. Compute the first derivative $f'(x)$.
2. Calculate the derivative of $f'(x)$; the result is $f''(x)$.

The second derivative of $y = f(x)$ is written in Leibniz notation as

$$\frac{d^2y}{dx^2}.$$

The second derivative of position with respect to time is called acceleration.

If we plot the graph $y = f(x)$, we know that $f'(x)$ represents the slope of the tangent line. Thus, if the second derivative is positive, the slope must be increasing as we move to the right, as in Fig. 2.1.7(a). Likewise, a negative second derivative means that the slope is decreasing as we increase x, as in Fig. 2.1.7(b).

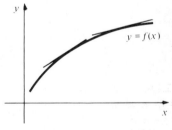

Figure 2.1.7. The rate of change of slope with respect to x is the second derivative $f''(x)$. (a) $f'' > 0$. (b) $f'' < 0$.

(a) increasing slope means positive second derivative

(b) decreasing slope means negative second derivative

Example 9 Calculate the second derivative of

(a) $f(x) = x^4 + 2x^3 - 8x$, (b) $f(x) = \dfrac{x+1}{\sqrt{x}}$.

(c) $\dfrac{d^2}{dx^2}(3x^2 - 2x + 1)$, (d) $\dfrac{d^2}{dr^2}(8r^2 + 2r + 10)$.

Solution (a) By our rules for differentiating polynomials from Section 1.4,

$$f'(x) = 4x^3 + 6x^2 - 8.$$

Now we differentiate this new polynomial:

$$f''(x) = 12x^2 + 12x - 0 = 12(x^2 + x).$$

(b) By the quotient rule from Section 1.5,

$$f'(x) = \frac{\sqrt{x} - (x+1) \cdot 1/2\sqrt{x}}{x} = \frac{x-1}{2\sqrt{x} \cdot x} = \frac{x-1}{2x^{3/2}}.$$

By the quotient rule again,

$$f''(x) = \frac{1}{2}\left(\frac{x^{3/2} - (x-1) \cdot \frac{3}{2}x^{1/2}}{x^3}\right) = \frac{3\sqrt{x} - \sqrt{x^3}}{4x^3}.$$

(c) Here, $\dfrac{d(3x^2 - 2x + 1)}{dx} = 6x - 2,$

and so $\dfrac{d^2(3x^2 - 2x + 1)}{dx^2} = \dfrac{d}{dx}(6x - 2) = 6,$

a constant function.

(d) $\dfrac{d^2}{dr^2}(8r^2 + 2r + 10) = \left(\dfrac{d}{dr}\right)\left[\left(\dfrac{d}{dr}\right)(8r^2 + 2r + 10)\right]$

$\qquad\qquad\qquad\qquad = \left(\dfrac{d}{dr}\right)[16r + 2] = 16.$ ▲

Next we consider a word problem involving second derivatives.

Example 10   A race car travels $\frac{1}{4}$ mile in 6 seconds, its distance from the start in feet after t seconds being $f(t) = 44t^2/3 + 132t$.

(a) Find its velocity and acceleration as it crosses the finish line.
(b) How fast was it going halfway down the track?

Solution   (a) The velocity at time t is $v = f'(t) = 88t/3 + 132$, and the acceleration is $a = f''(t) = \frac{88}{3}$. Substituting $t = 6$, we get $v = 308$ feet per second ($= 210$ miles per hour) and $a \approx 29.3$ feet per second^2.

(b) To find the velocity halfway down, we do *not* substitute $t = 3.00$ in $v = f'(t)$—that would be its velocity after half the time has elapsed. The total distance covered is $f(6) = (44)(36)/3 + (132)(6) = 1320$ feet ($= \frac{1}{4}$ mile). Thus, half the distance is 660 feet. To find the time t corresponding to the distance 660, we write $f(t) = 660$ and solve for t using the quadratic formula:

$$\frac{44t^2}{3} + 132t = 660,$$

$$t^2 + 9t - 45 = 0 \qquad \text{(multiply by } \tfrac{3}{44}\text{)},$$

$$t = \frac{-9 \pm \sqrt{81 + 180}}{2} \approx -12.58, 3.58 \qquad \text{(quadratic formula)}.$$

Since the time during the race is positive, we discard the negative root and retain $t = 3.58$. Substituting into $v = f'(t) = 88t/3 + 132$ gives $v \approx 237$ feet per second (≈ 162 miles per hour). ▲

We end this section with a discussion of some concepts from economics, where special names are given to certain rates of change.

Imagine a factory in which x worker-hours of labor can produce $y = f(x)$ dollars worth of output. First, suppose that y changes proportionally with x.

Then $\Delta y = f(x_0 + \Delta x) - f(x_0)$ represents the amount of extra output produced if Δx extra worker-hours of labor are employed. Thus, $\Delta y / \Delta x$ is the output per worker-hour. This average rate of change is called the *productivity of labor*.

Next, suppose that $f(x)$ is not necessarily linear. Then $\Delta y / \Delta x$ is the extra output per extra worker-hour of extra labor when Δx extra worker-hours are employed. The limiting value, as Δx becomes very small, is $f'(x_0)$. This instantaneous rate of change is called the *marginal productivity* of labor at the level x_0.

In Fig. 2.1.8 we sketch a possible productivity curve $y = f(x)$. Notice that as x_0 becomes larger and larger, the marginal productivity $f'(x_0)$ (= dollars of output per worker-hour at level x_0) becomes smaller. One says that the *law of diminishing returns* applies.

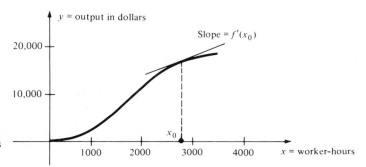

Figure 2.1.8. A possible productivity curve; the slope of the tangent line is the marginal productivity.

Example 11   A bagel factory produces $30x - 2x^2 - 2$ dollars worth of bagels for each x worker hours of labor. Find the marginal productivity when 5 workers are employed.

Solution   The output is $f(x) = 30x - 2x^2 - 2$ dollars. The marginal productivity at $x_0 = 5$ is $f'(5) = 30 - 4 \cdot 5 = 10$ dollars per worker-hour. Thus, at $x_0 = 5$, productivity would increase by 10 dollars per additional worker-hour. ▲

Next we discuss marginal cost and marginal revenue. Suppose that a company makes x calculators per week and that the management is free to adjust x. Define the following quantities:

$C(x) =$ the *cost* of making x calculators (labor, supplies, etc.)
$R(x) =$ the *revenue* obtained by producing x calculators (sales).
$P(x) = R(x) - C(x) =$ the *profit*.

Even though $C(x)$, $R(x)$, and $P(x)$ are defined only for integers x, economists find it useful to imagine them defined for all real x. This works nicely if x is so large that a change of one unit, $\Delta x = 1$, can legitimately be called "very small."

The derivative $C'(x)$ is called the *marginal cost* and $R'(x)$ is the *marginal revenue*:

$$C'(x) = \text{marginal cost} = \begin{cases} \text{the cost per calculator for producing} \\ \text{additional calculators at production level } x. \end{cases}$$

$$R'(x) = \text{marginal revenue} = \begin{cases} \text{the revenue per calculator obtained by} \\ \text{producing additional calculators at} \\ \text{production level } x. \end{cases}$$

Since $P(x) = R(x) - C(x)$, we get $P'(x) = R'(x) - C'(x)$, the profit per additional calculator at production level x. This is the *marginal profit*. If the price per unit is $f(x)$ and x calculators are sold, then $R(x) = xf(x)$. By the product rule, the marginal revenue is $R'(x) = xf'(x) + f(x)$.

Example 12   Suppose that it costs $(30x + 0.04x^2)/(1 + 0.0003x^3)$ dollars if x calculators are made, where $0 \leqslant x \leqslant 100$, and that calculators are priced at $100 - 0.05x$ dollars. If all x calculators are sold, what is the marginal profit?

Solution   The revenue is $R = x(100 - 0.05x)$, so the profit is $P = x(100 - 0.05x) - (30x + 0.04x^2)/(1 + 0.0003x^3)$. The marginal profit is therefore dP/dx, which may be calculated using the sum and quotient rules as follows:

$$\frac{dP}{dx} = \frac{d}{dx}(100x - 0.05x^2) - \frac{d}{dx}\left(\frac{30x + 0.04x^2}{1 + 0.0003x^3}\right)$$

$$= 100 - 0.1x - \frac{(30 + 0.08x)(1 + 0.0003x^3) - (30x + 0.04x^2)(0.0009x^2)}{(1 + 0.0003x^3)^2}$$

$$= 100 - 0.1x - \frac{30 + 0.08x - .018x^3 - 0.000012x^4}{(1 + 0.0003x^3)^2} \cdot \blacktriangle$$

Exercises for Section 2.1

In Exercises 1–4, assume that y changes proportionally with x and the rate of change is r. In each case, find y as a function of x, as in Example 1.
1. $r = 5$, $y = 1$ when $x = 4$.
2. $r = -2$, $y = 10$ when $x = 15$.
3. $r = \frac{1}{2}$, $y = 1$ when $x = 3$.
4. $r = 10$, $y = 4$ when $x = -1$.

5. If the price of electricity changes proportionally with time, and if the price goes from 2 cents per kilowatt-hour in 1982 to 3.2 cents per kilowatt-hour in 1984, what is the rate of change of price with respect to time? When will the price be 5 cents per kilowatt-hour? What will the price be in 1991?
6. It will take a certain woman seven bags of cement to build a 6-meter-long sidewalk of uniform width and thickness. Her husband offers to contribute enough of his own labor to extend the sidewalk to 7 meters. How much more cement do they need?
7. A rock is thrown straight down the face of a vertical cliff with an initial velocity of 3 meters per second. Two seconds later, the rock is falling at a velocity of 22.6 meters per second. Assuming that the velocity v changes proportionally with time t, find the equation relating v to t. How fast is the rock falling after 15 seconds?
8. In November 1980, Mr. B used 302 kilowatt-hours of electricity and paid \$18.10 to do so. In December 1980, he paid \$21.30 for 366 kilowatt-hours. Assuming that the cost of electricity changes linearly with the amount used, how much would Mr. B pay if he used no electricity at all? Suppose that Mr. B can reduce his bill to zero by selling solar-generated electricity back to

the company. How much must he sell? Interpret your answers on a graph.

Find the average rate of change of the functions in Exercises 9–12 on the specified interval.
9. $f(t) = 400 - 20t - 16t^2$; t between $t_0 = 1$ and $t_1 = \frac{3}{2}$.
10. $g(t) = 18t^2 + 2t + 3$; t between $t_0 = 2$ and $t_1 = 3.5$.
11. $f(x) = (x + \frac{1}{2})^2$; $x_0 = 2$, $\Delta x = 0.5$.
12. $g(s) = (3s + 2)(s - 1) - 3s^2$; $s_0 = 0$, $\Delta s = 6$.

13. The volume of a cone is $\frac{1}{3}$(area of base) $\times$ height. If the base has radius always equal to the height, find the rate of change of the volume with respect to this radius.
14. Find the rate of change of the area of an equilateral triangle with respect to the length of one of its sides.
15. During takeoff, a 747 has $25,000 - 80t + 2t^2 + 0.2t^3$ gallons of fuel in its tanks t seconds after starting its takeoff, $0 \leqslant t \leqslant 10$. How many gallons per second are being burned 2 seconds into the takeoff?
16. A space shuttle's external tank contains $10^5 - 10^4t - 10^3t^3$ liters of fuel t minutes after blastoff. How many liters per minute are being burned two minutes after blastoff?
17. If the height H in feet of a certain species of tree depends on its base diameter d in feet through the formula $H = 56d - 3d^2$, find the rate of change of H with respect to d at $d = 0.5$.
18. Suppose that tension T of a muscle is related to the time t of exertion by $T = 5 + 3t - t^2$, $0 \leqslant t \leqslant \frac{3}{2}$. Find the rate of change of T with respect to t at $t = 1$.

19. A flu epidemic has infected $P = 30t^2 + 100t$ people by t days after its outbreak. How fast is the epidemic spreading (in people per day) after 5 days?

20. Find the rate of change of the area of a circle with respect to its diameter when the diameter is 10. Compare with Example 4.

21. A sphere of radius r has volume $V = \frac{4}{3}\pi r^3$. What is the rate of change of the volume of the sphere with respect to its radius? Give a geometric interpretation of the answer.

22. A balloon being blown up has a volume $V = 3t^3 + 8t^2 + 16t$ cubic centimeters after t minutes. What is the rate of change of volume (in cubic centimeters per minute) at $t = 0.5$?

23. Let $x(t) = t^3 + ct$ be the position of a particle at time t. For which values of c does the particle reverse direction, and at what times does the reversal take place for each such value of c? Does the value of c affect the particle's acceleration?

24. An evasive moth has position $t^3 - t + 2$ at time t. A hungry bat has position $y(t) = -\frac{2}{3}t^2 + t + 2$ at time t. How many chances does the bat have to catch the moth? How fast are they going and what are their accelerations at these times?

25. If the position of a moving object at time t is $(t^5 + 1)(t + 2)$, find its velocity and acceleration when $t = 0.1$.

26. Let $h(t) = 2t^3$ be the position of an object moving along a straight line at time t. What are the velocity and acceleration at $t = 3$?

Compute the second derivatives in Exercises 27–32.

27. $\dfrac{d^2}{dx^2}(x^4 - 3x^2)$

28. $\dfrac{d^2}{dx^2}(3x^2 - 8x + 10)$

29. $\dfrac{d^2}{dx^2}\dfrac{x^2 + 1}{x + 2}$

30. $\dfrac{d^2}{dx^2}\dfrac{x^3 - 1}{x^4 + 8}$

31. $\dfrac{d^2}{dr^2}(3r^8 - 8r^7 + 10)$

32. $\dfrac{d^2}{ds^2}(s^9 - 10s^8 + 5)$

Find the second derivative of the functions in Exercises 33–40.

33. $f(x) = x^2 - 5$

34. $f(x) = x - 2$

35. $y = x^5 + 7x^4 - 2x + 3$

36. $y = [(x - 1) + x^2][x^3 - 1]$

37. $y = \dfrac{x^2}{x - 1}$

38. $y = x^5 + \dfrac{1}{x} + \dfrac{2}{x^3}$

39. $\dfrac{t^2 + 1}{t^2 - 1}$

40. $\dfrac{s}{s + 1}$

In Exercises 41–46, find the velocities and accelerations at the indicated times of the particles whose positions y (in meters) on a line are given by the following functions of time t (in seconds):

41. $y = 3t + 2$; $t_0 = 1$

42. $y = 5t - 1$; $t_0 = 0$

43. $y = 8t^2 + 1$; $t_0 = 0$

44. $y = 18t^2 - 2t + 5$; $t_0 = 2$

45. $y = 10 - 2t - 0.01t^4$; $t_0 = 0$

46. $y = 20 - 8t - 0.02t^6$; $t_0 = 1$

47. The height of a pebble dropped off a building at time $t = 0$ is $h(t) = 44.1 - 4.9t^2$ meters at time t. The pebble strikes the ground at $t = 3.00$ seconds.
    (a) What is its velocity and acceleration when it strikes the ground?
    (b) What is its velocity when it is halfway down the building?

48. The amount of rain y in inches at time x in hours from the start of the September 3, 1975 Owens Valley thunderstorm was given by $y = 2x - x^2$, $0 \leqslant x \leqslant 1$.
    (a) Find how many inches of rain per hour were falling halfway through the storm.
    (b) Find how many inches of rain per hour were falling after half an inch of rain has fallen.

49. A shoe repair shop can produce $20x - x^2 - 3$ dollars of revenue when x worker-hours of labor are used. Find the marginal productivity in dollars when 5 workers are employed.

50. The owners of a restaurant find that they can serve $300w - 2w^2 - 14$ dinners when w worker-weeks of labor are employed. If an average dinner is worth $7.50, what is the marginal productivity (in dollars) of a worker when 10 workers are employed?

51. A factory employing w workers produces $100w + w^2/100 - (1/5000)w^4$ dollars worth of tools per day. Find the marginal productivity of labor when $w = 20$.

52. A farm can grow $10000x - 35x^3$ dollars worth of tomatoes if x tons of fertilizer are used. Find the marginal productivity of the fertilizer when $x = 10$. Interpret the sign of your answer.

53. In a boot factory, the cost in dollars of making x boots is $(4x + 0.02x^2)/(1 + 0.002x^3)$. If boots are priced at $25 - 0.02x$ dollars, what is the marginal profit, assuming that x boots are sold?

54. In a pizza parlor, the cost in dollars of making x pizzas is $(5x + 0.01x^2)/(1 + 0.001x^3)$. The price per pizza sold is set by the rule: price $= 7 - 0.05x$ if x pizzas are made. If all x pizzas are sold, what is the marginal profit?

In each of Exercises 55–58, what name would you give to the rate of change of y with respect to x? In what units could this rate be expressed?

55. $x =$ amount of fuel used; $y =$ distance driven in an automobile.

56. $x =$ distance driven in an automobile; $y =$ amount of fuel used.

57. x = amount of fuel purchased; y = amount of money paid for fuel.

58. x = distance driven in an automobile; y = amount of money paid for fuel.

59. The cost c of fuel for driving, measured in cents per kilometer, can be written as the product $c = rp$, where r is the fuel consumption rate in liters per kilometer and p is the price of fuel in cents per liter. If r and p depend on time (the car deteriorates, price fluctuates), so does c. The rates of change are connected by the product rule

$$\frac{dc}{dt} = r\frac{dp}{dt} + \frac{dr}{dt}p.$$

Interpret in words each of the terms on the right-hand side of this equation, and explain why dc/dt should be their sum.

60. If $f(x)$ represents the cost of living at time x, then $f'(x) > 0$ means that there is inflation.
    (a) What does $f''(x) > 0$ mean?
    (b) A government spokesman says, "The rate at which inflation is getting worse is decreasing." Interpret this statement in terms of $f'(x)$, $f''(x)$, and $f'''(x)$.

61. Let $y = 4x^2 - 2x + 7$. Compute the average rate of change of y with respect to x over the interval from $x_0 = 0$ to $x_1 = \Delta x$ for the following values of Δx: $0.1, 0.001, 0.000001$. Compare with the derivative at $x_0 = 0$.

62. Repeat Exercise 61 with $\Delta x = -0.1, -0.001,$ and -0.000001.

63. Find the average rate of change of the following functions on the given interval. Compare with the derivative at the midpoint.
    (a) $f(x) = (x - \frac{1}{2})(x + 1)$ between $x = -\frac{1}{2}$ and $x = 0$.
    (b) $g(t) = 3(t + 5)(t - 3)$ on $[2, 6]$.
    (c) $h(r) = 10r^2 - 3r + 6$ on $[-0.1, 0.4]$.
    (d) $r(t) = (2 - t)(t + 4)$; t in $[3, 7]$.

64. (a) Let $y = ax^2 + bx + c$, where a, b, and c are constant. Show that the average rate of change of y with respect to x on any interval $[x_1, x_2]$ equals the instantaneous rate of change at the midpoint; i.e., at $(x_1 + x_2)/2$.
    (b) Let $f(x) = ax^2 + bx + c$, where a, b, and c are constant. Prove that, for any x_0,

    $$f(x) = f(x_0) + f'(m)(x - x_0)$$

    where $m = (x + x_0)/2$.

65. The length l and width w of a rectangle are functions of time given by $l = (3 + t^2 + t^3)$ centimeters and $w = (5 - t + 2t^2)$ centimeters at time t (in seconds). What is the rate of change of area with respect to time at time t?

66. If the height and radius of a right circular cylin-

der are functions of time given by $h = (1 + t^2 + t^3 + t^4)$ centimeters and $r = (1 + 2t - t^2 + t^5)$ centimeters at time t (in seconds), what is the rate of change with respect to time of the lateral surface area, i.e., the total surface area minus the top and bottom?

67. Let $f(t) = 2t^2 - 5t + 2$ be the position of object A and let $h(t) = -3t^2 + t + 3$ be the position of object B.
    (a) When is A moving faster than B?
    (b) How fast is B going when A stops?
    (c) When does B change direction?

68. Repeat Example 7 with the same data but with the following two conventions changed. First, the origin is now chosen at a point 20 meters above the bridge. Second, we designate down as "+" and up as "−" rather than vice versa.

69. For which functions $f(x) = ax^2 + bx + c$ is the second derivative equal to the zero function?

70. How do the graphs of functions $ax^2 + bx + c$ whose second derivative is positive compare with those for which the second derivative is negative and those for which the second derivative is zero?

71. A particle is said to be *accelerating* (or *decelerating*) if the sign of its acceleration is the same as (or opposite to) the sign of its velocity. (a) Let $f(t) = -t^3$ be the position of a particle on a straight line at time t. When is the particle accelerating and when is it decelerating? (b) If the position of a particle on a line is given as a quadratic function of time and the particle is accelerating at time t_0, does the particle ever decelerate?

72. One summer day in Los Angeles, the pollution index at 7:00 AM was 20 parts per million, increasing linearly 15 parts per million each hour until 5:00 PM. Let y be the amount of pollutants in the air x hours after 7:00 AM.
    (a) Find a linear equation relating y and x.
    (b) The slope is the increase in pollution for each hour increase in time. Find it.
    (c) Find the pollution level at 5:00 PM.

73. Straight-line depreciation means that the difference between current value and original value is directly proportional to the time t. Suppose a home office is presently furnished for $4000 and salvaged for $500 after ten years. Assume straight line depreciation.
    (a) Find a linear equation for the value V of the office furniture after t years, for tax purposes.
    (b) The slope of the line indicates the decrease in value each year of the office furniture, to be used in preparing a tax return. Find it.

★74. Suppose that the acceleration of an object is constant and equal to 9.8 meters/second² and that its velocity at time $x = 0$ is 2 meters/second.
(a) Express the velocity as a function of x.
(b) What is the velocity when $x = 3$?
(c) Express the position y of the object as a function of x, if $y = 4$ when $x = 0$.
(d) How far does the object travel between $x = 2$ and $x = 5$?

★75. Let

$$f(x) = \begin{cases} 0, & x \leq 0, \\ x^2, & x \geq 0. \end{cases}$$

(a) Sketch a graph of $f(x)$.
(b) Find $f'(x)$. Sketch its graph.
(c) Find $f''(x)$ for $x \neq 0$. Sketch its graph. What happens when $x = 0$?
(d) Suppose that $f(x)$ is the position of an object at time x. What might have happened at $x = 0$?

2.2 The Chain Rule

The derivative of $f(g(x))$ is a product of derivatives.

None of the rules which we have derived so far tell us how to differentiate $\sqrt{x^3 - 5} = (x^3 - 5)^{1/2}$. The chain rule will. Before deriving this rule, though, we shall look at what happens when we differentiate a function raised to an integer power.

If $g(x)$ is any function, we can use the product rule to differentiate $[g(x)]^2$:

$$\frac{d}{dx}\left[g(x)\right]^2 = \frac{d}{dx}\left[g(x)g(x)\right] = g'(x)g(x) + g(x)g'(x) = 2g(x)g'(x).$$

If we write $u = g(x)$, this can be expressed in Leibniz notation as

$$\frac{d}{dx}(u^2) = 2u\frac{du}{dx}.$$

In the same way, we may differentiate u^3:

$$\frac{d}{dx}(u^3) = \frac{d}{dx}(u^2 \cdot u) = \frac{d}{dx}(u^2) \cdot u + u^2\frac{du}{dx}$$

$$= 2u\frac{du}{dx} \cdot u + u^2\frac{du}{dx} = 3u^2\frac{du}{dx}.$$

Similarly, $(d/dx)(u^4) = 4u^3(du/dx)$ (check it yourself); and, for a general positive integer n, we have $(d/dx)u^n = nu^{n-1}(du/dx)$. (This may be formally proved by induction—see Exercise 52.)

Power of a Function Rule

To differentiate the nth power $[g(x)]^n$ of a function $g(x)$, where n is a positive integer, take out the exponent as a factor, reduce the exponent by 1, and multiply by the derivative of $g(x)$:

$$(g^n)'(x) = n[g(x)]^{n-1}g'(x),$$

$$\frac{d}{dx}(u^n) = nu^{n-1}\frac{du}{dx}$$

If $u = x$, then $du/dx = 1$, and the power of a function rule reduces to the ordinary power rule.

A common mistake made by students in applying the power of a function rule is to forget the extra factor of $g'(x)$—that is, du/dx.

Example 1 Find the derivative of $[g(x)]^3$, where $g(x) = x^4 + 2x^2$, first by using the power of a function rule and then by expanding the cube and differentiating directly. Compare the answers.

Solution By the power of a function rule, with $u = x^4 + 2x^2$ and $n = 3$,

$$\frac{d}{dx}(x^4 + 2x^2)^3 = 3(x^4 + 2x^2)^2 \frac{d}{dx}(x^4 + 2x^2) = 3(x^4 + 2x^2)^2(4x^3 + 4x).$$

If we expand the cube first, we get $(x^4 + 2x^2)^3 = x^{12} + 6x^{10} + 12x^8 + 8x^6$, so

$$\frac{d}{dx}(x^4 + 2x^2)^3 = \frac{d}{dx}(x^{12} + 6x^{10} + 12x^8 + 8x^6)$$

$$= 12x^{11} + 60x^9 + 96x^7 + 48x^5.$$

To compare the two answers, we expand the first one:

$$3(x^4 + 2x^2)^2(4x^3 + 4x) = 3(x^8 + 4x^6 + 4x^4) \cdot 4(x^3 + x)$$

$$= 12(x^{11} + 5x^9 + 8x^7 + 4x^5)$$

$$= 12x^{11} + 60x^9 + 96x^7 + 48x^5,$$

which checks. ▲

Example 2 Find $\dfrac{d}{ds}(s^4 + 2s^3 + 3)^8$.

Solution We apply the power of a function rule, with $u = s^4 + 2s^3 + 3$ (and the variable x replaced by s):

$$\frac{d}{ds}(s^4 + 2s^3 + 3)^8 = 8(s^4 + 2s^3 + 3)^7 \frac{d}{ds}(s^4 + 2s^3 + 3)$$

$$= 8(s^4 + 2s^3 + 3)^7(4s^3 + 6s^2).$$

(You could also do this problem by expanding the eighth power and then differentiating; obviously, this practice is not recommended.) ▲

Example 3 If $y = (x^2 + 1)^{27}(x^4 + 3x + 1)^8$, find the rate of change of y with respect to x.

Solution First of all, by the power of a function rule,

$$\frac{d}{dx}(x^2 + 1)^{27} = 27(x^2 + 1)^{26} \cdot 2x$$

and

$$\frac{d}{dx}(x^4 + 3x + 1)^8 = 8(x^4 + 3x + 1)^7(4x^3 + 3)$$

Thus, by the product rule, the rate of change of y with respect to x is

$$\frac{dy}{dx} = 27(x^2 + 1)^{26} \cdot 2x \cdot (x^4 + 3x + 1)^8$$

$$+ (x^2 + 1)^{27} \cdot 8(x^4 + 3x + 1)^7(4x^3 + 3).$$

To simplify this, we can factor out the highest powers of $x^2 + 1$ and $x^4 + 3x + 1$ to get

$$(x^2 + 1)^{26}(x^4 + 3x + 1)^7 \big[27 \cdot 2x(x^4 + 3x + 1) + (x^2 + 1) \cdot 8(4x^3 + 3)\big].$$

We can consolidate the expression in square brackets to a single polynomial of degree 5, getting $2(x^2 + 1)^{26}(x^4 + 3x + 1)^7(43x^5 + 16x^3 + 93x^2 + 27x + 12)$ as our rate of change. [*Note*: Consult your instructor regarding the amount of simplification required.] ▲

The power of a function rule is a special case of an important differentiation rule called the *chain rule*. To understand this more general rule, we begin by noting that the process of forming the power $[g(x)]^n$ can be broken into two successive operations: first find $u = g(x)$, and then find $f(u)$, where $y = f(u) = u^n$. The chain rule will help us to differentiate any function formed from two functions in this way.

If f and g are functions defined for all real numbers, we define their *composition* to be the function which assigns to x the number $f(g(x))$. The composition is often denoted by $f \circ g$. Thus $(f \circ g)(x) = f(g(x))$. To evaluate $y = (f \circ g)(x)$, we introduce an intermediate variable u and write $u = g(x)$ and $y = f(u)$. To evaluate y, we *substitute* $g(x)$ for u in $f(u)$. (If $f(x)$ and $g(x)$ are not defined for all x, then $(f \circ g)(x)$ is defined only when x is in the domain of g and $g(x)$ is in the domain of f.)

Example 4   (a) If $f(u) = u^3 + 2$ and $g(x) = (x^2 + 1)^2$, what is $h = f \circ g$?
(b) Let $f(x) = \sqrt{x}$ and $g(x) = x^3 - 5$. Find $f \circ g$ and $g \circ f$.
(c) Write $\sqrt{1 + x^2}/[2 + (1 + x^2)^3]$ as a composition of simpler functions.

Solution   (a) We calculate $h(x) = f(g(x))$ by writing $u = g(x)$ and substituting in $f(u)$. We get $u = (x^2 + 1)^2$ and so

$$h(x) = f(u) = u^3 + 2 = \left((x^2 + 1)^2\right)^3 + 2 = (x^2 + 1)^6 + 2.$$

(b) $(f \circ g)(x) = f(g(x)) = \sqrt{g(x)} = \sqrt{x^3 - 5}$.

$$(g \circ f)(x) = g(f(x)) = \left[f(x)\right]^3 - 5$$

$$= \left(\sqrt{x}\right)^3 - 5 = x^{3/2} - 5.$$

The functions $f \circ g$ and $g \circ f$ are certainly different.
(c) Let $g(x) = 1 + x^2$ and $f(u) = \sqrt{u}/(2 + u^3)$. Then the given function can be written as $f \circ g$. ▲

▦ Calculator Discussion

On electronic calculators, several functions, such as $1/x$, x^2, $\sqrt{x}$, and $\sin x$, are evaluated by the push of a single key. To evaluate the composite function $f \circ g$ on x, you first enter x, then push the key for g to get $g(x)$, then push the key for f to get $f(g(x))$. For instance, let $f(x) = x^2$, $g(x) = \sin x$. To calculate $(f \circ g)(x) = f(g(x)) = (\sin x)^2$ for $x = 32$ (degrees), we enter 32, then press the sin key, then the x^2 key. The result is $32 \rightarrow 0.52991926 \rightarrow 0.28081442$. Notice that $(g \circ f)(x) = \sin(x^2)$ is quite different: entering 32 and pressing the x^2 key followed by the sin key, we get $32 \rightarrow 1024 \rightarrow -0.82903756$. ▲

Do not confuse the composition of functions with the product. We have

$$(fg)(x) = f(x)g(x),$$

while

$$(f \circ g)(x) = f(g(x)).$$

In the case of the product we evaluate $f(x)$ and $g(x)$ separately and then multiply the results; in the case of the composition, we evaluate $g(x)$ first and then apply f to the result. While the order of f and g does not matter for the product, it does for composition.

Composition of Functions

The composition $f \circ g$ is obtained by writing $u = g(x)$ and evaluating $f(u)$. To break up a given function $h(x)$ as a composition, find an intermediate expression $u = g(x)$ such that $h(x)$ can be written in terms of u.

The derivative of a composite function turns out to be the product of the derivatives of the separate functions. The exact statement is given in the following box.

Chain Rule

To differentiate a composition $f(g(x))$, differentiate g at x, differentiate f at $g(x)$, and multiply the results:

$$(f \circ g)'(x) = f'(g(x)) \cdot g'(x).$$

In Leibniz notation,

$$\frac{dy}{dx} = \frac{dy}{du} \cdot \frac{du}{dx}$$

if y is a function of u and u is a function of x.

A complete proof of the chain rule can be given by using the theory of limits (see Review Exercise 99, Chapter 11). The basic argument, however, is simple and goes as follows. If x is changed by a small amount Δx and the corresponding change in $u = g(x)$ is Δu, we know that

$$g'(x) = \frac{du}{dx} = \lim_{\Delta x \to 0} \frac{\Delta u}{\Delta x}.$$

Corresponding to the small change Δu is a change Δy in $y = f(u)$, and

$$f'(u) = \frac{dy}{du} = \lim_{\Delta u \to 0} \frac{\Delta y}{\Delta u}.$$

To calculate the rate of change dy/dx, we write

$$\frac{dy}{dx} = \lim_{\Delta x \to 0} \frac{\Delta y}{\Delta x} = \lim_{\Delta x \to 0} \frac{\Delta y}{\Delta u}\frac{\Delta u}{\Delta x} = \left(\lim_{\Delta x \to 0} \frac{\Delta y}{\Delta u}\right)\left(\lim_{\Delta x \to 0} \frac{\Delta u}{\Delta x}\right)$$

$$= \left(\lim_{\Delta u \to 0} \frac{\Delta y}{\Delta u}\right)\left(\lim_{\Delta x \to 0} \frac{\Delta u}{\Delta x}\right) = \frac{dy}{du}\frac{du}{dx} = f'(u) \cdot g'(x).$$

In going to the second line, we replace $\Delta x \to 0$ by $\Delta u \to 0$ because the differentiable function g is continuous, i.e., $\Delta x \to 0$ implies $\Delta u \to 0$, as we saw in Section 1.3.

There is a flaw in this proof: the Δu determined by Δx could well be zero, and division by zero is not allowed. This difficulty is fortunately not an essential one, and the more technical proof given in Chapter 11 avoids it.

Notice that the chain rule written in Leibniz notation is closely related to our argument and is easy to remember. Although du does not really have an independent meaning, one may "cancel" it informally from the product $(dy/du) \cdot (du/dx)$ to obtain dy/dx.

A physical model illustrating the chain rule is given at the end of this section.

Example 5   Verify the chain rule for $f(u) = u^2$ and $g(x) = x^3 + 1$.

Solution   Let $h(x) = f(g(x)) = [g(x)]^2 = (x^3 + 1)^2 = x^6 + 2x^3 + 1$. Thus $h'(x) = 6x^5 + 6x^2$. On the other hand, since $f'(u) = 2u$ and $g'(x) = 3x^2$,

$$f'(g(x)) \cdot g'(x) = (2 \cdot (x^3 + 1))3x^2 = 6x^5 + 6x^2.$$

Hence the chain rule is verified in this case. ▲

Let us check that the power of a function rule follows from the chain rule: If $y = [g(x)]^n$, we may write $u = g(x)$, $y = f(u) = u^n$. Since $dy/du = nu^{n-1}$, the chain rule gives

$$\frac{dy}{dx} = f'(g(x))g'(x) = n[g(x)]^{n-1}g'(x).$$

This calculation applies to negative or zero powers as well as positive ones. Thus *the power of a function rule holds for all integer powers*.

Example 6   Let $f(x) = 1/[(3x^2 - 2x + 1)^{100}]$. Find $f'(x)$.

Solution   We write $f(x)$ as $(3x^2 - 2x + 1)^{-100}$. Thus

$$f'(x) = -100(3x^2 - 2x + 1)^{-101}(6x - 2). ▲$$

The chain rule also solves the problem which began the section.

Example 7   Differentiate $\sqrt{x^3 - 5}$.

Solution   In Example 4(b) we saw that $\sqrt{x^3 - 5} = \sqrt{u}$ if $u = x^3 - 5$. Thus, if $y = \sqrt{x^3 - 5} = \sqrt{u}$, then $dy/du = 1/2\sqrt{u}$ (Example 4, Sect. 1.3), and $du/dx = 3x^2$, so

$$\frac{dy}{dx} = \frac{dy}{du}\frac{du}{dx} = \frac{1}{2\sqrt{u}} \cdot 3x^2 = \frac{3x^2}{2\sqrt{x^3 - 5}} . ▲$$

Example 8   If $h(x) = f(x^2)$, find a formula for $h'(x)$. Check your formula in the case $f(u) = u^3$.

Solution   Let $u = g(x) = x^2$, so $h(x) = f(u)$. Then $h'(x) = f'(u) \cdot g'(x) = f'(x^2) \cdot 2x$. Thus $h'(x) = f'(x^2) \cdot 2x$.

If $f(u) = u^3$, then $f'(u) = 3u^2$ and $f'(x^2) = 3x^4$. Thus, $h'(x) = 3x^4 \, 2x = 6x^5$. In fact, $h(x) = (x^2)^3 = x^6$ in this case, so differentiating h directly gives the same result. ▲

Example 9   Use the chain rule to differentiate $f(x) = ((x^2 + 1)^{20} + 1)^4$. (Do not expand!)

Solution   Let $u = (x^2 + 1)^{20} + 1$ and $y = u^4$, so $y = f(x)$. By the chain rule,

$$\frac{dy}{dx} = \frac{dy}{du}\frac{du}{dx} = 4u^3\frac{du}{dx} = 4\left((x^2 + 1)^{20} + 1\right)^3\frac{du}{dx} .$$

To calculate du/dx we use the chain rule again (or the power of a function rule):

$$\frac{du}{dx} = \frac{d}{dx}\left[(x^2 + 1)^{20} + 1\right] = 20(x^2 + 1)^{19} \cdot 2x.$$

Thus

$$\frac{dy}{dx} = 4\big((x^2 + 1)^{20} + 1\big)^3 \cdot 20(x^2 + 1)^{19} \cdot 2x$$

$$= 160x\big((x^2 + 1)^{20} + 1\big)^3 (x^2 + 1)^{19}. \ \blacktriangle$$

The following examples require us to translate words into equations before using the chain rule.

Example 10 The population of Thin City is increasing at the rate of 10,000 people per day on March 30, 1984. The area of the city grows to keep the ratio of 1 square mile per 1000 people. How fast is the area increasing per day on this date?

Solution Let A = area, p = population, t = time (days). The rate of increase of area with respect to time is $dA/dt = (dA/dp)(dp/dt) = \frac{1}{1000} \cdot 10000 = 10$ square miles per day. ▲

Example 11 A dog 2 feet high trots proudly away from a 10-foot-high light post at 3 feet per second. When he is 8 feet from the post's base, how fast is the tip of his shadow moving?

Solution Refer to Fig. 2.2.1. By similar triangles, $y/(y - x) = 10/2$; solving, $y = 5x/4$. Then $dy/dt = (dy/dx)(dx/dt) = (\frac{5}{4})3 = 3\frac{3}{4}$ feet per second. ▲

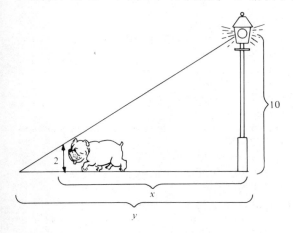

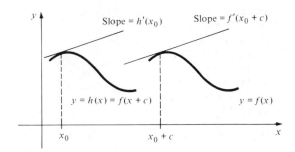

Figure 2.2.1. Dog trotting proudly away from lamp post.

Figure 2.2.2. The geometric interpretation of the shifting rule.

Another special case of the chain rule may help you to understand it. Consider $h(x) = f(x + c)$, c a constant. If we let $u = g(x) = x + c$, we get $g'(x) = 1$, so

$$h'(x) = f'(g(x)) \cdot g'(x) = f'(x + c) \cdot 1 = f'(x + c).$$

Note that the graph of h is the same as that of f except that it is shifted c units to the *left* (see Fig. 2.2.2). It is reasonable, then, that the tangent line to the graph of h is obtained by shifting the tangent line to the graph of f. Thus, in this case, the chain rule is telling us something geometrically obvious. One might call this formula the *shifting rule*. In Leibniz notation it reads

$$\frac{d}{dx} f(x + c)\Big|_{x_0} = \frac{d}{dx} f(x)\Big|_{x_0 + c}.$$

Supplement to Section 2.2
A Physical Model for the Chain Rule

When you change altitude rapidly, as in a moving car or plane, a pressure difference develops between the inside and outside of your eardrums, and your ears "pop." Three variables relevant to this phenomenon are the time t, the altitude u, and the air pressure p. Ear popping occurs when the rate of change dp/dt is too large.

The rate dp/dt is hard to measure directly. On the other hand, du/dt can be determined if we know the altitude as a function of t. For instance, if we are rolling down a hill at 100 kilometers per hour, we could have $du/dt = -3$ meters per second. The rate of change dp/du is known to meteorologists; near sea level, it is about -0.12 gsc per meter. (The unit "gsc" of pressure is "grams per square centimeter"; the rate of change is negative because pressure decreases as altitude increases.)

Now the chain rule enables us to calculate how fast the pressure is changing with time:

$$\frac{dp}{dt} = \frac{dp}{du} \cdot \frac{du}{dt} = (-0.12)(-3) \text{ gsc per second} = 0.36 \text{ gsc per second.}$$

This rate of pressure increase is fast enough so that the ears' internal pressure control system cannot keep up with it, and they "pop."

Exercises for Section 2.2

Find the derivatives of the functions in Exercises 1–10.
1. $(x + 3)^4$
2. $(x^2 + 3x + 1)^5$
3. $(x^3 + 10x)^{100}$
4. $(s^4 + 4s^3 + 3s^2 + 2s + 1)^8$
5. $(x^2 + 8x)^3 \cdot x$
6. $(x^2 + 2)^3(x^9 + 8)$
7. $(x^2 + 2)(x^9 + 8)^3$
8. $(x^3 + 2)^9(x^8 + 2x + 1)^{10}$
9. $(y + 1)^3(y + 2)^2(y + 3)$
10. $((x^2 - 1)^2 + 3)^{10}$

11. Let $g(x) = x + 1$ and $f(u) = u^2$. Find $f \circ g$ and $g \circ f$.
12. Let $h(x) = x^{24} + 3x^{12} + 1$. Write $h(x)$ as a composite function $f(g(x))$ with $g(x) = x^{12}$.

Find $f \circ g$ and $g \circ f$ in each of Exercises 13–16.
13. $g(x) = x^3$; $f(x) = (x - 2)^3$.
14. $g(x) = x^n$; $f(x) = x^m$.
15. $g(x) = \dfrac{1}{1 - x}$; $f(x) = \dfrac{1}{2} - \sqrt{3}\, x$.
16. $g(x) = \dfrac{3x - 2}{4x + 1}$; $f(x) = \dfrac{2x - 7}{9x + 3}$.

Write the functions in Exercises 17–20 as compositions of simpler functions.
17. $h(x) = \sqrt{4x^3 + 5x + 3}$.
18. $h(r) = \sqrt{1 + \sqrt{r}}$.
19. $h(u) = \left(\dfrac{1 - u}{1 + u} \right)^3$.
20. $h(x) = ((x^2 + 1) + (x^2 + 1)^2 + 1)^2$.

Verify the chain rule for $f(u)$ and $g(x)$ given in Exercises 21–24.
21. $f(u) = u^2$, $g(x) = x^2 - 1$
22. $f(u) = u^3$, $g(x) = x + 1$
23. $f(u) = u^2$, $g(x) = \sqrt{x}$
24. $f(u) = \sqrt{u}$, $g(x) = x^2$

Use the chain rule rule to differentiate the functions in Exercises 25–34.
25. $(x^2 - 6x + 1)^3$
26. $(x - 2x^2)^3$
27. $\dfrac{9 + 2x^5}{3 + 5x^5}$
28. $\dfrac{1}{(x^3 + 5x)^4}$
29. $((x^2 + 2)^2 + 1)^2$
30. $\dfrac{3}{\left[(x + 2)^2 + 4\right]^4}$
31. $\dfrac{(x^2 + 3)^5}{\left[1 + (x^2 + 3)\right]^8}$
32. $\dfrac{1}{2x + 1}[(2x + 1)^3 + 5]$
33. $\sqrt{4x^5 + 5x^2}$
34. $\sqrt{1 + \sqrt{x}}$

35. If $h(x) = x^3 f(2x^2)$, find a formula for $h'(x)$.
36. If $h(x) = f(g(x^2))$, find a formula for $h'(x)$.
37. Given three functions, f, g, and h:
    (a) How would you define the composition $f \circ g \circ h$?
    (b) Use the chain rule twice to obtain a formula for the derivative of $f \circ g \circ h$.
38. If $h(x) = f(g(x^3 + 2)) + g(f(x^2))$, find a formula for $h'(x)$.
39. Fat City occupies a circular area 10 miles in diameter and contains 500,000 inhabitants. If the population is growing now at the rate of 20,000 inhabitants per year, how fast should the diame-

ter be increasing now in order to maintain the circular shape and the same population density (= number of people per square mile). If the population continues to grow at the rate of 20,000 per year, how fast should the diameter be increasing in 5 years? Give an intuitive explanation of the relation between the two answers.

40. The radius at time t of a sphere S is given by $r = t^2 - 2t + 1$. How fast is the volume V of S changing at time $t = \frac{1}{2}, 1, 2$?

41. The kinetic energy K of a particle of mass m moving with speed v is $K = \frac{1}{2}mv^2$. A particle with mass 10 grams has, at a certain moment, velocity 30 centimeters per second and acceleration 5 centimeters per second per second. At what rate is the kinetic energy changing?

42. (a) At a certain moment, an airplane is at an altitude of 1500 meters and is climbing at the rate of 5 meters per second. At this altitude, pressure decreases with altitude at the rate of 0.095 gsc per meter. What is the rate of change of pressure with respect to time?

    (b) Suppose that the airplane in (a) is descending rather than climbing at the rate of 5 meters per second. What is the rate of change of pressure with respect to time?

43. At a certain moment, your car is consuming gasoline at the rate of 15 miles per gallon. If gasoline costs 75 cents per gallon, what is the cost per mile? Set the problem up in terms of functions and apply the chain rule.

44. The price of eggs, in cents per dozen, is given by the formula $p = 55/(s - 1)^2$, where s is the supply of eggs, in units of 10,000 dozen, available to the wholesaler. Suppose that the supply on July 1, 1986 is $s = 2.1$ and is falling at a rate of 0.03 per month. How fast is the price rising?

45. If an object has position $(t^2 + 4)^5$ at time t, what is its velocity when $t = -1$?

46. If an object has position $(t^2 + 1)/(t^2 - 1)$ at time t, what is its velocity when $t = 2$?

Find the second derivatives of the functions in Exercises 47–50.

47. $(x + 1)^{13}$
48. $(x^3 - 1)^8$
49. $(x^4 + 10x^2 + 1)^{98}$
50. $(x^2 + 1)^3(x^3 + 1)^2$

★51. (a) Find a "stretching rule" for the derivative of $f(cx)$, c a constant.

    (b) Draw the graphs of $y = 1 + x^2$ and of $y = 1 + (4x)^2$ and interpret the stretching rule geometrically.

★52. Prove that $(d/dx)(u^n) = nu^{n-1}du/dx$ for all natural numbers n as follows:

    (a) Note that this is established for $n = 1, 2, 3$ at the beginning of this section.

    (b) Assume that the result is true for $n - 1$, and write $u^n = u(u^{n-1})$. Now differentiate using the product rule to establish the result for n.

    (c) Use induction to conclude the result for all n. (See Exercise 65, p. 69.)

★53. Find a general formula for $(d^2/dx^2)(u^n)$, where $u = f(x)$ is any function of x.

★54. (a) Let i be the "identity function" $i(x) = x$. Show that $i \circ f = f$ and $f \circ i = f$ for any function f. (b) Verify the chain rule for $f = f \circ i$.

★55. Let f and g be functions such that $f \circ g = i$, where i is the function in Problem 54. Find a formula for $f'(x)$ in terms of the derivative of g.

★56. Use the result of Exercise 55 to find the derivative of $f(x) = \sqrt[3]{x}$ by letting $g(x) = x^3$.

★57. Find a formula for the second derivative of $f \circ g$ in terms of the first and second derivatives of f and g.

★58. Show that the power of a function rule for negative powers follows from that rule for positive powers and the reciprocal rule.

★59. For reasons which will become clear in Chapter 6, the quotient $f'(x)/f(x)$ is called the *logarithmic derivative* of $f(x)$.

    (a) Show that the logarithmic derivative of the product of two functions is the sum of the logarithmic derivatives of the functions.

    (b) Show that the logarithmic derivative of the quotient of two functions is the difference of their logarithmic derivatives.

    (c) Show that the logarithmic derivative of the nth power of a function is n times the logarithmic derivative of the function.

    (d) Develop a formula for the logarithmic derivative of

    $$[f_1(x)]^{n_1}[f_2(x)]^{n_2} \cdots [f_k(x)]^{n_k}$$

    in terms of the logarithmic derivatives of f_1 through f_k.

    (e) Using your formula in part (d), find the ordinary (*not* logarithmic) derivative of

    $$f(x) = \frac{(x^2 + 3)(x + 4)^8(x + 7)^9}{(x^4 + 3)^{17}(x^4 + 2x + 1)^5}.$$

    If you have enough stamina, compute $f'(x)$ without using the formula in part (d).

★60. Differentiate $(1 + (1 + (1 + x^2)^8)^8)^8$.

2.3 Fractional Powers and Implicit Differentiation

The power rule still holds when the exponent is a fraction.

In this section we extend the power rule to include fractional exponents by using a method called implicit differentiation, which can be applied to many other problems as well.

Let us begin by trying to find dy/dx when $y = x^{1/n} = \sqrt[n]{x}$, where n is a positive integer.[1] At the moment, we shall simply assume that this derivative exists and try to calculate its value. This assumption will be justified in Section 5.3, in connection with inverse functions.

We may rewrite the relation $y = x^{1/n}$ as $y^n = x$, so we must have

$$\frac{d}{dx}(y^n) = \frac{d}{dx}(x). \tag{1}$$

Recalling that y is a function of x, we may evaluate the left-hand side of (1) by the chain rule (or the power of a function rule) to get

$$\frac{d}{dx}(y^n) = ny^{n-1}\frac{dy}{dx}. \tag{2}$$

The right-hand side of (1) is simply

$$\frac{dx}{dx} = 1. \tag{3}$$

Substituting (2) and (3) into (1) gives

$$ny^{n-1}\frac{dy}{dx} = 1$$

which we may solve for dy/dx to obtain

$$\frac{dy}{dx} = \frac{1}{ny^{n-1}} = \frac{1}{n}y^{1-n} = \frac{1}{n}\left(x^{1/n}\right)^{1-n} = \frac{1}{n}x^{(1-n)/n} = \frac{1}{n}x^{(1/n)-1}.$$

Thus

$$\frac{d}{dx}(x^{1/n}) = \frac{1}{n}x^{(1/n)-1}. \tag{4}$$

Note that this rule reads the same as the ordinary power rule: "Bring down the exponent as a multiplier and then decrease the exponent by one." The special case $(d/dx)(x^{1/2}) = \frac{1}{2}x^{-1/2}$ has already been considered in Example 4, Section 1.3.

Example 1   Differentiate $f(x) = 3\sqrt[5]{x}$.

Solution $\dfrac{d}{dx}3\sqrt[5]{x} = 3\dfrac{d}{dx}x^{1/5} = \dfrac{3}{5}x^{(1/5)-1} = \dfrac{3}{5}x^{-4/5} = \dfrac{3}{5x^{4/5}}.$ ▲

Next, we consider a general rational power $f(x) = x^r$, where $r = p/q$ is a ratio of integers. Thinking of $x^{p/q}$ as $(x^{1/q})^p$, we set $g(x) = x^{1/q}$, so that $f(x) = [g(x)]^p$. Then, by the (integer) power of a function rule,

$$\frac{d}{dx}\left[g(x)\right]^p = p\left[g(x)\right]^{p-1}g'(x),$$

---

[1] Note that $x^{1/n}$ is defined for all x if n is odd but only for nonnegative x if n is even. A brief review of fractional exponents may be found in Section R.3.

so by formula (4) with $1/n$ replaced by $1/q$, we have

$$\frac{d}{dx}(x^{p/q}) = \frac{d}{dx}(x^{1/q})^p = p(x^{1/q})^{p-1} \cdot \frac{1}{q} x^{(1/q)-1}$$

$$= \frac{p}{q} x^{(p-1)/q} x^{(1-q)/q} = \frac{p}{q} x^{(p-q)/q} = \frac{p}{q} x^{(p/q)-1}.$$

We conclude that differentiation of rational powers follows the same rule as integer powers.

Rational Power Rule

To differentiate a power x^r (r a rational number), take out the exponent as a factor and then reduce the exponent by 1:

$$\frac{d}{dx}(x^r) = rx^{r-1}.$$

(The formula is valid for all x for which the right-hand side makes sense.)

Example 2   Differentiate $f(x) = 3x^2 + (x^2 + x^{1/3})/\sqrt{x}$.

Solution $f'(x) = \dfrac{d}{dx}(3x^2 + x^{3/2} + x^{-1/6})$

$$= 6x + \tfrac{3}{2}x^{1/2} - \tfrac{1}{6}x^{-7/6}$$

$$= 6x + \tfrac{3}{2}\sqrt{x} - \frac{1}{6x^{7/6}} \;\blacktriangle$$

We can combine the rational power rule with the chain rule to prove a rational power of a function rule. Let $y = [f(x)]^r$ and let $u = f(x)$ so that $y = u^r$. Then

$$\frac{dy}{dx} = \frac{dy}{du}\frac{du}{dx} = ru^{r-1}\frac{du}{dx} = r[f(x)]^{r-1}f'(x).$$

Rational Power of a Function Rule

To differentiate a power $[f(x)]^r$ (r a rational number), take out the exponent as a factor, reduce the exponent by 1, and multiply by $f'(x)$:

$$\frac{d}{dx}[f(x)]^r = r[f(x)]^{r-1}f'(x).$$

Example 3   Differentiate $g(x) = (9x^3 + 10)^{5/3}$.

Solution   Here $f(x) = 9x^3 + 10$, $r = \tfrac{5}{3}$, and $f'(x) = 27x^2$. Thus

$$g'(x) = \tfrac{5}{3}(9x^3 + 10)^{2/3} \cdot 27x^2 = 45x^2(9x^3 + 10)^{2/3}. \;\blacktriangle$$

The rules for rational powers can be combined with the quotient rule of differentiation, as in the next example.

Example 4 Differentiate $\dfrac{x^{1/2} + x^{3/2}}{x^{3/2} + 1}$.

Solution We use the quotient and rational power rules:

$$\frac{d}{dx}\left[\frac{x^{1/2} + x^{3/2}}{x^{3/2} + 1} \right]$$

$$= \frac{\left[x^{3/2} + 1 \right](d/dx)\left[x^{1/2} + x^{3/2} \right] - \left[x^{1/2} + x^{3/2} \right](d/dx)\left[x^{3/2} + 1 \right]}{\left[x^{3/2} + 1 \right]^2}$$

$$= \frac{\left[x^{3/2} + 1 \right]\left[\frac{1}{2}x^{-1/2} + \frac{3}{2}x^{1/2} \right] - \left[x^{1/2} + x^{3/2} \right]\left[\frac{3}{2}x^{1/2} \right]}{x^3 + 2x^{3/2} + 1}$$

$$= \frac{\left(\frac{1}{2}x + \frac{3}{2}x^2 + \frac{1}{2}x^{-1/2} + \frac{3}{2}x^{1/2} \right) - \left(\frac{3}{2}x + \frac{3}{2}x^2 \right)}{x^3 + 2x^{3/2} + 1}$$

$$= \frac{3x^{1/2} + x^{-1/2} - 2x}{2(x^3 + 2x^{3/2} + 1)} \; . \; \blacktriangle$$

The method which we used to differentiate $y = x^{1/n}$, namely differentiating the relation $y^n = x$ and then solving for dy/dx, is called *implicit differentiation*. This method can be applied to more complicated relationships such as $x^2 + y^2 = 1$ or $x^4 + xy + y^5 = 2$ which define y as a function of x *implicitly* rather than *explicitly*. In general, such a relationship will not define y uniquely as a function of x; it may define two or more functions. For example, the circle $x^2 + y^2 = 1$ is not the graph of a function, but the upper and lower semicircles are graphs of functions (see Fig. 2.3.1).

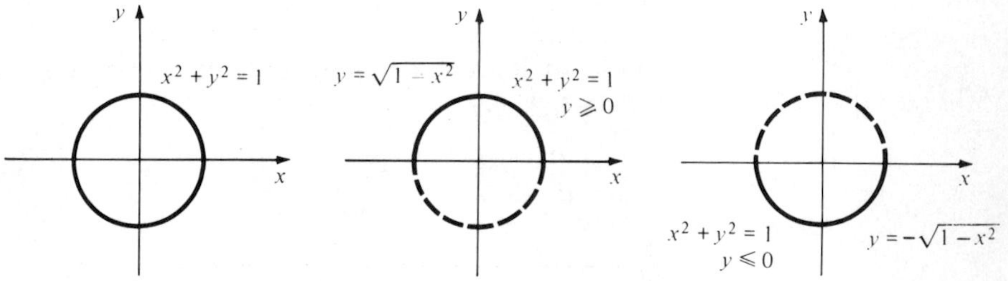

Figure 2.3.1. Parts of the circle $x^2 + y^2 = 1$ are the graphs of functions.

Example 5 If $y = f(x)$ and $x^2 + y^2 = 1$, express dy/dx in terms of x and y.

Solution Thinking of y as a function of x, we differentiate both sides of the relation $x^2 + y^2 = 1$ with respect to x. The derivative of the left-hand side is

$$\frac{d}{dx}\left(x^2 + y^2\right) = \frac{d}{dx}x^2 + \frac{d}{dx}y^2 = 2x + 2y\frac{dy}{dx} ,$$

while the right-hand side has derivative zero. Thus

$$2x + 2y\frac{dy}{dx} = 0 \quad \text{and so} \quad \frac{dy}{dx} = -\frac{x}{y} \; . \; \blacktriangle$$

The result of Example 5 can be checked, since in this case we can solve for y directly:

$$y = \pm\sqrt{1 - x^2} \; .$$

Notice that the given relation then defines *two* functions: $f_1(x) = \sqrt{1 - x^2}$ and

$f_2(x) = -\sqrt{1 - x^2}$. Taking the plus case, with $u = 1 - x^2$ and $y = \sqrt{u}$, the chain rule gives

$$\frac{dy}{dx} = \frac{dy}{du}\frac{du}{dx} = \frac{1}{2\sqrt{u}}(-2x)$$

$$= \frac{1}{2\sqrt{1 - x^2}}(-2x) = \frac{-x}{\sqrt{1 - x^2}} = \frac{-x}{y},$$

so it checks. The minus case gives the same answer.

From the form of the derivative given by implicit differentiation, $dy/dx = -x/y$, we see that the tangent line to a circle at (x, y) is perpendicular to the line through (x, y) and the origin, since their slopes are negative reciprocals of one another. (See Fig. 2.3.2.) Implicit differentiation often leads directly to such striking results, and for this reason it is sometimes preferable to use this method even when y could be expressed in terms of x.

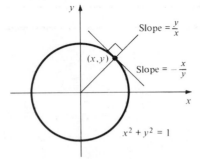

Figure 2.3.2. If $x^2 + y^2 = 1$, the formula $dy/dx = -x/y$ means that the tangent line to a circle at a point on the circle is perpendicular to the line from that point to the center of the circle.

There is a device which may help you to remember that the chain rule must be used. In Example 5, if we keep the notation $f(x)$ for y, then the relation $x^2 + y^2 = 1$ becomes $x^2 + [f(x)]^2 = 1$, and differentiating with respect to x gives $2x + 2f(x)f'(x) = 0$. Now we solve for $f'(x)$ to get $f'(x) = -x/f(x)$ or, in Leibniz notation, $dy/dx = -x/y$, just as before. Once you have done a few examples in this long-winded way, you should be able to go back to y and dy/dx without the f.

The following is an example in which we cannot solve for y in terms of x.

Example 6 Find the equation of the tangent line to the curve $2x^6 + y^4 = 9xy$ at the point $(1, 2)$.

Solution We note first that $(1, 2)$ lies on the curve, since $2(1)^6 + 2^4 = 9(1)(2)$. Now suppose that $y = f(x)$ and differentiate both sides of the defining relation. The left-hand side gives

$$\frac{d}{dx}(2x^6 + y^4) = 12x^5 + 4y^3\frac{dy}{dx},$$

while the right-hand side gives

$$\frac{d}{dx}(9xy) = 9y + 9x\frac{dy}{dx}.$$

Equating both sides and solving for dy/dx, we have

$$12x^5 + 4y^3\frac{dy}{dx} = 9y + 9x\frac{dy}{dx},$$

$$(4y^3 - 9x)\frac{dy}{dx} = 9y - 12x^5,$$

$$\frac{dy}{dx} = \frac{9y - 12x^5}{4y^3 - 9x}.$$

When $x = 1$ and $y = 2$,

$$\frac{dy}{dx} = \frac{9(2) - 12(1)^3}{4(2)^3 - 9(1)} = \frac{18 - 12}{32 - 9} = \frac{6}{23}.$$

Thus, the slope of the tangent line is $\frac{6}{23}$; by the point-slope formula, the equation of the tangent line is $y - 2 = \frac{6}{23}(x - 1)$, or $y = \frac{6}{23}x + \frac{40}{23}$. ▲

Implicit Differentiation

To calculate dy/dx if x and y are related by an equation:

1. Differentiate both sides of the equation with respect to x, thinking of y as a function of x and using the chain rule.
2. Solve the resulting equation for dy/dx.

Exercises for Section 2.3

Differentiate the functions in Exercises 1–24.

1. $10x^{1/8}$
2. $x^{3/5}$
3. $8x^{1/4} - x^{-2/3}$
4. $8x^4 - 3x^{-5/4}$
5. $3x^{2/3} - (5x)^{1/2}$
6. $x^2 - 3x^{1/2}$
7. $x^2(x^{1/3} + x^{4/3})$
8. $(x + 2)^{3/2}\sqrt{x + 2}$
9. $(x^5 + 1)^{7/9}$
10. $(x^{1/3} + x^{2/3})^{1/3}$
11. $\dfrac{1}{\sqrt{x}}$
12. $\dfrac{1}{\sqrt[4]{x^3 + 5x + 2}}$
13. $\dfrac{\sqrt{x^2 + 1}}{\sqrt{x^2 - 1}}$
14. $\dfrac{x^{1/7}}{(3x^{5/3} + x)}$
15. $\sqrt{(x + 3)^2 - 4}$
16. $3\sqrt{x} - \left(\dfrac{1}{x}\right) + 5$
17. $\dfrac{\sqrt{x}}{3 + x + x^3}$
18. $\dfrac{x}{\sqrt{x^2 + 2}}$
19. $\dfrac{\sqrt{x}}{1 + \sqrt{x}}$
20. $\dfrac{\sqrt{x}}{\sqrt[3]{x^4 + 2}}$
21. $\dfrac{\sqrt[3]{x}}{x^2 + 2}$
22. $\dfrac{(3x^2 - 6x)}{3\sqrt[4]{x}}$
23. $\sqrt{\dfrac{6x^2 + 2x + 1}{\sqrt{x} + 2x^3}}$
24. $\sqrt{\dfrac{\sqrt[3]{x}}{\sqrt{3x^2 + 1} + x}}$

Find the indicated derivatives in Exercises 25–28.

25. $\left(\dfrac{d}{dx}\right)(x^2 + 5)^{7/8}$

26. $\left(\dfrac{d^2}{dx^2}\right)(x^r)$, where r is rational

27. $f'(7)$, where $f(x) = 7\sqrt[3]{x}$

28. $\left(\dfrac{d}{dx}\right)\left(x^{1/4} + \dfrac{4}{\sqrt{x}}\right)\Big|_{x = 81}$

Find the derivatives of each of the functions in Exercises 29–34.

29. $f(x) = x^{3/11} - x^{1/5}$
30. $k(s) = \dfrac{1}{s^{3/5} - s}$
31. $h(y) = \dfrac{y^{1/8}}{y - 2}$
32. $g(t) = t(t^{2/3} + t^7)$

33. $l(x) = \left[\dfrac{x^{1/2} + 1}{x^{1/2} - 1}\right]^{1/2}$

34. $m(u) = (u^9 - 1)^{-6/7}$

35. If $x^2 + y^2 = 3$, compute dy/dx when $x = 0$ and $y = \sqrt{3}$.

36. If $x^3 + y^3 = xy$, compute dx/dy in terms of x and y.

37. Suppose that $x^4 + y^2 + y - 3 = 0$.
    (a) Compute dy/dx by implicit differentiation.
    (b) What is dy/dx when $x = 1, y = 1$?
    (c) Solve for y in terms of x (by the quadratic formula) and compute dy/dx directly. Compare with your answer in part (a).

38. Suppose that $xy + \sqrt{x^2 - y} = 7$.
    (a) Find dy/dx.
    (b) Find dx/dy.
    (c) What is the relation between dy/dx and dx/dy?

39. Suppose that $x^2/(x + y^2) = y^2/2$.
    (a) Find dy/dx when $x = 2, y = \sqrt{2}$.
    (b) Find dy/dx when $x = 2, y = -\sqrt{2}$.

40. Let $(u^2 + 6)(v^2 + 1) = 10uv$. Find du/dv and dv/du when $u = 2$ and $v = 1$.

41. Find the equation of the tangent line to the curve $x^4 + y^4 = 2$ when $x = y = 1$.

42. Find the equation of the tangent line to the curve $2x^2 + 2xy + y^2 = 8$ when $x = 2$ and $y = 0$.

43. Find $(d^2/dx^2)(x^{1/2} - x^{2/3})$.

44. Find $(d^2/dx^2)(x/\sqrt{1 + x^2})$.

45. Find the equation of the tangent line to the graph of $y = \sqrt{1 - x^2}$ at the point $(\sqrt{3}/2, 1/2)$.

46. Find the equation of the line tangent to $y = (x^{1/2} + x^{1/3})^{1/3}$ at $x = 1$.

47. Let $x^4 + y^4 = 1$. Find dy/dx as a function of x in two ways: by implicit differentiation and by solving for y in terms of x.

48. Differentiate the function $(x^3 + 2)/\sqrt{x^3 + 1}$ in at least two different ways. Be sure that the answers you get are equivalent. (Doing the same problem in several ways is a good method for checking your calculations, useful on examinations as well as in scientific work.)

Find linear approximations for the expressions in Exercises 49–52.

49. $\sqrt[4]{15.97}$

50. $(4.02)^{3/2}$

51. $(-26.98)^{2/3}$

52. $\sqrt[2]{122}$

53. The mass M of the first x meters of a concrete beam is $M = 24(2 + x^{1/3})^4$ kilograms. Find the density dM/dx.

54. The average pulse rate for persons 30 inches to 74 inches tall can be approximated by $y = 589/\sqrt{x}$ beats per minute for a person x inches tall.
    (a) Find dy/dx, and show that it is always negative.
    (b) The value of $|dy/dx|$ at $x = 65$ is the decrease in beats per minute expected for a 1-inch increase in height. Explain.
    (c) Do children have higher pulse rates than adults according to this model?

55. Let $y = 24\sqrt{x}$ be the learning curve for learning y items in x hours, $0 \leqslant x \leqslant 5$. Apply the linear approximation to estimate the number of new items learned in the 12-minute period given by $1 \leqslant x \leqslant 1.2$.

56. The daily demand function for a certain manufactured good is $p(x) = 75 + 4\sqrt{2x} - (x^2/2)$, where x is the production level.
    (a) Find dp/dx. Interpret.
    (b) Find the production level x for which $dp/dx = 0$. Interpret.

57. The fundamental period for vibration P and the tension T in a certain string are related by $P = \sqrt{32/T}$ seconds. Find the rate of change of period with respect to tension when $T = 9$ lbs.

58. Lions in a small district in an African game preserve defend an exclusive region of area A which depends on their body weight W by the formula $A = W^{1.31}$.
    (a) Find dA/dW.
    (b) By what percentage should the defended area increase after a 200-lb lion undergoes a 20-lb weight gain? (Use the linear approximation.)

59. The object distance x and image distance y satisfy the thin lens equation $1/x + 1/y = 1/f$, where f is the focal length.
    (a) Solve for y as a function of x when $f = 50$ millimeters.
    (b) Find dy/dx.
    (c) Find all (x, y) such that $(d/dx)(x + y) = 0$.

★60. Using implicit differentiation, find the equation of the tangent line at the point (x_0, y_0) on the circle $(x - a)^2 + (y - b)^2 = r^2$. Interpret your result geometrically. (a, b, and r are constants.)

2.4 Related Rates and Parametric Curves

If two quantities satisfy an equation, their rates of change can be related by implicit differentiation.

Suppose that we have two quantities, x and y, each of which is a function of time t. We know that the rates of change of x and y are given by dx/dt and dy/dt. If x and y satisfy an equation, such as $x^2 + y^2 = 1$ or $x^2 + y^6 + 2y = 5$, then the *rates* dx/dt and dy/dt can be *related* by differentiating the equation with respect to t and using the chain rule.

Example 1 Suppose that x and y are functions of t and that $x^4 + xy + y^4 = 1$. Relate dx/dt and dy/dt.

Solution Differentiate the relation between x and y *with respect to t*, thinking of x and y as functions of t:

$$\frac{d}{dt}(x^4 + xy + y^4) = 0,$$

$$4x^3 \frac{dx}{dt} + \frac{dx}{dt} y + x \frac{dy}{dt} + 4y^3 \frac{dy}{dt} = 0.$$

We can simplify this to

$$\frac{dy}{dt} = -\left(\frac{4x^3 + y}{x + 4y^3} \right) \frac{dx}{dt}$$

which is the desired relation. ▲

If you have trouble remembering to use the chain rule, you can use a device like that following Example 5 in Section 2.3. Namely, write $f(t)$ for x and $g(t)$ for y, then differentiate the relation (such as $[f(t)]^4 + f(t)g(t) + [g(t)]^4 = 1$) with respect to t. This will give a relation between $f'(t)$, $g'(t)$, $f(t)$, and $g(t)$. Once you have done a few examples in this long-winded way, you should be ready to go back to the d/dt's.

Related Rates

To relate the rates dx/dt and dy/dt if x and y satisfy a given equation:

1. Differentiate both sides of the equation with respect to t, thinking of x and y as functions of t.
2. Solve the resulting equation for dy/dt in terms of dx/dt (or vice versa if called for).

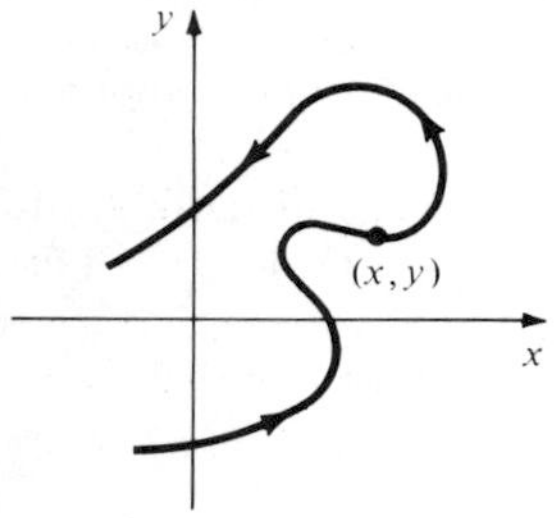

Figure 2.4.1. If x and y are functions of t, the point (x, y) follows a curve as t varies.

There is a useful geometric interpretation of related rates. (This topic is treated in more detail in Section 10.4.) If x and y are each functions of t, say $x = f(t)$ and $y = g(t)$, we can plot the points (x, y) for various values of t. As t varies, the point (x, y) will move along a curve. When a curve is described this way, it is called a *parametric curve* (see Fig. 2.4.1).

Example 2   If $x = t^4$ and $y = t^2$, what curve does (x, y) follow for $-\infty < t < \infty$?

Solution   We notice that $y^2 = x$, so the point (x, y) lies on a parabola. As t ranges from $-\infty$ to ∞, y goes from $+\infty$ to zero and back to $+\infty$, so (x, y) stays on the half of the parabola with $y \geqslant 0$ and traverses it twice (see Fig. 2.4.2). ▲

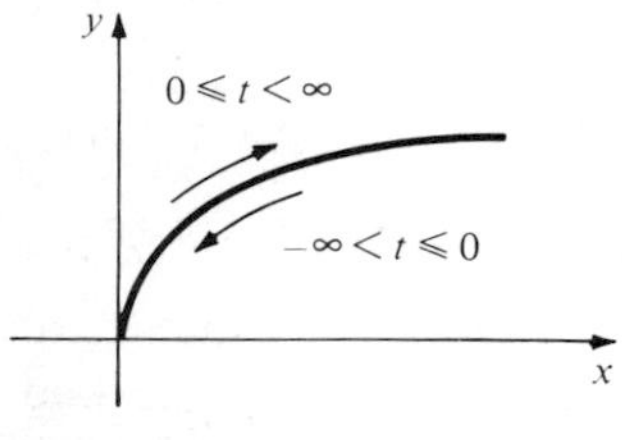

Figure 2.4.2. As t ranges from $-\infty$ to $+\infty$, the point (t^4, t^2) traverses the parabola twice in the directions shown.

It may be possible to describe a parametric curve in other ways. For instance, it may be described by a relation between x and y. Specifically, suppose that the parametric curve $x = f(t)$, $y = g(t)$ can be described by an equation $y = h(x)$ (the case $x = k(y)$ will be similar). Then we can differentiate by the chain rule. Using Leibniz notation:

$$\frac{dy}{dt} = \frac{dy}{dx} \frac{dx}{dt} \quad \text{so} \quad \frac{dy}{dx} = \frac{dy/dt}{dx/dt}.$$

This shows that the slope of the tangent line to a parametric curve is given by $(dy/dt)/(dx/dt)$.

Parametric Curves

As t varies, two equations $x = f(t)$ and $y = g(t)$ describe a curve in the plane called a parametric curve. The slope of its tangent line is given by

$$\frac{dy}{dx} = \frac{dy/dt}{dx/dt} \quad \text{if} \quad \frac{dx}{dt} \neq 0.$$

Example 3   Find the equation of the line tangent to the parametric curve given by the equations $x = (1 + t^3)^4 + t^2$, $y = t^5 + t^2 + 2$ at $t = 1$.

Solution   Here the relation between x and y is not clear, but we do not need to know it. (We tacitly assume that the path followed by (x, y) can be described by a function $y = h(x)$.) We have

$$\frac{dx}{dt} = 4(1 + t^3)^3 \cdot 3t^2 + 2t = 12(1 + t^3)^3 t^2 + 2t \quad \text{and} \quad \frac{dy}{dt} = 5t^4 + 2t,$$

so the slope of the tangent line is

$$\frac{dy}{dx} = \frac{dy/dt}{dx/dt} = \frac{5t^4 + 2t}{12t^2(1 + t^3)^3 + 2t} = \frac{5t^3 + 2}{12t(1 + t^3)^3 + 2}.$$

At $t = 1$, we get

$$\frac{dy}{dx} = \frac{7}{12 \cdot 2^3 + 2} = \frac{7}{98} = \frac{1}{14}.$$

Since $x = 17$ and $y = 4$ at $t = 1$, the equation of the tangent line is given by the point-slope formula:

$$y - 4 = \tfrac{1}{14}(x - 17),$$

$$y = \frac{x}{14} + \frac{39}{14}. \ \blacktriangle$$

Example 4   Show that the parametric equations $x = at + b$ and $y = ct + d$ describe a straight line if a and c are not both zero. What is its slope?

Solution   Multiplying $x = at + b$ by c, multiplying $y = ct + d$ by a, and subtracting, we get

$$cx - ay = bc - ad,$$

so $y = (c/a)x + (1/a)(ad - bc)$, which is the equation of a line with slope c/a. (If $a = 0$, x is constant and the line is vertical; if c were also zero the line would degenerate to a point.) Note that the slope can also be obtained as $(dy/dt)/(dx/dt)$, since $dy/dt = c$ and $dx/dt = a$. $\blacktriangle$

Example 5   Suppose that x and y are functions of time and that (x, y) moves on the circle $x^2 + y^2 = 1$. If x is increasing at 1 centimeter per second, what is the rate of change of y when $x = 1/\sqrt{2}$ and $y = 1/\sqrt{2}$?

Solution   Differentiating $x^2 + y^2 = 1$ gives $2x(dx/dt) + 2y(dy/dt) = 0$; so $dy/dt = (-x/y)(dx/dt)$. If $x = y = 1/\sqrt{2}$, $dy/dt = -dx/dt = -1$ centimeter per second. $\blacktriangle$

In word problems involving related rates, the hardest job may be to translate the verbal problem into mathematical terms. You need to identify the variables which are changing with time and to find relations between them. If some geometry is involved, drawing a figure is essential and will often help you to spot the important relations. Similar triangles and Pythagoras' theorem are frequently useful in these problems.

Example 6   A light L is being raised up a pole (see Fig. 2.4.3). The light shines on the object Q, casting a shadow on the ground. At a certain moment the light is 40 meters off the ground, rising at 5 meters per minute. How fast is the shadow shrinking at that instant?

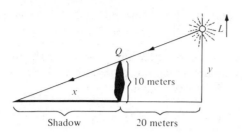

Figure 2.4.3. At what rate is the shadow shrinking?

Solution Let the height of the light be y at time t and the length of the shadow be x. By similar triangles, $x/10 = (x + 20)/y$; i.e., $xy = 10(x + 20)$. Differentiating with respect to t, $x(dy/dt) + (dx/dt)y = 10(dx/dt)$. At the moment in question $y = 40$, and so $x \cdot 40 = 10(x + 20)$ or $x = \frac{20}{3}$. Also, $dy/dt = 5$ and so $\frac{20}{3} \cdot 5 + 40(dx/dt) = 10(dx/dt)$. Solving for (dx/dt), we get $dx/dt = -\frac{10}{9}$. Thus the shadow is shrinking at $\frac{10}{9}$ meters per minute. ▲

Example 7 A spherical balloon is being blown up by a child. At a certain instant during inflation, air enters the balloon to make the volume increase at a rate of 50 cubic centimeters a second. At the same instant the balloon has a radius of 10 centimeters. How fast is the radius changing with time?

Solution Let the radius of the balloon be denoted by r and the volume by V. Thus $V = \frac{4}{3}\pi r^3$ and so

$$\frac{dV}{dt} = 4\pi r^2 \frac{dr}{dt}.$$

At the instant in question, $dV/dt = 50$ and $r = 10$. Thus

$$50 = 4\pi(10)^2 \frac{dr}{dt},$$

and so

$$\frac{dr}{dt} = \frac{50}{400\pi} = \frac{1}{8\pi} \approx 0.04 \text{ centimeters per second.} \ ▲$$

Example 8 A thunderstorm is dropping rain at the rate of 2 inches per hour into a conical tank of diameter 15 feet and height 30 feet. At what rate is the water level rising when the water is 20 feet deep?

Solution Figure 2.4.4 shows top and side views of the partially filled tank, both of which will be useful for our solution.

We denote by h the height of the water in the tank, so that dh/dt is the rate to be found. To proceed, we need to use the fact that the rate of rainfall is 2 inches per hour. What this means is that the water level in a *cylindrical* tank would rise uniformly at the rate of 2 inches per hour, so that the *volume* of the water pouring every hour into a circle of diameter 15 feet is $\pi \cdot (7.5)^2 \cdot \frac{1}{6}$ cubic feet $= \frac{75}{8}\pi$ cubic feet. It is useful, then, to introduce the variable V representing the volume of water in the tank; we have $dV/dt = 75\pi/8$ cubic feet per hour.

Now V and h are related by the formula for the volume of a cone: $V = \frac{1}{3}\pi r^2 h$, where r is the radius of the "base" of the cone, in this case, the radius of the water surface. From Fig. 2.4.4, we see, using similar triangles, that $r/h = 7.5/30 = 1/4$, so $r = \frac{1}{4}h$, and hence $V = \frac{1}{48}\pi h^3$. Differentiating and using the chain rule gives $dV/dt = \frac{1}{16}\pi h^2 dh/dt$. Inserting the specific data $h = 20$ and $dV/dt = \frac{75}{8}\pi$ gives the equation $\frac{75}{8}\pi = \frac{1}{16}\pi \cdot 400 \, dh/dt = 25\pi \, dh/dt$, which we may solve for dh/dt to get $dh/dt = \frac{3}{8}$. This is in feet per hour, so the water level is rising at the rate of $4\frac{1}{2}$ inches per hour. ▲

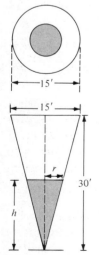

Figure 2.4.4. A conical tank partially filled with water.

Exercises for Section 2.4

In Exercises 1–8, assume that x and y are functions of t. Relate dx/dt and dy/dt using the given relation.

1. $x^2 - y^2 = 3$
2. $x + y = 4$
3. $x - y^2 - y^3 = 4$
4. $8x^2 + 9y^2 = 5$
5. $x + y^2 = y$
6. $\sqrt{x + 2y} = x$
7. $\sqrt{x} - \sqrt{y} = 5$
8. $(x^2 + y^2 + y^3)^{3/2} = 5$

9. Sketch the curve defined by the parametric equations $x = t^2$, $y = 1 - t$, $-\infty < t < \infty$.

10. Sketch the curve described by $x = 3t + 2$, $y = 4t - 8$, $-\infty < t < \infty$.

11. What curve do the parametric equations $x = t^2$ and $y = t^6$ describe?

12. If $x = (1 + t)^2$ and $y = (1 + t)^4$, what curve does (x, y) follow for $-\infty < t < \infty$?

13. Find the equation of the tangent line to the parametric curve $x = t^2$, $y = t^3$ at $t = 5$.

14. Find the equation of the line tangent to the parametric curve $x = t^2 + 1$, $y = 1/(t^4 + 1)$ at $t = 2$.

15. Find the equation of the tangent line to the parametric curve

$$x = \sqrt{t^4 + 6t^2 + 8t}, \qquad y = \frac{t^2 + 1}{\sqrt{t} - 1}$$

at $t = 3$.

16. (a) Find the slope of the parametric curve $y = t^4 + 2t$, $x = 8t$ at $t = 1$.
    (b) What relationship between x and y is satisfied by the points on this curve?
    (c) Verify that $dy/dx = (dy/dt)/(dx/dt)$ for this curve.

17. Suppose that $xy = 4$. Express dy/dt in terms of dx/dt when $x = 8$ and $y = \frac{1}{2}$.

18. If $x^2 + y^2 = x^5/y$ and $dy/dt = 3$ when $x = y = \sqrt{2}$, what is dx/dt at that point?

19. Suppose that $x^2 + y^2 = t$ and that $x = 3$, $y = 4$, and $dx/dt = 7$ when $t = 25$. What is dy/dt at that moment?

20. Let x and y depend on t in such a way that $(x + y)^2 + t^2 = 2t$ and such that $x = 0$ and $y = 1$ when $t = 1$. If $dx/dt = 4$ at that moment, what is dy/dt?

21. The radius and height of a circular cylinder are changing with time in such a way that the volume remains constant at 1 liter (= 1000 cubic centimeters). If, at a certain time, the radius is 4 centimeters and is increasing at the rate of $\frac{1}{2}$ centimeter per second, what is the rate of change of the height?

22. A hurricane is dropping 10 inches of rain per hour into a swimming pool which measures 40 feet long by 20 feet wide.
    (a) What is the rate at which the volume of water in the pool is increasing?
    (b) If the pool is 4 feet deep at the shallow end

and 8 feet deep at the deep end, how fast is the water level rising after 2 hours? (Suppose the pool was empty to begin with.) How fast after 6 hours?

23. Water is being pumped from a 20-meter square pond into a round pond with radius 10 meters. At a certain moment, the water level in the square pond is dropping by 2 inches per minute. How fast is the water rising in the round pond?

24. A ladder 25 feet long is leaning against a vertical wall. The bottom is being shoved along the ground, towards the wall at $1\frac{1}{2}$ feet per second. How fast is the top rising when it is 15 feet off the ground?

25. A point in the plane moves in such a way that it is always twice as far from $(0,0)$ as it is from $(0, 1)$.
    (a) Show that the point moves on a circle.
    (b) At the moment when the point crosses the segment between $(0,0)$ and $(0, 1)$, what is dy/dt?
    (c) Where is the point when $dy/dt = dx/dt$? (You may assume that dx/dt and dy/dt are not simultaneously zero.)

26. Two quantities p and q depending on t are subject to the relation $1/p + 1/q = 1$.
    (a) Find a relation between dp/dt and dq/dt.
    (b) At a certain moment, $p = \frac{4}{3}$ and $dp/dt = 2$. What are q and dq/dt?

27. Suppose the quantities x, y, and z are related by the equation $x^2 + y^2 + z^2 = 14$. If $dx/dt = 2$ and $dy/dt = 3$ when $x = 2$, $y = 1$, and $z = 3$, what is dz/dt?

28. The pressure P, volume V, and temperature T of a gas are related by the law $PV/T = \text{constant}$. Find a relation between the time derivatives of P, V, and T.

29. The area of a rectangle is kept fixed at 25 square meters while the length of the sides varies. Find the rate of change of the length of one side with respect to the other when the rectangle is a square.

30. The surface area of a cube is growing at the rate of 4 square centimeters per second. How fast is the length of a side growing when the cube has sides 2 centimeters long?

★31. (a) Give a rule for determining when the tangent line to a parametric curve $x = f(t)$, $y = g(t)$ is horizontal and when it is vertical.
    (b) When is the tangent line to the curve $x = t^2$, $y = t^3 - t$ horizontal? When is it vertical?

★32. (a) At which points is the tangent line to a parametric curve parallel to the line $y = x$?
    (b) When is the tangent line to the curve in part (b) of Exercise 31 parallel to the line $y = x$?
    (c) Sketch the curve of Exercise 31.

★33. Read Example 8. Show that for any conical tank, the ratio of dh/dt to the rate of rainfall is equal to the ratio between the area of the tank's open- ing to the area of the water surface. (In fact, this result is true for a tank of any shape; see Review Exercise 32, Chapter 9.)

2.5 Antiderivatives

An antiderivative of f is a function whose derivative is f.

Many applications of calculus require one to find a function whose derivative is given. In this section, we show how to solve simple problems of this type.

Example 1 Find a function whose derivative is $2x + 3$.

Solution We recall that the derivative of x^2 is $2x$ and that the derivative of $3x$ is 3, so the unknown function could be $x^2 + 3x$. We may check our answer by differentiating: $(d/dx)(x^2 + 3x) = 2x + 3$. ▲

The function $x^2 + 3x$ is not the only possible solution to Example 1; so are $x^2 + 3x + 1$, $x^2 + 3x + 2$, etc. In fact, since the derivative of a constant function is zero, $x^2 + 3x + C$ solves the problem for any number C.

A function F for which $F' = f$ is called an *antiderivative* of f. Unlike the derivative, the antiderivative of a function is never unique. Indeed, if F is an antiderivative of f, so is $F + C$ for an arbitrary constant C. In Section 3.6 we will show that *all* the antiderivatives are of this form. For now, we take this fact for granted. We can make the solution of an antidifferentiation problem unique by imposing an extra condition on the unknown function (see Fig. 2.5.1). The following example is a typical application of antidifferentiation.

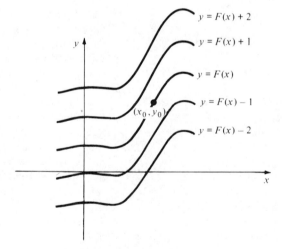

Figure 2.5.1. All these functions have the same derivative. Specifying $F(x_0) = y_0$ picks out one of them.

Example 2 The velocity of a particle moving along a line is $3t + 5$ at time t. At time 1, the particle is at position 4. Where is it at time 10?

Solution Let $F(t)$ denote the position of the particle at time t. We will determine the function F. Since velocity is the rate of change of position with respect to time, we must have $F'(t) = 3t + 5$; that is, F is an antiderivative of $f(t) = 3t + 5$. A function whose derivative is $3t$ is $\frac{3}{2}t^2$, since $(d/dt)\frac{3}{2}t^2 = \frac{3}{2}2t = 3t$. Similarly, a function whose derivative is 5 is $5t$. Therefore, we take

$$F(t) = \tfrac{3}{2}t^2 + 5t + C,$$

where C is a constant to be determined. To find the value of C, we use the information that the particle is at position 4 at time 1; that is, $F(1) = 4$. Substituting 1 for t and 4 for $F(t)$ in the equation $F(t) = \frac{3}{2}t^2 + 5t + C$ gives

$$4 = \frac{3}{2} + 5 + C = \frac{13}{2} + C,$$

or $C = -\frac{5}{2}$, and so $F(t) = \frac{3}{2}t^2 + 5t - \frac{5}{2}$. Finally, we substitute 10 for t, obtaining the position at time 10: $F(10) = \frac{3}{2} \cdot 100 + 5 \cdot 10 - \frac{5}{2} = 197\frac{1}{2}$. ▲

At this point in our study of calculus, we must solve antidifferentiation problems by guessing the answer and then checking and refining our guesses if necessary. More systematic methods will be given shortly.

Example 3   Find the general antiderivative for the function $f(x) = x^4 + 6$.

Solution   We may begin by looking for an antiderivative for x^4. If we guess x^5, the derivative is $5x^4$, which is five times too big, so we make a new guess, $\frac{1}{5}x^5$, which works. An antiderivative for 6 is $6x$. Adding our two results gives $\frac{1}{5}x^5 + 6x$; differentiating $\frac{1}{5}x^5 + 6x$ gives $x^4 + 6$, so $\frac{1}{5}x^5 + 6x$ is an antiderivative for $x^4 + 6$. We may add an arbitrary constant to get the general antiderivative $\frac{1}{5}x^5 + 6x + C$. ▲

Example 4   The acceleration of a falling body near the earth's surface is 9.8 meters per second per second. If the body has a downward velocity v_0 at $t = 0$, what is its velocity at time t? If the position is x_0 at time 0, what is the position at time t? (See Fig. 2.5.2.)

Solution   We measure the position x in the downward direction. Let v be the velocity. Then $dv/dt = 9.8$; since an antiderivative of 9.8 is $9.8t$, we have $v = 9.8t + C$. At $t = 0$, $v = v_0$, so $v_0 = (9.8)0 + C = C$, and so $v = 9.8t + v_0$. Now $dx/dt = v = 9.8t + v_0$. Since an antiderivative of $9.8t$ is $(9.8/2)t^2 = 4.9t^2$ and an antiderivative of v_0 is v_0t, we have $x = 4.9t^2 + v_0t + D$. At $t = 0$, $x = x_0$, so $x_0 = 4.9(0)^2 + v_0 \cdot 0 + D = D$, and so $x = 4.9t^2 + v_0t + x_0$. ▲

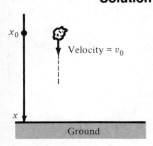

Figure 2.5.2. The body is moving downward at $t = 0$ with velocity v_0.

The most commonly used notation for the antiderivative is due to Leibniz. The symbol

$$\int f(x)\,dx$$

denotes the class of all antiderivatives of f; thus, if F is a particular antiderivative, we may write

$$\int f(x)\,dx = F(x) + C,$$

where C is an arbitrary constant. For instance, the result of Example 3 may be written

$$\int (x^4 + 6)\,dx = \frac{1}{5}x^5 + 6x + C.$$

The elongated S, called an *integral sign*, was introduced by Leibniz because antidifferentiation (also called *integration*) turns out to be a form of continuous ∫ummation. In Chapter 4, we will study this aspect of integration in detail. There and also in the supplement to this section, we explain the presence of the "dx" in the notation. For now, we simply think of dx as indicating that the independent variable is x. Its presence should also serve as a reminder that integrating is inverse to differentiating, where the dx occurs in the denominator of dy/dx.

The function $f(x)$ in $\int f(x)\,dx$ is called the *integrand*, and $\int f(x)\,dx$ is called the *indefinite integral* of $f(x)$. One traditionally refers to $f(x)\,dx$ as being "under" the integral sign, even though this is typographically inaccurate.

Antidifferentiation and Indefinite Integrals

An *antiderivative* for f is a function F such that $F'(x) = f(x)$. We write

$$F(x) = \int f(x)\, dx.$$

The function $\int f(x)\, dx$ is also called the *indefinite integral* of f, and f is called the *integrand*.

If $F(x)$ is an antiderivative of $f(x)$, the general antiderivative has the form $F(x) + C$ for an arbitrary constant C.

Some of the differentiation rules lead directly to systematic rules for antidifferentiation. The rules in the following box can be proved by differentiation of their right-hand sides (see Example 5 below).

Antidifferentiation Rules

Sum:
$$\int \left[f(x) + g(x) \right] dx = \int f(x)\, dx + \int g(x)\, dx$$

Constant multiple: $\int af(x)\, dx = a \int f(x)\, dx,$ where a is constant

Power:
$$\int x^n\, dx = \frac{x^{n+1}}{n+1} + C \quad (n \neq -1)$$

Polynomial:
$$\int (c_n x^n + \cdots + c_1 x + c_0)\, dx = \frac{c_n}{n+1} x^{n+1} + \cdots + \frac{c_1}{2} x^2 + c_0 x + C$$

Example 5   Prove the power rule $\int x^n\, dx = \dfrac{x^{n+1}}{n+1} + C$ $(n \neq -1)$

Solution   By definition, $F(x) = \int x^n\, dx$ is a function such that $F'(x) = f(x) = x^n$. However, $F(x) = [x^{n+1}/(n+1)] + C$ is such a function since its derivative is $F'(x) = (n+1) \cdot x^{n+1-1}/(n+1) = x^n$, by the power rule for derivatives. ▲

The exclusion $n \neq -1$ in the power rule arises because the formula $x^{n+1}/(n+1)$ makes no sense for $n = -1$; the denominator is zero. (It turns out that $1/x = x^{-1}$ does have an antiderivative, but it is a logarithm function rather than a power of x. We will study logarithms in Chapter 6.)

Example 6   Find $\displaystyle\int \left[\frac{1}{x^2} + 3x + 2 - \frac{8}{\sqrt{x}} \right] dx.$

Solution $\displaystyle\int \left(\frac{1}{x^2} + 3x + 2 - \frac{8}{\sqrt{x}} \right) dx$

$$= \int \frac{1}{x^2}\, dx + 3 \int x\, dx + 2 \int 1\, dx - 8 \int \frac{1}{\sqrt{x}}\, dx$$

$$= \int x^{-2}\, dx + 3 \int x^1\, dx + 2 \int x^0\, dx - 8 \int x^{-1/2}\, dx$$

$$= -1x^{-1} + \frac{3}{2}x^2 + 2x^1 - \frac{8}{1/2}x^{1/2} + C$$

$$= -\frac{1}{x} + \frac{3}{2}x^2 + 2x - 16\sqrt{x} + C.$$

We write C only once because the sum of four constants is a constant. ▲

Example 7   Find $\int \dfrac{dx}{(3x+1)^5}$.

Solution   We are looking for an antiderivative of $1/(3x+1)^5$. The power of a function rule suggests that we guess $1/(3x+1)^4$. Differentiating, we have

$$\frac{d}{dx}\frac{1}{(3x+1)^4} = \frac{-4}{(3x+1)^5}\frac{d}{dx}(3x+1) = \frac{-12}{(3x+1)^5}$$

Comparing with $1/(3x+1)^5$, we see that we are off by a factor of -12, so

$$\int \frac{dx}{(3x+1)^5} = \frac{-1}{12(3x+1)^4} + C. \; \blacktriangle$$

Using the same method as in Example 7, we find that

$$\int (ax+b)^n\, dx = \frac{1}{a(n+1)}(ax+b)^{n+1} + C,$$

where a and b are constants, $a \neq 0$, and n is a rational number, $n \neq -1$.

Example 8   Find $\int \sqrt{3x+2}\; dx$.

Solution   By the formula for $\int (ax+b)^n\, dx$ with $a = 3$, $b = 2$, and $n = \frac{1}{2}$, we get

$$\int \sqrt{3x+2}\; dx = \int (3x+2)^{1/2}\, dx$$

$$= \frac{1}{3(3/2)}(3x+2)^{3/2} + C = \frac{2}{9}(3x+2)^{3/2} + C. \; \blacktriangle$$

Example 9   Find $\int \dfrac{x^3-8}{x-2}\, dx$.

Solution   Here we simplify first. Dividing $x^3 - 8$ by $x - 2$ gives $(x^3 - 8)/(x - 2) = x^2 + 2x + 4$. Thus

$$\int \frac{x^3-8}{x-2}\, dx = \frac{x^3}{3} + x^2 + 4x + C. \; \blacktriangle$$

Example 10   Let x = position, v = velocity, a = acceleration, t = time. Express the relations between these variables by using the indefinite integral notation.

Solution   By the definitions of velocity and acceleration, we have $v = dx/dt$ and $a = dv/dt$. It follows that

$$v = \int a\, dt \quad \text{and} \quad x = \int v\, dt. \; \blacktriangle$$

Example 11   Water is flowing into a tub at $3t + 1/(t+1)^2$ gallons per minute after t minutes. How much water is in the tub after 2 minutes if it started out empty?

Solution Let $f(t)$ be the amount of water (in gallons) in the tub at time t. We are given

$$f'(t) = 3t + \frac{1}{(t+1)^2}.$$

This equation means that f is the antiderivative of $3t + 1/(t+1)^2$; thus,

$$f(t) = \int \left(3t + \frac{1}{(t+1)^2}\right) dt$$

$$= \int 3t\, dt + \int \frac{1}{(t+1)^2}\, dt = \frac{3t^2}{2} - \frac{1}{t+1} + C.$$

Since the tub started out empty, $f(0) = 0$; so $0 = -1 + C$, and thus $C = 1$. Therefore $f(t) = 3t^2/2 - 1/(t+1) + 1$. Setting $t = 2$ gives $f(2) = 3 \cdot 4/2 - \frac{1}{3} + 1 = 6\frac{2}{3}$ gallons after 2 minutes. ▲

Supplement to Section 2.5
The Notation $\int f(x)\, dx$

The Leibniz notation $\int f(x)\, dx$ for the antiderivative of a function $f(x)$ may seem strange at this point, but it is really rather natural and remarkably functional. To motivate it, let us study the velocity–distance relationship again. As in Section 1.1, we imagine a bus moving on a straight road with position $y = F(x)$ in meters from a designated starting point at time x in seconds (see Fig. 1.1.1). There we showed that $v = dy/dx$ is the velocity of the bus. As in Section 1.3, we may motivate this notation by writing the velocity as the limit

$$\frac{\Delta y}{\Delta x} = \frac{\text{distance travelled}}{\text{elapsed time}} \quad \text{as} \quad \Delta x \to 0.$$

Conversely, to reconstruct y from a given velocity function $v = f(x)$, we notice that in a short time interval from x to $x + \Delta x$, the bus has gone approximately $\Delta y \approx f(x) \Delta x$ meters (distance travelled = velocity × time elapsed). The total distance travelled is thus the $\int$um of $f(x) \Delta x$ over all the little Δx's making up the total time of the trip. This abbreviates to $\int f(x)\, dx$.

On the other hand, the distance travelled is $y = F(x)$, assuming $F(0) = 0$, and we know that $dy/dx = v = f(x)$, i.e., F is an antiderivative of f. Thus $\int f(x)\, dx$ is a reasonable notation for this antiderivative. The arbitrariness in the starting position $F(0)$ corresponds to the arbitrary constant that can be added to the antiderivative.

Exercises for Section 2.5

Find antiderivatives for each of the functions in Exercises 1–8.

1. $x + 2$
2. $x^6 + 9$
3. $s(s+1)(s+2)$
4. $4x^8 + 3x^2$
5. $\dfrac{1}{t^3}$
6. $x^5 + \dfrac{2}{x^4}$
7. $x^{3/2} - \sqrt{x}$
8. $x^4 - \dfrac{1}{\sqrt{x}} + x^{3/2}$

In Exercises 9–12, v is the velocity of a particle on the line, and $F(t)$ is the position at time t.

9. $v = 8t + 2$; $F(0) = 0$; find $F(1)$.
10. $v = -2t + 3$; $F(1) = 2$; find $F(3)$.
11. $v = t^2 + \sqrt{t}$; $F(1) = 1$; find $F(\frac{1}{2})$.
12. $v = t^{3/2} - t^2$; $F(2) = 1$; find $F(1)$.

Find the general antiderivatives for the functions f given in Exercises 13–20.

13. $f(x) = 3x$
14. $f(x) = 3x^4 + 4x^3$
15. $f(x) = \dfrac{x+1}{x^3}$
16. $f(t) = (t+1)^2$
17. $f(x) = \sqrt{x+1}$
18. $f(x) = (\sqrt{x} + 1)^2$
19. $f(t) = (t+1)^{3/2}$
20. $f(s) = (s+8)^{5/8}$

In Exercises 21–24, the velocity v_0 of a falling body (in meters per second) near the earth's surface is given at time $t = 0$. Find the velocity at time t and the position at time t with the given initial positions x_0. (The x axis is oriented downwards as in Fig. 2.5.2.)

21. $v_0 = 1$; $x_0 = 2$ 22. $v_0 = 3$; $x_0 = -1$
23. $v_0 = -2$; $x_0 = 0$ 24. $v_0 = -2$; $x_0 = -6$

25. Is it true that $\int f(x)g(x)\,dx$ is equal to $[\int f(x)\,dx]\,g(x) + f(x)[\int g(x)\,dx]$?

26. Prove the constant multiple rule for antidifferentiation.

27. Prove the sum rule for antiderivatives.

28. Prove that $\int f(x)f'(x)\,dx = \frac{1}{2}[f(x)]^2 + C$ for any function f.

Find the indefinite integrals in Exercises 29–40.

29. $\int (x^2 + 3x + 2)\,dx$ 30. $\int 4\pi r^2\,dr$

31. $\int (3t^2 + 2t + 1)\,dt$ 32. $\int (u^4 - 6u)\,du$

33. $\int (8t + 1)^{-2}\,dt$ 34. $\int \left(\dfrac{t^3 + t + 1}{t^5} \right) dt$

35. $\int \dfrac{4}{(3b + 2)^9}\,db$ 36. $\int \sqrt{2u + 5}\,du$

37. $\int \left(\dfrac{1}{x^4} + x^4 \right) dx$ 38. $\int \left(\dfrac{t^2 - t + 2}{t^6} \right) dt$

39. $\int \dfrac{x^2 + 3}{\sqrt{x}}\,dx$ 40. $\int \dfrac{t^3 - 8t + 1}{t^{3/2}}\,dt$

Find the indicated antiderivatives in Exercises 41–52.

41. $\int (x^3 + 3x)\,dx$ 42. $\int (t^3 + t^{-2})\,dt$

43. $\int \dfrac{1}{(t + 1)^2}\,dt$ 44. $\int \dfrac{w^2 + 2}{w^5}\,dw$

45. $\int \sqrt{8x + 3}\,dx$ 46. $\int \sqrt{10x - 3}\,dx$

47. $\int 10(8 - 3x)^{3/2}\,dx$ 48. $\int 3(x - 1)^{5/2}\,dx$

49. $\int \dfrac{\sqrt{x - 1} + 3}{(x - 1)^{1/2}}\,dx$ 50. $\int \dfrac{(3x - 2)^{3/2} - 8}{\sqrt{3x - 2}}\,dx$

51. $\int \dfrac{x^3 - 27}{x - 3}\,dx$ 52. $\int \dfrac{x^4 - 1}{x - 1}\,dx$

53. A ball is thrown downward with a velocity of 10 meters per second. How long does it take the ball to fall 150 meters?

54. A particle moves along a line with velocity $v(t) = \frac{1}{2}t^2 + t$. If it is at $x = 0$ when $t = 0$, find its position as a function of t.

55. The population of Booneville increases at a rate of $r(t) = (3.62)(1 + 0.8t^2)$ people per year, where t is the time in years from 1970. The population in 1976 was 726. What was it in 1984?

56. A car accelerates from rest to 55 miles per hour in 12 seconds. Assuming that the accerleration is constant, how far does the car travel during those 12 seconds?

57. A rock is thrown vertically upward with velocity 19.6 meters per second. After how long does it return to the thrower? (The acceleration due to gravity is 9.8 meters per second per second; see Example 4 and Fig. 2.5.3.)

58. Suppose that the marginal cost of producing grumbies at production level p is $100/(p + 20)^2$. If the cost of production is 100 when $p = 0$ (setup costs), what is the cost when $p = 80$?

59. (a) Find $\dfrac{d}{dx}\left(\dfrac{x^3 + 1}{x^3 - 1} \right)$.

  (b) Find $\int \dfrac{-6x^2}{(x^3 - 1)^2}\,dx$.

60. (a) Find $(d/dx)\sqrt{x^2 + 2x}$.

  (b) Find $\int \dfrac{z + 1}{\sqrt{z^2 + 2z}}\,dz$.

61. (a) Differentiate $(x^4 + 1)^{20}$.

  (b) Find $\int [(x^4 + 1)^{19}x^3 + 3x^{2/3}]\,dx$.

62. (a) Differentiate $\dfrac{1}{3 + x^{9/2}}$.

  (b) Find $\int \dfrac{x^{7/2}}{(3 + x^{9/2})^2}\,dx$.

63. Find a function $F(x)$ such that $x^5 F'(x) + x^3 + 2x = 3$.

64. Find a function $f(x)$ whose graph passes through $(1, 1)$ and such that the slope of its tangent line at $(x, f(x))$ is $3x + 1$.

65. Find the antiderivative $F(x)$ of the function $f(x) = x^3 + 3x^2 + 2$ which satisfies $F(0) = 1$.

66. Find the antiderivative $G(y)$ of $g(y) = (y + 4)^2$ which satisfies $G(1) = 0$.

★67. (a) What integration formula can you derive from the general power of a function rule? (See Exercises 60 and 61.) (b) Find $\int (x^3 + 4)^{-3}x^2\,dx$.

★68. (a) What integration formula can you derive from the chain rule?

  (b) Find $\int [\sqrt{x^2 + 20x} + (x^2 + 20x)](2x + 20)\,dx$.

Figure 2.5.3. The path of a rock thrown upwards from the earth.

Ground

Review Exercises for Chapter 2

Differentiate each of the functions in Exercises 1–10.

1. $(6x + 1)^3$
2. $(x^2 + 9x + 10)^8$
3. $(x^3 + x^2 - 1)^{10}$
4. $(x^2 + 1)^{-13}$
5. $6/x$
6. $\dfrac{9x^9 - x^8 + 14x^7 + x^6 + 5x^4 + x^2 + 2}{x^2}$
7. $\dfrac{x + 1}{x - 1}$
8. $\dfrac{x^2 - 1}{x^2 - 2}$
9. $\dfrac{(x^2 + 1)^{13}}{(x^2 - 1)^{14}}$
10. $\dfrac{\left[(x^3 + 6)^2 - (2x^4 + 1)^3\right]}{(x^5 + 8)}$

In Exercises 11–20, let

$$A(x) = x^3 - x^2 - 2x,$$

$$B(x) = x^2 + \tfrac{1}{4}x - \tfrac{3}{8},$$

$$C(x) = 2x^3 - 5x^2 + x + 2,$$

$$D(x) = x^2 + 8x + 16.$$

Differentiate the given functions in Exercises 11–16.

11. $[B(x)]^3$
12. $[A(x)]^2$
13. $\dfrac{A(x)}{D(x)}$
14. $C(x) - 3\left[\dfrac{D(x)}{A(x)}\right]$
15. $A(2x)$
16. $A(2B(2x))$

Find the equation of the tangent line to the graph of the given function in Exercises 17–20, where A, B, C, and D are given above.

17. $[A(x)]^{1/3}$ at $x = 1$
18. $[B(x)]^2$ at $x = 0$
19. $[C(x)]^2$ at $x = -2$
20. $\sqrt{D(x)}$ at $x = -1$

Differentiate the functions in Exercises 21–28.

21. $f(x) = x^{5/3}$
22. $h(x) = (1 + 2x^{1/2})^{3/2}$
23. $g(x) = \dfrac{x^{3/2}}{\sqrt{1 + x^2}}$
24. $l(y) = \dfrac{y^2}{\sqrt{1 - y^3}}$
25. $f(x) = \dfrac{1 + x^{3/2}}{1 - x^{3/2}}$
26. $f(y) = y^3 + \left(\dfrac{1 + y^3}{1 - y^3}\right)^{1/2}$
27. $f(x) = \dfrac{8\sqrt{x}}{1 + \sqrt{x}} + 3x\left(\dfrac{1 + \sqrt{x}}{1 - \sqrt{x}}\right)$
28. $f(y) = \dfrac{8y^4}{1 + \left[3/\left(1 + \sqrt{y}\right)\right]}$

Find the first and second derivatives of the functions in Exercises 29–40.

29. $f(x) = \dfrac{x - a}{x^2 + 2bx + c}$ (a, b, c constants).
30. $f(z) = \dfrac{az + b}{cz + d}$ (a, b, c, d constants).
31. $x(t) = \left(\dfrac{A}{1 - t}\right) + \left(\dfrac{B}{1 - t^2}\right) + \left(\dfrac{C}{1 - t^3}\right)$

   (A, B, C constants).
32. $s(t) = (t - 1)(t + 1)$.
33. $h(r) = r^{13} - \sqrt{2}\, r^4 - \left(\dfrac{r}{r^2 + 3}\right)$
34. $k(s) = s^{15} + \left(\dfrac{s^{12} - 2}{s - 1}\right) + s^2 - 1$.
35. $f(x) = (x - 1)^3 g(x)$ (here $g(x)$ is some differentiable function).
36. $V(r) = \tfrac{4}{3}\pi r^2 + 2\pi rh(r)$, where $h(r) = 2r - 1$.
37. $h(x) = (x - 2)^4(x^2 + 2)$
38. $f(x) = 1 - \dfrac{1}{x} - \dfrac{1}{x^2} + \dfrac{1}{x^3} + \dfrac{1}{x^{10}}$
39. $g(t) = \dfrac{t^5 - 3t^4}{t^3 - 1}$
40. $q(s) = s^5\left(\dfrac{1 - 2s}{1 - s}\right)$

41. The volume of a falling spherical raindrop grows at a rate which is proportional to the surface area of the drop. Show that the radius of the drop increases at a constant rate.
42. The temperature of the atmosphere decreases with altitude at the rate of 2°C per kilometer at the top of a certain cliff. A hang glider pilot finds that the outside temperature is rising at the rate of 10^{-4} degrees Centigrade per second. How fast is the glider falling?
43. For temperatures in the range $[-50, 150]$ (degrees Celsius), the pressure in a certain closed container of gas changes linearly with the temperature. Suppose that a 40° increase in temperature causes the pressure to increase by 30 millibars (a millibar is one thousandth of the average atmospheric pressure at sea level). (a) What is the rate of change of pressure with respect to temperature? (b) What change of temperature would cause the pressure to drop by 9 millibars?
44. Find the rate of change of the length of an edge of a cube with respect to its surface area.
45. The organism *amoebus rectilineus* always maintains the shape of a right triangle whose area is 10^{-6} square millimeters. Find the rate of change of the perimeter at a moment when the organism is isosceles and one of the legs is growing at 10^{-4} millimeters per second.

46. The price of calculus books rise at the rate of 75¢ per year. The price of books varies with weight at a rate of $2.00 per pound. How fast is the weight of books rising? (Ignore the effect of inflation).

47. Two ships, A and B, leave San Francisco together and sail due west. A sails at 20 miles per hour and B at 25 miles per hour. Ten miles out to sea, A turns due north and B continues due west. How fast are they moving away from each other 4 hours after departing San Francisco?

48. At an altitude of 2000 meters, a parachutist jumps from an airplane and falls $4.9t^2$ meters in t seconds. Suppose that the air pressure p decreases with altitude at the constant rate of 0.095 gsc per meter. The parachutist's ears pop when dp/dt reaches 2 gsc per second. At what time does this happen?

In Exercises 49–52, let A represent the area of the shaded region in Fig. 2.R.1.

49. Find dA/dx and d^2A/dx^2.
50. Find dA/dr and d^2A/dr^2.
51. Find dA/dy and d^2A/dy^2.
52. Find dA/dx and d^2A/dx^2.

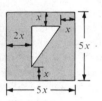 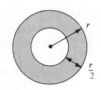

49 Find dA/dx and d^2A/dx^2 50 Find dA/dr and d^2A/dr^2

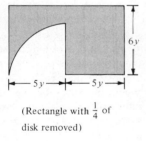 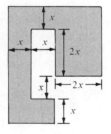

(Rectangle with $\frac{1}{4}$ of disk removed)

51 Find dA/dy and d^2A/dy^2 52 Find dA/dx and d^2A/dx^2

Figure 2.R.1. Find the indicated rates of change of the areas.

In Exercises 53–56 let P represent the perimeter of the shaded region in Figure 2.R.1.

53. For Exercise 49, find dA/dP and dP/dx.
54. For Exercise 50, find dA/dP and dP/dr.
55. For Exercise 51, find dA/dP and dP/dy.
56. For Exercise 52, find dA/dP and dP/dx.

57. The total cost C in dollars for producing x cases of solvent is given by $C(x) = 20 + 5x - (0.01)x^2$.

The number 20 in the formula represents the fixed cost for placing the order, regardless of size. The other terms represent the variable costs.
(a) Find the marginal cost.
(b) Find the cost for the 85th case of solvent, i.e., the marginal cost for a purchase of 84 cases.
(c) Explain in the language of marginal cost the statement "the more you buy, the cheaper it gets."
(d) Find a large value of x, beyond which it is unreasonable for the given formula for $C(x)$ to be applicable.

58. In Exercise 57, suppose that the solvent is priced at $8 - [(8x + 100)/(x + 300)]$ dollars per case at production level x. Calculate the marginal revenue and the marginal profit.

Find the equation of the line tangent to the graph of the function at the indicated point in Exercises 59–62.

59. $f(x) = (x^3 - 6x)^3$; $(0, 0)$

60. $f(x) = \dfrac{x^4 - 1}{6x^2 + 1}$; $(1, 0)$

61. $f(x) = \dfrac{x^3 - 7}{x^3 + 11}$; $(2, \frac{1}{19})$

62. $f(x) = \dfrac{x^5 - 6x^4 + 2x^3 - x}{x^2 + 1}$; $(1, -2)$

63. If $x^2 + y^2 + xy^3 = 1$, find dy/dx when $x = 0$, $y = 1$.

64. If x and y are functions of t, $x^4 + xy + y^4 = 2$, and $dy/dt = 1$ at $x = 1$, $y = 1$, find dx/dt at $x = 1$, $y = 1$.

65. Let a curve be described by the parametric equations

$$x = \sqrt{t} + t^2 + \frac{1}{t},$$
$$y = 1 + \sqrt[3]{t} + t, \qquad 1 \leqslant t \leqslant 3.$$

Find the equation of the tangent line at $t = 2$.

66. The *speed* of an object traveling on a parametric curve is given by $v = \sqrt{(dx/dt)^2 + (dy/dt)^2}$.
(a) Find the speed at $t = 1$ for the motion $x = t^3 - 3t^2 + 1$, $y = t^5 - t^7$.
(b) Repeat for $x = t^2 - 3$, $y = \frac{1}{3}t^3 - t$ at $t = 1$.

67. Find the linear approximations for: (a) $\sqrt[3]{27.11}$; (b) $\sqrt[3]{-63.01}$.

68. Find the linear approximations for (a) $\sqrt[5]{32.02}$ and (b) $\sqrt[6]{63.98}$.

69. (a) Find the linear approximation to the function $(x^{40} - 1)/(x^{29} + 1)$ at $x_0 = 1$.
(b) Calculate $[(1.021)^{40} - 1]/[(1.021)^{29} + 1]$ approximately.

70. Find an approximate value for $\sqrt{1 + (0.0036)^2}$.

71. Find a formula for $(d^2/dx^2)[f(x)g(x)]$.

72. If f is a given differentiable function and $g(x) = f(\sqrt{x})$, what is $g'(x)$?

73. Differentiate both sides of the equation

$$\left[f(x)^m \right]^n = f(x)^{mn}$$

where m and n are positive integers and show that you get the same result on each side.

74. Find a formula for $(d/dx)[f(x)^m g(x)^n]$, where f and g are differentiable functions and m and n are positive integers.

Find the antiderivatives in Exercises 75–94.

75. $\int 10 \, dx$

76. $\int (4.9t + 15) \, dt$

77. $\int (4x^3 + 3x^2 + 2x + 1) \, dx$

78. $\int (s^5 + 4s^4 + 9) \, ds$

79. $\int \frac{2}{3} x^{2/5} \, dx$

80. $\int 4x^{3/2} \, dx$

81. $\int \left[\frac{-1}{x^2} + \frac{-2}{x^3} + \frac{-3}{x^4} + \frac{-4}{x^5} \right] dx$

82. $\int \left(\frac{1}{z^2} + \frac{4}{z^3} \right) dz$

83. $\int (x^2 + \sqrt{x}) \, dx$

84. $\int (x + \sqrt{x} + \sqrt[3]{x}) \, dx$

85. $\int (x^{3/2} + x^{-1/2}) \, dx$

86. $\int \left(\frac{2}{\sqrt{x}} \right) dx$

87. $\int \frac{x^4 + x^6 + 1}{x^{1/2}} \, dx$

88. $\int \left(\sqrt{x} + \frac{1}{\sqrt{x}} \right) dx$

89. $\int \sqrt{x - 1} \, dx$

90. $\int (2x + 1)^{3/2} \, dx$

91. $\int \frac{1}{(x - 1)^2} \, dx$

92. $\int [(8x - 10)^{3/2} + 10x] \, dx$

93. $\int [(x - 1)^{1/2} - (x - 2)^{5/2}] \, dx$

94. $\int \frac{(2x - 1)^{3/2} - 1}{(10x - 5)^{1/2}} \, dx$

Differentiate each of the functions in Exercises 95–102 and write the corresponding antidifferentiation formula.

95. $f(x) = \dfrac{x^{1/3}}{(x^{1/2} - x^{1/3} + 1)^{1/2}}$

96. $f(x) = \sqrt{\left(\dfrac{8x^2 + 8x + 1}{x^{1/2} + x^{-1/2}} \right)}$

97. $f(x) = \sqrt{x} - \sqrt{\dfrac{x - 1}{x + 1}}$

98. $f(x) = \sqrt{\dfrac{\sqrt{x} - 1}{\sqrt{x} + 1}}$

99. $f(x) = \dfrac{\sqrt[4]{x} - 1}{\sqrt[4]{x} + 1}$

100. $f(x) = \sqrt[3]{\dfrac{x - 1}{3x + 1}}$

101. $f(x) = \sqrt{\dfrac{x^2 + 1}{x^2 - 1}}$

102. $f(x) = \sqrt{x^3 + 2\sqrt{x} + 1}$

103. The *lemniscate* $3(x^2 + y^2)^2 = 25(x^2 - y^2)$ is a planar curve which intersects itself at the origin.
(a) Show by use of symmetry that the entire lemniscate can be graphed by (1) reflecting the first quadrant portion through the x axis, and then (2) reflecting the right half-plane portion through the origin to the left half-plane.
(b) Find by means of implicit differentiation the value of dy/dx at $(2, 1)$.
(c) Determine the equation of the tangent line to the lemniscate through $(2, 1)$.

104. The *drag* on an automobile is the force opposing its motion down the highway, due largely to air resistance. The drag in pounds D can be approximated for velocities v near 50 miles per hour by $D = kv^2$. Using $k = 0.24$, find the rate of increase of drag with respect to time at 55 miles per hour when the automobile is undergoing an acceleration of 3 miles per hour each second.

105. The air resistance of an aircraft fuel tank is given approximately by $D = 980 + 7(v - 700)$ lbs for the velocity range of $700 \leqslant v \leqslant 800$ miles per hour. Find the rate of increase in air resistance with respect to time as the aircraft accelerates past the speed of sound (740 miles per hour) at a constant rate of 12 miles per hour each second.

106. A physiology experiment measures the heart rate $R(x)$ in beats per minute of an athlete climbing a vertical rope of length x feet. The experiment produces two graphs: one is the heart rate R versus the length x; the other is the length x versus the time t in seconds it took to climb the rope (from a fresh start, as fast as possible).
(a) Give a formula for the change in heart rate in going from a 12-second climb to a 13-second climb using the linear approximation.
(b) Explain how to use the tangent lines to the two graphs and the chain rule to compute the change in part (a).

★107. (a) Find a formula for the second and third derivatives of x^n.
(b) Find a formula for the rth derivative of x^n if $n > r$.
(c) Find a formula for the derivative of the product $f(x)g(x)h(x)$ of three functions.

★108. (a) Prove that if f/g is a rational function (i.e., a quotient of polynomials) with derivative zero, then f/g is a constant.
   (b) Conclude that if the rational functions F and G are both antiderivatives for a function h, then F and G differ by a constant.

★109. Prove that if the kth derivative of a rational function $r(x)$ is zero for some k, then $r(x)$ is a polynomial.

Chapter 3

Graphing and Maximum—Minimum Problems

Differential calculus provides tests for locating the key features of graphs.

Now that we know how to differentiate, we can use this information to assist us in plotting graphs. The signs of the derivative and the second derivative of a function will tell us which way the graph of the function is "leaning" and "bending."

Using the derivative to predict the behavior of graphs helps us to find the points where a function takes on its maximum and minimum values. Many interesting word problems requiring the "best" choice of some variable involve searching for such points.

In Section 3.1, we study the geometric aspects of continuity. This will provide a useful introduction to graphing. In Section 3.6, we use the ideas of maxima and minima to derive an important theoretical result—the mean value theorem. One consequence of this theorem is a fact which we used in connection with antiderivatives: a function whose derivative is zero must be constant.

3.1 Continuity and the Intermediate Value Theorem

If a continuous function on a closed interval has opposite signs at the endpoints, it must be zero at some interior point.

In Section 1.2, we defined continuity as follows: "A function $f(x)$ is said to be continuous at x_0 if $\lim_{x \to x_0} f(x) = f(x_0)$." A function is said to be *continuous on a given interval* if it is continuous at every point on that interval. If a function f is continuous on the whole real line, we just say that "f is continuous." An imprecise but useful guide is that a function is continuous when its graph can be drawn "without removing pencil from paper." In Figure 3.1.1, the curve on the left is continuous at x_0 while that on the right is not.

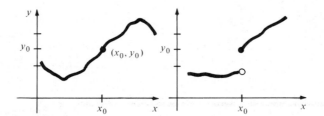

Figure 3.1.1. Illustrating a continuous curve (left) and a discontinuous curve (right).

Example 1   Decide where each of the functions whose graphs appear in Fig. 3.1.2 is continuous. Explain your answers.

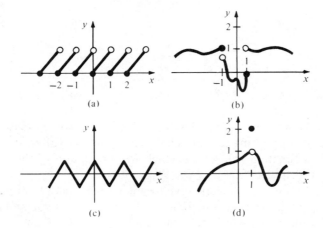

Figure 3.1.2. Where are these functions continuous?

Solution   (a) This function jumps in value at each of the points $x_0 = 0$, $x_0 = \pm 1$, $x_0 = \pm 2, \ldots$, so $\lim_{x \to x_0} f(x)$ does not exist at these points and thus f is not continuous there; however, f is continuous on each of the intervals between the jump points.

   (b) This function jumps in value at $x_0 = -1$ and $x_0 = +1$, and so $\lim_{x \to +1} f(x)$ and $\lim_{x \to -1} f(x)$ do not exist. Thus f is not continuous at $x_0 = \pm 1$; it is continuous on each of the intervals $(-\infty, -1)$, $(-1, 1)$, and $(1, \infty)$.

   (c) Even though this function has sharp corners on its graph, it *is* continuous; $\lim_{x \to x_0} f(x) = f(x_0)$ at each point x_0.

   (d) Here $\lim_{x \to 1} f(x) = 1$, so the limit exists. However, the limit does not equal $f(1) = 2$. Thus f is *not* continuous at $x_0 = 1$. It is continuous on the intervals $(-\infty, 1)$ *and* $(1, \infty)$. ▲

In Section 1.2, we used various limit theorems to establish the continuity of functions that are basic to calculus. For example, the rational function rule for limits says that a *rational function is continuous at points where its denominator does not vanish*.

Example 2   Show that the function $f(x) = (x - 1)/3x^2$ is continuous at $x_0 = 4$.

Solution   This is a rational function whose denominator does not vanish at $x_0 = 4$, so it is continuous by the rational function rule. ▲

Example 3   Let $g(x)$ be the *step function* defined by

$$g(x) = \begin{cases} 0 & \text{if } x \leqslant 0, \\ 1 & \text{if } x > 0. \end{cases}$$

Show that g is not continuous at $x_0 = 0$. Sketch.

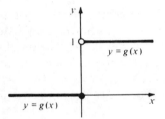

Figure 3.1.3. This step function is discontinuous at $x_0 = 0$.

Solution   The graph of g is shown in Fig. 3.1.3. Since g approaches (in fact, equals) 0 as x approaches 0 from the left, but approaches 1 as x approaches 0 from the right, $\lim_{x \to 0} g(x)$ does not exist. Therefore, g is not continuous at $x_0 = 0$. ▲

Example 4   Using the laws of limits, show that if f and g are continuous at x_0, so is fg.

Solution   We must show that $\lim_{x \to x_0}(fg)(x) = (fg)(x_0)$. By the product rule for limits, $\lim_{x \to x_0}[f(x)g(x)] = [\lim_{x \to x_0} f(x)][\lim_{x \to x_0} g(x)] = f(x_0)g(x_0)$, since f and g are continuous at x_0. But $f(x_0)g(x_0) = (fg)(x_0)$, and so $\lim_{x \to x_0}(fg)(x) = (fg)(x_0)$, as required. ▲

In Section 1.3, we proved the following theorem: *if f is differentiable at x_0, then f is continuous at x_0.* Using our knowledge of differential calculus, we can use this relationship to establish the continuity of additional functions or to confirm the continuity of functions originally determined using the laws of limits.

Example 5   (a) Show that $f(x) = 3x^2/(x^3 - 2)$ is continuous at $x_0 = 1$. Where else is it continuous?

(b) Show that $f(x) = \sqrt{x^2 + 2x + 1}$ is continuous at $x = 0$.

Solution   (a) By our rules for differentiation, we see that this function is differentiable at $x_0 = 1$; indeed, $x^3 - 2$ does not vanish at $x_0 = 1$. Thus f is also continuous at $x_0 = 1$. Similarly, f is continuous at each x_0 such that $x_0^3 - 2 \neq 0$, i.e., at each $x_0 \neq \sqrt[3]{2}$.

(b) This function is the composition of the square root function $h(u) = \sqrt{u}$ and the function $g(x) = x^2 + 2x + 1$; $f(x) = h(g(x))$. Note that $g(0) = 1 > 0$. Since g is differentiable at any x (being a polynomial), and h is differentiable at $u = 1$, f is differentiable at $x = 0$ by the chain rule. Thus f is continuous at $x = 0$. ▲

According to our previous discussion, a continuous function is one whose graph never "jumps." The definition of continuity is *local* since continuity at each point involves values of the function only near that point. There is a corresponding *global* statement, called the intermediate value theorem, which involves the behavior of a function over an entire interval $[a, b]$.

Intermediate Value Theorem (First Version)

Let f be continuous[1] on $[a, b]$ and suppose that, for some number c, $f(a) < c < f(b)$ or $f(a) > c > f(b)$. Then there is some point x_0 in (a, b) such that $f(x_0) = c$.

---

[1] Our definition of continuity on $[a, b]$ assumes that f is defined near each point $\bar{x}$ of $[a, b]$, including the endpoints, and that $\lim_{x \to \bar{x}} f(x) = f(\bar{x})$. Actually, at the endpoints, it is enough to assume that the one-sided limits (from inside the interval) exist, rather than the two-sided ones.

In geometric terms, this theorem says that for the graph of a continuous function to pass from one side of a horizontal line to the other, the graph must meet the line somewhere (see Fig. 3.1.4). The proof of the theorem depends on a careful study of properties of the real numbers and will be omitted. (See the references listed in the Preface.) However, by drawing additional graphs like those in Fig. 3.1.4, you should convince yourself that the theorem is reasonable.

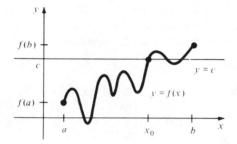

Figure 3.1.4. The graph of f must pierce the horizontal line $y = c$ if it is to get across.

Example 6   Show that there is a number x_0 such that $x_0^5 - x_0 = 3$.

Solution   Let $f(x) = x^5 - x$. Then $f(0) = 0$ and $f(2) = 30$. Since $0 < 3 < 30$, the intermediate value theorem guarantees that there is a number x_0 in $(0, 2)$ such that $f(x_0) = 3$. (The function f is continuous on $[0, 2]$ because it is a polynomial.) ▲

Notice that the intermediate value theorem does not tell us *how* to find the number x_0 but merely that it *exists*. (A look at Fig. 3.1.4 should convince you that there may be more than one possible choice for x_0.) Nevertheless, by repeatedly dividing an interval into two or more parts and evaluating $f(x)$ at the dividing points, we can solve the equation $f(x_0) = c$ as accurately as we wish. This *method of bisection* is illustrated in the next example.

Example 7   **(The method of bisection)** Find a solution of the equation $x^5 - x = 3$ in $(0, 2)$ to within an accuracy of 0.1 by repeatedly dividing intervals in half and testing each half for a root.

Solution   In Example 6 we saw that the equation has a solution in the interval $(0, 2)$. To locate the solution more precisely, we evaluate $f(1) = 1^5 - 1 = 0$. Thus $f(1) < 3 < f(2)$, so there is a root in $(1, 2)$. Now we bisect $[1, 2]$ into $[1, 1.5]$ and $[1.5, 2]$ and repeat: $f(1.5) \approx 6.09 > 3$, so there is a root in $(1, 1.5)$; $f(1.25) = 1.80 < 3$, so there is a root in $(1.25, 1.5)$; $f(1.375) \approx 3.54 > 3$, so there is a root in $(1.25, 1.375)$; thus $x_0 = 1.3$ is within 0.1 of a root. Further accuracy can be obtained by means of further bisections. (Related techniques for root finding are suggested in the exercises for this section. Other methods are presented in Section 11.4.) ▲

There is another useful way of stating the intermediate value theorem (the *contrapositive* statement).

Intermediate Value Theorem (Second Version)

Let f be a function which is continuous on $[a, b]$ and suppose that $f(x) \neq c$ for all x in $[a, b]$. If $f(a) < c$, then $f(b) < c$ as well. (See Fig. 3.1.5.) Similarly, if $f(a) > c$, then $f(b) > c$ as well.

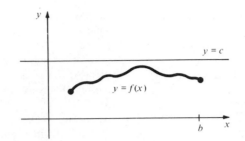

Figure 3.1.5. The graph starts below the line $y = c$ and never pierces the line, so it stays below the line.

In geometric language, this second version of the theorem says: "The graph of a continuous function which never meets a horizontal line must remain on one side of it." The first version says: "If the graph of a continuous function passes from one side of a horizontal line to the other, the graph must meet the line somewhere." You should convince yourself that these two statements really mean the same thing.

In practice, the second version of the intermediate value theorem can be useful for determining the sign of a function on intervals where it has no roots, as in the following example.

Example 8   Suppose that f is continuous on $[0, 3]$, that f has no roots on the interval, and that $f(0) = 1$. Prove that $f(x) > 0$ for *all* x in $[0, 3]$.

Solution   Apply the intermediate value theorem (version 2), with $c = 0$, to f on $[0, b]$ for each b in $(0, 3]$. Since f is continuous on $[0, 3]$, it is continuous on $[0, b]$; since $f(0) = 1 > 0$, we have $f(b) > 0$. But b was anything in $(0, 3]$, so we have proved what was asked. ▲

Exercises for Section 3.1

1. Decide where each of the functions whose graph is sketched in Fig. 3.1.6 is continuous.

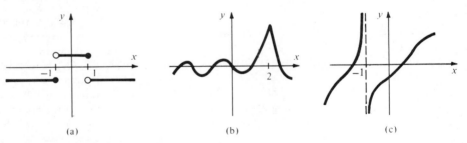

Figure 3.1.6. Where are these continuous?

(a)   (b)   (c)

2. Which of the functions in Fig. 3.1.7 are continuous at $x_0 = 1$?

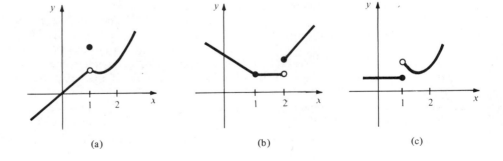

Figure 3.1.7. Which of these functions are continuous at $x_0 = 1$?

(a)   (b)   (c)

3. Show that $f(x) = (x^2 + 1)(x^2 - 1)$ is continuous at $x_0 = 0$.
4. Show that $f(x) = x^3 + 3x^2 - 2x$ is continuous at $x_0 = 3$.
5. Prove that $(x^2 - 1)/(x^3 + 3x)$ is continuous at $x_0 = 1$.
6. Prove that $(x^4 - 8)/(x^3 + 2)$ is continuous at $x_0 = 0$.
7. Where is $(x^2 - 1)/(x^4 + x^2 + 1)$ continuous?
8. Where is $(x^4 + 1)/(x^3 - 8)$ continuous?
9. Let $f(x) = (x^3 + 2)/(x^2 - 1)$. Show that f is continuous on $[-\frac{1}{2}, \frac{1}{2}]$.
10. Is the function $(x^3 - 1)/(x^2 - 1)$ continuous at 1? Explain your answer.
11. Let $f(x)$ be the step function defined by

$$f(x) = \begin{cases} -1 & \text{if } x < 0, \\ -2 & \text{if } x \geqslant 0. \end{cases}$$

Show that f is discontinuous at 0.
12. Let $f(x)$ be the *absolute value function*: $f(x) = |x|$; that is,

$$f(x) = \begin{cases} x & \text{if } x \geqslant 0, \\ -x & \text{if } x < 0. \end{cases}$$

Show that f is continuous at $x_0 = 0$.
13. Using the laws of limits, show that if f and g are continuous at x_0, so is $f + g$.
14. If f and g are continuous at x_0 and $g(x_0) \neq 0$, show that f/g is continuous at x_0.
15. Where is $f(x) = 8x^3/\sqrt{x^2 - 8}$ continuous?
16. Where is $f(x) = 9x^2 - 3x/\sqrt{x^4 - 2x^2 - 8}$ continuous?
17. Show that the equation $-s^5 + s^2 = 2s - 6$ has a solution.
18. Prove that the equation $x^3 + 2x - 1 = 7$ has a solution.
19. Prove that $f(x) = x^8 + 3x^4 - 1$ has at least two distinct zeros.
20. Show that $x^4 - 5x^2 + 1$ has at least two distinct zeros.
21. The roots of $f(x) = x^3 - 2x - x^2 + 2$ are $\sqrt{2}$, $-\sqrt{2}$, and 1. By evaluating $f(-3)$, $f(0)$, $f(1.3)$, and $f(2)$, determine the sign of $f(x)$ on each of the intervals between its roots.
22. Use the method of bisection to approximate $\sqrt{7}$ to within two decimal places. [*Hint:* Let $f(x) = x^2 - 7$. What should you use for a and b?]
23. Find a solution of the equation $x^5 - x = 3$ to an accuracy of 0.01.
24. Find a solution of the equation $x^5 - x = 5$ to an accuracy of 0.1.
25. Suppose that f is continuous on $[-1, 1]$ and that $f(x) - 2$ is never zero on $[-1, 1]$. If $f(0) = 0$, show that $f(x) < 2$ for all x in $[-1, 1]$.
26. Suppose that f is continuous on $[3, 5]$ and that $f(x) \neq 4$ for all x in $[3, 5]$. If $f(3) = 3$, show that $f(5) < 4$.

27. Let $f(x)$ be 1 if a certain sample of lead is in the solid state at temperature x; let $f(x)$ be 0 if it is in the liquid state. Define x_0 to be the melting point of this lead sample. Is there any way to define $f(x_0)$ so as to make f continuous? Give reasons for your answer, and supply a graph for f.
28. An empty bucket with a capacity of 10 liters is placed beneath a faucet. At time $t = 0$, the faucet is turned on, and water flows from the faucet at the rate of 5 liters per minute. Let $V(t)$ be the volume of the water in the bucket at time t. Present a plausible argument showing that V is a continuous function on $(-\infty, \infty)$. Sketch a graph of V. Is V differentiable on $(-\infty, \infty)$?
29. Let $f(x) = (x^2 - 4)/(x - 2)$, $x \neq 2$. Define $f(2)$ so that the resulting function is continuous at $x = 2$.
30. Let $f(x) = (x^3 - 1)/(x - 1)$, $x \neq 1$. How should $f(1)$ be defined in order that f be continuous at each point?
31. Let $f(x)$ be defined by

$$f(x) = \begin{cases} x^2 + 1 & \text{if } x < 1, \\ ? & \text{if } 1 \leqslant x \leqslant 3, \\ x - 6 & \text{if } 3 < x. \end{cases}$$

How can you define $f(x)$ on the interval $[1, 3]$ in order to make f continuous on $(-\infty, \infty)$? (A geometric argument will suffice.)
32. Let $f(x) = x + (4/x)$ for $x \leqslant -\frac{1}{2}$ and $x \geqslant 2$. Define $f(x)$ for x in $(-\frac{1}{2}, 2)$ in such a way that the resulting function is continuous on the whole real line.
33. Let $f(x)$ be defined by $f(x) = (x^2 - 1)/(x - 1)$ for $x \neq 1$. How should you define $f(1)$ to make the resulting function continuous? [*Hint:* Plot a graph of $f(x)$ for x near 1 by factoring the numerator.]
34. Let $f(x)$ be defined by $f(x) = 1/x$ for $x \neq 0$. Is there any way to define $f(0)$ so that the resulting function will be continuous?
35. Sketch the graph of a function which is continuous on the whole real line and differentiable everywhere except at $x = 0, 1, 2, 3, 4, 5, 6, 7, 8, 9, 10$.
36. Is a function which is continuous at x_0 necessarily differentiable there? Prove or give a counterexample.
37. The function $f(x) = 1/(x - 1)$ never takes the value zero, yet $f(0) = -1$ is negative and $f(2) = 1$ is positive. Why isn't this a counterexample to the intermediate value theorem?
38. "Prove" that you were once exactly one meter tall.

In Exercises 39–42, use the method of bisection to find a root of the given function, on the given interval, to the given accuracy.

▓ 39. $x^3 - 11$ on $[2, 3]$; accuracy $\frac{1}{100}$.

▓ 40. $x^3 + 7$ on $[-3, -1]$; accuracy $\frac{1}{100}$.

▓ 41. $x^5 + x^2 + 1$ on $[-2, -1]$; accuracy $\frac{1}{1000}$.

▓ 42. $x^4 - 3x - 2$ on $[1, 2]$; accuracy $\frac{1}{500}$.

★43. In the method of bisection, each estimate of the solution of $f(x) = c$ is approximately twice as accurate as the previous one. Examining the list of powers of 2: $2, 4, 8, 16, 32, 64, 128, 256, 512, 1024, 2048, 4096, 8192, 16384, 32768, 65536, 131072, \ldots$ —suggests that we get one more decimal place of accuracy for every three or four repetitions of the procedure. Explain.

★44. Using the result of Exercise 43, determine how many times to apply the bisection procedure to guarantee the accuracy A for the interval $[a, b]$ if:
   (a) $A = \frac{1}{100}$; $[a, b] = [3, 4]$.
   (b) $A = \frac{1}{1000}$; $[a, b] = [-1, 3]$.
   (c) $A = \frac{1}{700}$; $[a, b] = [11, 23]$.
   (d) $A = \frac{1}{15}$; $[a, b] = [0.1, 0.2]$.

★45. Can you improve the method of bisection by choosing a better point than the one halfway between the two previous points? [*Hint:* If $f(a)$ and $f(b)$ have opposite signs, choose the point where the line through $(a, f(a))$ and $(b, f(b))$ crosses the x axis.] Is there a method of division more appropriate to the decimal system?

   Do some experiments to see whether your method gives more accurate answers then the bisection method in the same number of steps. If so, does the extra accuracy justify the extra time involved in carrying out each step? You might wish to compare various methods on a competitive basis, either with friends, with yourself by timing the calculations, or by timing calculations done on a programmable calculator or microcomputer.

★46. Prove that if f is continuous on an interval I (not necessarily closed) and $f(x) \neq 0$ for all x in I, then the sign of $f(x)$ is the same for all x in I.

★47. Suppose that f is continuous at x_0 and that in any open interval I containing x_0 there are points x_1 and x_2 such that $f(x_1) < 0$ and $f(x_2) > 0$. Explain why $f(x_0)$ must equal 0.

★48. (a) Suppose that f and g are continuous on the real line. Show that $f - g$ is continuous. (b) Suppose that f and g are continuous functions on the whole real line. Prove that if $f(x) \neq g(x)$ for all x, and $f(0) > g(0)$, then $f(x) > g(x)$ for all x. Interpret your answer geometrically.

★49. Prove that any odd-degree polynomial has a root by following these steps.
   1. Reduce the problem to showing that $f(x) = x^n + a_{n-1}x^{n-1} + \cdots + a_1 x + a_0$ has a root, where a_i are constants and n is odd.
   2. Show that if $|x| > 1$ and $|x| > 2 \times \{|a_0| + \cdots + |a_{n-1}|\}$, then $f(x)/x^n$ is positive.
   3. Conclude that $f(x) < 0$ if x is large negative and $f(x) > 0$ if x is large positive.
   4. Apply the intermediate value theorem.

★50. Show that the polynomial $x^4 + bx + c = 0$ has a (real) root if $256c^3 < 27b^4$.

3.2 Increasing and Decreasing Functions

The sign of the derivative indicates whether a function is increasing or decreasing.

We begin this section by defining what it means for a function to be increasing or decreasing. Then we show that a function is increasing when its derivative is positive and is decreasing when its derivative is negative. Local maximum and minimum points occur where the derivative changes sign.

We can tell whether a function is increasing or decreasing at x_0 by seeing how its graph crosses the horizontal line $y = f(x_0)$ at x_0 (see Fig. 3.2.1). This

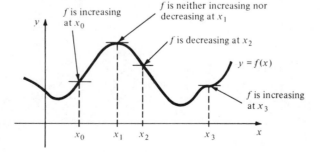

Figure 3.2.1. Places where the function f is increasing and decreasing.

geometric picture is the basis of the precise definition of increasing and decreasing. Note that at the point x_0 in Fig. 3.2.1, $f(x)$ is less than $f(x_0)$ for x just to the left of x_0, while $f(x)$ is larger than $f(x_0)$ for x just to the right. We cannot take x too far to the right, as the figure shows. The following paragraph gives the technical definition.

We say that a function f is *increasing* at x_0 if there is an interval (a, b) containing x_0 such that:

1. If $a < x < x_0$, then $f(x) < f(x_0)$.
2. If $x_0 < x < b$, then $f(x) > f(x_0)$.

Similarly, f is *decreasing* at x_0 if there is an interval (a, b) containing x such that:

1. If $a < x < x_0$, then $f(x) > f(x_0)$.
2. If $x_0 < x < b$, then $f(x) < f(x_0)$.

The purpose of the interval (a, b) is to limit our attention to a small region about x_0. Indeed, the notions of increasing and decreasing at x_0 are local; they depend only on the behavior of the function near x_0. In examples done "by hand," such as Example 1 below, we must actually find the interval (a, b). We will soon see that calculus provides an easier method of determining where a function is increasing and decreasing.

Example 1 Show that $f(x) = x^2$ is increasing at $x_0 = 2$.

Solution Choose (a, b) to be, say, $(1, 3)$. If $1 < x < 2$, we have $f(x) = x^2 < 4 = x_0^2$. If $2 < x < 3$, then $f(x) = x^2 > 4 = x_0^2$. We have verified conditions 1 and 2 of the definition, so f is increasing at $x_0 = 2$. ▲

Of special interest is the case $f(x_0) = 0$. If f is increasing at such an x_0, we say that f *changes sign from negative to positive at* x_0. By definition this occurs when $f(x_0) = 0$ and there is an open interval (a, b) containing x_0 such that $f(x) < 0$ when $a < x < x_0$ and $f(x) > 0$ when $x_0 < x < b$. (See Fig. 3.2.2.) Similarly if $f(x_0) = 0$ and f is decreasing at x_0, we say that f *changes sign from positive to negative at* x_0. (See Fig. 3.2.3.)

Notice that the chosen interval (a, b) may have to be small, since a function which changes sign from negative to positive may later change back from positive to negative (see Fig. 3.2.4).

Changes of sign can be significant in everyday life as, for instance, when the function $b = f(t)$ representing your bank balance changes from positive to negative. Changes of sign will be important for mathematical reasons in the next few sections of this chapter; it will then be useful to be able to determine when a function changes sign by looking at its formula.

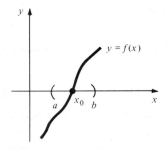

Figure 3.2.2. This function changes sign from negative to positive at x_0.

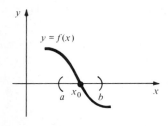

Figure 3.2.3. This function changes sign from positive to negative at x_0.

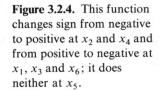

Figure 3.2.4. This function changes sign from negative to positive at x_2 and x_4 and from positive to negative at x_1, x_3 and x_6; it does neither at x_5.

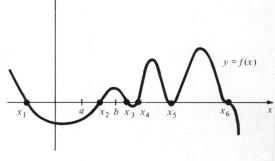

Example 2   Where does $f(x) = 3x - 5$ change sign?

Solution   We begin by finding those x for which $f(x)$ is negative and those for which it is positive. First, f is negative when

$$3x - 5 < 0,$$
$$3x < 5,$$
$$x < \tfrac{5}{3}.$$

(If you had difficulty following this argument, you may wish to review the material on inequalities in Section R.1.) Similarly, f is positive when $x > \tfrac{5}{3}$. So f changes sign from negative to positive at $x = \tfrac{5}{3}$. (See Fig. 3.2.5.) Here the chosen interval (a, b) can be arbitrarily large. ▲

If a function is given by a formula which factors, this often helps us to find the changes of sign.

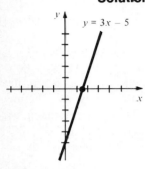

Figure 3.2.5. This function changes sign from negative to positive at $x = 5/3$.

Example 3   Where does $f(x) = x^2 - 5x + 6$ change sign?

Solution   We write $f(x) = (x - 3)(x - 2)$. The function f changes sign whenever one of its factors does. This occurs at $x = 2$ and $x = 3$. The factors have opposite signs for x between 2 and 3, and the same sign otherwise, so f changes from positive to negative at $x = 2$ and from negative to positive at $x = 3$. (See Fig. 3.2.6.) ▲

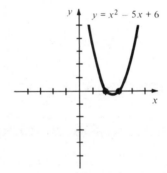

Figure 3.2.6. This function changes sign from positive to negative at $x = 2$ and from negative to positive at $x = 3$.

Given a more complicated function, such as $x^5 - x^3 - 2x^2$, it may be difficult to tell directly whether it is increasing or decreasing at a given point. The derivative is a very effective tool for helping us answer such questions.

A linear function $l(x) = mx + b$ is increasing everywhere when the slope m is positive. If the derivative $f'(x)$ of a function f is positive when $x = x_0$, the linear approximation to f at x_0 is increasing, so we may expect f to be increasing at x_0 as well. (See Fig. 3.2.7.)

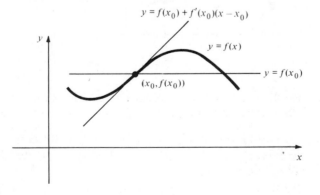

Figure 3.2.7. The function f is increasing where its derivative is positive.

We can verify that a function is increasing where its derivative is positive by using difference quotients. Let $y = f(x)$ and assume that $f'(x_0) > 0$. Then

$$\frac{\Delta y}{\Delta x} = \frac{f(x_0 + \Delta x) - f(x_0)}{\Delta x} > 0$$

for Δx sufficiently small, since $\Delta y / \Delta x$ approaches the positive number $f'(x_0)$ as Δx approaches zero. Thus, there is an interval (a, b) about x_0 such that $\Delta x = x - x_0$ and $\Delta y = f(x) - f(x_0)$ have the same sign when x is in the interval. Thus, when $a < x < x_0$, Δx is negative and so is Δy; hence $f(x) < f(x_0)$. Similarly, $f(x) > f(x_0)$ when $x_0 < x < b$. Thus f is increasing at x_0. If $f'(x_0) < 0$, a similar argument shows that f is decreasing at x_0.

The functions $y = x^3$, $y = -x^3$, and $y = x^2$ show that many kinds of behavior are possible when $f'(x_0) = 0$ (see Fig. 3.2.8).

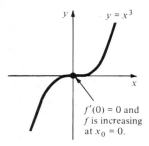

$f'(0) = 0$ and f is increasing at $x_0 = 0$.

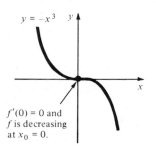

$f'(0) = 0$ and f is decreasing at $x_0 = 0$.

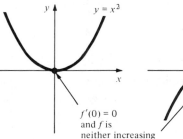

$f'(0) = 0$ and f is neither increasing nor decreasing at $x_0 = 0$.

Figure 3.2.8. If $f'(x_0) = 0$, you cannot tell if f is increasing or decreasing without further information.

Thus we arrive at the following test (see Fig. 3.2.9).

Increasing–Decreasing Test

1. If $f'(x_0) > 0$, f is increasing at x_0.
2. If $f'(x_0) < 0$, f is decreasing at x_0.
3. If $f'(x_0) = 0$, the test is inconclusive.

Figure 3.2.9. $f'(x_1) > 0$; f is increasing at x_1. $f'(x_2) < 0$; f is decreasing at x_2. $f'(x_3) = f'(x_4) = 0$; f is neither increasing nor decreasing at x_3 and x_4. $f'(x_5) = 0$; f is decreasing at x_5. $f'(x_6) = 0$; f is increasing at x_6.

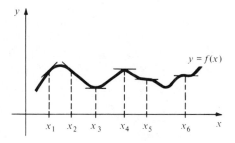

Example 4 (a) Is $x^5 - x^3 - 2x^2$ increasing or decreasing at -2? (b) Is $g(s) = \sqrt{s^2 - s}$ increasing or decreasing at $s = 2$?

Solution (a) Letting $f(x) = x^5 - x^3 - 2x^2$, we have $f'(x) = 5x^4 - 3x^2 - 4x$, and $f'(-2) = 5(-2)^4 - 3(-2)^2 - 4(-2) = 80 - 12 + 8 = 76$, which is positive. Thus $x^5 - x^3 - 2x^2$ is increasing at -2.

(b) By the chain rule, $g'(s) = (2s - 1)/2\sqrt{s^2 - s}$, so $g'(2) = 3/2\sqrt{3} > 0$. Thus g is increasing at $s = 2$. ▲

Example 5   Let $f(x) = x^3 - x^2 + x + 3$. How does f change sign at $x = -1$?

Solution   Notice that $f(-1) = 0$. Also, $f'(-1) = 3(-1)^2 - 2(-1) + 1 = 6 > 0$, so f is increasing at $x = -1$. Thus f changes sign from negative to positive. ▲

We can also interpret the increasing–decreasing test in terms of velocities and other rates of change. For instance, if $f(t)$ is the position of a particle on the real-number line at time t, and $f'(t_0) > 0$, then the particle is moving to the right at time t_0; if $f'(t_0) < 0$, the particle is moving to the left (see Fig. 3.2.10).

Figure 3.2.10. Positive velocity means that the motion is to the right, and negative velocity means that the motion is to the left.

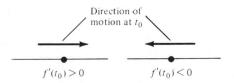

Example 6   The temperature at time t is given by $f(t) = (t + 1)/(t - 1)$ for $t < 1$. Is it getting warmer or colder at $t = 0$?

Solution   We calculate $f'(t)$ by the quotient rule:

$$f'(t) = \frac{(t - 1) - (t + 1)}{(t - 1)^2} = -\frac{2}{(t - 1)^2}.$$

Since $f'(0) = -2$ is negative, f is decreasing, so it is getting colder. ▲

Instead of focusing our attention on the small intervals used so far, one can also consider the idea of increasing and decreasing functions on general intervals, which could be large.

Let f be a function defined on an interval I. If $f(x_1) < f(x_2)$ for all $x_1 < x_2$ in I, we say that f is *increasing on I*. If $f(x_1) > f(x_2)$ for all $x_1 < x_2$ in I, we say that f is *decreasing on I*.

It is plausible that *if f is increasing at each point of an interval, then f is increasing on the whole interval* in this new sense. We shall use this important fact now, deferring the formal proof until Section 3.6.

Example 7   On what intervals is $f(x) = x^3 - 2x + 6$ increasing or decreasing?

Solution   We consider the derivative $f'(x) = 3x^2 - 2$. This is positive when $3x^2 - 2 > 0$, i.e., when $x^2 > 2/3$, i.e., either $x > \sqrt{2/3}$ or $x < -\sqrt{2/3}$. Similarly, $f'(x) < 0$ when $x^2 < 2/3$, i.e., $-\sqrt{2/3} < x < \sqrt{2/3}$. Thus, f is increasing on the intervals $(-\infty, -\sqrt{2/3})$ and $(\sqrt{2/3}, \infty)$, and f is decreasing on $(-\sqrt{2/3}, \sqrt{2/3})$. ▲

The result of Example 7 enables us to make a good guess at what the graph $y = f(x)$ looks like. We first plot the points where $x = \pm\sqrt{2/3}$ as in Fig. 3.2.11(a). When $x = \pm\sqrt{2/3}$, we get $y = 6 \mp 4\sqrt{6}/9$. We also plot the point $x = 0$, $y = 6$. Since f is increasing on $(-\infty, -\sqrt{2/3})$ and on $(\sqrt{2/3}, \infty)$, and decreasing elsewhere, the graph must look something like that in Fig. 3.2.11(b). Later in this chapter, we will use techniques like this to study graphing more systematically.

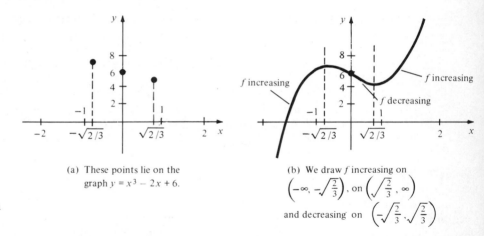

(a) These points lie on the graph $y = x^3 - 2x + 6$.

(b) We draw f increasing on $\left(-\infty, -\sqrt{\frac{2}{3}}\right)$, on $\left(\sqrt{\frac{2}{3}}, \infty\right)$ and decreasing on $\left(-\sqrt{\frac{2}{3}}, \sqrt{\frac{2}{3}}\right)$

Figure 3.2.11. First steps in sketching a graph.

Example 8 Match each of the functions in the left-hand column of Fig. 3.2.12 with its derivative in the right-hand column.

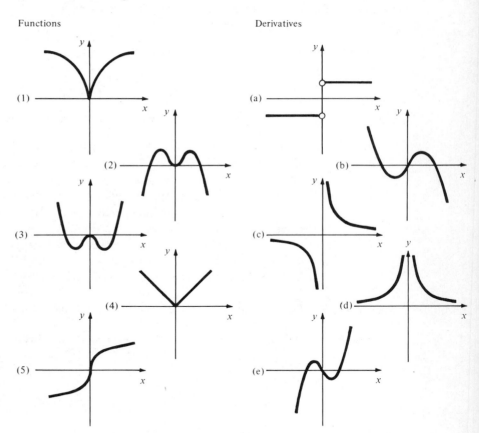

Figure 3.2.12. Matching functions and their derivatives.

Solution Function (1) is decreasing for $x < 0$ and increasing for $x > 0$. The only functions in the right-hand column which are negative for $x < 0$ and positive for $x > 0$ are (a) and (c). We notice, further, that the derivative of function 1 is not constant for $x < 0$ (the slope of the tangent is constantly changing), which eliminates (a). Similar reasoning leads to the rest of the answers, which are: (1)-(c), (2)-(b), (3)-(e), (4)-(a), (5)-(d). ▲

We now turn our attention to the points which separate the intervals on which a function is increasing or decreasing, such as the points $x = \pm\sqrt{2/3}$ in Fig.

3.2.11. In this figure, we see that these points are places where f has a maximum and a minimum. Here are the formal definitions: A point x_0 is called a *local minimum point* for f if there is an open interval (a, b) around x_0 such that $f(x) \geqslant f(x_0)$ for all x in (a, b). Similarly, a point x_0 is called a *local maximum point* for f if there is an open interval (a, b) around x_0 such that $f(x) \leqslant f(x_0)$ for all x in (a, b).[2] (See Fig. 3.2.13.)

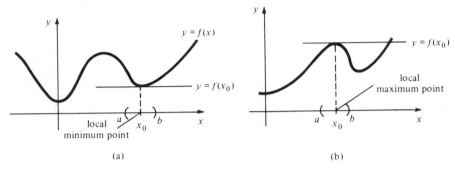

Figure 3.2.13. The graph of f lies above the line $y = f(x_0)$ near a local minimum point (a) and below that line near a local maximum point (b).

(a)   (b)

Example 9   For each of the functions in Fig. 3.2.14, tell whether x_0 is a local minimum point, a local maximum point, or neither.

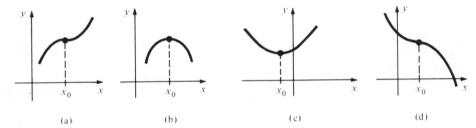

Figure 3.2.14. Is x_0 a local maximum? Local minimum? Neither?

(a)   (b)   (c)   (d)

Solution   Comparing each graph in Fig. 3.2.14 with the horizontal line through the heavy dot, we find that x_0 is a local maximum point in (b), a local minimum point in (c), and neither in (a) and (d). ▲

At a local maximum or minimum point x_0, a function f can be neither increasing nor decreasing, as we see from a comparison of the definitions or graphic interpretations of these concepts. It follows that the derivative $f'(x_0)$ (if it exists) can be neither positive nor negative at such a point; hence, it must be zero. Points x_0 where $f'(x_0) = 0$ are called *critical points* of f.

The critical point test described in the following display is very important. (Some people remember nothing else after a year of calculus.) A good portion of this chapter explores the applications of the test and its limitations.

Critical Point Test

If x_0 is a local maximum or minimum point of a (differentiable) function f, then x_0 is a critical point, i.e., $f'(x_0) = 0$.

In Figs. 3.2.8 and 3.2.9, you will observe that not every critical point is a local maximum or minimum. In the remainder of this section we shall develop a test for critical points to be maxima or minima. This will be useful for both

---

[2] Sometimes the phrase "strict local minimum point" is used when $f(x) > f(x_0)$ for all x in (a, b) other than x_0. Likewise for strict local maximum points. Here, and elsewhere, we use the term "point" to refer to a number in the domain of f rather than to a point in the plane.

graphing and problem solving. To lead to the test, we ask this question: How does a function f behave just to the right of a critical point x_0? The two simplest possibilities are that $f'(x) > 0$ for all x in some interval (x_0, b) or that $f'(x) < 0$ in such an interval. In the first case, f is increasing on (x_0, b), and the second case is decreasing there. These possibilities are illustrated in Fig. 3.2.15.

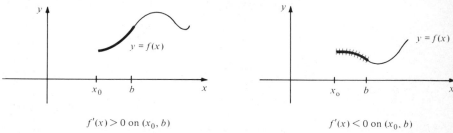

Figure 3.2.15. Behavior to the right of a critical point.

$f'(x) > 0$ on (x_0, b) $f'(x) < 0$ on (x_0, b)

Likewise, the behavior of f just to the left of x_0 can be determined if we know the sign of $f'(x)$ on an interval (a, x_0), as shown in Fig. 3.2.16.

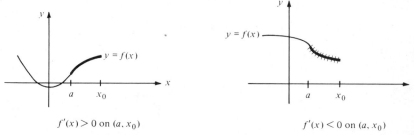

Figure 3.2.16. Behavior to the left of a critical point.

$f'(x) > 0$ on (a, x_0) $f'(x) < 0$ on (a, x_0)

The two possibilities in Fig. 3.2.15 can be put together with the two possibilities in Fig. 3.2.16 to give four different ways in which a function may behave near a critical point, as shown in Fig. 3.2.17.

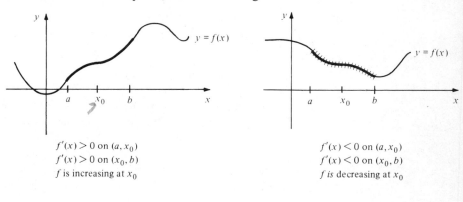

$f'(x) > 0$ on (a, x_0) $f'(x) < 0$ on (a, x_0)
$f'(x) > 0$ on (x_0, b) $f'(x) < 0$ on (x_0, b)
f is increasing at x_0 f is decreasing at x_0

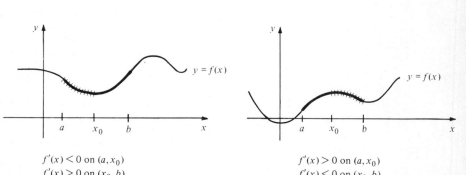

Figure 3.2.17. Four ways in which a function can behave near a critical point.

$f'(x) < 0$ on (a, x_0) $f'(x) > 0$ on (a, x_0)
$f'(x) > 0$ on (x_0, b) $f'(x) < 0$ on (x_0, b)
x_0 is a local minimum point x_0 is a local maximum point

The situations shown in Fig. 3.2.17 are not the only possibilities, but they are the most common. For example, $f'(x)$ might be zero on one side of x_0 or the other, or it might oscillate wildly. For most of the functions we encounter in this course, though, the classification of critical points in Fig. 3.2.17 will suffice. As we have already seen in Fig. 3.2.8, the functions x^3, $-x^3$, x^2, and $-x^2$ provide examples of all four possibilities. Notice that in the case where x_0 is a local maximum or minimum point, f' changes sign at x_0, while f' has the same sign on both sides of a critical point where f is increasing or decreasing.

We summarize our analysis of critical points in the form of a test.

First Derivative Test

Suppose that x_0 is a critical point for $f(x)$, i.e., $f'(x_0) = 0$.

1. If $f'(x)$ changes sign from negative to positive at x_0, then x_0 is a (strict) local minimum point for f.
2. If $f'(x)$ changes sign from positive to negative at x_0, then x_0 is a (strict) local maximum point for f.
3. If $f'(x)$ is negative for all $x \neq x_0$ near x_0, then f is decreasing at x_0.
4. If $f'(x)$ is positive for all $x \neq x_0$ near x_0, then f is increasing at x_0.

Sign of $f'(x)$ near x_0 to the left	Sign of $f'(x)$ near x_0 to the right	Behavior of f at x_0
$-$	$+$	local minimum
$+$	$-$	local maximum
$-$	$-$	decreasing
$+$	$+$	increasing

This test should *not* be literally memorized. If you understand Fig. 3.2.17, you can reproduce the test accurately.

Example 10 Find the critical points of the function $f(x) = 3x^4 - 8x^3 + 6x^2 - 1$. Are they local maximum or minimum points?

Solution We begin by finding the critical points: $f'(x) = 12x^3 - 24x^2 + 12x = 12x(x^2 - 2x + 1) = 12x(x - 1)^2$; the critical points are thus 0 and 1. Since $(x - 1)^2$ is always nonnegative, the only sign change is from negative to positive at 0. Thus 0 is a local minimum point, and f is increasing at 1. ▲

Example 11 Find and classify the critical points of the function

$$f(x) = x^3 + 3x^2 - 6x.$$

Solution The derivative $f'(x) = 3x^2 + 6x - 6$ has roots at $-1 \pm \sqrt{3}$; it is positive on $(-\infty, -1 - \sqrt{3})$ and $(-1 + \sqrt{3}, \infty)$ and is negative on $(-1 - \sqrt{3}, -1 + \sqrt{3})$. Changes of sign occur at $-1 - \sqrt{3}$ (positive to negative) and $-1 + \sqrt{3}$ (negative to positive), so $-1 - \sqrt{3}$ is a local maximum point and $-1 + \sqrt{3}$ is a local minimum point. ▲

Example 12   Discuss the critical points of $y = x^4$ and $y = -x^4$.

Solution   If $f(x) = x^4$, $f'(x) = 4x^3$, and the only critical point is at $x_0 = 0$. We know that x^3 changes sign from negative to positive at 0, so the same is true for $4x^3$, and hence x^4 has a local minimum at 0. Similarly, the only critical point of $-x^4$ is 0, which is a local maximum point. (See Fig. 3.2.18.) ▲

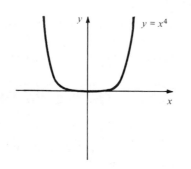

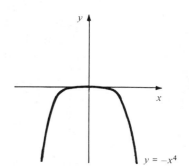

Figure 3.2.18. Critical behavior of $y = x^4$ and $y = -x^4$.

Exercises for Section 3.2

1. Using algebra alone, show that $f(x) = x^2$ is increasing at $x_0 = 3$.
2. Using algebra alone, show that $f(x) = x^2$ is decreasing at $x_0 = -1$.
3. Show by algebra that $f(x) = mx + b$ is increasing for all x_0 if $m > 0$.
4. Show by algebra that $f(x) = mx + b$ is decreasing for all x_0 if $m < 0$.

Using only algebra, find the sign changes of the functions at the indicated points in Exercises 5–8.

5. $f(x) = 2x - 1$; $x_0 = \frac{1}{2}$.
6. $f(x) = x^2 - 1$; $x_0 = -1$.
7. $f(x) = x^5$; $x_0 = 0$.
8. $h(z) = z(z - 2)$; $z_0 = 2$.

In Exercises 9–12, determine whether the functions are increasing, decreasing, or neither at the indicated points.

9. $x^3 + x + 1$; $x_0 = 0$.
10. $x^4 + x + 5$; $x_0 = 0$.
11. $\dfrac{x^2 - 1}{\sqrt{x^2 + 1}}$; $x_0 = 0$.
12. $\dfrac{1}{(x^4 - x^3 + 1)^{3/2}}$; $x_0 = 1$.
13. If $f(t) = t^5 - t^4 + 2t^2$ is the position of a particle on the real-number line at time t, is it moving to the left or right at $t = 1$?
14. A ball is thrown upward with an initial velocity of 30 meters per second. The ball's height above the ground at time t is $h(t) = 30t - 4.9t^2$. When is the ball rising? When is it falling?
15. The annual inflation rate in Uland during 1968 was approximately $r(t) = 20[1 + (t^2 - 6t)/500]$ percent *per year*, where t is the time *in months* from the beginning of the year. During what months was the inflation rate decreasing? What are the max–min points of $r(t)$? Explain their (political) significance.

Derivatives   Functions

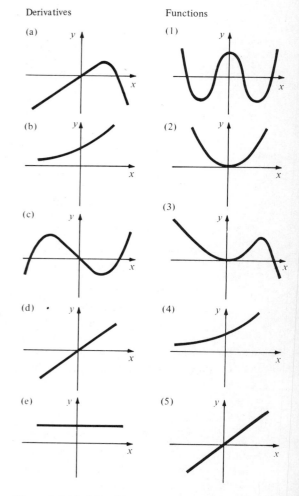

Figure 3.2.19. Matching derivatives and functions (Exercise 21).

16. The rate of a chemical reaction between $t = 0$ and $t = 10$ is given by $r(t) = 2t^3 - 3t^2 + 1$. When is the reaction slowing down? When is it speeding up?

17. Find the intervals on which $f(x) = x^2 - 1$ is increasing or decreasing.

18. Find the intervals on which $x^3 - 3x^2 + 2x$ is increasing or decreasing.

19. Find the points at which $f(x) = 2x^3 - 9x^2 + 12x + 5$ is increasing or decreasing.

20. Find all points in which $f(x) = x^2 - 3x + 2$ is increasing, and at which it changes sign.

21. Match each derivative on the left in Fig. 3.2.19 with the function on the right.

22. Sketch functions whose derivatives are shown in Fig. 3.2.20.

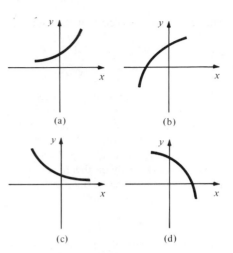

(a) (b)

(c) (d)

Figure 3.2.21. Are f and f' increasing or decreasing?

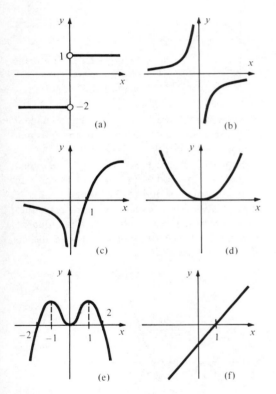

(a) (b)

(c) (d)

(e) (f)

Figure 3.2.20. Sketch functions that have these derivatives.

23. For each of the functions shown in Fig. 3.2.20, state: (1) where it is increasing: (2) where it is decreasing; (3) its local maximum and minimum points; (4) where it changes sign.

24. For each of the functions in Fig. 3.2.21 tell whether the function is increasing or decreasing and whether the derivative of the function is increasing or decreasing.

25. In Fig. 3.2.22, which points are local maxima and which are local minima?

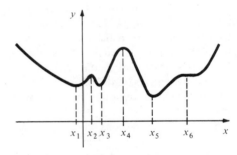

Figure 3.2.22. Find the local maxima and minima.

26. Tell where the function in Fig. 3.2.22 is increasing and where it is decreasing.

Find the critical points of the functions in Exercises 27–34 and decide whether they are local maxima, local minima, or neither.

27. $f(x) = x^2 - 2$

28. $f(x) = \dfrac{x^2 + 1}{x^2 - 1}$

29. $f(x) = x^3 + x^2 - 2$

30. $f(x) = 3x^4$

31. $g(y) = \dfrac{y}{1 + y^2}$

32. $h(z) = \dfrac{z^2}{1 + z^2}$

33. $l(r) = (r^4 - r^2)^2$

34. $m(q) = (q^3 - q)^{347}$

35. Is $f(x) = 1/(x^2 + 1)$ increasing or decreasing at $x = 1, -3, \frac{3}{4}, 25, -36$?

36. Let $f(x) = 4x^2 + (1/x)$. Determine whether f is increasing or decreasing at each of the following points: (a) 1; (b) $-\frac{1}{2}$; (c) -5; (d) $2\frac{1}{3}$.

37. Describe the change of sign at $x = 0$ of the function $f(x) = mx$ for $m = -2, 0, 2$.

38. Describe the change of sign at $x = 0$ of the function $f(x) = mx - x^2$ for $m = -1, -\frac{1}{2}, 0, \frac{1}{2}, 1$.

Find the sign changes (if any) for each of the functions in Exercises 39–42.

39. $x^2 + 3x + 2$

40. $x^2 + x - 1$

41. $x^2 - 4x + 4$

42. $\dfrac{x^3 - 1}{x^2 + 1}$

43. Using only algebra, determine the sign change of $f(x) = (x - r_1)(x - r_2)$ at $x = r_1$, where $r_1 < r_2$.

44. For an observer standing on the Earth, let $f(t)$ denote the angle from the horizon line to the sun at time t. When does $f(t)$ change sign?

Find the intervals on which each of the functions in Exercises 45–48 is increasing or decreasing:

45. $2x^3 - 5x + 7$

46. $x^5 - x^3$

47. $\dfrac{x^2 + 1}{x - 3}$

48. $x^4 - 2x^2 + 1$

49. Find a quadratic polynomial which is zero at $x = 1$, is decreasing if $x < 2$, and is increasing if $x > 2$.

50. Herring production T (in grams) is related to the number N of fish stocked in a storage tank by the equation $T = 500N - 50N^2$.
    (a) Find dT/dN.
    (b) Unless too many fish are stocked, an increase in the number of fish stocked will cause an increase in production at the expense of a reduction in the growth of each fish. (The weight for each fish is T/N.) Explain this statement mathematically in terms of derivatives and the level N^* of stocking which corresponds to maximum production.

★51. Prove the following assertions concerning the function $f(x) = (x^3 - 1)/(x^2 - 1)$:
    (a) f can be defined at $x = 1$ so that f becomes continuous and differentiable there, but cannot be so defined at $x = -1$.
    (b) f is increasing on $(-\infty, -2]$ and decreasing on $[-2, -1)$.
    (c) If $a < -1$, then $(a^3 - 1)/(a^2 - 1) \leqslant -3$.
    (d) f is increasing on $[0, \infty)$ and decreasing on $(-1, 0]$. Make up an equality based on this fact.

★52. Prove that f is increasing at x_0 if and only if the function $f(x) - f(x_0)$ changes sign from negative to positive at x_0.

★53. Using the definition of an increasing function, prove that if f and g are increasing at x_0, then so is $f + g$.

★54. Prove that if f and g are increasing and positive on an interval I, then fg is increasing on I.

★55. Let $f(x) = a_0 + a_1 x + a_2 x^2 + \cdots + a_n x^n$. Under what conditions on the a_i's is f increasing at $x_0 = 0$?

★56. Under what conditions on a, b, c, and d is the cubic polynomial $ax^3 + bx^2 + cx + d$ strictly increasing or strictly decreasing on $(-\infty, \infty)$? (Assume $a \neq 0$.)

★57. If g and h are positive functions, find criteria involving $g'(x)/g(x)$ and $h'(x)/h(x)$ to tell when (a) the product $g(x)h(x)$ and (b) the quotient $g(x)/h(x)$ are increasing or decreasing.

★58. Let f be a function, and $a > 0$ a positive real number. Discuss the relation between the critical points of $f(x)$, $af(x)$, $a + f(x)$, $f(ax)$, and $f(a + x)$.

★59. Find a cubic polynomial with a graph like the one shown in Fig. 3.2.23.

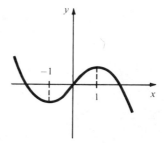

Figure 3.2.23. This is the graph of what cubic polynomial?

★60. (a) Show that there is no quartic polynomial whose graph is consistent with the information shown in Fig. 3.2.20(e).
    (b) Show that if ± 2 is replaced by $\pm\sqrt{2}$, then there is a quartic polynomial consistent with the information in Fig. 3.2.20(e).

★61. Find a relationship between the (positive) values of a and b which insures that there is a quartic polynomial with a graph consistent with the information in Fig. 3.2.24.

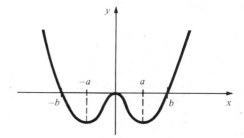

Figure 3.2.24. When is this the graph of a quartic?

3.3 The Second Derivative and Concavity

The sign of the second derivative indicates which way the graph of a function is bending.

In the last section we saw that the classification of critical points of a function $f(x)$ depends on the sign changes of the derivative $f'(x)$. On the other hand, the sign changes of $f'(x)$ at a critical point of f are determined by the sign of the derivative of $f'(x)$, i.e., by the sign of the *second* derivative $f''(x)$. (See Fig. 3.3.1.) After exploring the consequences of this idea, we shall see that the sign of $f''(x)$ is important even when x is not a critical point of f.

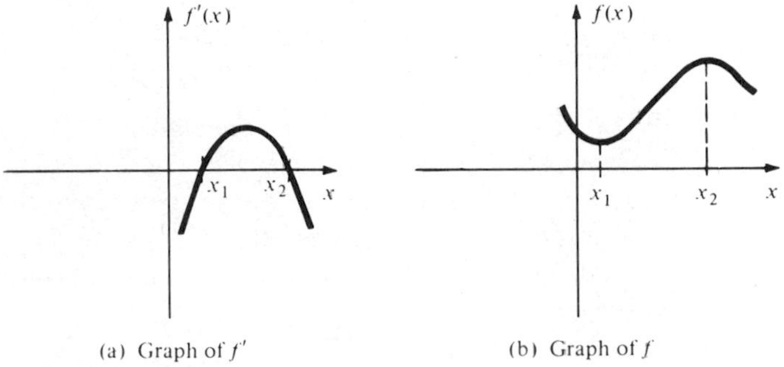

Figure 3.3.1. $f''(x_1) > 0$ so f' is increasing at x_1 and changes sign from negative to positive; thus f has a local minimum at x_1. Likewise $f''(x_2) < 0$ and x_2 is a local maximum.

(a) Graph of f' (b) Graph of f

Recall that if $g(x_0) = 0$ and $g'(x_0) > 0$, then $g(x)$ changes sign from negative to positive. Applying this to the case where g is the derivative f' of a function f, we find that for a critical point x_0 of $f(x)$, $f'(x)$ changes sign from negative to positive if $f''(x_0) > 0$. Thus, by the first derivative test, x_0 is a local minimum point. Similar reasoning when $f''(x_0) < 0$ leads to the following test.

Second Derivative Test

Suppose that $f'(x_0) = 0$.

1. If $f''(x_0) > 0$, then x_0 is a (strict) local minimum point.
2. If $f''(x_0) < 0$, then x_0 is a (strict) local maximum point.
3. If $f''(x_0) = 0$, the test is inconclusive.

We will discuss the case $f''(x_0) = 0$ shortly. For now, notice that the functions $y = x^3$, $y = -x^3$, $y = x^4$, $y = -x^4$ (Figs. 3.2.8 and 3.2.18) show that various things can happen in this case.

Example 1 Use the second derivative test to analyze the critical points of the function $f(x) = x^3 - 6x^2 + 10$.

Solution Since $f'(x) = 3x^2 - 12x = 3x(x - 4)$, the critical points are at 0 and 4. Since $f''(x) = 6x - 12$, we find that $f''(0) = -12 < 0$ and $f''(4) = 12 > 0$. By the second derivative test, 0 is a local maximum point and 4 is a local minimum point. ▲

Example 2   Analyze the critical points of $f(x) = x^3 - x$.

Solution $f'(x) = 3x^2 - 1$ and $f''(x) = 6x$. The critical points are zeros of $f'(x)$; that is, $x = \pm(1/\sqrt{3})$; $f''(-1/\sqrt{3}) = -(6/\sqrt{3}) < 0$ and $f''(1/\sqrt{3}) = 6/\sqrt{3} > 0$. By the second derivative test, $-(1/\sqrt{3})$ is a local maximum point and $1/\sqrt{3}$ is a local minimum point. ▲

When $f''(x_0) = 0$, the second derivative test is inconclusive. We may sometimes use the first derivative test to analyze the critical points, however.

Example 3   Analyze the critical point $x_0 = -1$ of $f(x) = 2x^4 + 8x^3 + 12x^2 + 8x + 7$.

Solution   The derivative is $f'(x) = 8x^3 + 24x^2 + 24x + 8$, and $f'(-1) = -8 + 24 - 24 + 8 = 0$, so -1 is a critical point. Now $f''(x) = 24x^2 + 48x + 24$, so $f''(-1) = 24 - 48 + 24 = 0$, and the second derivative test is inconclusive. If we factor f', we find $f'(x) = 8(x^3 + 3x^2 + 3x + 1) = 8(x + 1)^3$. Thus -1 is the only root of f', $f'(-2) = -8$, and $f'(0) = 8$, so f' changes sign from negative to positive at -1; hence, -1 is a local minimum point for f. ▲

Whether or not $f'(x_0)$ is zero, the sign of $f''(x_0)$ has an important geometric interpretation: it tells us which way the tangent line to the graph of f turns as the point of tangency moves to the right along the graph (see Fig. 3.3.2). The two graphs in Fig. 3.3.2 are bent in opposite directions. The graph in part (a) is said to be *concave upward*; the graph in part (b) is said to be *concave downward*.

Figure 3.3.2. (a) The slope of the tangent line is increasing; $f''(x) > 0$.
(b) The slope of the tangent line is decreasing; $f''(x) < 0$.

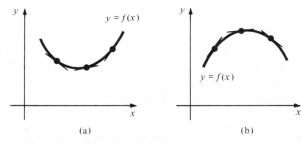

(a)   (b)

We can give precise definitions of upward and downward concavity by considering how the graph of f lies in relation to one of its tangent lines. To accomplish this, we compare $f(x)$ with its linear approximation at x_0: $l(x) = f(x_0) + f'(x_0)(x - x_0)$. If there is an open interval (a, b) about x_0 such that $f(x) > l(x)$ for all x in (a, b) other than x_0, then f is called *concave upward* at x_0. If, on the other hand, there is an open interval (a, b) about x_0 such that $f(x) < l(x)$ for all x in (a, b) other than x_0, then f is called *concave downward* at x_0.

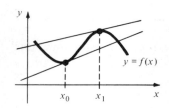

Figure 3.3.3. The function f is concave upward at x_0 and concave downward at x_1.

Geometrically, f is concave upward (downward) at x_0 if the graph of f lies locally above (below) its tangent line at x_0, as in Fig. 3.3.3.

Notice that the difference $h(x) = f(x) - l(x)$ is positive or negative according to whether $f(x) > l(x)$ or $f(x) < l(x)$. Since $h(x_0) = f(x_0) - l(x_0) = 0$, we see that f is concave upward or downward at x_0 according to whether $h(x)$ has a local minimum or maximum at x_0. (See Fig. 3.3.4.)

If we differentiate $h(x) = f(x) - [f(x_0) + f'(x_0)(x - x_0)]$ twice, we obtain $h'(x) = f'(x) - f'(x_0)$ and $h''(x) = f''(x)$ (x_0 is treated as a constant). Notice that $h'(x_0) = 0$, so x_0 is a critical point for h. Next, observe that $h''(x_0) = f''(x_0)$, so we may conclude from the second derivative test for local maxima and minima that x_0 is a local minimum for h if $f''(x_0) > 0$ and a local maximum if $f''(x_0) < 0$. Thus we have the test in the next box. (Once again, the functions $x^3, -x^3, x^4, -x^4$ at $x_0 = 0$ illustrate the possibilties in case 3.)

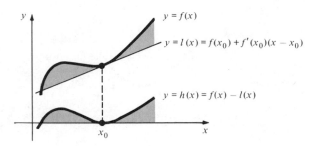

Figure 3.3.4. The function f is concave upward at x_0 when the difference $h(x)$ between $f(x)$ and its linear approximation at x_0 has a local minimum at x_0.

The Second Derivative Test for Concavity

1. If $f''(x_0) > 0$, then f is concave upward at x_0.
2. If $f''(x_0) < 0$, then f is concave downward at x_0.
3. If $f''(x_0) = 0$, then f may be either concave upward at x_0, concave downward at x_0, or neither.

Example 4 Discuss the concavity of $f(x) = 4x^3$ at the points $x = -1$, $x = 0$, and $x = 1$.

Solution We have $f'(x) = 12x^2$ and $f''(x) = 24x$, so $f''(-1) = -24$, $f''(0) = 0$, and $f''(1) = 24$. Therefore, f is concave downward at -1 and concave upward at 1. At zero the test is inconclusive; we can see, however, that f is neither concave upward nor downward by noticing that f is increasing at zero, so that it crosses its tangent line at zero (the x axis). That is, f is neither above nor below its tangent line near zero, so f is neither concave up nor concave down at zero. ▲

Example 5 Find the intervals on which $f(x) = 3x^3 - 8x + 12$ is concave upward and on which it is concave downward. Make a rough sketch of the graph.

Solution Differentiating f, we get $f'(x) = 9x^2 - 8$, $f''(x) = 18x$. Thus f is concave upward when $18x > 0$ (that is, when $x > 0$) and concave downward when $x < 0$. The critical points occur when $f'(x) = 0$, i.e., at $x = \pm \sqrt{8/9} = \pm \frac{2}{3}\sqrt{2}$. Since $f''(-\frac{2}{3}\sqrt{2}) < 0$, $-\frac{2}{3}\sqrt{2}$ is a local maximum, and since $f''(\frac{2}{3}\sqrt{2}) > 0$, $\frac{2}{3}\sqrt{2}$ is a local minimum. This information is sketched in Fig. 3.3.5. ▲

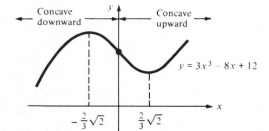

Figure 3.3.5. The critical points and concavity of $3x^3 - 8x + 12$.

We have just seen that a function f is concave upward where $f''(x) > 0$ and concave downward where $f''(x) < 0$. Points which separate intervals where f has the two types of concavity are of special interest and are called *inflection points*. More formally, we say that the point x_0 is a *inflection point* for the function f if f is twice differentiable near x_0 and f'' changes sign at x_0. (See Fig. 3.3.6.)

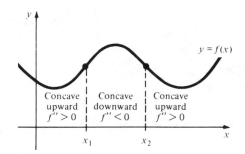

Figure 3.3.6. The inflection points of f are x_1 and x_2.

To see where f'' changes sign, we begin by looking for the points where it is zero. Then we look at the next derivative, $f'''(x_0)$, to see whether a sign change actually takes place.

Inflection Points

An inflection point for f is a point where f'' changes sign. If x_0 is an inflection point for f, then $f''(x_0) = 0$. If $f''(x_0) = 0$ and $f'''(x_0) \neq 0$, then x_0 is an inflection point for f.

Example 6 Find the inflection points of $f(x) = x^2 + (1/x)$.

Solution The first derivative is $f'(x) = 2x - (1/x^2)$, and so the second derivative is $f''(x) = 2 + (2/x^3)$. The only possible inflection points occur where

$$0 = f''(x) = 2 + \frac{2}{x^3}.$$

That is, $x^3 = -1$; hence, $x = -1$. To test whether this is an inflection point, we calculate the third derivative: $f'''(x) = -6/x^4$, so $f'''(-1) = -6 \neq 0$; hence, -1 is a inflection point. ▲

If $f''(x_0) = 0$ and $f'''(x_0) = 0$, then x_0 may or may not be an inflection point. For example, $f(x) = x^4$ does not have an inflection point at $x_0 = 0$ (since $f''(x) = 12x^2$ does not change sign at 0), whereas $f(x) = x^5$ does have an inflection point at $x_0 = 0$ (since $f''(x) = 20x^3$ does change sign at 0). In both cases, $f''(x_0) = f'''(x_0) = 0$, so the test in the preceding box fails in this case.

We can also detect sign changes of f'' by examining the sign of $f''(x)$ in each interval between its roots.

Example 7 Find the inflection points of the function $f(x) = 24x^4 - 32x^3 + 9x^2 + 1$.

Solution We have $f'(x) = 96x^3 - 96x^2 + 18x$, so $f''(x) = 288x^2 - 192x + 18$ and $f'''(x) = 576x - 192$. To find inflection points, we begin by solving $f''(x) = 0$; the quadratic formula gives $x = (4 \pm \sqrt{7})/12$. Using our knowledge of parabolas, we can conclude that f'' changes from positive to negative at $(4 - \sqrt{7})/12$ and from negative to positive at $(4 + \sqrt{7})/12$; thus both are inflection points. One could also evaluate $f'''((4 \pm \sqrt{7})/12)$, but this requires more computation. ▲

Some additional insight into the meaning of inflection points can be obtained by considering the motion of a moving object. If $x = f(t)$ is its position at time t, then the second derivative $d^2x/dt^2 = f''(t)$ is the rate of change of the velocity $dx/dt = f'(t)$ with respect to time—the *acceleration*. We assume that $dx/dt > 0$, so that the object is moving to the right on the number line. If $d^2x/dt^2 > 0$, the velocity is increasing; that is, the object is *accelerating*. If $d^2x/dt^2 < 0$, the velocity is decreasing; the object is *decelerating*. Therefore, a point of inflection occurs when the object switches from accelerating to decelerating or vice versa.

Example 8   In Fig. 3.3.7, tell whether the points x_1 through x_6 are local maxima, local minima, inflection points, or none of these.

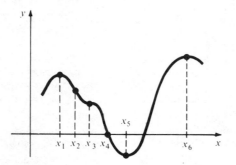

Figure 3.3.7. Locate the local maxima and minima and inflection points.

Solution x_1 is a local maximum point; x_4 is none of these (it is a zero of f);
x_2 is an inflection point; x_5 is a local minimum point;
x_3 is an inflection point; x_6 is a local maximum point. ▲

Exercises for Section 3.3

Use the second derivative test to analyze the critical points of the functions in Exercises 1–8.

1. $f(x) = 3x^2 + 2$
2. $f(x) = x^3 - 6x - 3$
3. $f(x) = 6x^5 - x + 20$
4. $f(x) = x^4 - x^2$
5. $f(x) = \dfrac{x^2 - 1}{x^2 + 1}$
6. $f(x) = \dfrac{x^3 - 1}{x^2 + 1}$
7. $g(s) = \dfrac{s}{1 + s^2}$
8. $h(p) = p + \dfrac{1}{p}$

Find the intervals on which the functions in Exercises 9–16 are concave upward and those on which they are concave downward:

9. $f(x) = 3x^2 + 8x + 10$
10. $f(x) = x^3 + 3x + 8$
11. $f(x) = x^4$
12. $f(x) = \dfrac{1}{x}$
13. $f(x) = \dfrac{x}{x - 1}$
14. $f(x) = \dfrac{1}{1 + x^2}$
15. $f(x) = x^3 + 4x^2 - 8x + 1$.
16. $f(x) = (x - 2)^3 + 8$.

Find the inflection points for the functions in Exercises 17–24.

17. $f(x) = x^3 - x$
18. $f(x) = x^4 - x^2 + 1$
19. $f(x) = x^7$
20. $f(x) = x^6$
21. $f(x) = (x - 1)^4$
22. $f(x) = 2x^3 + 3x$
23. $f(x) = \dfrac{1}{1 + x^2}$
24. $f(x) = \dfrac{x^2 - 1}{x^2 + 1}$

25. In each of the graphs of Fig. 3.3.8, tell whether x_0 is a local maximum point, a local minimum point, an inflection point, or none of these.

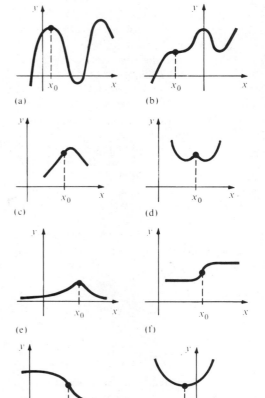

Figure 3.3.8. Is x_0 a local maximum? A local minimum? An inflection point? Neither?

26. Identify each of the points x_1 through x_6 in Fig. 3.3.9 as a local maximum point, local minimum point, inflection point, or none of these.

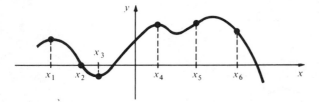

Figure 3.3.9. Classify the points $x_1, \ldots, x_6$.

Find the local maxima, local minima, and inflection points of each of the functions in Exercises 27–30. Also find the intervals on which each function is increasing, decreasing, and concave upward and downward.

27. $\frac{1}{4}x^2 - 1$ 28. $\dfrac{1}{x(x-1)}$

29. $x^3 + 2x^2 - 4x + \frac{3}{2}$ 30. $x^2 - x^4$

31. Find the inflection points for x^n, n a positive integer. How does the answer depend upon n?

32. Is an inflection point always, sometimes, or never a critical point? Explain.

33. Find a function with inflection points at 1 and 2. [*Hint*: Start by writing down $f''(x)$. Then figure out what $f'(x)$ and $f(x)$ should be.]

34. Find a function with local maxima or minima at 1 and 3 and an inflection point at 2.

35. (a) Relate the sign of the error made in the linear approximation of f to the second derivative of f.
    (b) Apply your conclusion to the linear approximation of $1/x$ at $x_0 = 1$.

36. (a) Use the second derivative to compare x^2 with $9 + 6(x - 3)$ for x near 3. (b) Show by algebra that $x^2 > 9 + 6(x - 3)$ for all $x \neq 3$.

37. Let $f(x) = x^3 - x$.
    (a) Find the linear approximations to f at $x_0 = -1, 0$, and 1.
    (b) For each such x_0, compare the value of $f(x_0 + \Delta x)$ with the linear approximation for $\Delta x = \pm 1, \pm 0.1, \pm 0.01$. How does the error depend upon $f''(x_0)$?

38. Show that if $f'(x_0) = f''(x_0) = 0$ and $f'''(x_0) \neq 0$, then x_0 is *not* a local maximum or local minimum point for f.

39. The power output of a battery is given by $P = EI - RI^2$, where E and R are positive constants.
    (a) For which current I is the power P a local maximum? Justify using the second derivative test.
    (b) What is the maximum power?

40. A generator of E volts is connected to an inductor of L henrys, a resistor of R ohms, and a second resistor of x ohms. Heat is dissipated from the second resistor, the power P being given by

$$P = \frac{E^2 x}{(2\pi L)^2 + (x + R)^2}.$$

(a) Find the resistance value x_0 which makes the power as large as possible. Justify with the second derivative test.
(b) Find the maximum power which can be achieved by adjustment of the resistance x.

41. A rock thrown upward attains a height $s = 3 + 40t - 16t^2$ feet in t seconds. Using the second derivative test, find the maximum height of the rock.

42. An Idaho cattle rancher owns 1600 acres adjacent to the Snake River. He wishes to make a three-sided fence from 2 miles of surplus fencing, the enclosure being set against the river to make a rectangular corral. If x is the length of the short side of the fence, then $A = x(2 - 2x)$ is the area enclosed by the fence (assuming the river is straight).
    (a) Show that the maximum area occurs when $x = \frac{1}{2}$, using the second derivative test.
    (b) Verify that the maximum area enclosed is 0.5 square mile.
    (c) Verify that the fence dimensions are $\frac{1}{2}, 1, \frac{1}{2}$ miles, when the area enclosed is a maximum.

★43. Sketch the graphs of continuous functions on $(-\infty, \infty)$ with the following descriptions. (If you think no such function can exist, state that as your answer.)
    (a) Three local maxima or minima and two points of inflection.
    (b) Two local maxima or minima and three points of inflection.
    (c) Four local maxima or minima and no points of inflection.
    (d) Two (strict) local maxima and no (strict) local minima.

★44. Suppose that $f'(x_0) = f''(x_0) = f'''(x_0) = 0$, but $f''''(x_0) \neq 0$. Is x_0 a local maximum point, a local minimum point, or an inflection point of f? Give examples to show that anything can happen if $f''''(x_0) = 0$.

★45. If $f(x)$ is positive for all x, do $f(x)$ and $1/f(x)$ have the same inflection points?

★46. Prove that no odd-degree polynomial can be everywhere concave upward. (As part of your solution, give a few simple examples and include a brief discussion of the possibilities for even-degree polynomials.)

★47. Prove the following theorem, which shows that the tangent line at a point of inflection crosses the graph:

Let x_0 be an inflection point for f and let $h(x) = f(x) - [f(x_0) + f'(x_0)(x - x_0)]$ be the difference between f and its linear approximation at x_0. If f'' changes sign from negative to positive [or positive to negative] at x_0 (for example, if $f'''(x_0)$ > 0 [or < 0]), then h changes sign from negative to positive [or positive to negative] at x_0.

The two cases are illustrated in Fig. 3.3.10.

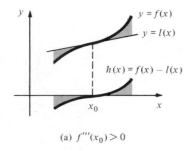

(a) $f'''(x_0) > 0$

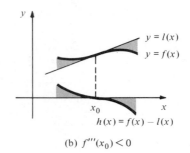

(b) $f'''(x_0) < 0$

Figure 3.3.10. The graph of f crosses its tangent line at a point of inflection.

3.4 Drawing Graphs

Using calculus to determine the principal features of a graph often produces better results than simple plotting.

One of the best ways to understand a function is to see its graph. The simplest way to draw a graph is by plotting some points and connecting them with a smooth curve, but this method can lead to serious errors unless we are sure that we have plotted enough points. The methods described in the first three sections of this chapter, combined with the techniques of differentiation, help us make a good choice of which points to plot and show us how to connect the points by a curve of the proper shape.

We begin by outlining a systematic procedure to follow in graphing any function.

Graphing Procedure

To sketch the graph of a function f:

1. Note any *symmetries* of f. Is $f(x) = f(-x)$, or $f(x) = -f(-x)$, or neither? In the first case, f is called *even*; in the second case, f is called *odd*. (See Fig. 3.4.1 and the remarks below.)
2. Locate any points where f is not defined and determine the behavior of f near these points. Also determine, if you can, the behavior of $f(x)$ for x very large positive and negative.
3. Locate the local maxima and minima of f, and determine the intervals on which f is increasing and decreasing.
4. Locate the inflection points of f, and determine the intervals on which f is concave upward and downward.
5. Plot a few other key points, such as x and y intercepts, and draw a small piece of the tangent line to the graph at each of the points you have plotted. (To do this, you must evaluate f' at each point.)
6. Fill in the graph consistent with the information gathered in steps 1 through 5.

Let us examine the graphing procedure beginning with step 1. If f is even—that is, $f(x) = f(-x)$—we may plot the graph for $x \geq 0$ and then reflect the result across the y axis to obtain the graph for $x \leqslant 0$. (See Fig. 3.4.1 (a).) If f is odd, that is, $f(x) = -f(-x)$ then, having plotted f for $x \geqslant 0$, we may reflect the graph in the y axis and then in the x axis to obtain the graph for $x \leqslant 0$. (See Fig. 3.4.1 (b).)

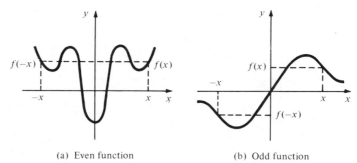

Figure 3.4.1. f is even when $f(-x) = f(x)$ and odd when $f(-x) = -f(x)$.

(a) Even function   (b) Odd function

To decide whether a function is even or odd, substitute $-x$ for x in the expression for $f(x)$ and see if the resulting expression is the same as $f(x)$, the negative of $f(x)$, or neither.

Example 1   Classify each of the following functions as even, odd, or neither:

(a) $f(x) = x^4 + 3x^2 + 12$;
(b) $g(x) = x/(1 + x^2)$;
(c) $h(x) = x/(1 + x)$;

Solution   (a) $f(-x) = (-x)^4 + 3(-x)^2 + 12 = x^4 + 3x^2 + 12 = f(x)$, so f is even.
(b) $g(-x) = (-x)/(1 + (-x)^2) = -x/(1 + x^2) = -g(x)$, so g is odd.
(c) $h(-x) = (-x)/(1 + (-x)) = -x/(1 - x)$, which does not appear to equal $h(x)$ or $-h(x)$. To be sure, we substitute $x = 2$, for which $h(x) = \frac{2}{3}$ and $h(-x) = 2$; thus, h is neither even nor odd. ▲

Step 2 is concerned with what is known as the *asymptotic* behavior of the function f and is best explained through an example. The asymptotic behavior involves infinite limits of the type $\lim_{x \to x_0} f(x) = \pm \infty$ and $\lim_{x \to \pm\infty} f(x) = l$, as were discussed in Section 1.2.

Example 2   Find the asymptotic behavior of $f(x) = x/(1 - x)$ (f is not defined for $x = 1$).

Solution   For x near 1 and $x > 1$, $1 - x$ is a small negative number, so $f(x) = x/(1 - x)$ is large and negative; for x near 1 and $x < 1$, $1 - x$ is small and positive, so $x/(1 - x)$ is large and positive. Thus we could sketch the part of the graph of f near $x = 1$ as in Fig. 3.4.2. The line $x = 1$ is called a *vertical asymptote* for $x/(1 - x)$. In terms of limits, we write $\lim_{x \to 1-} [x/(1 - x)] = +\infty$ and $\lim_{x \to 1+} [x/(1 - x)] = -\infty$.

Next we examine the behavior of $x/(1 - x)$ when x is large and positive and when x is large and negative. Since both the numerator and denominator also become large, it is not clear what the ratio does. We may note, however, that

$$\frac{x}{1 - x} = \frac{1}{(1/x) - 1}$$

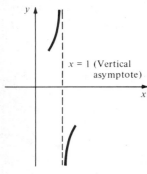

$x = 1$ (Vertical asymptote)

Figure 3.4.2. Pieces of the graph $y = x/(1 - x)$ are plotted near the vertical asymptote $x = 1$.

for $x \neq 0$. As x becomes large (positive or negative), $1/x$ becomes small, and $(1/x) - 1$ approaches -1, so $1/[(1/x) - 1]$ approaches $1/(-1) = -1$, i.e., $\lim_{x \to \infty}[x/(1 - x)] = -1$ and $\lim_{x \to -\infty}[x/(1 - x)] = -1$. Furthermore, when x is large and positive, $1/x > 0$, so $(1/x) - 1 > -1$, and therefore $1/[(1/x) - 1] < 1/(-1) = -1$, so the graph lies near, but below, the line

$y = -1$. Similarly, for x large and negative, $1/x < 0$, so $(1/x) - 1 < -1$, and therefore $1/[(1/x) - 1] > -1$, so the graph lies near, but above, the line $y = -1$. Thus, we could sketch the part of the graph for x large as in Fig. 3.4.3. The line $y = -1$ is called a *horizontal asymptote* for f. ▲

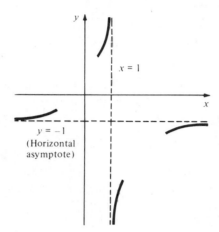

Figure 3.4.3. Pieces of the graph $y = x/(1 - x)$ are plotted near the horizontal asymptote $y = -1$.

Steps 3 and 4 were described in detail in Sections 3.2 and 3.3; step 5 increases the accuracy of plotting, and step 6 completes the job. These steps will be carried out in detail in the examples that follow. The graph $y = x/(1 - x)$ begun above is discussed again in Example 9.

Some words of advice: It is important to be systematic; follow the procedure step by step, and introduce the information on the graph as you proceed. A haphazard attack on a graph often leads to confusion and sometimes to desperation. Just knowing steps 1 through 6 is not enough—you must be able to employ them effectively. The only way to develop this ability is through practice.

Example 3   Sketch the graph of $f(x) = x - \dfrac{1}{x}$.

Solution   We carry out the six-step procedure:

1. $f(-x) = -x + (1/x) = -f(x)$; f is odd, so we need only study $f(x)$ for $x \geqslant 0$.
2. f is not defined for $x = 0$. For x small and positive, $-(1/x)$ is large in magnitude and negative in sign, so $x - (1/x)$ is large and negative as well; $x = 0$ is a vertical asymptote. For x large and positive, $-(1/x)$ is small and negative; thus the graph of $f(x) = x - (1/x)$ lies below the line $y = x$, approaching the line as x becomes larger. The line $y = x$ is again called an *asymptote* (see Fig. 3.4.4).

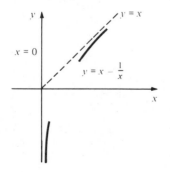

Figure 3.4.4. The lines $x = 0$ and $y = x$ are asymptotes.

3. $f'(x) = 1 + (1/x^2)$, which is positive for all $x \neq 0$. Thus f is always increasing and there are no maxima or minima.
4. $f''(x) = -(2/x^3)$, which is negative for all $x > 0$; f is concave downward on $(0, \infty)$.
5. The intercept occurs where $x - (1/x) = 0$; that is, $x = 1$. We have $f'(1) = 2$, $f(2) = \frac{3}{2}$, $f'(2) = \frac{5}{4}$.

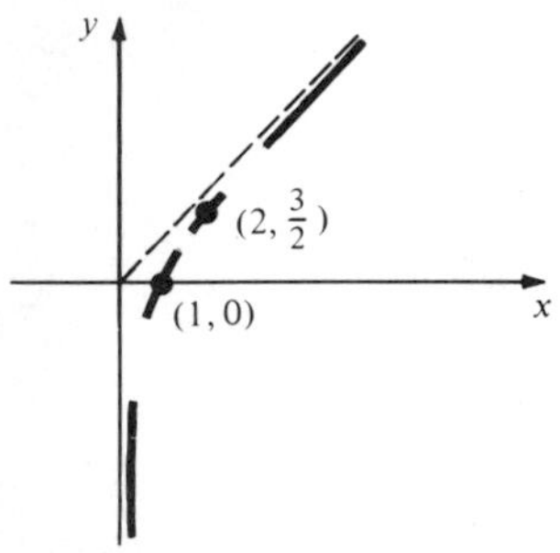

The information obtained in steps 1 through 5 is placed on the graph in Fig. 3.4.5.

6. We fill in the graph for $x > 0$ (Fig. 3.4.6). Finally, we use the fact that f is odd to obtain the other half of the graph by reflecting through the x and y axes. (See Fig. 3.4.7.) ▲

Figure 3.4.5. The information obtained from steps 1 to 5.

Figure 3.4.6. The graph for $x > 0$ is filled in (step 6).

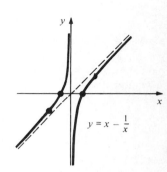

Figure 3.4.7. The complete graph is obtained by using the fact that f is odd.

▦ Calculator Remark

While calculators enable one to plot points relatively quickly, and computers will plot graphs from formulas, the use of calculus is still essential. A calculator can be deceptive if used alone, as we saw in Example 7, Section R.6. In Chapter 14 we will see how the computer can help us graph complicated surfaces in space, but it may be unwise to begin expensive computation before a thorough analysis using calculus. Of course, it may be even quicker to solve a simple problem by calculus than to go to a machine for plotting. ▲

Example 4   Sketch the graph of $f(x) = \dfrac{x}{1 + x^2}$.

Solution   Again we carry out the six-step procedure:

1. $f(-x) = -x/[1 + (-x)^2] = -x/[1 + x^2] = -f(x)$; f is odd, so its graph must by symmetric when reflected in the x and y axes.
2. Since the denominator $1 + x^2$ is never zero, the function is defined everywhere; there are no vertical asymptotes. For $x \neq 0$, we have

$$f(x) = \frac{x}{1 + x^2} = \frac{1}{x + (1/x)}.$$

Since $1/x$ becomes small as x becomes large, $f(x)$ looks like $1/(x + 0) = 1/x$ for x large. Thus $y = 0$ is a horizontal asymptote; the graph is below $y = 0$ for x large and negative and above $y = 0$ for x large and positive.

3. $$f'(x) = \frac{1 + x^2 - x(2x)}{(1 + x^2)^2} = \frac{(1 - x^2)}{(1 + x^2)^2},$$

which vanishes when $x = \pm 1$. To check the sign of $f'(x)$ on $(-\infty, -1)$, $(-1, 1)$, and $(1, \infty)$, we evaluate it at conveniently chosen points: $f'(-2) = -\frac{3}{25}$, $f'(0) = 1$, $f'(2) = -\frac{3}{25}$. Thus f is decreasing on $(-\infty, -1)$ and on $(1, \infty)$ and f is increasing on $(-1, 1)$. Hence -1 is a local minimum and 1 is a local maximum by the first derivative test.

4. $$f''(x) = \frac{(1 + x^2)^2(-2x) - (1 - x^2) \cdot 2(1 + x^2)2x}{(1 + x^2)^4} = \frac{2x(x^2 - 3)}{(1 + x^2)^3}.$$

This is zero when $x = 0$, $\sqrt{3}$, and $-\sqrt{3}$. Since the denominator of f'' is positive, we can determine the sign by evaluating the numerator. Evaluating at -2, -1, 1, and 2, we get -4, 4, -4, and 4, so f is concave downward on $(-\infty, -\sqrt{3})$ and $(0, \sqrt{3})$ and concave upward on $(-\sqrt{3}, 0)$ and $(\sqrt{3}, \infty)$; $-\sqrt{3}$, 0, and $\sqrt{3}$ are points of inflection.

5. $$f(0) = 0; \qquad f'(0) = 1,$$
$$f(1) = \tfrac{1}{2}; \qquad f'(1) = 0,$$
$$f(\sqrt{3}) = \tfrac{1}{4}\sqrt{3}; \qquad f'(\sqrt{3}) = -\tfrac{1}{8}.$$

The only solution of $f(x) = 0$ is $x = 0$.

The information obtained in steps 1 through 5 is placed tentatively on the graph in Fig. 3.4.8. As we said in step 1, we need do this only for $x \geqslant 0$.

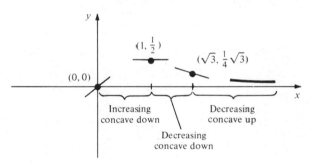

Figure 3.4.8. The graph $y = x/(1 + x^2)$ after steps 1 to 5.

6. We draw the final graph, remembering to obtain the left-hand side by reflecting the right-hand side in both axes. (You can get the same effect by rotating the graph $180°$, keeping the origin fixed.) The result is shown in Fig. 3.4.9. ▲

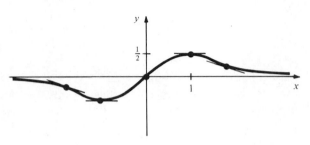

Figure 3.4.9. The complete graph $y = x/(1 + x^2)$.

Example 5 Sketch the graph of $f(x) = 2x^3 + 8x + 1$.

Solution The six steps are as follows:

1. $f(-x) = -2x^3 - 8x + 1$ is not equal to $f(x)$ or $-f(x)$, so f is neither even nor odd.

2. f is defined everywhere. We may write

$$f(x) = 2x^3\left(1 + \frac{4}{x^2} + \frac{1}{2x^3}\right).$$

For x large, the factor $1 + (4/x^2) + (1/2x^3)$ is near 1, so $f(x)$ is large and positive for x large and positive, and large and negative for x large and negative. There are no horizontal or vertical asymptotes.

3. $f'(x) = 6x^2 + 8$, which vanishes nowhere and is always positive. Thus f is increasing on $(-\infty, \infty)$ and has no critical points.

4. $f''(x) = 12x$, which is negative for $x < 0$ and positive for $x > 0$. Thus f is concave downward on $(-\infty, 0)$ and concave upward on $(0, \infty)$; zero is a point of inflection.

5.
$$f(0) = 1; \qquad f'(0) = 8,$$
$$f(1) = 11; \qquad f'(1) = 14,$$
$$f(-1) = -9; \qquad f'(-1) = 14.$$

The information obtained so far is plotted in Fig. 3.4.10.

6. A look at Fig. 3.4.10 suggests that the graph will be very long and thin. In fact, $f(2) = 33$, which is way off the graph. To get a useful picture, we may stretch the graph horizontally by changing units on the x axis so that a unit on the x axis is, say, four times as large as a unit on the y axis. We add a couple of additional points by calculating

$$f(\tfrac{1}{2}) = 5\tfrac{1}{4}; \qquad f'(\tfrac{1}{2}) = 9\tfrac{1}{2},$$
$$f(-\tfrac{1}{2}) = -3\tfrac{1}{4}; \qquad f'(-\tfrac{1}{2}) = 9\tfrac{1}{2}.$$

Then we draw a smooth curve as in Fig. 3.4.11. ▲

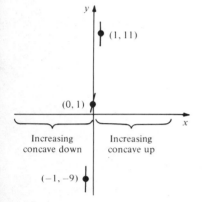

Figure 3.4.10. The graph $y = 2x^3 + 8x + 1$ after steps 1 to 5.

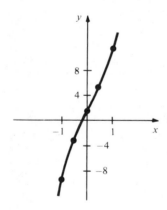

Figure 3.4.11. The completed graph $y = 2x^3 + 8x + 1$.

Any cubic function $y = ax^3 + bx^2 + cx + d$ may be plotted just as the one in the preceding example. The critical points are obtained by solving the quadratic equation

$$f'(x) = 3ax^2 + 2bx + c = 0$$

which may have one or two roots or none. In Example 5, $f'(x) = 0$ had no roots, and f' was always positive. If f' has two roots, $y = f(x)$ will have one local maximum and one local minimum. Let us do an example of this type.

Example 6   Sketch the graph of $f(x) = 2x^3 - 8x + 1$.

Solution   Again, we use the six-step procedure:

1. There are no symmetries.
2. There are no asymptotes. As in Example 5, $f(x)$ is large positive for x large positive and large negative for x large negative.
3. $f'(x) = 6x^2 - 8$, which is zero when $x = \pm\sqrt{4/3} = \pm 2/\sqrt{3} \approx \pm 1.15$. Also, $f'(-2) = f'(2) = 16$ and $f'(0) = -8$, so f is increasing on $(-\infty, -2/\sqrt{3}\,]$ and $[2/\sqrt{3}, \infty)$ and decreasing on $[-2/\sqrt{3}, 2/\sqrt{3}\,]$. Thus, $-2/\sqrt{3}$ is a local maximum point and $2/\sqrt{3}$ a local minimum point.
4. $f''(x) = 12x$, so f is concave downward on $(-\infty, 0)$ and concave upward on $(0, \infty)$. Zero is an inflection point.
5.
$$f(0) = 1; \qquad\qquad\qquad f'(0) = -8,$$
$$f(-2/\sqrt{3}\,) = 1 + 32/3\sqrt{3} \approx 7.16; \quad f'(-2/\sqrt{3}\,) = 0,$$
$$f(2/\sqrt{3}\,) = 1 - 32/3\sqrt{3} \approx -5.16; \quad f'(2/\sqrt{3}\,) = 0,$$
$$f(-\tfrac{1}{2}) = 4\tfrac{3}{4}; \qquad\qquad\qquad f'(-\tfrac{1}{2}) = -6\tfrac{1}{2},$$
$$f(\tfrac{1}{2}) = -2\tfrac{3}{4}; \qquad\qquad\qquad f'(\tfrac{1}{2}) = -6\tfrac{1}{2},$$
$$f(-2) = 1; \qquad\qquad\qquad f'(-2) = 16,$$
$$f(2) = 1; \qquad\qquad\qquad f'(2) = 16.$$

The data are plotted in Fig. 3.4.12. The scale is stretched by a factor of 4 in the x direction, as in Fig. 3.4.11.
6. We draw the graph (Fig. 3.4.13). ▲

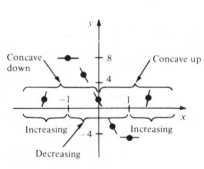

Figure 3.4.12. The graph $y = 2x^3 - 8x + 1$ after steps 1 to 5.

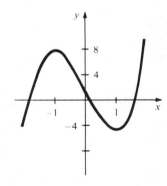

Figure 3.4.13. The completed graph $y = 2x^3 - 8x + 1$.

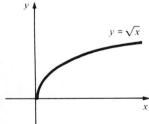

Figure 3.4.14. The graph $y = \sqrt{x}$.

Some interesting new features arise when we graph functions which involve fractional powers. For example, consider the graph of $y = \sqrt{x}$ for $x \geqslant 0$. Notice that the slope $dy/dx = \frac{1}{2}x^{-1/2} = 1/(2\sqrt{x}\,)$ becomes large and positive as $x \to 0$, while dy/dx approaches 0 as $x \to \infty$. Thus, the graph appears as in Fig. 3.4.14.

Something similar happens for the cube root $y = x^{1/3}$, so that $dy/dx = 1/(3x^{2/3})$. This time, the function is defined for all x; its derivative exists for $x \neq 0$ and is large and positive for x near 0, of both signs. (See Fig. 3.4.15.) We sometimes say that the graph $y = x^{1/3}$ has a *vertical tangent* at $x = 0$.

You may notice that the graphs $y = x^{1/2}$ and $y = x^{1/3}$ resemble $y = x^2$

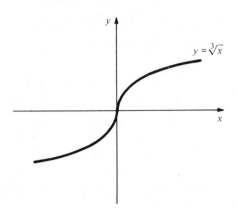

Figure 3.4.15. The graph $y = \sqrt[3]{x}$ has a "vertical tangent line" at the origin.

and $y = x^3$ turned on their sides. This relationship will be explored when we study inverse functions in Section 5.3.

Still more interesting is the graph of $y = x^{2/3}$, which is also defined for all x. The derivative is $dy/dx = \frac{2}{3}x^{-1/3} = 2/(3\sqrt[3]{x})$. For x near 0 and positive, dy/dx is large and positive, whereas for x near 0 and negative, dy/dx is large and negative. Thus, the graph has the appearance shown in Fig. 3.4.16. Again,

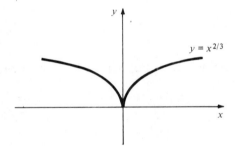

Figure 3.4.16. The graph $y = x^{2/3}$ has a cusp at the origin.

we can say that the graph has a vertical tangent at $x = 0$. However, the shape of the graph near $x = 0$ has not been encountered before. We call $x = 0$ a *cusp*. Note that $x = 0$ is a minimum point of $f(x) = x^{2/3}$, but that $x^{2/3}$ is not differentiable there.

In general, a continuous function f is said to have a *cusp* at x_0 if $f'(x)$ has opposite signs on opposite sides of x_0 but $f'(x)$ "blows up" at x_0 in the sense that $\lim_{x \to x_0}[1/f'(x)] = 0$; thus $\lim_{x \to x_0\pm} f'(x) = \pm\infty$ or $\mp\infty$.

Example 7   Let $f(x) = (x^2 + 1)^{3/2}$. (a) Where is f increasing? (b) Sketch the graph of f. Are there any cusps?

Solution   (a) Using the chain rule, we get $f'(x) = 3x\sqrt{x^2 + 1}$. Hence $f'(x) < 0$ (so f is decreasing) on $(-\infty, 0)$, and $f'(x) > 0$ (so f is increasing) on $(0, \infty)$.

(b) By the first derivative test, $x = 0$ is a local minimum point. Note that $f(x)$ is an even function and

$$f''(x) = 3\left(\frac{x^2}{\sqrt{x^2 + 1}} + \sqrt{x^2 + 1} \right) > 0,$$

so f is concave upward. Thus we can sketch its graph as in Fig. 3.4.17. There are no cusps. ▲

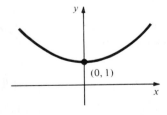

Figure 3.4.17. The graph $y = (x^2 + 1)^{3/2}$.

Example 8 Sketch the graph of $(x + 1)^{2/3}x^2$.

Solution Letting $f(x) = (x + 1)^{2/3}x^2$, we have

$$f'(x) = \tfrac{2}{3}(x + 1)^{-1/3}x^2 + (x + 1)^{2/3} \cdot 2x$$

$$= \left[2x/3(x + 1)^{1/3}\right](4x + 3).$$

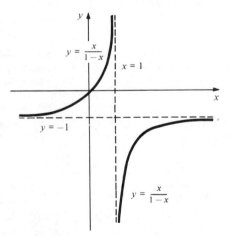

For x near -1, but $x > -1$, $f'(x)$ is large positive, while for $x < -1$, $f'(x)$ is large negative. Since f is continuous at -1, this is a local minimum and a cusp.

  The other critical points are $x = 0$ and $x = -\tfrac{3}{4}$. From the first derivative test (or second derivative test, if you prefer), $-\tfrac{3}{4}$ is a local maximum and zero is a local minimum. For $x > 0$, f is increasing since $f'(x) > 0$; for $x < -1$, f is decreasing since $f'(x) < 0$. Thus we can sketch the graph as in Fig. 3.4.18. (We located the inflection points at $(-33 \pm \sqrt{609}\,)/40 \approx -0.208$ and -1.442 by setting the second derivative equal to zero.) ▲

Figure 3.4.18. The graph $y = x^2(x + 1)^{2/3}$ has a cusp at $x = -1$.

Sometimes algebraic transformations simplify the job of drawing a graph.

Example 9 Sketch the graph $y = x/(1 - x)$ by (a) the six-step procedure and (b) by making the transformation $u = 1 - x$.

Solution (a) In Example 1 we carried out steps 1 and 2. To carry out step 3, we compute using the quotient rule:

$$\frac{dy}{dx} = \frac{(1 - x) - x(-1)}{(1 - x)^2} = \frac{1}{(1 - x)^2}.$$

Since dy/dx is always positive (undefined if $x = 1$), the graph has no maxima and minima and the function is increasing on the intervals $(-\infty, 1)$ and $(1, \infty)$. For step 4, we compute that

$$\frac{d^2y}{dx^2} = \frac{2}{(1 - x)^3},$$

so the graph is concave upward on $(-\infty, 1)$ and concave downward on $(1, \infty)$. For step 5 we note that $y = 0$ when $x = 0$ and $y = -2$ when $x = 2$. At both of these points dy/dx is 1. The graph is then plotted in Fig. 3.4.19.

Figure 3.4.19. The graph $y = x/(1 - x)$.

  (b) If we let $u = x - 1$, then $x = 1 + u$, so $y = (1 + u)/(-u) = -1 - 1/u$. Letting $z = y + 1$, we get $z = -1/u$. This graph is easy to sketch and may already be familiar to you as a hyperbola in the second and fourth quadrants of the uz plane. Translating axes (just as we did for parabolas in Section R.2) produces the same result as in Fig. 3.4.19, in which the dashed lines are the u and z axes. ▲

Supplement to Section 3.4
The General Cubic

In Section R.2, we saw how to plot the general linear function, $f(x) = ax + b$, and the general quadratic function, $f(x) = ax^2 + bx + c$ ($a \neq 0$). The methods of this section yield the same results: the graph of $ax + b$ is a straight line, while the graph of $f(x) = ax^2 + bx + c$ is a parabola, concave upward if $a > 0$ and concave downward if $a < 0$. Moreover, the maximum (or minimum) point of the parabola occurs when $f'(x) = 2ax + b = 0$, that is, $x = -(b/2a)$, which is the same result as was obtained by completing the square.

A more ambitious task is to determine the shape of the graph of the general cubic $f(x) = ax^3 + bx^2 + cx + d$. (We assume that $a \neq 0$; otherwise, we are dealing with a quadratic or linear function.) Of course, any specific cubic can be plotted by techniques already developed, but we wish to get an idea of what *all possible* cubics look like and how their shapes depend on a, b, c, and d.

We begin our analysis with some simplifying transformations. First of all, we can factor out a and obtain a new polynomial $f_1(x)$ as follows:

$$f(x) = a\left(x^3 + \frac{b}{a}x^2 + \frac{c}{a}x + \frac{d}{a}\right) = af_1(x).$$

The graphs of f and f_1 have the same basic shape; if $a > 0$, the y axis is just rescaled by multiplying all y values by a; if $a < 0$, the y axis is rescaled and the graph is flipped about the x axis. It follows that we not lose any generality by assuming that the coefficient of x^3 is 1.

Therefore, consider the simpler form

$$f_1(x) = x^3 + b_1x^2 + c_1x + d_1,$$

where $b_1 = b/a$, $c_1 = c/a$, and $d_1 = d/a$. In trying to solve cubic equations, mathematicians of the early Renaissance noticed a useful trick: if we replace x by $x - (b_1/3)$, then the quadratic term drops out; that is,[3]

$$f_1\left(x - \frac{b_1}{3}\right) = x^3 + c_2x + d_2,$$

---

[3] These algebraic ideas are related to a formula for the roots of a general cubic, discovered by Niccolo Tartaglia (1506–1559) but published by (without Tartaglia's permission) and often credited to Girolamo Cardano (1501–1576). Namely, the solutions of the cubic equation $x^3 + bx^2 + cx + d = 0$ are

$$x_1 = S + T - \frac{b}{3}, \qquad\qquad \text{where} \quad S = \sqrt[3]{R + \sqrt{Q^3 + R^2}}$$

$$x_2 = -\frac{1}{2}(S + T) - \frac{b}{3} + \frac{1}{2}\sqrt{-3}(S - T), \qquad\qquad T = \sqrt[3]{R - \sqrt{Q^3 + R^2}}$$

$$x_3 = -\frac{1}{2}(S + T) - \frac{b}{3} - \frac{1}{2}\sqrt{-3}(S - T), \quad Q = \frac{3c - b^2}{9}, \quad R = \frac{9bc - 27d - 2b^3}{54}$$

There is also a formula for the roots of a quartic equation, but a famous theorem (due to Abel and Ruffini in the nineteenth century) states that there can be no such algebraic formula for the general equation of degree $\geqslant 5$. Modern proofs of this theorem can be found in advanced textbooks on the algebra [such as L. Goldstein, *Abstract Algebra*, Prentice-Hall, (1973).] These proofs are closely related to the proof of the impossibility of trisecting angles with ruler and compass.

where c_2 and d_2 are new constants, depending on b_1, c_1, and d_1. (We leave to the reader the task of verifying this last statement and expressing c_2 and d_2 in terms of b_1, c_1, and d_1; see Exercise 56.)

The graph of $f_1(x - b_1/3) = f_2(x)$ is the same as that of $f_1(x)$ except that it is shifted by $b_1/3$ units along the x axis. This means that we lose no generality by assuming that the coefficient of x^2 is zero—that is, we only need to graph $f_2(x) = x^3 + c_2x + d_2$. Finally, replacing $f_2(x)$ by $f_3(x) = f_2(x) - d_2$ just corresponds to shifting the graph d_2 units along the y axis.

We have now reduced the graphing of the general cubic to the case of graphing $f_3(x) = x^3 + c_2x$. For simplicity let us write $f(x)$ for $f_3(x)$ and c for c_2. To plot $f(x) = x^3 + cx$, we go through steps 1 to 6:

1. f is odd.
2. f is defined everywhere. Since $f(x) = x^3(1 + c/x^2)$, $f(x)$ is large and positive (negative) when x is large and positive (negative); there are no horizontal or vertical asymptotes.
3. $f'(x) = 3x^2 + c$. If $c > 0$, $f'(x) > 0$ for all x, and f is increasing everywhere. If $c = 0$, $f'(x) > 0$ except at $x = 0$, so f is increasing everywhere even though the graph has a horizontal tangent at $x = 0$. If $c < 0$, $f'(x)$ has roots at $\pm\sqrt{-c/3}$; f is increasing on $(-\infty, -\sqrt{-c/3}\,]$ and $[\sqrt{-c/3}, \infty)$ and decreasing on $[-\sqrt{-c/3}, \sqrt{-c/3}\,]$. Thus, $-\sqrt{-c/3}$ is a local maximum point and $\sqrt{-c/3}$ is a local minimum point.
4. $f''(x) = 6x$, so f is concave downward for $x < 0$ and concave upward for $x > 0$. Zero is an inflection point.
5. $f(0) = 0$, $f'(0) = c$,

$$f(\pm\sqrt{-c}\,) = 0, \qquad\qquad f'(\pm\sqrt{-c}\,) = -2c \quad (\text{if } c < 0);$$

$$f\left(\pm\sqrt{\frac{-c}{3}}\,\right) = \pm\frac{2}{3}c\sqrt{\frac{-c}{3}}, \qquad f'\left(\pm\sqrt{\frac{-c}{3}}\,\right) = 0 \quad (\text{if } c < 0).$$

6. We skip the preliminary sketch and draw the final graphs. (See Fig. 3.4.20.)

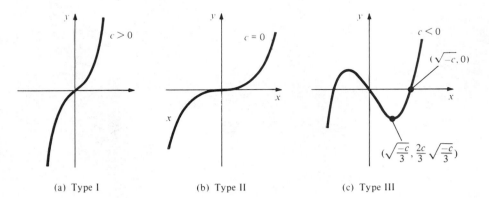

Figure 3.4.20. The graph of $y = x^3 + cx$ for: $c > 0$ (I); $c = 0$ (II); and $c < 0$ (III).

(a) Type I   (b) Type II   (c) Type III

Thus there are three types of cubics (with $a > 0$):

Type I: f is increasing at all points, $f' > 0$ everywhere.
Type II: f is increasing at all points; $f' = 0$ at one point.
Type III: f has a maximum and a minimum.

Type II is the transition type between types I and III. You may imagine the graph changing as c begins with a negative value and then moves toward zero. As c gets smaller and smaller, the turning points move in toward the origin and the bumps merge at the point where $f'(x) = 0$. As c passes zero to become positive, the bumps disappear completely.

Example 10   Convert $2x^3 + 3x^2 + x + 1$ to the form $x^3 + cx$ and determine whether the cubic is of type I, II, or III.

Solution $\frac{1}{2}(2x^3 + 3x^2 + x + 1) = x^3 + \frac{3}{2}x^2 + \frac{1}{2}x + \frac{1}{2}$. Substituting $x - \frac{1}{2}$ for x (since the coefficient of x^2 is $\frac{3}{2}$) gives

$$\left(x - \tfrac{1}{2}\right)^3 + \tfrac{3}{2}\left(x - \tfrac{1}{2}\right)^2 + \tfrac{1}{2}\left(x - \tfrac{1}{2}\right) + \tfrac{1}{2}$$

$$= x^3 - \tfrac{3}{2}x^2 + \tfrac{3}{4}x - \frac{1}{8} + \tfrac{3}{2}\left(x^2 - x + \tfrac{1}{4}\right) + \tfrac{1}{2}\left(x - \tfrac{1}{2}\right) + \tfrac{1}{2}$$

$$= x^3 - \tfrac{1}{4}x + \tfrac{1}{2}.$$

Thus, after being shifted along the x and y axes, the cubic becomes $x^3 - \frac{1}{4}x$. Since $c < 0$, it is of type III. ▲

Exercises for Section 3.4

In Exercises 1–4, is the function even, odd, or neither?

1. $f(x) = \dfrac{x^3 + 6x}{x^2 + 1}$ 2. $f(x) = x$

3. $f(x) = \dfrac{x}{x^3 + 1}$ 4. $f(x) = x^6 + 8x^2 + 3$

In Exercises 5–8, describe the behavior of the given functions near their vertical asymptotes:

5. $f(x) = \dfrac{x}{x^3 + 1}$ 6. $f(x) = \dfrac{x}{x^3 - 1}$

7. $f(x) = \dfrac{x^3 + 1}{x^2 - 1}$ 8. $f(x) = \dfrac{x^3 - 1}{x^2 - 1}$

Sketch the graphs of the functions in Exercises 9–30.

9. $\dfrac{x^2}{1 - x^2}$ 10. $\dfrac{3x^2 + 4}{x^2 - 9}$

11. $\dfrac{x(x - 2)}{x - 1}$ 12. $\dfrac{x^3 + 1}{x^2 - 1}$

[Hint: Let $u = x - 1$]

13. $x^4 - x^3$ 14. $x^4 + x^3$

15. $x^4 - x^2$ 16. $x^4 - 3x^2 + 2\sqrt{2}$

17. $-x^3 + x + 1$ 18. $-x^3 + \frac{3}{2}x^2 + x + 1$

19. $x^3 + \frac{1}{4}x^2 - \frac{7}{8}x - \frac{3}{8}$ 20. $x^3 + 3x^2 + 2x$

21. $\dfrac{x^3}{2 - x^2} + x$ 22. $\dfrac{x^2 + 1}{(x - 1)(x + 1)^2}$

23. $\dfrac{x^2}{2 + x}$ 24. $x^2 + \dfrac{1}{1 - x}$

25. $x^3 + 7x^2 - 2x + 10$ 26. $3x^3 + x^2 + 1$

27. $\dfrac{1 - x^2}{1 + x^2}$ 28. $\dfrac{x}{1 - x}$

29. $8x^3 - 3x^2 + 2x$ 30. $x^4 - 6x^2 + 5x + 2$
[*Hint:* Sketch the derivative first.]

31. Match the following functions with the graphs in Fig. 3.4.21.

(a) $\dfrac{x^2 + 1}{x^2 - 1}$,

(b) $x + \dfrac{1}{x}$,

(c) $x^3 - 3x^2 - 9x + 1$,

(d) $\dfrac{x^2 - 1}{x^2 + 1}$.

A 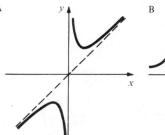   B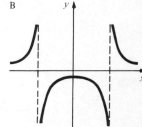

C 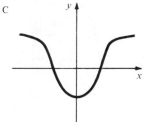   D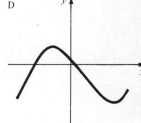

Figure 3.4.21. Match these graphs with the functions in Exercise 31.

(A)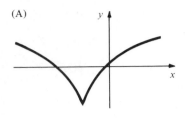

(B)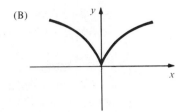

(C)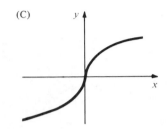

(D)

Figure 3.4.22. Match these graphs with the functions in Exercise 32.

32. Match the following functions with the graphs in Fig. 3.4.22: (a) $x^{2/7}$, (b) $x^{3/7}$, (c) $(1 + x)^{2/7} - 1$, (d) $(1 + x^2)^{1/7} - 1$.

33. Let $f(x) = (x^2 - 3)^{2/3}$. (a) Where is f increasing? Decreasing? (b) Sketch the graph of f, noting any cusps.

34. Show that the graph of

$$f(x) = (1 - x)^{2/3}(1 + x)^{2/3}$$

has two cusps. Sketch.

Sketch the graph of each of the functions in Exercises 35–40.

35. $(x^2 + 1)^{1/3}$ 36. $(x^2 + 5x + 4)^{2/7}$
37. $(x - 1)^{4/3}(x + 1)^{2/3}$ 38. $x + x^{2/3}$
39. $(x - 4)^{100/99}$ 40. $(x^3 + 2x^2 + x)^{4/9}$.

41. Let $f(x)$ be a polynomial. Show that $f(x)$ is an even (odd) function if only even (odd) powers of x occur with nonzero coefficients in $f(x)$.

42. Find a criterion for telling when the quotient $f(x)/g(x)$ of two polynomials is even, odd, or neither.

43. A simple model for the voting population in a certain district is given by $N(t) = 30 + 12t^2 - t^3$, $0 \leqslant t \leqslant 8$, where t is the time in years, N the population in thousands.
    (a) Graph N versus t on $0 \leqslant t \leqslant 8$.
    (b) At what time t will the population of voters increase most rapidly?
    (c) Explain the significance of the points $t = 0$ and $t = 8$.

44. The population P of mice in a wood varies with the number x of owls in the wood according to the formula $P = 30 + 10x^2 - x^3$, $0 \leqslant x \leqslant 10$. Graph P versus x.

★45. Suppose that $f(x)$ is defined on all of $(-\infty, \infty)$. Show that

$$f(x) = e(x) + o(x),$$

where e is an even function and o is an odd function. [*Hint*: Substitute $-x$ for x, use the fact that e is even and o is odd, and solve for $e(x)$ and $o(x)$.]

★46. There is one function which is both even and odd. What is it?

★47. If f is twice differentiable and x_0 is a critical point of f, must x_0 be either a local maximum, local minimum, or inflection point?

★48. What does the graph of $[ax/(bx + c)] + d$ look like if a, b, c, and d are positive constants?

★49. Prove that the graph of any cubic $f(x) = ax^3 + bx^2 + cx + d$ ($a \neq 0$) is symmetric about its inflection point in the sense that the function

$$g(x) = f\left(x - \frac{b}{3a}\right) - f\left(-\frac{b}{3a}\right)$$

is odd.

★50. A drug is injected into a person's bloodstream and the temperature increase T recorded one hour later. If x milligrams are injected, then

$$T(x) = \frac{x^2}{8}\left(1 - \frac{x}{16}\right), \qquad 0 \leqslant x \leqslant 16.$$

    (a) The rate of change T with respect to dosage x is called the *sensitivity* of the body to the dosage. Find it.
    (b) Use the techniques for graphing cubics to graph T versus x.
    (c) Find the dosage at which the sensitivity is maximum.

★51. Convert $x^3 + x^2 + 3x + 1$ to the form $x^3 + cx$ and determine whether the cubic is of type I, II, or III.

★52. Convert $x^3 - 3x^2 + 3x + 1$ to the form $x^3 + cx$ and determine whether the cubic is of type I, II, or III.

★53. Suppose that an object has position at time t given by $x = (t - 1)^{5/3}$. Discuss its velocity and acceleration near $t = 1$ with the help of a graph.

★54. Consider $g(x) = f(x)(x - a)^{p/q}$, where f is differentiable at $x = a$, p is even. q is odd and $p < q$. If $f(a) \neq 0$, show that g has a cusp at $x = a$. [*Hint*: Look at $g'(x)$ for x on either side of a.]

⋆55. For which values of a and p does the cubic $g(x) = ax^3 + px$ have zero, one, or two critical points? (Assume $a \neq 0$).

⋆56. This problem concerns the graph of the general cubic (see the Supplement to this Section):

(a) Find an explicit formula for the coefficient c_2 in $f(x - b_1/3)$ in terms of b_1, c_1, and d_1 and thereby give a simple rule for determining whether the cubic

$$x^3 + b_1 x^2 + c_2 x + d_1$$

is of type I, II, or III.

(b) Give a rule, in terms of a, b, c, d, for determining the type of the general cubic $ax^3 + bx^2 + cx + d$.

(c) Use the quadratic formula on the derivative of $ax^3 + bx^2 + cx + d$ to determine, in terms of a, b, c, and d, how many turning points there are. Compare with the result in part (b).

Exercises 57 to 64 concern the graph of the general quartic:

$$f(x) = ax^4 + bx^3 + cx^2 + dx + e, \qquad a \neq 0.$$

⋆57. Using the substitution $x - b/4a$ for x, show that one can reduce to the case

$$f(x) = x^4 + cx^2 + dx + e$$

(with a new c, d and e!).

⋆58. According to the classification of cubics, $f'(x) = 4x^3 + 2cx + d$ can be classified into three types: I $(c > 0)$, II $(c = 0)$, and III $(c < 0)$, so we may name each quartic by the type of its derivative. Sketch the graph of a typical quartic of type I.

⋆59. Divide type II quartics into three cases: II_1 $(d > 0)$, II_2 $(d = 0)$, II_3 $(d < 0)$. Sketch their graphs with $e = 0$.

⋆60. In case III $(c < 0)$, $f'(x)$ has two turning points and can have one, two, or three roots. By considering Fig. 3.4.23, show that the sign of $f'(x)$ at its critical points determines the number of roots. Obtain thereby a classification of type III quartics into five subtypes III_1, III_2, III_3, III_4, and III_5. Sketch the graphs for each case and determine the conditions on c and d which govern the cases.

⋆61. Using your results from Exercises 59 and 60, show that the (c, d) plane may be divided into regions, as shown in Fig. 3.4.24, which determine the type of quartic.[4]

Using the results of Exercise 61, classify and sketch the graphs of the quartics in Exercises 62–64.

⋆62. $x^4 - 3x^2 - 4x$.

⋆63. $x^4 + 4x^2 + 6x$.

⋆64. $x^4 + 7x$.

[4] For advanced (and sometimes controversial) applications of this figure, see T. Poston and I. Stuart, *Catastrophe Theory and Its Applications*, Pitman (1978).

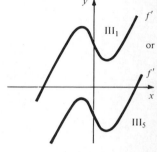

(a) One root: (f' has the same sign at its two critical points)

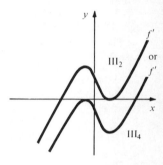

(b) Two roots: (f' is zero at one of its critical points)

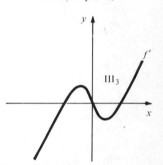

Figure 3.4.23. The five possible positions of the graph of f' in case III.

(c) Three roots: (f' has opposite signs at its critical points)

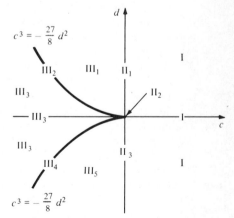

Figure 3.4.24. Locating the value of (c, d) in this graph tells the type of quartic.

3.5 Maximum–Minimum Problems

A step-by-step procedure aids in the solution of practical maximum–minimum problems.

This section is concerned with maximum–minimum problems of two types. First, we consider the problem of maximizing or minimizing a function on an interval. Then we apply these ideas to maximum–minimum problems that are presented in words rather than in formulas. Students are often overwhelmed by such "word problems," which appear to admit no systematic means of solution. Fortunately, guidelines do exist for attacking these problems; they are best learned through practice.

The maximum and minimum points discussed in Section 3.2 were *local*, since we compared $f(x_0)$ with $f(x)$ for x near x_0. For many applications of calculus, however, it is important to find the points where $f(x)$ has the largest or smallest possible value as x ranges over an entire given interval. In this section, we show how calculus helps us to locate these *global* maximum and minimum points, and we discuss how to translate word problems into calculus problems involving maxima and minima.

Global maxima and minima should be as familiar to you as the daily weather report. The statement on the 6 P.M. news that "today's high temperature was 26°C" means that:

1. At no time today was the temperature higher than 26°C.
2. At some time today, the temperature was exactly 26°C.

If we let f be the function which assigns to each t in the interval $[0, 18]$ the temperature in degrees at time t hours after midnight, then we may say that 26 is the (global) *maximum value of f on* $[0, 18]$. It is useful to have a formal definition.

Maximum and Minimum Values on Intervals

Let f be a function which is defined on an interval I of real numbers.
  If M is a real number such that:

1. $f(x) \leqslant M$ for all x in I
2. $f(x_0) = M$ for at least one x_0 in I

then we call M the *maximum value of f on I*.

  If m is a real number such that:

1. $f(x) \geqslant m$ for all x in I
2. $f(x_1) = m$ for at least one x_1 in I

then we call m the *minimum value of f on I*.

The numbers x_0 and x_1 in the definition of maximum and minimum values represent the points at which these values are attained. They are called maximum or minimum *points* for f on I. For the temperature function discussed above, x_0 might be 15.5, indicating that the high temperature occurred at 3:30 P.M. Of course, it might be possible that the temperature rose to 26° at 2 P.M. dipped due to a sudden rain shower, rose again to 26° at 3:30 P.M. and finally decreased toward evening. In that case both 14 and 15.5 would be acceptable values for x_0. (See Fig. 3.5.1.) Note that there were *local minima at $t = 11$ (11 A.M.), and $t = 15$ (3 P.M.).*

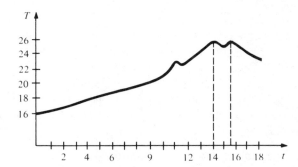

Figure 3.5.1. The maximum temperature (26°) occurred at 2 P.M. ($t = 14$) and also at 3:30 P.M. ($t = 15.5$).

Sometimes, to distinguish the points considered here from the local maxima and minima discussed in Section 3.2, we call them *global* maximum and minimum points.

We will see shortly that differential calculus provides a powerful technique for locating the maximum and minimum points of functions defined by formulas. For functions defined by other means, there are sometimes direct approaches to finding maxima and minima. For example:

1. The highest ring around a bathtub indicates the maximum water level achieved since the tub was last scrubbed. (Other rings indicate local maxima.)
2. On a maximum–minimum thermometer, one can read directly the maximum and minimum temperatures reached since the last time the thermometer was reset.
3. If the graph of a function is available, either from experimental data or by plotting from a formula as in Sections R.6 and 3.4, the maxima and minima can be seen as the high and low points on the graph.

Example 1 Find the maximum and minimum points and values of $f(x) = 1/(1 + x^2)$ on the interval $[-2, 2]$.

Solution $f(x)$ is largest where its denominator is smallest, and vice versa. The maximum point occurs, therefore, at $x = 0$; the maximum value is 1. The minimum points occur at -2 and 2; the minimum value is $\frac{1}{5}$.

We may verify these statements with the assistance of some calculus: $f'(x) = -2x/(1 + x^2)^2$, which is positive for $x < 0$ and negative for $x > 0$, so f is increasing on $[-2, 0]$ and decreasing on $[0, 2]$. It follows that

$$\tfrac{1}{5} = f(-2) < f(x) < f(0) = 1 \qquad \text{for} \quad -2 < x < 0$$

and

$$1 = f(0) > f(x) > f(2) = \tfrac{1}{5} \qquad \text{for} \quad 0 < x < 2,$$

and we have $\frac{1}{5} \leqslant f(x) \leqslant 1$ for $-2 \leqslant x \leqslant 2$. (See Fig. 3.5.2.) ▲

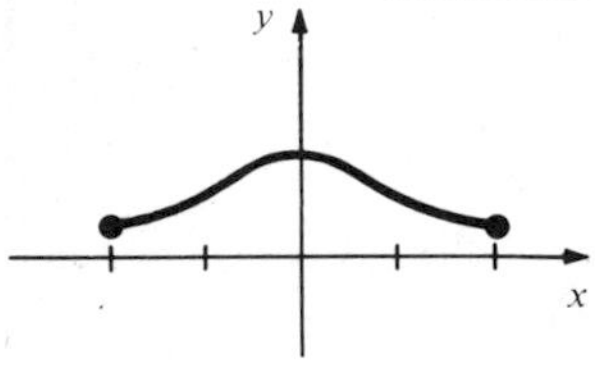

Figure 3.5.2. The function $f(x) = 1/(1 + x^2)$ on $[-2, 2]$ has a maximum point at $x = 0$ and minimum points at $x = \pm 2$.

Example 2 Find the maximum and minimum points and values, if they exist, for the function $f(x) = x^2 + 1$ on each of the following intervals:

(a) $(-\infty, \infty)$ (c) $(-1, 1)$ (e) $(0, 1]$ (g) $(-2, 1]$
(b) $(0, \infty)$ (d) $[-1, 1]$ (f) $[\frac{1}{2}, 1]$ (h) $[-2, 1)$

Solution   The graphs are indicated in Fig. 3.5.3. In case (a), there is no highest point on the graph; hence, no maximum. In case (c), one might be tempted to call 2 the maximum value, but it is not attained at any point of the interval $(-1, 1)$. Study this example well; it will be a useful test case for the general statements to be made later in this section. ▲

	l	Graph	*Maximum points*	*Maximum value*	*Minimum points*	*Minimum value*
(a)	$(-\infty, \infty)$		None	None	0	1
(b)	$(0, \infty)$		None	None	None	None
(c)	$(-1, 1)$		None	None	0	1
(d)	$[-1, 1]$		$-1, 1$	2	0	1
(e)	$(0, 1]$		1	2	None	None
(f)	$[\frac{1}{2}, 1]$		1	2	$\frac{1}{2}$	$\frac{5}{4}$
(g)	$(-2, 1]$	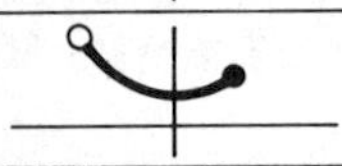	None	None	0	1
(h)	$[-2, 1)$	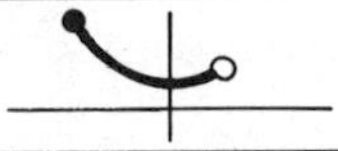	-2	5	0	1

Figure 3.5.3. Solutions to Example 2.

➤ means that the graph goes off to infinity

○ means that the endpoint does not belong to the graph

Example 3   Figure 3.5.4 shows the amount of solar energy received at various latitudes in the northern hemisphere on June 21 on a square meter of horizontal surface located at the top of the atmosphere. Find the maximum and minimum points and values.

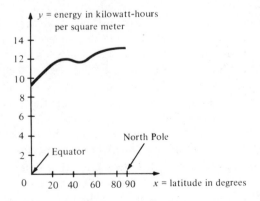

Figure 3.5.4. The solar energy y received on June 21 at the top of the atmosphere at various latitudes x. (See W. G. Kendrew, *Climatology*, Oxford University Press, 1949.)

Solution   The maximum value is about 13 kilowatt-hours, attained at $x = 90$ (the North Pole); the minimum value is about 9 kilowatt-hours, attained at $x = 0$ (the equator). (There are local maximum and minimum points at about $x = 30$ and $x = 50$, but these are not what we are looking for.) ▲

The explanation for the unexpected result in Example 3 is that although the sun's radiation is weakest at the pole, the summer day is longest there, resulting in a larger amount of accumulated energy. At the surface of the earth, absorption of solar energy by the atmosphere is greatest near the pole, since the low angle of the sun makes the rays pass through more air. If this absorption is taken into account, the resulting graph is more like the one in Fig. 3.5.5, which is in better correspondence with the earth's climate.

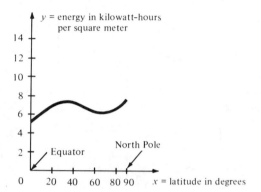

Figure 3.5.5. The solar energy received on June 21 at the surface of the earth, assuming clear skies. (See W. G. Kendrew, *Climatology*, Oxford University Press, 1949.)

If I is a closed interval, the problem of finding maxima and minima for a function f on I is guaranteed to have a solution by the following theoretical result.

Extreme Value Theorem

Let f be continuous on the closed interval $[a,b]$. Then f has both a maximum and a minimum value on $[a,b]$.

The proof of this theorem is omitted,[4] but the statement should be understood by everyone. It says that two conditions together are *sufficient* to ensure that f has both a maximum and a minimum value on I:

1. f is continuous on I.
2. I is a closed interval.

Cases (a), (b), (c), (e), and (g) of Example 2 show that condition 1 alone is not sufficient; that is, if I is not closed, the maxima and minima may not exist. Case (h) shows that the maxima and minima might happen to exist even if I is not closed; thus these two conditions are not *necessary* for the existence of maxima and minima.

Notice that, like the intermediate value theorem, the extreme value theorem is an *existence theorem* which tells you nothing about how to find the maxima and minima. However, combining the extreme value theorem with the critical point test does yield a practical test.

To obtain this test, we first note that by the extreme value theorem, the maximum and minimum points must exist; by the critical point test (Section 3.2), these points must be either critical points or endpoints. It remains, therefore, to determine *which* amongst the critical points and endpoints are the maximum and minimum points; to do this, it suffices to evaluate f at all of them and then compare the values. The following box summarizes the situation.

---

[4] The proof may be found in any of the references listed in the Preface.

Closed Interval Test

To find the maxima and minima for a function f which is continuous on $[a,b]$ and differentiable on (a,b):

1. Make a list $x_1, \ldots, x_n$ consisting of the critical points of f on (a,b) and the endpoints a and b of $[a,b]$.
2. Compute the values $f(x_1), \ldots, f(x_n)$.

The largest of the $f(x_i)$ is the maximum value of f on $[a,b]$. The maximum points for f on $[a,b]$ are those x_i for which $f(x_i)$ equals the maximum value.

The smallest of the $f(x_i)$ is the minimum value of f on $[a,b]$. The minimum points for f on $[a,b]$ are those x_i for which $f(x_i)$ equals the minimum value.

Example 4 Find the maximum and minimum points and values for the function $f(x) = (x^2 - 8x + 12)^4$ on the interval $[-10, 10]$.

Solution The list indicated by the closed interval test consists of $-10, 10$, and the critical points. To find the critical points we differentiate:

$$\frac{d}{dx}(x^2 - 8x + 12)^4 = 4(x^2 - 8x + 12)^3(2x - 8)$$

$$= 8(x - 6)^3(x - 2)^3(x - 4)$$

which is zero when $x = 2$, 4, or 6. We compute the value of f at each of these points and put the results in a table:

x	-10	2	4	6	10
$f(x)$	$(192)^4$	0	$(-4)^4$	0	$(32)^4$

The maximum value is $(192)^4 = 1358954496$; the maximum point is the endpoint -10. The minimum value is zero; the minimum points are the critical points 2 and 6. ▲

For intervals which are not closed, we may use graphing techniques to decide which critical points are maxima and minima.

Example 5 Find maximum and minimum points and values for $y = x^4 - x^2$ on $[-1, \infty)$.

Solution The critical points satisfy the equation $0 = 4x^3 - 2x = 2x(2x^2 - 1)$. The solutions of the equation are 0, $\sqrt{1/2}$, and $-\sqrt{1/2}$, all of which lie in the interval $[-1, \infty)$. We make a table of values of the function

x	-1	$-\sqrt{1/2}$	0	$\sqrt{1/2}$
$x^4 - x^2$	0	$-\frac{1}{4}$	0	$-\frac{1}{4}$

and its derivative

x	-1	$-\frac{1}{2}$	0	$\frac{1}{2}$	1
$4x^3 - 2x$	-2	$\frac{1}{2}$	0	$-\frac{1}{2}$	2

and draw a rough graph (Fig. 3.5.6).

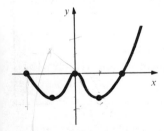

Figure 3.5.6. The graph of $y = x^4 - x^2$ on $[-1, \infty)$.

We see that the critical points $-\sqrt{1/2}$ and $\sqrt{1/2}$ are minimum points; the minimum value is $-\frac{1}{4}$. The endpoint -1 and the critical point 0 are not maximum or minimum points since $x^4 - x^2$ takes on values both greater and less than 0 on the interval $[-1, \infty)$. Thus, there are no maximum points (0 is only a *local* maximum). ▲

We now turn our attention to solving maximum–minimum problems given in words rather than by formulas. To illustrate a general approach to these problems, we will go through the solution of four sample problems step by step.[5]

1. A shepherd lives on a straight coastline and has 500 meters of fencing with which to enclose his sheep. Assuming that he uses the coastline as one side of a rectangular enclosure, what dimensions should the rectangle have in order that the sheep have the largest possible area in which to graze?[6]
2. Illumination from a point light source is proportional to the intensity of the source and inversely proportional to the square of the distance from the source to the point of observation. Given two point sources 10 meters apart, with one source four times as intense as the other, find the darkest point on the line segment joining the sources.
3. Given four numbers, a, b, c, d, find a number x which best approximates them in the sense that the sum of the squares of the differences between x and each of the four numbers is as small as possible.
4. Suppose that it costs $(x^2/100) + 10x$ cents to run your car for x days. Once you sell your car, it will cost you 50 cents a day to take the bus. How long should you keep the car?

The first thing to do is read the problem carefully. Then ask yourself: "What is given? What is required?"

Sometimes it appears that not enough data are given. In Problem 2, for instance, one may think: "Illumination is proportional to the intensity, but I'm not given the constant of proportionality, so the problem isn't workable." It turns out that in this problem, the answer does not depend upon the proportionality constant. (If it did, you could at least express your answer in terms of this unknown constant.) On the other hand, some of the data given in the statement of a problem may be irrelevant. You should do your best at the beginning of solving a problem to decide which data are relevant and which are not.

Here, in full, is the first step in attacking a maximum–minimum word problem.

Step 1: Setting up the Problem

(a) Read the problem carefully, give names to any unnamed relevant variables, and note any relations among the variables.
(b) Draw a figure, if one is appropriate.
(c) Identify the quantity to be maximized or minimized.
(d) Make sure that the relevant and irrelevant information are clearly distinguished.

---

[5] For a general discussion of how to attack a problem, we enthusiastically recommend *How to Solve It*, by G. Polya (Princeton University Press, Second Edition, 1957).

[6] This ancient Greek problem is a variant of a famous problem ingeniously solved by Dido, the daughter of the king of Tyre and founder of Carthage (see M. Kline, *Mathematics: A Cultural Approach*, Addison-Wesley, 1962, p. 114).

Example 6   Carry out step 1 for Problems 1, 2, 3, and 4.

Solution   *Problem* 1: We draw a picture (Fig. 3.5.7). Let l and w denote the length and width of the rectangle, and let A be the area enclosed. These quantities are related by the equations $lw = A$ and $l + 2w = 500$ (since there are 500 meters of fencing available). We want to maximize A.

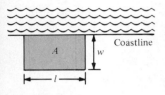

Figure 3.5.7. For which shape is A largest?

Problem 2: We place the first source at $x = 0$ and the second source at $x = 10$ on the real line (Fig. 3.5.8). Denote by I_1 and I_2 the intensities of the two sources. Let $L_1 =$ illumination at x from the first source and $L_2 =$ illumination at x from the second source. Then $L = L_1 + L_2$ is the total illumination.

The given relations are

$$I_2 = 4I_1,$$

$$L_1 = \frac{kI_1}{x^2} \qquad (k > 0 \text{ is a proportionality constant}),$$

$$L_2 = \frac{kI_2}{(10 - x)^2}.$$

We want to minimize L.

Figure 3.5.8. Light sources of intensities I_1 and I_2 are placed at $x = 0$ and $x = 10$. The observer is at x.

Problem 3: Call the unknown number x. We want to minimize

$$y = (x - a)^2 + (x - b)^2 + (x - c)^2 + (x - d)^2.$$

Problem 4: x is the number of days we run the car. At this point in solving the problem, it is completely legitimate to pace around the room, muttering "What should be minimized?" This is not clearly stated in the problem, so we must determine it ourselves. A reasonable objective is to minimize the total amount of money to be paid. How is this to be done? Well, as soon as the cost of running the car exceeds 50 cents per day, we should switch to the bus. So let $y =$ the cost per day of running the car at day x. We want the first x for which $y \geq 50$. The relation between the variables is

$$y = \frac{d}{dx} \left(\frac{x^2}{100} + 10x \right) = \frac{x}{50} + 10. \; \blacktriangle$$

Having set up a problem, we are ready to apply the methods of calculus.

Step 2: Solving the Problem

(a) Write the quantity to be maximized or minimized as a function of one of the other variables in the problem. (This is usually done by expressing all other variables in terms of the one chosen.)

(b) Note any restrictions on the chosen variable.

(c) Find the maxima and minima by the methods of this chapter.

(d) State the answer in words.

The main thing to be mastered in word problems is the technique of translating words into relevant mathematical symbols to which the tools of calculus can be applied. Once the calculus work is done, the answer must then be translated back into the terms of the original word problem.

Example 7 Carry out step 2 for Problems 1, 2, 3, and 4.

Solution *Problem* 1: We want to maximize A, so we write it as a function of l or w; we choose w. Now $l + 2w = 500$, so $l = 500 - 2w$. Thus $A = lw = (500 - 2w)w = 500w - 2w^2$.

The restriction on w is that $0 \leqslant w \leqslant 250$. (Clearly, only a non-negative w can be meaningful, and w cannot be more than 250 or else l would be negative.)

To maximize $A = 500w - 2w^2$ on $[0, 250]$, we compute $dA/dw = 500 - 4w$, which is zero if $w = 125$. Since $d^2A/dw^2 = -4$ for all w, the second derivative test tells us that 125 is a maximum point. Hence the maximum occurs when $w = 125$ and $l = 250$.

The rectangle should be 250 meters long in the direction parallel to the coastline and 125 meters in the direction perpendicular to the coastline in order to enclose the maximal area.

Problem 2: We want to minimize

$$L = L_1 + L_2 = kI_1 \left[\frac{1}{x^2} + \frac{4}{(10 - x)^2} \right].$$

Since the point is to be between the sources, we must have $0 < x < 10$. To minimize L on $(0, 10)$, we compute:

$$L'(x) = kI_1 \left[-\frac{2}{x^3} + \frac{4(20 - 2x)}{(10 - x)^4} \right]$$

$$= 2kI_1 \left[-\frac{1}{x^3} + \frac{4}{(10 - x)^3} \right].$$

The critical points occur when

$$\frac{4}{(10 - x)^3} = \frac{1}{x^3}.$$

Hence

$$\frac{10 - x}{x} = \sqrt[3]{4},$$

$$10 - x - \sqrt[3]{4} \cdot x = 0,$$

$$x = \frac{10}{1 + \sqrt[3]{4}}.$$

Thus there is one critical point; we use the first derivative test to determine if it is a maximum or minimum point. It suffices to check the sign of the derivative at a point on each side of the critical point.

Let $x_- = 0.0001$. Then

$$L'(x_-) = 2kI_1 \left[-\frac{1}{(0.0001)^3} + \frac{4}{(10 - 0.0001)^3} \right].$$

Without calculating this explicitly, we can see that the term $-(1/0.0001)^3$ is negative and much larger in size than the other term, so $L'(x_-) < 0$.

Let $x_+ = 9.9999$. Then

$$L'(x_+) = 2kI_1\left[-\frac{1}{(9.9999)^3} + \frac{4}{(0.0001)^3} \right].$$

Now it is the second term which dominates; since it is positive, $L'(x_+) > 0$. Thus L' changes from negative to positive at x_0, so $x_0 = 10/(1 + \sqrt[3]{4})$ must be a minimum point.

The darkest point is thus at a distance of $10/(1 + \sqrt[3]{4}) \approx 3.86$ meters from the smaller source. It is interesting to compare the distances of the darkest point from the two sources. The ratio

$$\frac{10 - \left[10/(1 + \sqrt[3]{4}) \right]}{10/(1 + \sqrt[3]{4})}$$

is simply $\sqrt[3]{4}$.

Problem 3: $y = (x - a)^2 + (x - b)^2 + (x - c)^2 + (x - d)^2$. There are no restrictions on x.

$$\frac{dy}{dx} = 2(x - a) + 2(x - b) + 2(x - c) + 2(x - d)$$
$$= 2(4x - a - b - c - d)$$

which is 0 only if $x = \frac{1}{4}(a + b + c + d)$.

Since $d^2y/dx^2 = 8$ is positive, $\frac{1}{4}(a + b + c + d)$ is the minimum point. The number required is thus the average, or arithmetic mean, of a, b, c, and d.

Problem 4: This is *not* a standard maximum–minimum problem. We minimize our expenses by selling the car at the time when the cost per day, $(x/50) + 10$, reaches the value 50. (See Fig. 3.5.9.) We solve $(x/50) + 10 = 50$, getting $x = 2000$. Thus the car should be kept for 2000 days. ▲

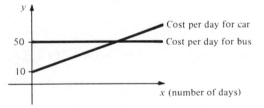

Figure 3.5.9. When does the car become more expensive than the bus?

In the process of doing a word problem, it is useful to ask general questions like, "Can I guess any properties of the answer? Is the answer reasonable?" Sometimes a clever or educated guess can carry one surprisingly far toward the solution of a problem.

For instance, consider the problem "Find the triangle with perimeter 1 which has the greatest area." In the statement of the problem, all three sides of the triangle enter in the same way—there is nothing to single out any side as special. Therefore, we guess that the answer must have the three sides equal; that is, the triangle should be equilateral. This is, in fact, correct. Such reasoning must be used with care: if we ask for the triangle with perimeter 1 and the *least* area, the answer is *not* an equilateral triangle; thus, reasoning by symmetry must often be supplemented by a more detailed analysis.

In Problems 1 and 3, the answers have as much symmetry as the data, and indeed these answers might have been guessed before any calculation had been done. In Problem 2, most people would have a hard time guessing the answer, but at least one can observe that the final answer has the darkest point nearer to the weaker of the two sources, which is reasonable.

Example 8   Of all rectangles inscribed in a circle of radius 1, guess which has the largest area; the largest perimeter. (See Fig. 3.5.10.)

Figure 3.5.10. What shape gives the rectangle the largest area? The largest perimeter?

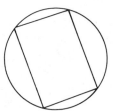

Solution   Since the length and width enter symmetrically into the formulas for area and perimeter, we may guess that the maximum of both area and perimeter occurs when the rectangle is a square. ▲

We can now solve word problems by bringing together all of the preceding techniques.

Example 9   Find the dimensions of a rectangular box of minimum cost if the manufacturing costs are 10 cents per square meter on the bottom, 5 cents per square meter on the sides, and 7 cents per square meter on the top. The volume is to be 2 cubic meters and the height is to be 1 meter.

Solution   Let the dimensions of the base be l and w; the height is 1. If the total cost is C, then

$$C = 10lw + 7lw + 2 \cdot 5 \cdot (l \cdot 1 + w \cdot 1) = 17lw + 10l + 10w.$$

Now $l \cdot w \cdot 1 = 2$ is the total volume. Eliminating w,

$$C = 34 + 10l + \frac{20}{l}.$$

We are to minimize C. Let $f(l) = 34 + 10l + (20/l)$ on $(0, \infty)$. Then

$$f'(l) = 10 - \frac{20}{l^2}$$

which is 0 when $l = \sqrt{2}$. (We are concerned only with $l > 0$.) Since $f''(l) = 40/l^3$ is positive at $l = \sqrt{2}$, $l = \sqrt{2}$ is a local minimum point. Since this is the only critical point, it is also the global minimum. Thus, the dimensions of minimum cost are $\sqrt{2}$ by $\sqrt{2}$ by 1. ▲

Example 10   The stiffness S of a wooden beam of rectangular cross-section is proportional to its breadth and the cube of its thickness. Find the stiffest rectangular beam that can be cut from a circular log of diameter d.

Solution   Let the breadth be b and thickness be t. Then S is proportional to bt^3. From Fig. 3.5.11, we have $(\frac{1}{2}b)^2 + (\frac{1}{2}t)^2 = (d/2)^2$, i.e., $b = \sqrt{d^2 - t^2}$ $(0 < t < d)$. To maximize $f(t) = bt^3 = t^3\sqrt{d^2 - t^2}$, we note that $f'(t) = t^2(3d^2 - 4t^2)/\sqrt{d^2 - t^2} = 0$ if $3d^2 - 4t^2 = 0$, i.e., $t = \sqrt{3}\,d/2$. Since $f'(t)$ has the same sign as $3d^2 - 4t^2$, which changes from positive to negative at $t = \sqrt{3}\,d/2$, $t = \sqrt{3}\,d/2$ is a maximum point; so the dimensions of the beam should be $t = \sqrt{3}\,d/2$ and $b = d/2$. ▲

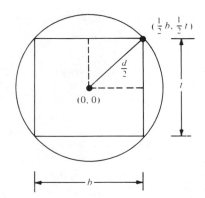

Figure 3.5.11. Cross section of a beam being cut from a log.

Example 11   The terms marginal revenue and marginal cost were defined in Section 2.1.

(a) Prove that at a production level x_0 that maximizes profit, the marginal revenue equals the marginal cost (see Fig. 3.5.12).

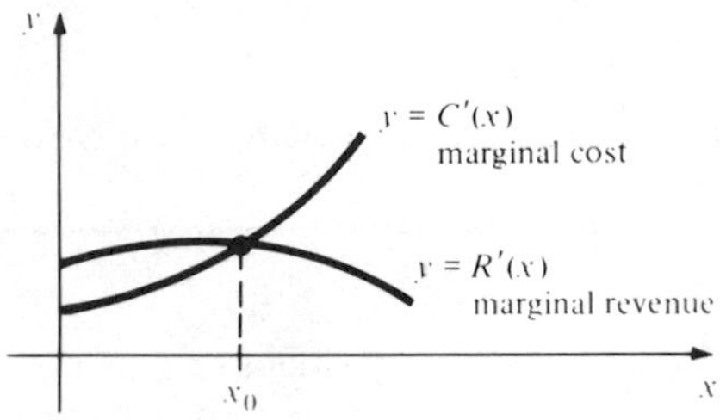

Figure 3.5.12. The production level x_0 that maximizes profit has $C'(x_0) = R'(x_0)$.

(b) Determine the number of units of a commodity that should be produced to maximize the profit when the cost and revenue functions are given by $C(x) = 800 + 30x - 0.01x^2$ and $R(x) = 50x - 0.02x^2$, $0 \leqslant x \leqslant 1500$.

Solution   (a) We have $P(x) = R(x) - C(x)$. At as maximum point x_0 for P, we must have $P'(x_0) = 0$ by the critical point test. Therefore, $R'(x_0) - C'(x_0) = 0$ or $R'(x_0) = C'(x_0)$ as required.

(b) The profit is

$$P(x) = R(x) - C(x)$$

$$= (50x - 0.02x^2) - (800 + 30x - 0.01x^2)$$

$$= 20x - 0.01x^2 - 800.$$

The limits on x are that $0 \leqslant x \leqslant 1500$. The critical points occur where $P'(x) = 0$; i.e., $20 - 0.02x = 0$, i.e., $x = 1000$. This is a maximum since $P''(x) = -0.02$ is negative; at $x = 1000$, $P(1000) = 9200$. The endpoint $x = 0$ is not a maximum since $P(0) = -800$ is negative. Also $P(1500) = 6700$, so the value at this endpoint is less than at $x = 1000$. Thus, the production level should be set at $x = 1000$. ▲

Example 12   Given a number $a > 0$, find the minimum value of $(a + x)/\sqrt{ax}$ where $x > 0$.

Solution   We are to minimize $f(x) = (a + x)/\sqrt{ax}$ on $(0, \infty)$. By the quotient rule,

$$f'(x) = \frac{\sqrt{ax} - (a + x)(a/2\sqrt{ax})}{ax}$$

$$= \frac{2ax - a(a + x)}{2(ax)^{3/2}} = \frac{a(x - a)}{2(ax)^{3/2}}.$$

Thus, $x = a$ is the only critical point in $(0, \infty)$. We observe that $f'(x)$ changes sign from negative to positive at $x = a$, so this is a minimum point. The minimum value is $f(a) = (a + a)/\sqrt{a^2} = 2$. ▲

The result of this example can be rephrased by saying that

$$\frac{a + x}{\sqrt{ax}} \geqslant 2 \quad \text{for every} \quad a > 0, \quad x > 0.$$

Writing b for x, this becomes

$$\frac{a + b}{2} \geqslant \sqrt{ab} \quad \text{for every} \quad a > 0, \quad b > 0.$$

That is, the *arithmetic mean* $(a + b)/2$ of a and b is larger than their *geometric mean* $\sqrt{ab}$. This inequality was proved by using calculus. It can also be proved by algebra alone. Indeed, since the square of any number is non-negative,

$$0 \leqslant \left(\sqrt{a} - \sqrt{b}\right)^2 = a - 2\sqrt{ab} + b,$$

so $2\sqrt{ab} \leqslant a + b$. Hence,

$$\frac{a + b}{2} \geqslant \sqrt{ab}.$$

This is sometimes called the *arithmetic–geometric mean inequality*.

Exercises for Section 3.5

Find the maximum and minimum points and values for each function on the given interval in Exercises 1–4.

1. $2x^3 - 5x + 2$ on $[1, \infty)$
2. $x^3 - 6x + 3$ on $(-\infty, \infty)$
3. $\dfrac{x^2 + 2}{x}$ on $(0, 5]$
4. $-3x^2 + 2x + 1$ on $(-\infty, \infty)$

Find the maximum and minimum points for each of the functions in Exercises 5–8.

5. $\dfrac{x^2 + 1}{x}$ on $(-\infty, 0)$
6. $\dfrac{x^2 + 2x + 3}{x^2 + 5}$ on $(-2, 2]$
7. $-x^3 + 5x + 4$ on $(-2, \frac{5}{3})$
8. $x^3 - 3x + 5$ on $(-3, \frac{3}{2})$

9. Figure 3.5.13 shows the annual inflation rate in Oxbridge for 1970–1980. Find the approximate maximum and minimum points and values.

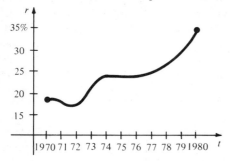

Figure 3.5.13. The annual inflation rate in Oxbridge.

10. Figure 3.5.14 shows the temperatures recorded in Goose Brow during a 24-hour period. When did the maxima and minima occur, and what were their values?

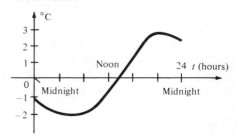

Figure 3.5.14. A cold day in Goose Brow.

In Exercises 11–30, find the critical points, endpoints, and global maximum and minimum points and values for each function on the given interval.

11. $f(x) = x^2 - x$ on $[0, 1)$.
12. $f(x) = x^3$ on $[-1, \infty)$.
13. $f(x) = x^4 - 4x^2 + 7$ on $[-4, 2]$.
14. $f(x) = 4x^4 - 2x^2 + 1$ (a) on $[-10, 20]$ and (b) on $[-0.2, -0.1]$.
15. $f(x) = \dfrac{1}{4x^4 - 2x^2 + 200{,}000}$ on $[-10, 20]$.
16. $f(x) = \dfrac{1 + x^2}{1 - x^2}$ on $[-\frac{1}{2}, \frac{3}{4}]$.
17. $f(x) = x^2 - 3x + 1$ on each of the following intervals.
    (a) $(2, \infty)$ (c) $(-3, 2]$
    (b) $(-\infty, \frac{1}{2}]$ (d) $(-\frac{3}{2}, 2]$

(e) $(-2, 2)$ (g) $[-1, 1]$
(f) $(-\infty, \infty)$ (h) $[-8, 8]$

18. $f(x) = x^3 - 2x + 1$ on each of the intervals in Exercise 17.
19. $f(x) = 7x^2 + 2x + 4$ on (a) $[-1, 1]$, (b) $(0, \infty)$, (c) $[-4, 2)$.
20. $f(x) = x^3 + 6x^2 - 12x + 7$ on (a) $(-\infty, \infty)$, (b) $(-2, 6]$, (c) $[-2, 1)$, (d) $[-4, 4]$.
21. $f(x) = x^3 - 3x^2 + 3x + 1$ on $(-\infty, \infty)$.
22. $f(x) = x^3 + 3x^2 - 3x + 1$ on $[-1, 2]$.
23. $f(x) = x^3 - x$ on $[-2, 3]$.
24. $f(x) = x^4 + 8x^3 + 3$ on $[-1, 1]$.
25. $f(x) = \dfrac{x^3 - 1}{x^2 + 1}$ on $[-10, 10]$.
26. $f(x) = \dfrac{x^2 + 1}{x^4 + 1}$ on $(-\infty, \infty)$.
27. $f(x) = -\dfrac{x}{1 + x^2}$ on $[-1, 6]$.
28. $f(x) = \dfrac{x}{1 + x^2}$ on $(-\infty, \infty)$.
29. $f(x) = \dfrac{x^3}{1 + x^2}$ on $(-\infty, \infty)$.
30. $f(x) = \dfrac{x^n}{1 + x^2}$ on $(-\infty, \infty)$, n a positive integer.

Carry out step 1 for Exercises 31–34, (see Example 6).

31. Of all rectangles with area 1, which has the smallest perimeter?
32. Find the point on the arc of the parabola $y = x^2$ for $0 \leqslant x \leqslant 1$ which is nearest to the point $(0, 1)$. [*Hint*: Consider the *square* of the distance between points.]
33. Two point masses which are a fixed distance apart attract one another with a force which is proportional to the product of the masses. Assuming that the sum of the two masses is M, what must the individual masses be so that the force of attraction is as large as possible?
34. Ten miles from home you remember that you left the water running, which is costing you 10 cents an hour. Driving home at speed s miles per hour costs you $6 + (s/10)$ cents per mile. At what speed should you drive to minimize the total cost of gas and water?
35. Carry out step 2 for Exercise 31.
36. Carry out step 2 for Exercise 32.
37. Carry out step 2 for Exercise 33.
38. Carry out step 2 for Exercise 34.
39. A rectangular box, open at the top, is to be constructed from a rectangular sheet of cardboard 50 centimeters by 80 centimeters by cutting out equal squares in the corners and folding up the sides. What size squares should be cut out for the container to have maximum volume?
40. A window in the shape of a rectangle with a semicircle on the top is to be made with a perimeter of 4 meters. What is the largest possible area for such a window?

41. (a) A can is to be made to hold 1 liter ($= 1000$ cubic centimeters) of oil. If the can is in the shape of a circular cylinder, what should the radius and height be in order that the surface area of the can (top and bottom and curved part) be as small as possible?
    (b) What is the answer in part (a) if the total capacity is to be V cubic centimeters?
    (c) Suppose that the surface area of a can is fixed at A square centimeters. What should the dimensions be so that the capacity is maximized?
42. Determine the number of units of a commodity that should be produced to maximize the profit when the cost and revenue functions are given by $C(x) = 700 + 40x - 0.01x^2$, $R(x) = 80x - 0.03x^2$.
43. Determine the number of units of a commodity that should be produced to maximize the profit for the following cost and revenue functions: $C(x) = 360 + 80x + .002x^2 + .00001x^3$, $R(x) = 100x - .0001x^2$.
44. If the cost of producing x calculators is $C(x) = 100 + 10x + 0.01x^2$ and the price per calculator at production level x is $P(x) = 26 - 0.1x$ (this is called the *demand* equation), what production level should be set in order to maximize profit?
45. The U.S. Post Office will accept rectangular boxes only if the sum of the length and girth (twice the width plus twice the height) is at most 72 inches. What are the dimensions of the box of maximum volume the Post Office will accept? (You may assume that the width and height are equal.)
46. Given n numbers, $a_1, \ldots, a_n$, find a number x which best approximates them in the sense that the sum of the squares of the differences between x and the n numbers is as small as possible.
47. One positive number plus the square of another equals 48. Choose the numbers so that their product is as large as possible.
48. Find the point or points on the arc of the parabola $y = x^2$ for $0 \leqslant x \leqslant 1$ which are nearest to the point $(0, q)$. Express your answer in terms of q. [*Hint*: Minimize the *square* of the distance between points.]
49. One thousand feet of fencing is to be used to surround two areas, one square and one circular. What should the size of each area be in order that the total area be (a) as large as possible and (b) as small as possible?
50. A forest can support up to 10,000 rabbits. If there are x rabbits in the forest, each female can be expected to bear $\frac{1}{1000}(10{,}000 - x)$ bunnies in a year. What total population will give rise to the

greatest number of newborn bunnies in a year? (Assume that exactly half the rabbits are female; ignore the fact that the bunnies may themselves give birth to more young during the year; and remember that the total population including new bunnies is not to exceed 10,000).

51. (a) In Fig. 3.5.15, for which value of y does the line segment PQ have the shortest length? Express your answer in terms of a and b. [*Hint*: Minimize the square of the length.]

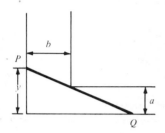

Figure 3.5.15. The "ladder" PQ just fits into the corner of the corridor.

   (b) What is the length of the longest ladder which can be slid along the floor around the corner from a corridor of width a to a corridor of width b?

52. One hundred feet of fencing is to be used to enclose two pens, one square and one triangular. What dimensions should the pens have to enclose the largest possible area?

53. A conical dunce cap is to be made from a circular piece of paper of circumference c by cutting out a pie-shaped piece whose curved outer edge has length l. What should l be so that the resulting dunce cap has maximum volume?

54. (a) Suppose that you drive from coast to coast on Interstate Route 80 and your altitude above sea level is $f(x)$ when you are x miles from San Francisco. Discuss the critical points, endpoints, global maximum and minimum points and values, and local maximum and minimum points for $f(x)$.

   (b) Do as in part (a) for a hike to the top of Mt. Whitney, where x is the distance walked from your starting point.

In Exercises 55–59, try to guess the answer, or some part of the answer, by using some symmetry of the data.

55. Of all rectangles of area 1, which has the smallest perimeter?

56. Of all geometric figures with perimeter 1, which has the greatest area?

57. In Problem 1 (see Example 6), suppose that we allow the fencing to assume any shape, not necessarily rectangular, but still with one side along the shore requiring no fencing. What shape gives

the maximum area? A formal proof is not required.

58. What is the answer to Problem 2 (see Example 6) if the two intensities are equal? If one of the intensities is eight times the other?

59. Of all *right* triangles of area 1, guess which one has the shortest perimeter. Which one, if any, has the longest perimeter?

60. Use calculus to show that the answer in Example 8 is correct.

★61. Find a function on $[-1, 1]$ which is continuous but which is not differentiable at its maximum point.

★62. Find a function defined on $[0, 1]$ which does not have a maximum value on $[0, 1]$.

★63. Find a function defined on $[-2, 2]$ which has neither a maximum value nor a minimum value on $[-2, 2]$.

★64. Find a function defined on $[-3, -1]$ which is continuous at -3 and -1, has a maximum value on $[-3, -1]$, but has no minimum value on $[-3, -1]$.

★65. Prove that the maximum value of f on I is unique; that is, show that if M_1 and M_2 both satisfy conditions 1 and 2 of the definition at the beginning of this section, then $M_1 = M_2$. [*Hint*: Show that $M_1 \leqslant M_2$ and $M_2 \leqslant M_1$.]

★66. Let the set I be contained in the set J, and suppose that f is defined on J (and therefore defined on I). Show that if m_I and m_J are the minimum values of f on I and J, respectively, then $m_J \leqslant m_I$.

In Exercises 67–70, find the maximum and minimum values of the given function on the given interval.

★67. $f(x) = px + (q/x)$, on $(0, \infty)$ where p and q are nonzero numbers.

★68.
$$f(x) = \begin{cases} 5x, & x \leqslant 0 \\ -3x, & x \geqslant 0 \end{cases} \quad \text{on } [-1, 1].$$

★69. $f(x) = x/(1 - x^2)$ for $-1 < x \leqslant 0$ and $f(x) = x^3 - x$ for $0 \leqslant x < 2$, on $(-1, 2)$.

★70.
$$f(x) = \begin{cases} -x^4 - 1 & \text{for } -1 < x \leqslant 0 \\ -1 & \text{for } 0 \leqslant x \leqslant 1 \\ x - 2 & \text{for } 1 \leqslant x < \infty \end{cases}$$

on $(-1, \infty)$.

★71. Three equal light sources are spaced at $x = 0, 1, 2$, along a line. At what points between the sources is the total illumination least? See Problem 2, Example 6. You should get a cubic equation for $(x - 1)^2$. Solve the equation numerically by using the method of bisection.

★72. What happens in Exercise 71 if there are four light sources instead of three? Can you guess the correct answer by using a symmetry argument without doing any calculation?

★73. (This minimization problem involves no calculus.) You ran out of milk the day before your weekly visit to the supermarket, and you must pick up a container at the corner grocery store. At the corner store, a quart costs Q cents and a half gallon costs G cents. At the supermarket, milk costs q cents a quart and g cents a half gallon.

(a) If $Q = 45$, $G = 80$, $q = 38$, and $g = 65$, what size container should you buy at the corner grocery to minimize your eventual milk expense? (Assume that a quart will get you through the day.)

(b) Under what conditions on Q, G, q, and g should you buy a half gallon today?

★74. The cost of running a boat is $10v^3$ dollars per mile where v is its speed in still water. What is the most economical speed to run the boat upstream against a current of 5 miles per hour?

★75. Let f and g be defined on I. Under what conditions is the maximum value of $f + g$ on I equal to the sum of the maximum values of f and g on I?

★76. Let I be the set consisting of the whole numbers from 1 to 1000, and let $f(x) = 45x - x^2$. Find the maximum and minimum points and values for f on I.

★77. Suppose that f is continuous on $[a, b]$ and is differentiable and concave upward on the interval (a, b). Show that the maximum point of f is an endpoint.

3.6 The Mean Value Theorem

If the derivative of a function is everywhere zero, then the function is constant.

The mean value theorem is a technical result whose applications are more important than the theorem itself. We begin this section with a statement of the theorem, proceed immediately to the applications, and conclude with a proof which uses the idea of global maxima and minima.

The mean value theorem is, like the intermediate value and extreme value theorems, an *existence* theorem. It asserts the existence of a point in an interval where a function has a particular behavior, but it does not tell you how to find the point.

Mean Value Theorem

If f is continuous on $[a, b]$ and is differentiable on (a, b), then there is a point x_0 in (a, b) at which

$$f'(x_0) = \frac{f(b) - f(a)}{b - a}.$$

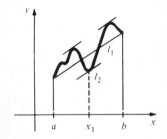

Figure 3.6.1. The slope $[f(b) - f(a)]/(b - a)$ of the secant line l_1 is equal to the slope of the tangent line l_2.

In physical terms, the mean value theorem says that the average velocity of a moving object during an interval of time is equal to the instantaneous velocity at some moment in the interval. Geometrically, the theorem says that a secant line drawn through two points on a smooth graph is parallel to the tangent line at some intermediate point on the curve. There may be more than one such point, as in Fig. 3.6.1. Consideration of these physical and geometric interpretations should make the theorem believable.

We will prove the mean value theorem at the end of this section. For now, we will concentrate on some applications. These tell us that if we know something about $f'(x)$ for all x in $[a, b]$, then we can conclude something about the relation between values of $f(x)$ at different points in $[a, b]$.

Consequences of the Mean Value Theorem

Suppose that f is differentiable on (a, b) and continuous on $[a, b]$.

1. Assume that there are two numbers A and B such that

$$A \leq f'(x) \leq B \qquad \text{for all } x \text{ in } (a, b).$$

Then for any two distinct points x_1 and x_2 in $[a, b]$,

$$A \leq \frac{f(x_2) - f(x_1)}{x_2 - x_1} \leq B.$$

2. If $f'(x) = 0$ on (a, b), then f is constant on $[a, b]$.
3. Let $F(x)$ and $G(x)$ be functions such that $F'(x) = G'(x)$ for all x in an open interval (a, b). Then there is a constant C such that $F(x) = G(x) + C$ for all x in (a, b).

The first consequence holds since $[f(x_2) - f(x_1)]/[x_2 - x_1] = f'(x_0)$ for some x_0 between x_1 and x_2, by the mean value theorem applied to f on the interval with endpoints x_1 and x_2. Since x_0 is in (a, b), $f'(x_0)$ lies between A and B, so does the difference quotient. In particular, if $f'(x) = 0$, we can choose $A = B = 0$, which implies that $f(x_1) = f(x_2)$. Hence f is constant. To obtain the third consequence, observe that $F(x) - G(x)$ has zero derivative, so it is a constant. This consequence of the mean value theorem is a fact about antiderivatives which we used in Section 2.5.

Example 1 Suppose that f is differentiable on the whole real line and that $f'(x)$ is constant. Prove that f is linear.

Solution Let m be the constant value of f'.

Method 1. We may apply the first consequence of the mean value theorem with $x_1 = 0$, $x_2 = x$, $A = m = B$, to conclude that $[f(x) - f(0)]/(x - 0) = m$. But then $f(x) = mx + f(0)$ for all x, so f is linear.

Method 2. Let $g(x) = f(x) - mx$. Then $g'(x) = m - m = 0$, so g is a constant. Setting $x = 0$, $g(x) = f(0)$. Thus $f(x) = mx + f(0)$ so f is linear. ▲

Example 2 Let f be continuous on $[1, 3]$ and differentiable on $(1, 3)$. Suppose that for all x in $(1, 3)$, $1 \leq f'(x) \leq 2$. Prove that $2 \leq f(3) - f(1) \leq 4$.

Solution Apply the first consequence with $A = 1$, $B = 2$. Then we have the inequalities $1 \leq [f(3) - f(1)]/(3 - 1) \leq 2$, and so $2 \leq f(3) - f(1) \leq 4$. ▲

Example 3 Let $f(x) = (d/dx)|x|$.
(a) Find $f'(x)$.
(b) What does the second consequence of the mean value theorem tell you about f? What does it not tell you?

Solution (a) Since $|x|$ is linear on $(-\infty, 0)$ and $(0, \infty)$, its second derivative $d^2|x|/dx^2 = f'(x)$ is identically zero for all $x \neq 0$.
(b) By consequence 2, f is constant *on any open interval on which it is differentiable*. It follows that f is constant on $(-\infty, 0)$ and $(0, \infty)$. The corollary does *not* say that f is constant on $(-\infty, \infty)$. In fact, $f(-2) = -1$, while $f(2) = +1$. ▲

Example 4   Suppose that $F'(x) = x$ for all x and that $F(3) = 2$. What is $F(x)$?

Solution   Let $G(x) = \frac{1}{2}x^2$. Then $G'(x) = x = F'(x)$, so $F(x) = G(x) + C = \frac{1}{2}x^2 + C$. To evaluate C, set $x = 3$: $2 = F(3) = \frac{1}{2}(3^2) + C = \frac{9}{2} + C$. Thus $C = 2 - \frac{9}{2} = -\frac{5}{2}$ and $F(x) = \frac{1}{2}x^2 - \frac{5}{2}$. ▲

Example 5   The velocity of a train is kept between 40 and 50 kilometers per hour during a trip of 200 kilometers. What can you say about the duration of the trip?

Solution   Before presenting a formal solution using the mean value theorem, let us use common sense. If the velocity is at least 40 kilometers per hour, the trip takes at most $\frac{200}{40} = 5$ hours. If the velocity is at most 50 kilometers per hour, the trip takes at least $\frac{200}{50} = 4$ hours. Thus, the trip takes between 4 and 5 hours.

To use the mean value theorem, let $f(t)$ be the position of the train at time t; let a and b be the beginning and ending times of the trip. By consequence 1 with $A = 40$ and $B = 50$, we have $40 \leqslant [f(b) - f(a)]/(b - a) \leqslant 50$. But $f(b) - f(a) = 200$, so

$$40 \leqslant \frac{200}{b - a} \leqslant 50,$$

$$\frac{1}{5} \leqslant \frac{1}{b - a} \leqslant \frac{1}{4},$$

$$5 \geqslant b - a \geqslant 4.$$

Hence the trip takes somewhere between 4 and 5 hours, as we found above. ▲

Our proof of the mean value theorem will use two results from Sections 3.1 and 3.4, which we recall here:

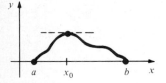

Figure 3.6.2. Rolle's theorem: If f is zero at the ends of the interval, its graph must have a horizontal tangent line somewhere between.

1. If x_0 lies in the open interval (a, b) and is a maximum or minimum point for a function f on an interval $[a, b]$, and if f is differentiable at x_0, then $f'(x_0) = 0$ (critical point test).
2. If f is continuous on a closed interval $[a, b]$, then f has a maximum and a minimum point in $[a, b]$ (extreme value theorem).

We now proceed with the proof in three steps.

Step 1 (Rolle's Theorem[7])   *Let f be continuous on $[a, b]$ and differentiable on (a, b), and assume that $f(a) = f(b) = 0$. Then there is a point x_0 in (a, b) at which $f'(x_0) = 0$.*

Indeed, if $f(x) = 0$ for all x in $[a, b]$, we can choose any x_0 in (a, b). So assume that f is not everywhere zero. By the extreme value theorem, f has a maximum point x_1 and a minimum point x_2. Since f is zero at the ends of the interval but is not identically zero, at least one of x_1, x_2 lies in (a, b), and not at an endpoint. Let x_0 be this point. By the critical point test, $f'(x_0) = 0$, so Rolle's theorem is proved.

Rolle's theorem has a simple geometric interpretation (see Fig. 3.6.2).

Step 2 (Horserace Theorem)   *Suppose that f_1 and f_2 are continuous on $[a, b]$ and differentiable on (a, b), and assume that $f_1(a) = f_2(a)$ and $f_1(b) = f_2(b)$. Then there exists a point x_0 in (a, b) such that $f_1'(x_0) = f_2'(x_0)$.*

[7] Michel Rolle (1652–1719) (pronounced "roll") was actually best known for his attacks on the calculus. He was one of the critics of the newly founded theory of Newton and Leibniz. It is an irony of history that he has become so famous for "Rolle's theorem" when he did not even prove the theorem but used it only as a remark concerning the location of roots of polynomials. (See D. E. Smith, *Source Book in Mathematics*, Dover, 1929, pp. 251–260, for further information.)

Let $f(x) = f_1(x) - f_2(x)$. Since f_1 and f_2 are differentiable on (a, b) and continuous on $[a, b]$, so is f. By assumption, $f(a) = f(b) = 0$, so from Step 1, $f'(x_0) = 0$ for some x_0 in (a, b). Thus $f_1'(x_0) = f_2'(x_0)$ as required (see Fig. 3.6.3).

We call this the "horserace theorem" because it has the following interpretation. Suppose that two horses run a race starting together and ending in a tie. Then, at some time during the race, they must have had the same velocity.

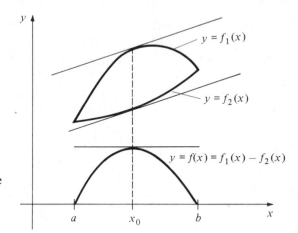

Figure 3.6.3. The curves $y = f_1(x)$ and $y = f_2(x)$ have parallel tangent lines when $y = f_1(x) - f_2(x)$ has a horizontal tangent line.

Step 3.

We apply Step 2 to a given function f and a linear function l that matches f at its endpoints, namely,

$$l(x) = f(a) + (x - a)\left[\frac{f(b) - f(a)}{b - a}\right].$$

Note that $l(a) = f(a)$, $l(b) = f(b)$, and $l'(x) = [f(b) - f(a)]/(b - a)$. By Step 2, $f'(x_0) = l'(x_0) = [f(b) - f(a)]/(b - a)$ for some point x_0 in (a, b). This completes our proof of the mean value theorem. ∎

Example 6 Let $f(x) = x^4 - 9x^3 + 26x^2 - 24x$. Note that $f(0) = 0$ and $f(2) = 0$. Show without calculating that $4x^3 - 27x^2 + 52x - 24$ has a root somewhere strictly between 0 and 2.

Solution Since $f(0) = 0$ and $f(2) = 0$, Rolle's theorem shows that f' is zero at some x_0 in $(0, 2)$; that is, $0 < x_0 < 2$. ▲

Example 7 Suppose that f is a differentiable function such that $f(0) = 0$ and $f(1) = 1$. Show that $f'(x_0) = 2x_0$ for some x_0 in $(0, 1)$.

Solution Use the horserace theorem with $f_1(x) = f(x)$, $f_2(x) = x^2$, and $[a, b] = [0, 1]$. ▲

Exercises for Section 3.6

1. Suppose that f is continuous on $[0, \frac{1}{2}]$, and differentiable on $(0, \frac{1}{2})$, and that $0.3 \leq f'(x) \leq 1$ for $0 < x < \frac{1}{2}$. Prove that $0.15 \leq [f(\frac{1}{2}) - f(0)] \leq 0.5$.

2. Suppose that f is continuous on $[3, 5]$ and differentiable on $(3, 5)$, and that $\frac{1}{2} \leq f'(x) \leq \frac{2}{3}$ for $3 < x < 5$. Show that $1 \leq [f(5) - f(3)] \leq \frac{4}{3}$.

3. Suppose that $(d/dx)[f(x) - 2g(x)] = 0$. What can you say about the relationship between f and g?

4. If $f''(x) = 0$ on (a, b), what can you say about f?

5. Let $f(x) = x^5 + 8x^4 - 5x^2 + 15$. Prove that somewhere between -1 and 0 the tangent line to the graph of f has slope -2.

6. Let $f(x) = 5x^4 + 9x^3 - 11x^2 + 10$. Prove that the graph of f has slope 9 somewhere between -1 and 1.

7. Let $f(x) = \sqrt{x^3 - 8}$. Show that somewhere between 2 and 3 the tangent line to the graph of f has slope $\sqrt{19}$.

8. Let $f(x) = x^7 - x^5 - x^4 + 2x + 1$. Prove that the graph of f has slope 2 somewhere between -1 and 1.

9. Suppose that an object lies at $x = 4$ when $t = 0$ and that the velocity dx/dt is 35 with a possible error of ± 1, for all t in $[0, 2]$. What can you say about the object's position when $t = 2$?

10. The fuel consumption of an automobile varies between 17 and 23 miles per gallon, according to the conditions of driving. Let $f(x)$ be the number of gallons of fuel in the tank after x miles have been driven. If $f(100) = 15$, give upper and lower estimates for $f(200)$.

11. Directly verify the validity of the mean value theorem for $f(x) = x^2 - x + 1$ on $[-1, 2]$ by finding the point(s) x_0. Sketch.

12. Let $f(x) = x^3$ on the interval $[-2, 3]$. Find explicitly the value(s) of x_0 whose existence is guaranteed by the mean value theorem. Sketch.

13. Suppose that $F'(x) = -(1/x^2)$ for all $x \neq 0$. Is $F(x) = 1/x + C$, where C is a constant?

14. Suppose that $f'(x) = x^2$ and $f(1) = 0$. What is $f(x)$?

15. Let $f(x) = |x| - 1$. Then $f(-1) = f(1) = 0$, but $f'(x)$ is never equal to zero on $[-1, 1]$. Does this contradict Rolle's theorem? Explain.

16. Suppose that the horses in a race cross the finish line with equal velocities. Must they have had the same acceleration at some time during the race?

Find the antiderivatives of the functions in Exercises 17–20.

17. $f(x) = \frac{1}{2}x - 4x^2 + 21$.

18. $f(x) = 6x^5 - 12x^3 + 15x - 11$.

19. $f(x) = \dfrac{1}{x^2} + 2x$.

20. $f(x) = 2x(x^2 + 7)^{100}$.

Find the antiderivative $F(x)$ for the given function $f(x)$ satisfying the given condition in Exercises 21–24.

21. $f(x) = 2x^4$; $F(1) = 2$.

22. $f(x) = 4 - x$; $F(2) = 1$.

23. $f(x) = x^4 + x^3 + x^2$; $F(1) = 1$.

24. $f(x) = \dfrac{1}{x^5}$; $F(1) = 3$.

25. Let f be twice differentiable on (a, b), and suppose that $f(x) = 0$ at three distinct points in (a, b). Prove that there is a point x_0 in (a, b) at which $f''(x_0) = 0$.

★26. Use the mean value theorem to prove *the increasing function theorem*: If f is continuous on $[a, b]$ and differentiable on (a, b), and $f'(x) > 0$ for x in (a, b), then f is increasing on $[a, b]$.

★27. Suppose that f and g are continuous on $[a, b]$ and that f' and g' are continuous on (a, b). Assume that

$$f(a) = g(a) \quad \text{and} \quad f(b) = g(b).$$

Prove that there is a number c in (a, b) such that the line tangent to the graph of f at $(c, f(c))$ is parallel to the line tangent to the graph of g at $(c, g(c))$.

★28. Let f be a polynomial. Suppose that f has a double zero at a and b. (A polynomial f has a *double zero* at $x = a$ if $f(x) = (x - a)^2 g(x)$ for some polynomial g). Show that $f'(x)$ has at least three roots in $[a, b]$.

★29. The coyote population in Nevada was the same at three consecutive times t_1, t_2, t_3. Assume that the population $N(t)$ is a differentiable function of time t and is nonconstant on $[t_1, t_2]$ and on $[t_2, t_3]$. Establish by virtue of the mean value theorem the existence of two times T, T^* in (t_1, t_2) and (t_2, t_3), respectively, for which the coyote population decreases.

★30. (a) Let f be differentiable on (a, b) [and continuous on $[a, b]$]. Suppose that, for all x in the open interval (a, b), the derivative $f'(x)$ belongs to a certain set S of real numbers. Show that for any two distinct points x_1 and x_2 in (a, b) [in $[a, b]$], the difference quotient

$$\frac{f(x_2) - f(x_1)}{x_2 - x_1}$$

belongs to S as well.

(b) Use (a) to prove consequence 1 of the mean value theorem.

(c) What does (a) tell you if $S = [a, b]$? If $S = (0, \infty)$?

Review Exercises for Chapter 3

On what intervals are the functions in Exercises 1–4 continuous?

1. $f(x) = \dfrac{1}{\sqrt{x^2 - 1}}$

2. $f(x) = \dfrac{1}{\sqrt{x^3 - 1}}$

3. $f(x) = \dfrac{1}{x - 1} + \dfrac{1}{x - 2}$

4. $f(x) = \dfrac{x - 5}{x^2 - 3x + 2}$

5. Explain why the function h given by

$$h(x) = \begin{cases} -x^3 + 5 & \text{for } x \leq 2 \\ x^2 - 7 & \text{for } x > 2 \end{cases}$$

is continuous at 2. Is h continuous on the whole real line?

6. Let $f(x) = [1/x] + [(x^2 - 1)/x]$. Can you define $f(0)$ so that the resulting function is continuous at all x?

7. Find a function which is continuous on the whole real line and differentiable for all x except 1, 2, and 3. (A sketch will do.)

8. Sketch the graph of a function that has one vertical asymptote at $x = 1$, two maximum points, and no other critical points.

9. Show that $f(x) = x^4 + x^2 + x - 2$ has a root between $x = 0$ and $x = 1$.

10. Show that $f(x) = x^5 + x^3 + 2x + 1$ has a root between $x = -1$ and $x = 0$.

▤ 11. Find a solution to the equation in Exercise 9 to within two decimal places using the method of bisection.

▤ 12. Find a solution to the equation in Exercise 10 to within two decimal places using the method of bisection.

In Exercises 13–16, determine the intervals on which the given function is increasing and decreasing.

13. $8x^3 - 3x^2 + 2$.

14. $5x^3 + 2x^2 - 3x + 10$.

15. $\dfrac{x}{1 + (x - 1)^2}$.

16. $\dfrac{x}{x^2 - 3x + 2}$, $x \neq 1, 2$.

17. Suppose that the British inflation rate (in percent per year) from 1990 to 2000 is given by the function $B(t) = 10(t^3/100 - t^2/20 - t/4 + 1)$, where t is the time in years from January 1, 1993. When is the inflation rate increasing? How might a politician react to minimum points? Maximum points? Inflection points?

18. A paint can is kicked off the roof of a 156-foot building, its distance S from the ground being given by $S = 156 + 4t - 16t^2$ (S in feet, t in seconds).

   (a) How high up does the can go before it falls downward?

   (b) At what time does the can return to roof level?

   (c) Find the velocity of the can when it collides with the ground.

19. The rate of growth of a tree between $t = 0$ and $t = 100$ years is given by the formula $r(t) = 10^{-6}(t^3 - 75t^2) + 10$ inches per year. When is the growth slowing down? speeding up?

20. A bicycle is moving at $(\frac{1}{16}t^3 + 4t^2 - 3t + 2)$ miles per hour, where t is the time in hours, $0 \leq t \leq 2$. Does the bicycle change direction during this period?

In Exercises 21–24, find the critical points of the given function and determine if they are local maxima, minima, or neither.

21. $2x^3 - 5x^2 + 4x + 3$

22. $-8x^2 + 2x - 3$

23. $(x^2 + 2x - 3)^2$

▤ 24. $(x^3 - 12x + 1)^6$

For each of the functions in Exercises 25–30, answer the questions below and draw a graph:

(i) Where is f continuous?

(ii) Where is f differentiable?

(iii) On which intervals is f increasing? Decreasing?

(iv) Where is f concave upward or downward?

(v) What are the critical points, endpoints, local maximum and minimum points, and inflection points for f?

25. $f(x) = -7x^3 + 2x^2 + 15$ on $(-\infty, \infty)$.

26. $f(x) = 6x^2 + 3x + 4$ on $(-\infty, \infty)$.

27. $f(x) = x^3 - 2x + 1$ on $[-1, 2]$.

28. $f(x) = x^4 - 3x^3 + x^2$ on $[-6, 6]$.

29. $f(x) = \dfrac{3x}{2 - 5x}$ on its domain.

30. $f(x) = \dfrac{x}{1 + x^{2n}}$ on $(-\infty, \infty)$, n is a positive integer.

Explain why each statement in Exercises 31–40 is true or false: (Justify if true, give a counterexample if false.)

31. Every continuous function is differentiable.

32. If f is a continuous function on $[0, 2]$, and $f(1) = -1$ and $f(2) = 1$, then there must be a point x in $[0, 1]$ where $f(x) = 0$.

33. If a function is increasing at $x = 1$ and at $x = 2$, it must be increasing at every point between 1 and 2.

34. If a differentiable function f on $(-\infty, \infty)$ has a local maximum point at $x = 0$, then $f'(0) = 0$.

35. $f(x) = x^4 - x^3$ is increasing at zero.

36. $f(x) = x^3 + 3x$ takes its largest value on $[-1, 1]$ at an endpoint.

37. Parabolas never have inflection points.

38. All cubic functions $y = ax^3 + bx^2 + cx + d$, $a \neq 0$ have exactly one inflection point.

39. $f(x) = 3/(5x^2 + 1)$ on $(-\infty, \infty)$ has a local maximum at $x = 0$.

40. $y = x^5$ has an inflection point at $x = 0$.

Sketch the graphs of the functions in Exercises 41–50.

41. $x^3 + 3x + 2$

42. $x^5 - 3x^4$

43. $\dfrac{1}{x + 3}$

44. $\dfrac{1}{1 + x^4}$

45. $\dfrac{1 - x^3}{1 + 2x^3}$

▤ 46. $\dfrac{x - 5}{x^2 - 3x + 2}$

47. $x^{2/3} + (x - 1)^{2/3}$

48. $x^{4/3}$

49. $x(x - 1)^{3/2}$

50. $x(x - 1)^{2/3}$

In Exercises 51–56, find the maximum and minimum *value* of each function on the designated interval.

51. $f(x) = 3x^3 + x^2 - x + 5$; $[-2, 2]$

52. $f(x) = 2x^3 + 5x^2 - 4x$; $[-2, 2]$

53. $f(x) = \dfrac{5x^4 + 1}{x^2 + 1}$; $[-1, 1]$

54. $f(x) = \dfrac{x^3}{x^2 - 4}$; $[-1, 1]$

55. $f(x) = \dfrac{x^3}{x^2 - 4}$; $(-2, 2)$

56. $f(x) = -x^4 + 8x^2 + 2$; $(-\infty, \infty)$

In Exercises 57–62, sketch the graph of the indicated function on the designated interval.

57. The function in Exercise 51.

58. The function in Exercise 52.

59. The function in Exercise 53.

60. The function in Exercise 54.

61. The function in Exercise 55.

62. The function in Exercise 56.

63. A wooden picture frame is to be 2 inches wide on top and bottom and 1 inch wide on the sides. Assuming the cost of a frame to be proportional to its front surface area, find the dimensions of the cheapest frame which will surround an area of 100 square inches.

64. At time t, a rectangle has sides given by $l(t) = 2 + t$ and $w(t) = 1 + t^2$, for $-1 \leqslant t \leqslant 1$. (a) When does this rectangle have minimum area? (b) When is the area shrinking the fastest?

65. Find the dimensions of the right circular cylinder of greatest volume that can be inscribed in a given right circular cone. Express your answer in terms of the height h of the cone and the radius r of the base of the cone.

66. A rectangular box with square bottom is to have a volume of 648 cubic centimeters. The top and bottom are to be padded with foam and pressboard, which costs three times as much per square centimeter as the fiberboard used for the sides. Which dimensions produce the box of least cost?

67. A box company paints its open-top square-bottom boxes white on the bottom and two sides and red on the remaining two sides. If red paint costs 50% more than white paint, what are the dimensions of the box with volume V which costs least to paint? In what sense is the "shape" of this box independent of V?

68. Find the maximum area an isosceles triangle can have if each of its equal sides has a length of 10 centimeters.

69. The material for the top, bottom, and lateral surface of a tin can costs $\frac{1}{20}$ of a cent per square centimeter. The cost of sealing the top and bottom to the lateral surface is p cents times the total length in centimeters of the rims (see Fig. 3.R.1) which are to be sealed. Find (in terms of p) the dimensions of the cheapest can which will hold a volume of V cubic centimeters. Express your answer in terms of the solution of a cubic equation; do not solve it.

Rims

Figure 3.R.1. Minimize the cost of making the can.

70. We quote from V. Belevitch, *Classical Network Theory* (Holden-Day, 1968, p. 159): "When a generator of e.m.f. e and internal *positive* resistance R_i is connected to a positive load resistance R, the active power dissipated in the load

$$w = |e|^2 R / (R + R_i)^2, \qquad e \text{ a constant,}$$

is maximum with respect to R for $R = R_i$, that is, when the load resistance is *matched* to the internal resistance of the generator."

Verify this statement. (You do not need to know any electrical engineering.)

71. A fixed-frequency generator producing 6 volts is connected to a coil of 0.05 henry, in parallel with a resistor of 0.5 ohm and a second resistor of x ohms. The power P is given by

$$P = \frac{36x}{(\pi/10)^2 + (x + 1/2)^2}.$$

Find the maximum power and the value of x which produces it.

72. The bloodstream drug concentration $C(t)$ in a certain patient's bloodstream t hours after injection is given by

$$C(t) = \frac{16t}{(10t + 20)^2}.$$

Find the maximum concentration and the number of hours after injection at which it occurs.

73. In a drug-sensitivity problem, the change T in body temperature for an x-milligram injection is given by

$$T(x) = x^2 \left(1 - \frac{x}{4}\right), \qquad 0 \leqslant x \leqslant 4.$$

(a) Find the *sensitivity* $T'(x)$ when $x = 2$.
(b) Graph T versus x.

74. The power P developed by the engine of an aircraft flying at a constant (subsonic) speed v is given by

$$P = \left(cv^2 + \frac{d}{v^2}\right)v,$$

where $c > 0, d > 0$.
(a) Find the speed v_0 which minimizes the power.
(b) Let $Q(t)$ denote the amount of fuel (in gallons) the aircraft has at time t. Assume the power is proportional to the rate of fuel consumption. At what speed will the flight time from takeoff to fuel exhaustion be maximized?

75. Prove that for any positive numbers a, b, c,

$$\frac{a + b + c}{3} \geqslant \sqrt[3]{abc}$$

by minimizing $f(x) = (a + b + x)/\sqrt[3]{abx}$.

76. Find the number x which best approximates 1, 2, 3, and 5 in the sense that the sum of the *fourth* powers of the differences between x and each number is minimized. Compute x to within 0.1 by using the bisection method.

77. A manufacturer of hand calculators can produce up to 50,000 units with a wholesale price

of $16 for a fixed cost of $9000 plus $11.50 for each unit produced. Let x stand for the number of units produced.

(a) Explain why the *total revenue R*, *total cost C* and *profit P* must satisfy the equations

$$R = 16x, \qquad C = 9000 + (11.5)x,$$
$$P = R - C = (4.5)x - 9000.$$

(b) The *break-even point* is the production level x for which the profit is zero. Find it.

(c) Determine the production level x which corresponds to a $4500 profit.

(d) If more than 50,000 units are produced, the revenue is $R = 16x - \frac{4}{9} \times 10^{-6}x^2$, but the cost formula is unchanged. Find the x that maximizes profits.

78. A manufacturer sells x hole punchers per week. The weekly cost and revenue equations are

$$C(x) = 5000 + 2x,$$
$$R(x) = 10x - \frac{x^2}{1000}, \qquad 0 \le x \le 8000$$

(a) Find the minimum cost.

(b) Find the maximum revenue.

(c) Define the profit P by the formula $P(x) = R(x) - C(x)$. Find the maximum profit.

79. A homeowner plans to construct a rectangular vegetable garden with a fence around it. The garden requires 800 square feet, and one edge is on the property line. Three sides of the fence will be chain-link costing $2.00 per linear foot, while the property line side will be inexpensive screening costing $.50 per linear foot. Which dimensions will cost the least?

80. A rental agency for compact cars rents 96 cars each day for $16.00 per day. Each dollar increase in the rental rate results in four fewer cars being rented.

(a) How should the rate be adjusted to maximize the income?

(b) What is the maximum income?

81. The Smellter steel works and the Green Copper Corporation smelter are located about 40 miles apart. Particulate matter concentrations in parts per million theoretically decrease by an inverse square law, giving, for example,

$$C(x) = \frac{k}{x^2} + \frac{3k}{(40 - x)^2}$$

$$(1 \le x \le 39), \quad k \text{ a constant,}$$

as the concentration x miles from Smellter. This model assumes that Green emits three times more particulate matter than Smellter.

(a) Find $C'(x)$ all critical points of C.

(b) Assuming you wished to build a house between Smellter and Green at the point of least particulate concentration, how far would you be from Smellter?

82. Saveway checkers have to memorize the sale prices from the previous day's newspaper advertisement. A reasonable approximation for the percentage P of the new prices memorized after t hours of checking is $P(t) = 96t - 24t^2$, $0 \le t \le 3$.

(a) Is a checker who memorizes the whole list in 30 minutes above or below average?

(b) What is the maximum percentage of the list memorized after 3 hours?

83. If $f'''(x) = 0$ on $(-\infty, \infty)$, show that there are constants A, B, and C such that $f(x) = Ax^2 + Bx + C$.

84. Verify the horserace theorem for $f_1(x) = x^2 + x - 2$ and $f_2(x) = x^3 + 3x^2 - 2x - 2$ on $[0, 1]$.

85. Persons between 30 and 75 inches in height h have average weight $W = \frac{1}{2}(h/10)^3$ lbs.

(a) What is the average weight of a person 5 feet 2 inches in height?

(b) A second grade child grows from 48 inches to 50 inches. Use the linear approximation to estimate his approximate weight gain and compare with a direct calculation.

86. A storage vessel for a chemical bleach mixture is manufactured by coating the inside of a thin hollow plastic cube with fiberglass. The cube has 12-inch sides, and the coating is $\frac{1}{5}$ inch thick. Use the volume formula $V = x^3$ and the mean value theorem to approximate the volume of the fiberglass coating. [*Hint*: The fiberglass volume is $V(12) - V(11.8)$.]

87. (a) Show that the quotient f/g has a critical point when the ratios f'/f and g'/g are equal.

(b) Find a similar criterion for a product fg to have a critical point.

88. (Refer to Review Problem 30 of Chapter 1 and Fig. 1.R.2.) Let P, with coordinates (x_1, y_1), be inside the parabola $y = x^2$; that is, $y_1 > x_1^2$. Show that the path consisting of two straight lines joining P to a point (x, x^2) on the parabola and then (x, x^2) to $(0, \frac{1}{4})$ has minimum length when the first segment is vertical.

89. The graph of a factored polynomial (such as $y = x(x - 1)^2(x - 2)^3(x - 3)^7(x - 4)^{12}$) near a root r appears similar to that of the function $y = c(x - r)^n$, where c and n are chosen appropriately for the root r.

(a) Let $y = x(x + 1)^2(x - 2)^4$. For values of x near 2, the factor $x(x + 1)^2$ is approximately $2(2 + 1)^2 = 18$, so y is approximately $18(x - 2)^4$. Sketch the graph near $x = 2$.

(b) Argue that near the roots $0, 1, 3$ the equation $y = 10x(x - 1)^3(x - 3)^2$ looks geometrically like $y = -90x$, $y = 40(x - 1)^3$, $y = 240(x - 3)^2$, respectively. Use this information to help sketch the graph from $x = 0$ to $x = 4$.

★90. The *astroid* $x^{2/3} + y^{2/3} = 1$ is a planar curve which admits no self-intersections, but it has four cusp points.
   (a) Apply symmetry methods to graph the astroid.
   (b) Divide the astroid into two curves, each of which is the graph of a continuous function. Find equations for these functions.
   (c) Find the points where dy/dx is not defined and compare with the cusp points. Use implicit differentiation.
   (d) Explain why no tangent line exists at the cusp points.

★91. A function f is said to majorize a function g on $[a,b]$ if $f(x) \geqslant g(x)$ for all $a \leqslant x \leqslant b$.
   (a) Show by means of a graph that x majorizes x^2 on $[0,1]$.
   (b) Argue that "f majorizes g on $[a,b]$" means that the curve $y = f(x)$, $a \leqslant x \leqslant b$, lies on or above the curve $y = g(x)$, $a \leqslant x \leqslant b$.
   (c) If $m > n > 0$, then x^n majorizes x^m on $[0,1]$. Explain fully.
   (d) If $m > n > 0$, $(x-a)^n$ majorizes $(x-a)^m$ on $[a, a+1]$. Why?
   (e) Given that $m \neq n$, $m > 0$, and $n > 0$, determine on which intervals $(x-a)^n$ majorizes $(x-a)^m$ (or conversely).
   (f) Graph the functions $y = x - 1$, $y = (x-1)^2$, $y = (x-1)^3$, $y = (x-1)^4$, $y = (x-1)^5$ on the same set of axes for $-1 \leqslant x \leqslant 3$.

★92. Let $f(x) = 1/(1+x^2)$.
   (a) For which values of c is the function $f(x) + cx$ increasing on the whole real line? Sketch the graph of $f(x) + cx$ for one such c.
   (b) For which values of c is the function $f(x) - cx$ decreasing on the whole real line? Sketch the graph of $f(x) - cx$ for one such c.
   (c) How are your answers in parts (a) and (b) related to the inflection points of f?

★93. Let f be a nonconstant polynomial such that $f(0) = f(1)$. Prove that f has a local minimum or a local maximum point somewhere in the interval $(0, 1)$.

★94. Prove that, given any n numbers $a_1, \ldots, a_n$, there is a uniquely determined number x for which the sum of the fourth powers of the differences between x and the a_i's is minimized. [*Hint*: Use the second derivative.]

★95. Prove the following intermediate value theorem for derivatives: If f is differentiable at all points

of $[a,b]$, and if $f'(a)$ and $f'(b)$ have opposite signs, then there is a point $x_0 \in (a,b)$ such that $f'(x_0) = 0$. [The example given in Review Exercise 84, Chapter 5, shows that this theorem does not follow from the intermediate value theorem of Section 3.1.]

★96. Let f be differentiable function on $(0, \infty)$ such that all the tangent lines to the graph of f pass through the origin. Prove that f is linear. [*Hint*: Consider the function $f(x)/x$.]

★97. Let f be continuous on $[3,5]$ and differentiable on $(3,5)$, and suppose that $f(3) = 6$ and $f(5) = 10$. Prove that, for some x_0 in the interval $(3,5)$, the tangent line to the graph of f at x_0 passes through the origin. [*Hint*: Consider the function $f(x)/x$.] Illustrate your result with a sketch.

★98. A function defined on (a,b) is called *convex* when the following inequality holds for x, y in (a,b) and t in $[0,1]$:

$$f(tx + (1-t)y) \leqslant tf(x) + (1-t)f(y).$$

Demonstrate by the following graphical argument that if f'' is continuous and positive on (a,b), then f is convex:
   (a) Show that f is concave upward at each point of (a,b).
   (b) By drawing a graph, convince yourself that the straight line joining two points on the graph lies above the graph between those points.
   (c) Use the fact in (b) to deduce the desired inequality.

★99. Suppose that f is continuous on $[a,b]$, $f(a) = f(b) = 0$, and $x^2f''(x) + 4xf'(x) + 2f(x) \geqslant 0$ for x in (a,b). Prove that $f(x) \leqslant 0$ for x in $[a,b]$.

★100. A rubber cube of incompressible material is pulled on all faces with a force T. The material stretches by a factor v in two directions and contracts by a factor v^{-2} in the other. By balancing forces, one can establish *Rivlin's equation*:

$$v^3 - \frac{T}{2\alpha} v^2 + 1 = 0,$$

where α is a constant (analogous to the spring constant for a spring). Show that Rivlin's equation has one (real) solution if $T < 3\sqrt[3]{2}\,\alpha$ and has three solutions if $T > 3\sqrt[3]{2}\,\alpha$.

Chapter 4

The Integral

Integration, defined as a continuous summation process, is linked to differentiation by the fundamental theorem of calculus.

In everyday language, the word *integration* refers to putting things together, while *differentiation* refers to separating, or distinguishing, things.

The simplest kind of "differentiation" in mathematics is subtraction, which tells us the difference between two numbers. We may think of differentiation in calculus as telling the difference between the values of a function at nearby points in its domain.

By analogy, the simplest kind of "integration" in mathematics is addition. Given two or more numbers, we can put them together to obtain their sum. Integration in calculus is an operation on functions, giving a "continuous sum" of all the values of a function on an interval. This process can be applied whenever a physical quantity is built up from another quantity which is spread out over space or time. For example, in this chapter, we shall see that the distance travelled by an object moving on a line is the integral of its velocity with respect to time, generalizing the formula "distance = velocity × time," which is valid when the velocity is constant. Other examples are that the volume of a wire of variable cross-sectional area is obtained by integrating this area over the length of the wire, and the total electrical energy consumed in a house during a day is obtained by integrating the time-varying power consumption over the day.

4.1 Summation

The symbol $\sum_{i=1}^{n} a_i$ is shorthand for $a_1 + a_2 + \cdots + a_n$.

To illustrate the basic ideas and properties of integration, we shall reexamine the relationship between distance and velocity. In Section 1.1 we saw that velocity is the time derivative of distance travelled, i.e.,

$$\text{velocity} \approx \frac{\Delta d}{\Delta t} = \frac{\text{change in distance}}{\text{change in time}} . \tag{1}$$

In this chapter, it will be more useful to look at this relationship in the form

$$\Delta d \approx \text{velocity} \times \Delta t. \tag{2}$$

To be more specific, suppose that a bus is travelling on a straight highway and that its position is described by a function $y = F(t)$, where y is the position of the bus measured in meters from a designated starting position, and t is the time measured in seconds. (See Fig. 4.1.1.) We wish to obtain the

Figure 4.1.1. What is the position of the bus in terms of its velocity?

Starting position

position y in terms of velocity v. In Section 2.5 we did this by using the formula $v = dy/dt$ and the notion of an antiderivative. This time, we shall go back to basic principles, starting with equation (2).

If the velocity is constant over an interval of length Δt, then the approximation ($\approx$) in equation (2) becomes equality, i.e., $\Delta d = v\,\Delta t$. This suggests another easily understood case: suppose that our time interval is divided into two parts with durations Δt_1 and Δt_2 and that the velocity during these time intervals equals the constants v_1 and v_2, respectively. (This situation is slightly unrealistic, but it is a convenient idealization.) The distance travelled during the first interval is $v_1 \Delta t_1$ and that during the second is $v_2 \Delta t_2$; thus, the total distance travelled is

$$\Delta d = v_1 \Delta t_1 + v_2 \Delta t_2.$$

Continuing in the same way, we arrive at the following result:

Summation, Distance, and Velocity

If a particle moves with a constant velocity v_1 for a time interval Δt_1, v_2 for a time interval Δt_2, v_3 for a time interval Δt_3, . . . , and velocity v_n for time interval Δt_n, then the total distance travelled is

$$\Delta d = v_1 \Delta t_1 + v_2 \Delta t_2 + v_3 \Delta t_3 + \cdots + v_n \Delta t_n. \tag{3}$$

In (3), the symbol "$+ \cdots +$" is interpreted as "continue summing until the last term $v_n \Delta t_n$ is reached."

Example 1   The bus in Fig. 4.1.1 moves with the following velocities:

> 4 meters per second for the first 2.5 seconds,
> 5 meters per second for the second 3 seconds,
> 3.2 meters per second for the third 2 seconds, and
> 1.4 meters per second for the fourth 1 second.

How far does the bus travel?

Solution   We use formula (3) with $n = 4$ and

$$v_1 = 4, \qquad \Delta t_1 = 2.5,$$
$$v_2 = 5, \qquad \Delta t_2 = 3,$$
$$v_3 = 3.2, \qquad \Delta t_3 = 2,$$
$$v_4 = 1.4, \qquad \Delta t_4 = 1$$

to get

$$\Delta d = 4 \times 2.5 + 5 \times 3 + 3.2 \times 2 + 1.4 \times 1$$
$$= 10 + 15 + 6.4 + 1.4$$
$$= 32.8 \text{ meters. } \blacktriangle$$

Integration involves a summation process similar to (3). To prepare for the development of these ideas, we need to develop a systematic notation for summation. This notation is not only useful in the discussion of the integral but will appear again in Chapter 12 on infinite series.

Given n numbers, a_1 through a_n, we denote their sum $a_1 + a_2 + \cdots + a_n$ by

$$\sum_{i=1}^{n} a_i \tag{4}$$

Here $\sum$ is the capital Greek letter *sigma*, the equivalent of the Roman S (for *sum*). We read the expression above as "the sum of a_i, as i runs from 1 to n."

Example 2 (a) Find $\sum_{i=1}^{4} a_i$, if $a_1 = 2$, $a_2 = 3$, $a_3 = 4$, $a_4 = 6$. (b) Find $\sum_{i=1}^{4} i^2$.

Solution (a) $\sum_{i=1}^{4} a_i = a_1 + a_2 + a_3 + a_4 = 2 + 3 + 4 + 6 = 15$.

(b) Here $a_i = i^2$, so

$$\sum_{i=1}^{4} i^2 = 1^2 + 2^2 + 3^2 + 4^2 = 1 + 4 + 9 + 16 = 30. \ \blacktriangle$$

Notice that formula (3) can be written in summation notation as

$$\Delta d = \sum_{i=1}^{n} v_i \Delta t_i. \tag{3'}$$

The letter i in (4) is called a *dummy index*; we can replace it everywhere by any other letter without changing the value of the expression. For instance,

$$\sum_{k=1}^{n} a_k \quad \text{and} \quad \sum_{i=1}^{n} a_i$$

have the same value, since both are equal to $a_1 + \cdots + a_n$.

A summation need not start at 1; for instance

$$\sum_{i=2}^{6} b_i \quad \text{means} \quad b_2 + b_3 + b_4 + b_5 + b_6$$

and

$$\sum_{j=-2}^{3} c_j \quad \text{means} \quad c_{-2} + c_{-1} + c_0 + c_1 + c_2 + c_3.$$

Example 3 Find $\sum_{k=2}^{5} (k^2 - k)$.

Solution $\sum_{k=2}^{5} (k^2 - k) = (2^2 - 2) + (3^2 - 3) + (4^2 - 4) + (5^2 - 5) = 2 + 6 + 12 + 20$
$$= 40. \ \blacktriangle$$

Summation Notation

To evaluate

$$\sum_{i=m}^{n} a_i ,$$

where $m \leqslant n$ are integers, and a_i are real numbers, let i take each integer value such that $m \leqslant i \leqslant n$. For each such i, evaluate a_i and add the resulting numbers. (There are $n - m + 1$ of them.)

We list below some general properties of the summation operation:

Properties of Summation

1. $$\sum_{i=m}^{n} (a_i + b_i) = \sum_{i=m}^{n} a_i + \sum_{i=m}^{n} b_i .$$

2. $$\sum_{i=m}^{n} ca_i = c \sum_{i=m}^{n} a_i , \qquad \text{where } c \text{ is a constant.}$$

3. If $m \leqslant n$ and $n + 1 \leqslant p$, then

$$\sum_{i=m}^{p} a_i = \sum_{i=m}^{n} a_i + \sum_{i=n+1}^{p} a_i .$$

4. If $a_i = C$ for all i with $m \leqslant i \leqslant n$, where C is some constant, then

$$\sum_{i=m}^{n} a_i = C(n - m + 1).$$

5. If $a_i \leqslant b_i$ for all i with $m \leqslant i \leqslant n$, then

$$\sum_{i=m}^{n} a_i \leqslant \sum_{i=m}^{n} b_i .$$

These are just basic properties of addition extended to sums of many numbers at a time. For instance, property 3 says that $a_m + a_{m+1} + \cdots + a_p = (a_m + \cdots + a_n) + (a_{n+1} + \cdots + a_p)$, which is a generalization of the associative law. Property 2 is a distributive law; property 1 is a commutative law. Property 4 says that repeated addition of the same number is the same as multiplication; property 5 is a generalization of the basic law of inequalities: if $a \leqslant b$ and $c \leqslant d$, then $a + c \leqslant b + d$.

A useful formula gives the sum of the first n integers:

Sum of the First n Integers

$$\sum_{i=1}^{n} i = \frac{1}{2} n(n + 1) \tag{5}$$

To prove this formula, let $S = \sum_{i=1}^{n} i = 1 + 2 + \cdots + n$. Then write S again with the order of the terms reversed and add the two sums:

$$
\begin{aligned}
S &= 1 &+ 2 \quad &+ 3 \quad &+ \cdots + (n-2) + (n-1) + n \\
S &= n &+ (n-1) &+ (n-2) &+ \cdots + 3 \quad\quad + 2 \quad\quad + 1 \\
\hline
2S &= (n+1) &+ (n+1) &+ (n+1) &+ \cdots + (n+1) + (n+1) + (n+1)
\end{aligned}
$$

Since there are n terms in the sum, the right-hand side is $n(n + 1)$, so $2S = n(n + 1)$, and $S = \frac{1}{2}n(n + 1)$.

Example 4   Find the sum of the first 100 integers.[1]

Solution   We substitute $n = 100$ into $S = \frac{1}{2}n(n + 1)$, giving $\frac{1}{2} \cdot 100 \cdot 101 = 50 \cdot 101 = 5050.$ ▲

Example 5   Find the sum $4 + 5 + 6 + \cdots + 29$.

Solution   This sum is $\sum_{i=4}^{29} i$. We may write it as a difference $\sum_{i=1}^{29} i - \sum_{i=1}^{3} i$ using either "common sense" or summation property 3. Using formula (5) twice gives

$$\sum_{i=4}^{29} i = \frac{1}{2} \cdot 29 \cdot 30 - \frac{1}{2} \cdot 3 \cdot 4 = 29 \cdot 15 - 3 \cdot 2 = 435 - 6 = 429. \blacktriangle$$

Example 6   Find $\sum_{j=3}^{102}(j - 2)$.

Solution   We use the summation properties as follows:

$$\sum_{j=3}^{102}(j - 2) = \sum_{j=3}^{102} j - \sum_{j=3}^{102} 2 \qquad \text{(property 1)}$$

$$= \sum_{j=1}^{102} j - \sum_{j=1}^{2} j - 2(100) \qquad \text{(properties 3 and 4)}$$

$$= \frac{1}{2}(102)(103) - 3 - 200 \qquad \text{(formula (5))}$$

$$= 5050.$$

We can also do this problem by making the substitution $i = j - 2$. As j runs from 3 to 102, i runs from 1 to 100, and we get

$$\sum_{j=3}^{102}(j - 2) = \sum_{i=1}^{100} i = \frac{1}{2} \cdot 100 \cdot 101 = 5050. \blacktriangle$$

The second method used in Example 6 is usually best carried out by thinking about the meaning of the notation in a given problem. However, for reference, we record the general formula for *substitution of an index*: With the substitution $i = j + q$,

$$\sum_{j=m}^{n} a_{j+q} = \sum_{i=m+q}^{n+q} a_i. \tag{6}$$

The following example illustrates a trick that utilizes cancellation.

Example 7   Show that $\sum_{i=1}^{n}[i^3 - (i - 1)^3] = n^3$.

Solution   The easiest way to do this is by writing out the sum:

$$\sum_{i=1}^{n}\left[i^3 - (i - 1)^3\right] = \left[1^3 - 0^3\right] + \left[2^3 - 1^3\right] + \left[3^3 - 2^3\right] + \left[4^3 - 3^3\right]$$

$$+ \cdots + \left[(n - 1)^3 - (n - 2)^3\right] + \left[n^3 - (n - 1)^3\right]$$

and observing that we can cancel 1^3 with -1^3, 2^3 with -2^3, 3^3 with -3^3, and so on up to $(n - 1)^3$ with $-(n - 1)^3$. This leaves only the terms

$$-0^3 + n^3 = n^3. \blacktriangle$$

---

[1] A famous story about the great mathematician C. F. Gauss (1777–1855) concerns a task his class had received from a demanding teacher in elementary school. They were to add up the first 100 numbers. Gauss wrote the answer 5050 on his slate immediately; had he derived $S = \frac{1}{2}n(n + 1)$ in his head at age 10?

The kind of sum encountered in Example 7 is called a *telescoping*, or *collapsing*, sum. A similar argument proves the result in the following box.

Telescoping Sum

$$\sum_{i=1}^{n} (a_i - a_{i-1}) = a_n - a_0 \tag{7}$$

The next example uses summation notation to retrieve a result which may already be obvious, but the idea will reappear later in the fundamental theorem of calculus.

Example 8   Suppose that the bus in Fig. 4.1.1 is at position y_i at time t_i, $i = 0, \ldots, n$, and that during time interval (t_{i-1}, t_i), the velocity is a constant

$$v_i = \frac{y_i - y_{i-1}}{t_i - t_{i-1}} = \frac{\Delta y_i}{\Delta t_i}, \qquad i = 1, \ldots, n.$$

Using a telescoping sum, confirm that the distance travelled equals the difference between the final and initial position.

Solution   By formula (3′), the distance travelled is

$$\Delta d = \sum_{i=1}^{n} v_i \, \Delta t_i .$$

Since $v_i = \Delta y_i / \Delta t_i$, we get

$$\Delta d = \sum_{i=1}^{n} \left(\frac{\Delta y_i}{\Delta t_i} \right) \Delta t_i = \sum_{i=1}^{n} \Delta y_i = \sum_{i=1}^{n} (y_i - y_{i-1}).$$

This is a telescoping sum which, by (7), equals $y_n - y_0$; i.e., the final position minus the initial position (see Fig. 4.1.2 where $n = 3$). ▲

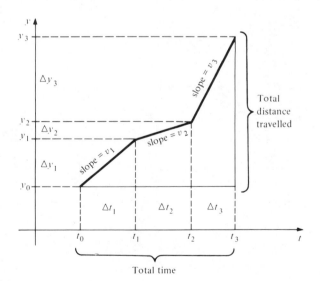

Figure 4.1.2. Motion of the bus in Example 8 ($n = 3$).

Exercises for Section 4.1

In Exercises 1–4, a particle moves along a line with the given velocities for the given time intervals. Compute the total distance travelled.

1. 2 meters per second for the first 3 seconds,
   1.8 meters per second for the second 2 seconds,
   2.1 meters per second for the third 3 seconds,
   3 meters per second for the fourth 1.5 seconds.

2. 3 meters per second for the first 1.5 seconds,
   1.2 meters per second for the second 3 seconds,
   2.1 meters per second for the third 2.4 seconds,
   4 meters per second for the fourth 3 seconds.

3. 8 meters per second for the first 1.2 seconds,
   10 meters per second for the second 3.1 seconds,
   12 meters per second for the third 4.2 seconds.

4. 2 meters per second for the first 8.1 seconds,
   3.2 meters per second for the second 2 seconds,
   4.6 meters per second for the third 1.1 seconds.

Find the sums in Exercises 5–8.

5. $\sum_{i=1}^{4}(i^2 + 1)$

6. $\sum_{i=1}^{3} i^3$

7. $\sum_{i=1}^{5} i(i - 1)$

8. $\sum_{i=1}^{6} i(i - 2)$

Find the sums in Exercises 9–12.

9. $1 + 2 + \cdots + 25$

10. $3 + 4 + \cdots + 39$

11. $\sum_{i=1}^{45} i$

12. $\sum_{i=3}^{99} i$

Find the sums in Exercises 13–16.

13. $\sum_{j=4}^{80}(j - 3)$

14. $\sum_{j=8}^{108}(j - 7)$

15. $\sum_{i=1}^{99}[(i + 1)^2 - i^2]$

16. $\sum_{i=1}^{100}[(i + 1)^5 - i^5]$

17. Find $\sum_{j=-2}^{2} j^3$.

18. Find $\sum_{j=-1000}^{1000} j^5$.

19. Find $\sum_{j=1}^{102}(j + 6)$.

20. Find $\sum_{k=-20}^{10} k$.

21. Find a formula for $\sum_{i=m}^{n} i$, where m and n are positive integers.

22. Find $\sum_{i=1}^{12} a_i$, where a_i is the number of days in the ith month of 1987.

23. Show that $\sum_{k=1}^{1000} 1/(1 + k^2) \leqslant 1000$.

24. Show that $\sum_{i=1}^{300} 3/(1 + i) \leqslant 300$.

Find the telescoping sums in Exercises 25–28.

25. $\sum_{i=1}^{100}[i^4 - (i - 1)^4]$

26. $\sum_{i=1}^{32}\{(3i)^2 - [3(i - 1)]^2\}$

27. $\sum_{i=1}^{100}[(i + 2)^2 - (i + 1)^2]$

28. $\sum_{i=1}^{41}[(i + 3)^2 - (i + 2)^2]$

29. Draw a graph like Fig. 4.1.2 for the data in Exercise 1.

30. Draw a graph like Fig. 4.1.2 for the data in Exercise 2.

Find the sums in Exercises 31–40.

31. $\sum_{k=0}^{100}(3k - 2)$

32. $\sum_{i=0}^{n}(2i + 1)$

33. $\sum_{k=1}^{n}[(k + 1)^4 - k^4]$

34. $\sum_{k=1}^{300}[(k + 1)^8 - k^8]$

35. $\sum_{i=1}^{100}[(i + 2)^2 - (i - 1)^2]$

36. $\sum_{i=1}^{50}[(2i + 2)^3 - (2i)^3]$

37. $\sum_{i=-30}^{30}[i^5 + i + 2]$

38. $\sum_{l=-75}^{75}[l^9 + 5l^7 - 13l^5 + l]$

39. $\sum_{j=2}^{6} 2^j$

40. $\sum_{k=1}^{3} 3^k$

★41. By the method of telescoping sums, we have

$$\sum_{i=1}^{n}\left[(i + 1)^3 - i^3\right] = (n + 1)^3 - 1.$$

(a) Write $(i + 1)^3 - i^3 = 3i^2 + 3i + 1$ and use properties of summation to prove that

$$\sum_{i=1}^{n} i^2 = \frac{n(n + 1)(2n + 1)}{6}.$$

(b) Find a formula for

$$\sum_{i=m}^{n} i^2$$

in terms of m and n.

(c) Using the *method* and result of (a), find a formula for $\sum_{i=1}^{n} i^3$. (You may wish to try guessing an answer by experiment.)

★42. (a) Prove that

$$\sum_{i=1}^{n} i(i + 1) = \frac{1}{3} n(n + 1)(n + 2)$$

by writing

$$i(i + 1) = \frac{1}{3}[i(i + 1)(i + 2) - (i - 1)i(i + 1)]$$

and using a telescoping sum.

(b) Find $\sum_{i=1}^{n} i(i + 1)(i + 2)$.

(c) Find $\sum_{i=1}^{n}[1/i(i + 1)]$.

4.2 Sums and Areas

Areas under graphs can be approximated by sums.

In the last section, we saw that the formula for distance in terms of velocity is $\Delta d = \sum_{i=1}^{n} v_i \Delta t_i$ when the velocity is a constant v_i during the time interval (t_{i-1}, t_i). In this section we shall discuss a geometric interpretation of this fact which will be important in the study of integration.

Let us plot the velocity of a bus as a function of time. Suppose that the total time interval in question is $[a,b]$; i.e., t runs from a to b, and this interval is divided into n smaller intervals so that $a = t_0 < t_1 < \cdots < t_n = b$. The ith interval is (t_{i-1}, t_i), and v is a constant v_i on this interval.[2] The form of v is shown in Fig. 4.2.1 for $n = 5$.

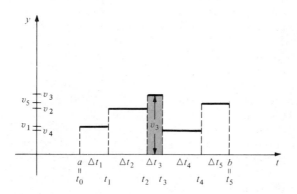

Figure 4.2.1. The velocity of the bus.

We notice that $v_i \Delta t_i$ is exactly the area of the rectangle over the ith interval with base Δt_i and height v_i (the rectangle for $i = 3$ is shaded in the figure). Thus,

$$\Delta d = \sum_{i=1}^{n} v_i \Delta t_i \text{ is the total area of the rectangles under the graph of } v.$$

This suggests that the problem of finding distances in terms of velocities should have something to do with areas, even when the velocity changes smoothly rather than abruptly. Turning our attention to areas then, we go back to the usual symbol x (rather than t) for the independent variable.

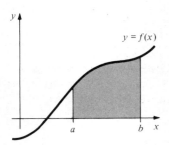

Figure 4.2.2. The region under the graph of f on $[a,b]$.

The *area under the graph* of a function f on an interval $[a,b]$ is defined to be the area of the region in the plane enclosed by the graph $y = f(x)$, the x axis, and the vertical lines $x = a$ and $x = b$. (See Fig. 4.2.2.) Here we assume that $f(x) \geqslant 0$ for x in $[a,b]$. (In the next section, we shall deal with the possibility that f might take negative values.)

Let us examine certain similarities between properties of sums and areas. To the property $\sum_{i=m}^{p} a_i = \sum_{i=m}^{n} a_i + \sum_{i=n+1}^{p} a_i$ of sums, there corresponds the additive property of areas: if a plane region is split into two parts which overlap only along their edges, the area of the region is the sum of the areas of the parts. (See Fig. 4.2.3.) Another property of sums is that if $a_i \leqslant b_i$ for

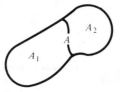

Figure 4.2.3. Area $(A) =$ Area $(A_1) +$ Area (A_2).

Figure 4.2.4. Area $(A) \geqslant$ Area (B).

[2] We are deliberately vague about the value of v at the end points, when the bus must suddenly switch velocities. The value of Δd does not depend on what v is at each t_i, so we can safely ignore these points.

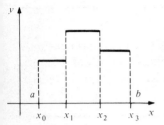

Figure 4.2.5. The graph of a step function on $[a, b]$ with $n = 3$.

$i = m, m + 1, \ldots, n$, then $\sum_{i=m}^{n} a_i \leqslant \sum_{i=m}^{n} b_i$; the counterpart for areas is the inclusion property: if one plane region is contained in another, the containing region has more area. (See Fig. 4.2.4.)

The connection between areas and sums becomes more explicit if we consider step functions. A function g on the interval $[a, b]$ is called a *step function* if $[a, b]$ can be broken into smaller intervals (called subintervals) with g constant on each part. More precisely, there should be numbers x_0, $x_1, \ldots, x_n$, with $a = x_0 < x_1 < x_2 < \cdots < x_{n-1} < x_n = b$, such that g is constant on each of the intervals $(x_0, x_1), (x_1, x_2), \ldots, (x_{n-1}, x_n)$, as in Fig. 4.2.5. The values of g at the endpoints of these intervals will not affect any of our calculations. The list $(x_0, x_1, \ldots, x_n)$ is called a *partition* of $[a, b]$.

Example 1   Draw a graph of the step function g defined on $[2, 4]$ by

$$g(x) = \begin{cases} 1 & \text{if} \quad 2 \leqslant x \leqslant 2.5, \\ 3 & \text{if} \quad 2.5 < x < 3.5, \\ 2 & \text{if} \quad 3.5 \leqslant x \leqslant 4. \end{cases}$$

Solution   The graph of g on $[2, 2.5]$ is a horizontal line with height 1 on this interval. The endpoints on the graph are drawn as solid dots to indicate that g takes the value 1 at the endpoints $x = 2$ and $x = 2.5$. Continuing through the remaining subintervals and using open dots to indicate endpoints which do not belong to the graph, we obtain Fig. 4.2.6. ▲

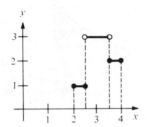

Figure 4.2.6. The graph of the step function g in Example 1.

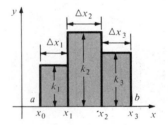

Figure 4.2.7. The shaded area is the sum of $k_1 \Delta x_1$, $k_2 \Delta x_2$ and $k_3 \Delta x_3$.

If a step function is non-negative, then the region under its graph can be broken into rectangles, and the area of the region can be expressed as a sum. It is common to write Δx_i for length $x_i - x_{i-1}$ of the ith partition interval; if the value of g on this interval is $k_i \geqslant 0$, then the area of the rectangle from x_{i-1} to x_i with height k_i is $k_i \Delta x_i$. Thus the total area under the graph is $k_1 \Delta x_1 + k_2 \Delta x_2 + \cdots + k_n \Delta x_n = \sum_{i=1}^{n} k_i \Delta x_i$, as in Fig. 4.2.7.

Example 2   What are the x_i's, Δx_i's, and k_i's for the step function in Example 1? Compute the area of the region under its graph.

Solution   Looking at Figs. 4.2.6 and 4.2.7, we begin by labelling the left endpoint as x_0; i.e., $x_0 = 2$. The remaining partition points are $x_1 = 2.5$, $x_2 = 3.5$, and $x_3 = 4$. The Δx_i's are the widths of the intervals: $\Delta x_1 = x_1 - x_0 = 0.5$, $\Delta x_2 = 1$, and $\Delta x_3 = 0.5$. Finally, the k_i's are the heights of the rectangles: $k_1 = 1$, $k_2 = 3$, and $k_3 = 2$. The area under the graph is

$$\sum_{i=1}^{3} k_i \Delta x_i = (1)(0.5) + (3)(1) + (2)(0.5) = 4.5. \quad ▲$$

Step Functions

A function g on $[a, b]$ is a step function when $[a, b]$ can be broken up into intervals of width Δx_i, on each of which g equals a constant k_i.

If each k_i is positive (or zero), the area under the graph of g is $\sum_{i=1}^{n} k_i \Delta x_i$.

In deriving our formula for the area under the graph of a step function, we used the fact that the area of a rectangle is its length times width, and the additive property of areas. By using the inclusion property, we can find the areas under graphs of *general* functions by comparison with step functions—this idea, which goes back to the ancient Greeks, is the key to defining the integral.

Given a non-negative function f, we wish to compute the area A under its graph on $[a, b]$. A *lower sum* for f on $[a, b]$ is defined to be the area under the graph of a non-negative step function g for which $g(x) \leqslant f(x)$ on $[a, b]$. If $g(x) = k_i$ on the ith subinterval, then the inclusion property of areas tells us that $\sum_{i=1}^{n} k_i \Delta x_i \leqslant A$. (Fig. 4.2.8).

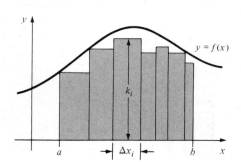

Figure 4.2.8. The shaded area $\sum_{i=1}^{n} k_i \Delta x_i$ is a lower sum for f on $[a, b]$.

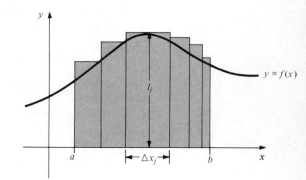

Figure 4.2.9. The shaded area $\sum_{j=1}^{m} l_j \Delta x_j$ is an upper sum for f on $[a, b]$.

Similarly, an *upper sum* for f on $[a, b]$ is defined to be $\sum_{j=1}^{m} l_j \Delta x_j$, where h is a step function with $f(x) \leqslant h(x)$ on $[a, b]$, and $h(x) = l_j$ on the jth subinterval of a partition of $[a, b]$ (Fig. 4.2.9). By the inclusion property for areas, $A \leqslant \sum_{j=1}^{m} l_j \Delta x_j$, so the area lies between the upper and lower sums.

Example 3 Let $f(x) = x^2 + 1$ for $0 \leqslant x \leqslant 2$. Let

$$g(x) = \begin{cases} 0 & 0 \leqslant x \leqslant 1 \\ 2 & 1 < x \leqslant 2 \end{cases} \quad \text{and} \quad h(x) = \begin{cases} 2 & 0 \leqslant x \leqslant \frac{2}{3}, \\ 4 & \frac{2}{3} < x \leqslant \frac{4}{3}, \\ 5 & \frac{4}{3} < x \leqslant 2. \end{cases}$$

Draw a graph showing $f(x)$, $g(x)$, and $h(x)$. What upper and lower sums for f can be obtained from g and h?

Solution The graphs are shown in Fig. 4.2.10.

For g we have $\Delta x_1 = 1$, $k_1 = 0$ and $\Delta x_2 = 1$, $k_2 = 2$. Since $g(x) \leqslant f(x)$ for all x in the open interval $(0, 2)$ (the graph of g lies below that of f), we have as a lower sum,

$$\sum_{i=1}^{2} k_i \Delta x_i = 0 \cdot 1 + 2 \cdot 1 = 2.$$

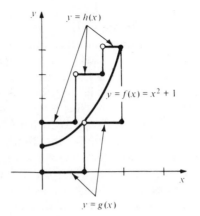

Figure 4.2.10. The area under the graph of h is an upper sum for f; the area under the graph of g is a lower sum.

For h we have $\Delta x_1 = \frac{2}{3}$, $l_1 = 2$, $\Delta x_2 = \frac{2}{3}$, $l_2 = 4$, and $\Delta x_3 = \frac{2}{3}$, $l_3 = 5$. Since the graph of h lies above that of f, $h(x) \geqslant f(x)$ for all x in the interval $(0, 2)$, we get the upper sum

$$\sum_{i=1}^{3} l_i \Delta x_i = 2 \cdot \frac{2}{3} + 4 \cdot \frac{2}{3} + 5 \cdot \frac{2}{3} = \frac{22}{3} = 7\frac{1}{3} . \ \blacktriangle$$

Using partitions with sufficiently small subintervals, we hope to find step functions below and above f such that the corresponding lower and upper sums are as close together as we wish. Notice that the difference between lower and upper sums is the area *between* the graphs of the step functions (Fig. 4.2.11). We expect this area to be very small if the subintervals are small enough and the values of the step functions are close to the values of f.

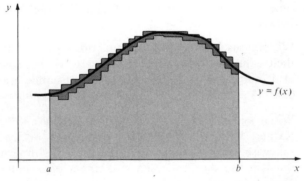

Figure 4.2.11. The dark shaded area is the difference between upper and lower sums for f on $[a, b]$. The area under the graph is between the upper and lower sums.

Suppose that there are lower sums and upper sums which are arbitrarily close to one another. Then there can only be one number A such that $L \leqslant A \leqslant U$ for every lower sum L and every upper sum U, and this number must be the area under the graph.

Area Under a Graph

To calculate the area under the graph of a non-negative function f, we try to find upper and lower sums (areas under graphs of step functions lying below and above f) which are closer and closer together. (See Example 6 below for a specific instance of what is meant by "closer and closer.") The area A is the number which is above all the lower sums and below all the upper sums.

What we have done here for areas has a counterpart in our distance–velocity problem. Suppose that $v = f(t)$ defined for $a \leqslant t \leqslant b$ gives the velocity of a moving bus as a function of time, and that there is a partition $(t_0, t_1, \ldots, t_n)$ of $[a, b]$ and numbers $k_1, \ldots, k_n$ such that $k_i \leqslant f(t)$ for t in the ith interval (t_{i-1}, t_i). Taking for granted that a faster moving object travels further in a given time interval, we may conclude that the bus travels a distance at least $k_i(t_i - t_{i-1})$ in the ith time interval. Thus the total distance travelled must be *at least* $k_1 \Delta t_1 + \cdots + k_n \Delta t_n = \sum_{i=1}^{n} k_i \Delta t_i$ (where, as usual, we write Δt_i for $t_i - t_{i-1}$), so we have a *lower estimate* for the distance travelled between $t = a$ and $t = b$. Similarly, if we know that $f(t) \leqslant l_i$ on (t_{i-1}, t_i) for some numbers $l_1, \ldots, l_n$, we get an *upper estimate* $\sum_{i=1}^{n} l_i \Delta t_i$ for the distance travelled. By making the time intervals short enough we hope to be able to find k_i and l_i close together, so that we can estimate the distance travelled as accurately as we wish.

Example 4 The velocity of a moving bus (in meters per second) is observed over periods of 10 seconds, and it is found that

$$4 \leqslant v \leqslant 5 \qquad \text{when} \quad 0 < t < 10,$$
$$5.5 \leqslant v \leqslant 6.5 \qquad \text{when} \quad 10 < t < 20,$$
$$5 \leqslant v \leqslant 5.7 \qquad \text{when} \quad 20 < t < 30.$$

Estimate the distance travelled during the interval $0 \leqslant t \leqslant 30$.

Solution A lower estimate is $4 \cdot 10 + 5.5 \cdot 10 + 5 \cdot 10 = 145$, and an upper estimate is $5 \cdot 10 + 6.5 \cdot 10 + 5.7 \cdot 10 = 172$, so the distance travelled is between 145 and 172 meters. ▲

Example 5 The velocity of a snail at time t seconds is $(0.001)(t^2 + 1)$ meters per second at time t. Use the calculations in Example 3 to estimate how far the snail crawled between $t = 0$ and $t = 2$.

Solution We may use the comparison functions g and h in Example 3 if we multiply their values by 0.001 (and change x to t). The lower sum and upper sum are also multiplied by 0.001, and so the distance crawled is between 0.002 and $0.00733\ldots$ meters, i.e., between 2 and $7\frac{1}{3}$ millimeters. ▲

When we calculate derivatives, we seldom use the definition in terms of limits. Rather, we use the rules for derivatives, which are much more efficient. Likewise, we will not usually calculate areas in terms of upper and lower sums but will use the fundamental theorem of calculus once we have learned it. Now, however, to reinforce the idea of upper and lower sums, we shall do one area problem "the hard way."

Example 6 Use upper and lower sums to find the area under the graph of $f(x) = x$ on $[0, 1]$.

Solution The area is shaded in Fig. 4.2.12.

We will look for upper and lower sums which are close together. The simplest way to do this is to divide the interval $[0, 1]$ into equal parts with a partition of the form $(0, 1/n, 2/n, \ldots, (n-1)/n, 1)$. A step function $g(x)$ below $f(x)$ is given by setting $g(x) = (i-1)/n$ on the interval $[(i-1)/n, i/n)$, while the step function with $h(x) = i/n$ on $((i-1)/n, i/n]$ is above $f(x)$ (Fig. 4.2.13).

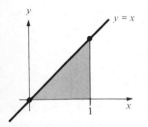

Figure 4.2.12. The region under the graph of $f(x) = x$ on $[0, 1]$.

The difference between the upper and lower sums is equal to the total area of the chain of boxes in Fig. 4.2.14, on which both $g(x)$ and $h(x)$ are graphed. Each of the n boxes has area $(1/n) \cdot (1/n) = 1/n^2$, so their total area is $n \cdot (1/n^2) = 1/n$, which becomes arbitrarily small as $n \to \infty$, so we know that the area under our graph will be precisely determined. To find the area, we compute the upper and lower sums. For the lower sum, $g(x) = (i - 1)/n = k_i$ on the ith subinterval, and $\Delta x_i = 1/n$ for all i, so

$$\sum_{i=1}^{n} k_i \Delta x_i = \sum_{i=1}^{n} \frac{i-1}{n} \cdot \frac{1}{n} = \frac{1}{n^2} \sum_{i=1}^{n} (i-1) = \frac{1}{n^2} \left(\sum_{i=1}^{n} i - \sum_{i=1}^{n} 1 \right)$$

$$= \frac{1}{n^2} \left[\frac{n(n+1)}{2} - n \right] = \frac{n+1}{2n} - \frac{1}{n} = \frac{1}{2} - \frac{1}{2n}.$$

The upper sum is

$$\sum_{i=1}^{n} \frac{i}{n} \cdot \frac{1}{n} = \frac{1}{n^2} \sum_{i=1}^{n} i = \frac{1}{n^2} \frac{n(n+1)}{2} = \frac{1}{2} + \frac{1}{2n}.$$

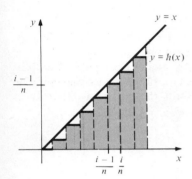

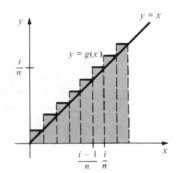

Figure 4.2.13. Lower and upper sums for $f(x) = x$ on $[0, 1]$.

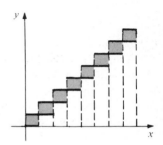

Figure 4.2.14. Difference between the upper and lower sums.

The area under the graph is the unique number A which satisfies the inequalities $1/2 - 1/2n \leqslant A \leqslant 1/2 + 1/2n$ for all n (see Fig. 4.2.15). Since the number $\frac{1}{2}$ satisfies the condition, we must have $A = \frac{1}{2}$. ▲

Figure 4.2.15. The area lies in the interval $[1/2 - 1/2n, 1/2 + 1/2n]$ for all n. The length of this interval $\to 0$ as $n \to \infty$.

The result of Example 6 agrees with the rule from elementary geometry that the area of a triangle is half the base times the height. The advantage of the method used here is that it can applied to more general graphs. (Another case is given in Exercise 20.) This method was used extensively during the century before the invention of calculus, and is the basis for the definition of the integral.

Exercises for Section 4.2

Draw the graphs of the step functions in Exercises 1–4.

1. $g(x) = \begin{cases} 0 & \text{if } 0 \leqslant x \leqslant 1, \\ 2 & \text{if } 1 < x < 2, \\ 1 & \text{if } 2 \leqslant x \leqslant 3. \end{cases}$

2. $g(x) = \begin{cases} 1 & \text{if } 0 \leqslant x \leqslant 0.5, \\ 3 & \text{if } 0.5 < x < 2, \\ 2 & \text{if } 2 \leqslant x \leqslant 3, \\ 4 & \text{if } 3 < x \leqslant 4. \end{cases}$

3. $g(x) = \begin{cases} 0 & \text{if } 0 \leqslant x < 1, \\ 1 & \text{if } 1 \leqslant x < 2, \\ 2 & \text{if } 2 \leqslant x \leqslant 3. \end{cases}$

4. $g(x) = \begin{cases} 1 & \text{if } 2 \leqslant x < 2.5, \\ 3 & \text{if } 2.5 \leqslant x \leqslant 3, \\ 4 & \text{if } 3 < x < 4, \\ 0 & \text{if } 4 \leqslant x \leqslant 4.5. \end{cases}$

In Exercises 5–8 compute the x_i's, Δx_i's, and k_i's for the indicated step function and compute the area of the region under its graph.

5. For g in Exercise 1.

6. For g in Exercise 2.

7. For g in Exercise 3.

8. For g in Exercise 4.

In Exercises 9 and 10, draw a graph showing f, g, and h and compute the upper and lower sums for f obtained from g and h.

9. $f(x) = x^2$, $1 \le x \le 3$;

$$g(x) = \begin{cases} 1, & 1 \le x \le 2, \\ 4, & 2 < x \le 3; \end{cases}$$

$$h(x) = \begin{cases} 4, & 1 \le x < 2, \\ 9, & 2 \le x \le 3. \end{cases}$$

10. $f(x) = x^3 + 1$, $1 \le x \le 3$;

$$g(x) = \begin{cases} 2, & 1 \le x \le 1.5, \\ 4, & 1.5 < x < 2, \\ 9, & 2 \le x \le 3; \end{cases}$$

$$h(x) = \begin{cases} 9, & 1 \le x \le 2, \\ 28, & 2 < x \le 3. \end{cases}$$

11. The velocity of a moving bus (in meters per second) is observed over periods of 5 seconds and it is found that

$$\begin{array}{lll} 5.0 \le v \le 6.0 & \text{when} & 0 < t < 5, \\ 4.0 \le v \le 5.5 & \text{when} & 5 < t < 10, \\ 6.1 \le v \le 7.2 & \text{when} & 10 < t < 15, \\ 3.2 \le v \le 4.7 & \text{when} & 15 < t < 20. \end{array}$$

Estimate the distance travelled during the interval $t = 0$ to $t = 20$.

12. The velocity of a moving bus (in meters per second) is observed over periods of 7.5 seconds and it is found that

$$\begin{array}{lll} 4.0 \le v \le 5.1 & \text{when} & 0 < t < 7.5, \\ 3.0 \le v \le 5.0 & \text{when} & 7.5 < t < 15, \\ 4.4 \le v \le 5.5 & \text{when} & 15 < t < 22.5, \\ 3.0 \le v \le 4.1 & \text{when} & 22.5 < t < 30. \end{array}$$

Estimate the distance travelled during the interval $t = 0$ to $t = 30$.

13. The velocity of a snail at time t is $(0.002)t^2$ meters per second at time t. Use the functions g and h in Exercise 9 to estimate how far the snail crawled between $t = 1$ and $t = 3$.

14. The velocity of a snail at time t is given by $(0.0005)(t^3 + 1)$ meters per second at time t. Use the functions g and h in Exercise 10 to estimate how far the snail crawled between $t = 1$ and $t = 3$.

In Exercises 15–18, use upper and lower sums to find the area under the graph of the given function.

15. $f(x) = x$ for $1 \le x \le 2$.

16. $f(x) = 2x$ for $0 \le x \le 1$.

17. $f(x) = 5x$ for $a \le x \le b$, $a > 0$.

18. $f(x) = x + 3$ for $a \le x \le b$, $a > 0$.

19. Using upper and lower sums, find the area under the graph of $f(x) = 1 - x$ between $x = 0$ and $x = 1$.

★20. Using upper and lower sums, show that the area under the graph of $f(x) = x^2$ between $x = a$ and $x = b$ is $\frac{1}{3}(b^3 - a^3)$. (You will need to use the result of Exercise 41(a) from Section 4.1.)

★21. Let $f(x) = \begin{cases} x^2, & 0 \le x < 1, \\ x, & 1 \le x \le 2. \end{cases}$

Find the area under the graph of f on $[0, 2]$, using the results of the Exercises 15 and 20.

★22. Let

$$f(x) = \begin{cases} 1 - x, & 0 \le x < 1, \\ 5x, & 1 \le x \le 4. \end{cases}$$

Using the results of Exercises 17 and 19, find the area under the graph of f on $[0, 4]$.

★23. By combining the results of Example 6 and Exercise 20, find the area of the shaded region in Fig. 4.2.16. (*Hint:* Write the area as a difference of known areas.)

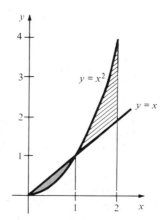

Figure 4.2.16. Find the shaded and striped areas.

★24. Using the results of previous exercises, find the area of the striped region in Fig. 4.2.16.

4.3 The Definition of the Integral

The integral of a function is a "signed" area.

In the previous section, we saw how areas under graphs could be approximated by the areas under graphs of step functions. Now we shall extend this idea to functions that need not be positive and shall give the formal definition of the integral.

Recall that if g is a step function with constant value $k_i \geqslant 0$ on the interval (x_{i-1}, x_i) of width $\Delta x_i = x_i - x_{i-1}$, then the area under the graph of g is

$$\text{Area} = \sum_{i=1}^{n} k_i \Delta x_i.$$

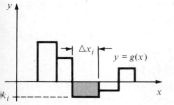

Figure 4.3.1. The product $k_i \Delta x_i$ is the negative of the shaded area.

This formula is analogous to the formula for distance travelled when the velocity is a step function; see formula (3), Section 4.1. In that situation, it is reasonable to allow negative velocity (reverse motion). Likewise, in the area formula we wish to allow negative k_i. To do so, we shall have to interpret "area" correctly. Suppose that $g(x)$ is a *negative* constant k_i on an interval of width Δx_i. Then $k_i \Delta x_i$ is the *negative* of the area between the graph of g and the x axis on that interval. (See Fig. 4.3.1.)

To formalize this idea, we introduce the notion of signed area. If R is any region in the xy plane, its *signed area* is defined to be the area of the part of R lying above the x axis, minus the area of the part lying below the axis.

If f is a function defined on the interval $[a, b]$, the region between the graph of f and the x axis consists of those points (x, y) for which x is in $[a, b]$ and y lies betwen 0 and $f(x)$. It is natural to consider the signed area of such a region, as illustrated in Fig. 4.3.2. For a step function g with values k_i on intervals of length Δx_i, the sum $\sum_{i=1}^{n} k_i \Delta x_i$ gives the signed area of the region between the graph of g and the x axis.

Figure 4.3.2. The signed area between the graph of f and the x axis on $[a, b]$ is the area of the + regions minus the area of the − regions.

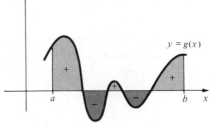

Signed Area

The signed area of a region is the area of the portion above the x axis minus the area of the portion below the x axis.

For the region between the x axis and the graph of a step function g, this signed area is $\sum_{i=1}^{n} k_i \Delta x_i$.

Example 1 Draw a graph of the step function g on $[0, 1]$ defined by

$$g(x) = \begin{cases} -2 & \text{if } 0 \leqslant x < \frac{1}{3}, \\ 3 & \text{if } \frac{1}{3} \leqslant x \leqslant \frac{3}{4}, \\ 1 & \text{if } \frac{3}{4} < x \leqslant 1. \end{cases}$$

Compute the signed area of the region between its graph and the x axis.

Solution The graph is shown in Fig. 4.3.3. There are three intervals, with $\Delta x_1 = \frac{1}{3}$, $\Delta x_2 = \frac{3}{4} - \frac{1}{3} = \frac{5}{12}$, and $\Delta x_3 = 1 - \frac{3}{4} = \frac{1}{4}$; $k_1 = -2$, $k_2 = 3$, and $k_3 = 1$. Thus the signed area is

$$\sum_{i=1}^{3} k_i \Delta x_i = (-2)\left(\tfrac{1}{3}\right) + (3)\left(\tfrac{5}{12}\right) + (1)\left(\tfrac{1}{4}\right) = -\tfrac{2}{3} + \tfrac{5}{4} + \tfrac{1}{4} = \tfrac{5}{6}. \quad \blacktriangle$$

The counterpart of signed area for our distance–velocity problem is directed distance, explained as follows: If the bus is moving to the right, then $v > 0$ and distances are increasing. If the bus is moving to the left, then $v < 0$ and the distances are decreasing. In the formula $\Delta d = \sum_{i=1}^{n} v_i \Delta t_i$, Δd is the displacement, or the net distance the bus has moved, not the total distance travelled, which would be $\sum_{i=1}^{\infty} |v_i| \Delta t$. Just as with signed areas, movement to the left is considered negative and is subtracted from movement to the right. (See Fig. 4.3.4.)

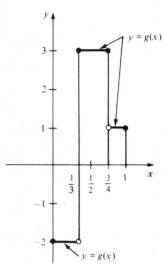

Figure 4.3.3. The graph of the step function in Example 1.

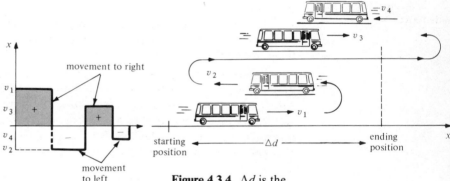

Figure 4.3.4. Δd is the displacement; i.e., net distance travelled.

To find the signed area between the graph and the x axis for a function which is not a step function, we can use upper and lower sums. Just as with positive functions, if g is a step function lying below f, i.e., $g(x) \leqslant f(x)$ for x in $[a, b]$, we call

$$L = \sum_{i=1}^{n} k_i \Delta x_i$$

a *lower sum* for f. Likewise, if h is a step function lying above f, with values l_j on intervals of width Δx_j for $j = 1, \ldots, m$, then

$$U = \sum_{j=1}^{m} l_j \Delta x_j$$

is an *upper sum* for f. If we can find L's and U's arbitrarily close together, lying on either side of a number A, then A must be the *signed area* between the graph of f and the x axis on $[a, b]$. (See Fig. 4.3.5.)

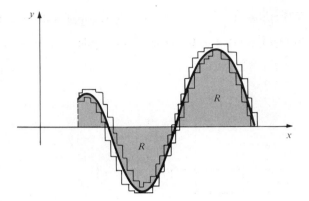

Figure 4.3.5. The signed area of the region R lies between the upper and lower sums.

We are now ready to define the integral of a function f.

Definition   Let f be a function defined on $[a, b]$. We say that f *has an integral* or that f *is integrable* if upper and lower sums for f can be found which are arbitrarily close together. The number I such that $L \leqslant I \leqslant U$ for all lower sums L and upper sums U is called the *integral* of f and is denoted

$$\int_a^b f(x)\,dx.$$

We call $\int$ the *integral sign*, a, b, the *endpoints* or *limits of integration*, and f the *integrand*.

The precise meaning of "arbitrarily close together" is the same as in Example 6, Section 4.2, namely, that there should exist sequences L_n and U_n of lower and upper sums such that $\lim_{n\to\infty}(U_n - L_n) = 0$. (Limits of sequences will be treated in detail in Chapter 11.)

The Integral

Given a function f on $[a, b]$, the integral of f, if it exists, is the number

$$\int_a^b f(x)\,dx$$

which separates the upper and lower sums. This number is the signed area of the region between the graph of f and the x axis.

The notation for the integral is derived from the notation for sums. The Greek letter $\sum$ has turned into an elongated S; k_i and l_j have turned into function values $f(x)$; Δx_i has become dx; and the limits of summation (e.g., i goes from 1 to n) have become limits of integration:

$$\sum_{i=1}^n k_i\,\Delta x_i$$

$$\downarrow$$

$$\int_a^b f(x)\,dx$$

Just as with antiderivatives, the "x" in "dx" indicates that x is the variable of integration.

Example 2   Compute $\int_{-2}^{3} f(x)\,dx$ for the function f sketched in Fig. 4.3.6. How is the integral related to the area of the shaded region in the figure?

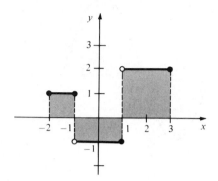

Figure 4.3.6. The signed area of the shaded region is an integral.

Solution   The integral $\int_{-2}^{3} f(x)\,dx$ is the signed area of the shaded region.

$$\int_{-2}^{3} f(x)\,dx = (1)(1) + (-1)(2) + (2)(2) = 1 - 2 + 4 = 3. \;\blacktriangle$$

Example 3   Write the signed area of the region in Fig. 4.3.7 as an integral.

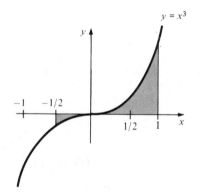

Figure 4.3.7. The signed area of this region equals what integral?

Solution   The region is that between the graph of $y = x^3$ and the x axis from $x = -\frac{1}{2}$ to $x = 1$, so the signed area is

$$\int_{-1/2}^{1} x^3\,dx. \;\blacktriangle$$

The next example shows how upper and lower sums can be used to approximate an integral. (In Chapter 6, we will learn how to compute this integral exactly.)

Example 4   Using a division of the interval $[1, 2]$ into three equal parts, find $\int_{1}^{2}(1/x)\,dx$ to within an error of no more than $\frac{1}{10}$.

Solution   To estimate the integral within $\frac{1}{10}$, we must find lower and upper sums which are within $\frac{2}{10}$ of one another. We divide the interval into three equal parts and use the step functions which give us the lowest possible upper sum and highest possible lower sum, as shown in Fig. 4.3.8. For a lower sum we have

$$\int_{1}^{2} g(x)\,dx = \frac{1}{4/3}\left(\frac{4}{3} - 1\right) + \frac{1}{5/3}\left(\frac{5}{3} - \frac{4}{3}\right) + \frac{1}{2}\left(2 - \frac{5}{3}\right)$$

$$= \frac{3}{4}\cdot\frac{1}{3} + \frac{3}{5}\cdot\frac{1}{3} + \frac{1}{2}\cdot\frac{1}{3} = \frac{1}{3}\left(\frac{3}{4} + \frac{3}{5} + \frac{1}{2}\right) = \frac{1}{3}\left(\frac{37}{20}\right) = \frac{37}{60}.$$

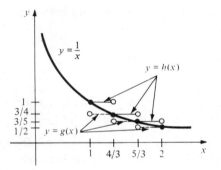

Figure 4.3.8. Illustrating upper and lower sums for $1/x$ on $[1, 2]$.

For an upper sum we have

$$\int_1^2 h(x)\,dx = \frac{1}{1}\cdot\frac{1}{3} + \frac{1}{4/3}\cdot\frac{1}{3} + \frac{1}{5/3}\cdot\frac{1}{3}$$

$$= \frac{1}{3}\left(1 + \frac{3}{4} + \frac{3}{5}\right)$$

$$= \frac{1}{3}\left(\frac{47}{20}\right) = \frac{47}{60}.$$

It follows that

$$\frac{37}{60} \leqslant \int_1^2 \frac{1}{x}\,dx \leqslant \frac{47}{60}.$$

Since the integral lies in the interval $[\frac{37}{60}, \frac{47}{60}]$, whose length is $\frac{1}{6}$, we may take the midpoint $\frac{42}{60} = \frac{7}{10}$ as our estimate; it will differ from the true integral by no more than $\frac{1}{2}\cdot\frac{1}{6} = \frac{1}{12}$, which is less than $\frac{1}{10}$. ▲

We have been calculating approximations to integrals without knowing whether some of those integrals actually exist or not. Thus it may be reassuring to know the following fact whose proof is given in more advanced courses.[3]

Existence Theorem   *If f is continuous on $[a, b]$, then it has an integral.*

In particular, all differentiable functions have integrals, but so do step functions and functions whose graphs have corners (such as $y = |x|$); thus, integrability is a more easily satisfied requirement than differentiability or even continuity. The reader should note, however, that there do exist some "pathological" functions that are not integrable. (See Exercise 36).

It is possible to calculate integrals of functions which are not necessarily positive by the method used in Example 6 of the previous section, but this is a tedious process. Rather than doing any such examples here, we shall wait until we have developed the machinery of the fundamental theorem of calculus to assist us.

Let us now interpret the integral in terms of the distance–velocity problem. We saw in our previous work that the upper and lower sums represent the displacement of vehicles whose velocities are step functions and which are faster or slower than the one we are studying. Thus, the displace-

---

[3] See, for instance, *Calculus Unlimited* by J. Marsden and A. Weinstein, Benjamin/Cummings (1981), p. 159, or one of the other references given in the Preface.

ment, like the integral, is sandwiched between upper and lower sums for the velocity function, so we must have

$$\text{displacement} = \int_a^b f(t)\, dt.$$

Example 5   A bus moves on the line with velocity $v = (t^2 - 4t + 3)$ meters per second. Write formulas in terms of integrals for:

(a) the displacement of the bus between $t = 0$ and $t = 3$;
(b) the actual distance the bus travels between $t = 0$ and $t = 3$.

Solution   (a) The displacement is $\int_0^3 (t^2 - 4t + 3)\, dt$.
(b) We note that v can be factored as $(t - 1)(t - 3)$, so it is positive on $(0, 1)$ and negative on $(1, 3)$. The total distance travelled is thus

$$\int_0^1 (t^2 - 4t + 3)\, dt - \int_1^3 (t^2 - 4t + 3)\, dt. \; \blacktriangle$$

We close this section with a discussion of a different approach to the integral, called the *method of Riemann sums*. Later we shall usually rely on the step function approach, but Riemann sums are also widely used, and so you should have at least a brief exposure to them.

The idea behind Riemann sums is to use step functions to *approximate* the function to be integrated, rather than bounding it above and below. Given a function f defined on $[a, b]$ and a partition $(x_0, x_1, \ldots, x_n)$ of that interval, we choose points $c_1, \ldots, c_n$ such that c_i lies in the interval $[x_{i-1}, x_i]$. The step function which takes the constant value $f(c_i)$ on (x_{i-1}, x_i) is then an approximation to f; the signed area under its graph, namely,

$$S_n = \sum_{i=1}^n f(c_i)\, \Delta x_i,$$

is called a Riemann sum.[4] It lies above all the lower sums and below all the upper sums constructed using the same partition, so it is a good approximation to the integral of f on $[a, b]$ (see Fig. 4.3.9). Notice that the Riemann sum

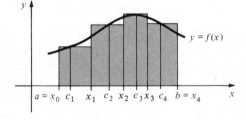

Figure 4.3.9. The area of the shaded region is a Riemann sum for f on $[a, b]$.

is formed by "sampling" the values of f at points $c_1, \ldots, c_n$, "weighting" the samples according to the lengths of the intervals from which the c_i's are chosen, and then adding.

If we choose a sequence of partitions, one for each n, such that the lengths Δx_i approach zero as n becomes larger, then the Riemann sums approach the integral $\int_a^b f(x)\, dx$ in the limit as $n \to \infty$. From this and Fig. 4.3.9, we again see the connection between integrals and areas.

Just as the derivatives may be defined as a limit of difference quotients, so the integral may be defined as a limit of Riemann sums; the integral as defined this way is called the *Riemann integral*.

[4] After the German mathematician Bernhard Riemann (1826–1866).

Riemann Sums

Choose, for each n, a partition of $[a, b]$ into n subintervals such that the maximum of Δx_i in the nth partition approaches zero as $n \to \infty$. If c_i is a point chosen in the interval $[x_{i-1}, x_i]$, then

$$\lim_{n \to \infty} \sum_{i=1}^{n} f(c_i) \Delta x_i = \int_a^b f(x)\, dx.$$

Example 6   Write $\int_0^1 x^3 \, dx$ as a limit of sums.

Solution   As in Example 6, Section 4.2, divide $[0, 1]$ into n equal parts by the partition $(0, 1/n, 2/n, \ldots, (n-1)/n, 1)$. Choose $c_i = i/n$, the right endpoint of the interval $[(i-1)/n, i/n]$. (We may choose any point we wish; the left endpoint or midpoint would have been just as good.) Then with $f(x) = x^3$, we get

$$S_n = \sum_{i=1}^{n} f(c_i) \Delta x_i = \sum_{i=1}^{n} c_i^3 \cdot \frac{1}{n} = \sum_{i=1}^{n} \left(\frac{i}{n}\right)^3 \left(\frac{1}{n}\right) = \sum_{i=1}^{n} \frac{i^3}{n^4}.$$

Therefore,

$$\lim_{n \to \infty} \frac{1}{n^4} \sum_{i=1}^{n} i^3 = \int_0^1 x^3 \, dx.$$

Thus, we can find $\int_0^1 x^3 \, dx$ if we can evaluate this limit, or vice versa. ▲

Supplement to Section 4.3
Solar Energy

Besides the distance–velocity and area problems, which we used to introduce the integral, there are other physical problems that could be used in the same way. Here, we consider the problem of computing solar energy and shall see how it, too, leads naturally to the integral in terms of upper and lower sums.

Consider a solar cell attached to an energy storage unit (such as a battery) as in Fig. 4.3.10. When light shines on the solar cell, it is converted into electrical energy which is stored in the battery (as electrical–chemical energy) for later use.

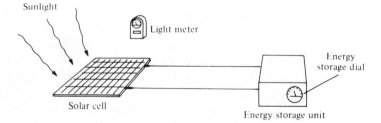

Figure 4.3.10. The storage unit accumulates the power received by the solar cell.

We will be interested in the relation between the amount E of energy stored and the intensity I of the sunlight. The number E can be read off a dial on the energy storage device; I can be measured with a photographer's light meter. (The units in which E and I are measured are unimportant for this discussion.)

Experiments show that when the solar cell is exposed to a steady source of sunlight, the change ΔE in the amount of energy stored is proportional to the product of the intensity I and the length Δt of the period of exposure. Thus

$$\Delta E = \kappa I \Delta t,$$

where κ is a constant depending on the apparatus and on the units used to measure energy, time, and intensity. (We can imagine κ being told to us as a manufacturer's specification.)

The intensity I can change—for example, the sun can move behind a cloud. If during two periods, Δt_1 and Δt_2, the intensity is, respectively, I_1 and I_2, then the total change in energy is the sum of the energies stored over each individual period. That is,

$$\Delta E = \kappa I_1 \Delta t_1 + \kappa I_2 \Delta t_2 = \kappa (I_1 \Delta t_1 + I_2 \Delta t_2).$$

Likewise, if there are n periods, $\Delta t_1, \ldots, \Delta t_n$, during which the intensity is $I_1, \ldots, I_n$ (as in Fig. 4.3.11(a)), the energy stored will be the sum of n terms,

$$\Delta E = \kappa (I_1 \Delta t_1 + I_2 \Delta t_2 + \cdots + I_n \Delta t_n) = \kappa \sum_{i=1}^{n} I_i \Delta t_i .$$

Notice that this sum is exactly κ times the integral of the step function g, where $g(t) = I_i$ on the interval of length Δt_i.

In practice, as the sun moves gradually behind the clouds and its elevation in the sky changes, the intensity I of sunlight does not change by jumps but varies continuously with t (Fig. 4.3.11(b)). The change in stored

Figure 4.3.11. The intensity of sunlight varying with time.

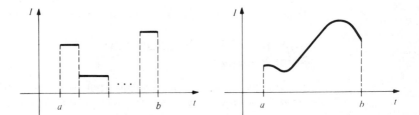

energy ΔE can still be measured on the energy storage meter, but it can no longer be represented as a sum in the ordinary sense. In fact, the intensity now takes on infinitely many values, but it does not stay at a given value for any length of time.

If $I = f(t)$, the true change in stored energy is given by the integral

$$\Delta E = \kappa \int_a^b f(t)\, dt,$$

which is κ times the area under the graph $I = f(t)$. If $g(t)$ is a step function with $g(t) \leqslant f(t)$, then the integral of g is less than or equal to the integral of $f(t)$. This is in accordance with our intuition: the less the intensity, the less the energy stored.

The passage from step functions to general functions in the definition of the integral and the interpretation of the integral can be carried out in many contexts; this gives integral calculus a wide range of applications.

Exercises for Section 4.3

In Exercises 1–4, draw a graph of the given step function, and compute the signed area of the region between its graph and the x axis.

1. $g(x) = \begin{cases} 1 & \text{if } 0 \leq x \leq 1, \\ -3 & \text{if } 1 < x \leq 2. \end{cases}$

2. $g(x) = \begin{cases} -4 & \text{if } -1 \leq x < 0, \\ 2 & \text{if } 0 \leq x \leq 1, \\ 3 & \text{if } 1 < x \leq 2. \end{cases}$

3. $g(x) = \begin{cases} 1 & \text{if } -2 \leq x < -1, \\ -1 & \text{if } -1 \leq x \leq 0. \end{cases}$

4. $g(x) = \begin{cases} -3 & \text{if } -3 \leq x \leq -2, \\ -2 & \text{if } -2 < x \leq -1, \\ -1 & \text{if } -1 < x \leq 0. \end{cases}$

In Exercises 5–8, compute the indicated integrals.

5. $\int_0^2 g(x)\,dx$, g as in Exercise 1.

6. $\int_{-1}^0 g(x)\,dx$, g as in Exercise 2.

7. $\int_{-2}^0 g(x)\,dx$, g as in Exercise 3.

8. $\int_{-2}^0 g(x)\,dx$, g as in Exercise 4.

In Exercises 9–12, write the signed areas of the shaded regions in terms of integrals. (See Figure 4.3.12.)

9.

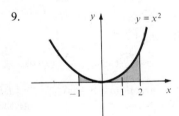

10.

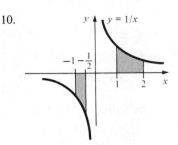

11.

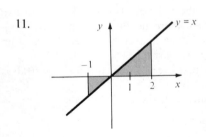

12.

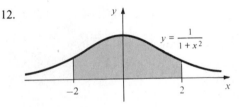

Figure 4.3.12. Graphs for Exercises 9–12.

13. Find $\int_2^4 (1/x)\,dx$ to within an error of no more than $\frac{1}{10}$.

14. If you used the method in Example 4 to calculate $\int_1^2 (1/x)\,dx$ to within $\frac{1}{100}$, how many subintervals would you need?

15. Estimate $\int_1^2 (1/x^2)\,dx$ to within $\frac{1}{10}$.

16. Estimate $\int_1^2 (1/x^2)\,dx$ to within $\frac{1}{100}$.

17. A bus moves on the line with velocity given by $v = 5(t^2 - 5t + 6)$. Write a formula in terms of integrals for:
    (a) the displacement of the bus between $t = 0$ and $t = 3$;
    (b) the actual distance the bus travels between $t = 0$ and $t = 3$.

18. A bus moves on the line with velocity given by $v = 6t^2 - 30t + 24$. Write a formula in terms of integrals for:
    (a) the displacement of the bus between $t = 0$ and $t = 5$;
    (b) the actual distance the bus travels between $t = 0$ and $t = 5$.

In Exercises 19–22, write the given integral as a limit of sums.

19. $\int_0^1 x^5\,dx.$

20. $\int_0^1 9x^3\,dx.$

21. $\int_2^4 \frac{1}{1+x^2}\,dx.$

22. $\int_3^4 \frac{x^2}{1+x}\,dx.$

23. Show that $-3 \leq \int_1^2 (t^3 - 4)\,dt \leq 4$.

24. Show that $\int_0^1 t^{10}\,dt \leq 1$.

25. Let $f(t)$ be defined by
$$f(t) = \begin{cases} 2 & \text{if } 0 \leq t < 1, \\ 0 & \text{if } 1 \leq t < 3, \\ -1 & \text{if } 3 \leq t \leq 4. \end{cases}$$

For any number x in $(0, 4]$, $f(t)$ is a step function on $[0, x]$.

(a) Find $\int_0^x f(t)\,dt$ as a function of x. (You will need to use different formulas on different intervals.)

(b) Let $F(x) = \int_0^x f(t)\,dt$, for x in $(0, 4]$. Draw a graph of F.

(c) At which points is F differentiable? Find a formula for $F'(x)$.

26. Let f be the function defined by

$$f(x) = \begin{cases} 2, & 1 \leqslant x < 4, \\ 5, & 4 \leqslant x < 7, \\ 1, & 7 \leqslant x \leqslant 10. \end{cases}$$

(a) Find $\int_1^{10} f(x)\,dx$.

(b) Find $\int_2^9 f(x)\,dx$.

(c) Suppose that g is a function on $[1, 10]$ such that $g(x) \leqslant f(x)$ for all x in $[1, 10]$. What inequality can you derive for $\int_1^{10} g(x)\,dx$?

(d) With $g(x)$ as in part (c), what inequalities can you obtain for $\int_1^{10} 2g(x)\,dx$ and $\int_1^{10} -g(x)\,dx$? [*Hint:* Find functions like f with which you can compare $2g$ and $-g$.]

27. Let $f(t)$ be the "greatest integer function"; that is, $f(t)$ is the greatest integer which is less than or equal to t—for example, $f(n) = n$ for any integer, $f(5\frac{1}{2}) = 5$, $f(-5\frac{1}{2}) = -6$, and so on.

(a) Draw a graph of $f(t)$ on the interval $[-4, 4]$.

(b) Find $\int_0^1 f(t)\,dt$, $\int_0^6 f(t)\,dt$, $\int_{-2}^2 f(t)\,dt$, and $\int_0^{4.5} f(t)\,dt$.

(c) Find a general formula for $\int_0^n f(t)\,dt$, where n is any positive integer.

(d) Let $F(x) = \int_0^x f(t)\,dt$, where $x > 0$. Draw a graph of F for $x \in [0, 4]$, and find a formula for $F'(x)$, where it is defined.

28. A rod 1 meter long is made of 100 segments of equal length such that the linear density of the kth segment is $30k$ grams per meter. What is the total mass of the rod?

29. The volume of a rod of uniform shape is $A\,\Delta x$, where A is the cross-sectional area and Δx is the length.

(a) Suppose that the rod consists of n pieces, with the ith piece having cross-sectional area A_i and length Δx_i. Write a formula for the volume.

(b) Suppose that the cross-sectional area is $A = f(x)$, where f is a function on $[0, L]$, L being the total length of the rod. Write a formula for the volume of the rod, using integral notation.

★30. Suppose that $f(x)$ is a step function on $[a, b]$, and let $g(x) = f(x) + k$, where k is a constant.

(a) Show that $g(x)$ is a step function.

(b) Find $\int_a^b g(x)\,dx$ in terms of $\int_a^b f(x)\,dx$.

★31. Let $h(x) = kf(x)$, where $f(x)$ is a step function on $[a, b]$.

(a) Show that $h(x)$ is a step function.

(b) Find $\int_a^b h(x)\,dx$ in terms of $\int_a^b f(x)\,dx$.

★32. For $x \in [0, 1]$ let $f(x)$ be the first digit after the decimal point in the decimal expansion of x.

(a) Draw a graph of f. (b) Find $\int_0^1 f(x)\,dx$.

★33. Define the functions f and g on $[0, 3]$ as follows:

$$f(x) = \begin{cases} 4, & 0 \leqslant x < 1, \\ -1, & 1 \leqslant x < 2, \\ 2, & 2 \leqslant x \leqslant 3, \end{cases}$$

$$g(x) = \begin{cases} 2, & 0 \leqslant x < 1\frac{1}{2}, \\ 1, & 1\frac{1}{2} \leqslant x \leqslant 3. \end{cases}$$

(a) Draw the graph of $f(x) + g(x)$ and compute $\int_0^3 [f(x) + g(x)]\,dx$.

(b) Compute $\int_1^2 [f(x) + g(x)]\,dx$.

(c) Compare $\int_0^3 2f(x)\,dx$ with $2\int_0^3 f(x)\,dx$.

(d) Show that

$$\int_0^3 [f(x) - g(x)]\,dx = \int_0^3 f(x)\,dx - \int_0^3 g(x)\,dx.$$

(e) Is the following true?

$$\int_0^3 f(x) \cdot g(x)\,dx \overset{?}{=} \int_0^3 f(x)\,dx \cdot \int_0^3 g(x)\,dx.$$

★34. Suppose that f is a continuous function on $[a, b]$ and that $f(x) \neq 0$ for all x in $[a, b]$. Assume that $a \neq b$ and that $f((a + b)/2) = 1$. Prove that $\int_a^b f(x)\,dx > 0$. [*Hint:* Find a lower sum.]

★35. Compute the exact value of $\int_0^1 x^5\,dx$ by using Riemann sums and the formula

$$1^5 + 2^5 + 3^5 + \cdots + N^5 = \frac{N^6}{6} + \frac{N^5}{2} + \frac{5N^4}{12} - \frac{N^2}{12}.$$

★36. Let the function f be defined on $[0, 3]$ by

$$f(x) = \begin{cases} 0 & \text{if } x \text{ is a rational number,} \\ 2 & \text{if } x \text{ is irrational.} \end{cases}$$

(a) Using the fact that between every two real numbers there lie both rationals and irrationals, show that every upper sum for f on $[0, 3]$ is at least 6.

(b) Show that every lower sum for f on $[0, 3]$ is at most 0.

(c) Is f integrable on $[0, 3]$? Explain.

4.4 The Fundamental Theorem of Calculus

The processes of integration and differentiation are inverses to one another.

We now know two ways of expressing the solution of the distance–velocity problem. Let us recall the problem and these two ways.

Problem   A bus moves on a straight line with given velocity $v = f(t)$ for $a \leqslant t \leqslant b$. Find the displacement Δd of the bus during this time interval.

First Solution   The first solution uses antiderivatives and was presented in Section 2.5. Let $y = F(t)$ be the position of the bus at time t. Then since $v = dy/dt$, i.e., $f = F'$, F is an antiderivative of f. The displacement is the final position minus the initial position; i.e.,

$$\Delta d = F(b) - F(a), \tag{1}$$

the difference between the values of the antiderivative at $t = a$ and $t = b$.

Second Solution   The second solution uses the integral as defined in the previous section. We saw that

$$\Delta d = \int_a^b f(t)\, dt. \tag{2}$$

We arrived at formulas (1) and (2) by rather different routes. However, the displacement must be the same in each case. Equating (1) and (2), we get

$$F(b) - F(a) = \int_a^b f(t)\, dt. \tag{3}$$

This equality is called the *fundamental theorem of calculus*. It expresses the integral in terms of an antiderivative and establishes the key link between differentiation and integration.

The argument by which we arrived at (3) was based on a physical model. Later, in this section, we shall also give a purely mathematical proof.

With a slight change of notation, we restate (3) in the following box.

Fundamental Theorem of Calculus

Suppose that the function F is differentiable everywhere on $[a,b]$ and that F' is integrable on $[a,b]$. Then

$$\int_a^b F'(x)\, dx = F(b) - F(a).$$

In other words, if f is integrable on $[a,b]$ and has an antiderivative F, then

$$\int_a^b f(x)\, dx = F(b) - F(a).$$

We may use this theorem to find the integral which we previously computed "by hand" (Example 6, Section 4.2).

Example 1   Using the fundamental theorem of calculus, compute $\int_0^1 x\,dx$.

Solution   By the power rule, an antiderivative for $f(x) = x$ is $F(x) = \frac{1}{2}x^2$. (You could also have found $F(x)$ by guessing, and you can always check the answer by differentiating $\frac{1}{2}x^2$.) The fundamental theorem gives

$$\int_0^1 x\,dx = \int_0^1 f(x)\,dx = F(1) - F(0) = \frac{1}{2}\cdot 1^2 - \frac{1}{2}\cdot 0^2 = \frac{1}{2},$$

which agrees with our earlier result. ▲

Next, we use the fundamental theorem to obtain a new result.

Example 2   Using the fundamental theorem of calculus, compute $\int_a^b x^2\,dx$.

Solution   Let $f(x) = x^2$; again by the power rule, we may take $F(x) = \frac{1}{3}x^3$. By the fundamental theorem, we have

$$\int_a^b x^2\,dx = \int_a^b f(x)\,dx = F(b) - F(a) = \frac{1}{3}b^3 - \frac{1}{3}a^3.$$

We conclude that $\int_a^b x^2\,dx = \frac{1}{3}(b^3 - a^3)$. This gives the area under a segment of the parabola $y = x^2$ (Fig. 4.4.1). ▲

We can summarize the integration method provided by the fundamental theorem as follows:

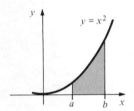

Figure 4.4.1. The shaded area is $\int_a^b x^2\,dx = \frac{1}{3}(b^3 - a^3)$.

Fundamental Integration Method

To integrate the function $f(x)$ over the interval $[a,b]$: find an antiderivative $F(x)$ for $f(x)$, then evaluate F at a and b and subtract the results:

$$\int_a^b f(x)\,dx = F(b) - F(a).$$

Notice that the fundamental theorem does not specify which antiderivative to use. However, if F_1 and F_2 are two antiderivatives of f on $[a,b]$, they differ by a constant (see Section 3.6); $F_1(t) = F_2(t) + C$, and so

$$F_1(b) - F_1(a) = \left[F_2(b) + C\right] - \left[F_2(a) + C\right] = F_2(b) - F_2(a).$$

(The C's cancel.) Thus all choices of F give the same result.

Expressions of the form $F(b) - F(a)$ occur so often that it is useful to have a special notation for them.

Notation for the Fundamental Theorem

$$F(x)\Big|_a^b \quad \text{means} \quad F(b) - F(a).$$

Example 3   Find $(x^3 + 5)\big|_2^3$.

Solution   Here $F(x) = x^3 + 5$ and

$$(x^3 + 5)\big|_2^3 = F(3) - F(2)$$
$$= (3^3 + 5) - (2^3 + 5)$$
$$= 32 - 13 = 19. \ \blacktriangle$$

In terms of this new notation, we can write the formula of the fundamental theorem of calculus in the form

$$\int_a^b f(x)\,dx = F(x)\Big|_a^b,$$

where F is an antiderivative of f on $[a,b]$.

Example 4   Find $\displaystyle\int_2^6 (x^2 + 1)\,dx$.

Solution   By the sum and power rules for antiderivatives, and antiderivative for $x^2 + 1$ is $\frac{1}{3}x^3 + x$. By the fundamental theorem,

$$\int_2^6 (x^2 + 1)\,dx = \left(\frac{1}{3}x^3 + x\right)\Big|_2^6$$

$$= \left(\frac{6^3}{3} + 6\right) - \left(\frac{2^3}{3} + 2\right)$$

$$= 78 - 4\tfrac{2}{3} = 73\tfrac{1}{3}. \ \blacktriangle$$

Example 5   Evaluate $\displaystyle\int_1^2 \frac{1}{x^4}\,dx$.

Solution   An antiderivative of $1/x^4 = x^{-4}$ is $-1/3x^3$, since

$$\frac{d}{dx}\left(-\frac{1}{3}x^{-3}\right) = -\frac{1}{3}\cdot(-3)x^{-4} = x^{-4}.$$

Hence

$$\int_1^2 \frac{1}{x^4}\,dx = -\frac{1}{3x^3}\Big|_1^2 = \left(-\frac{1}{3\cdot 2^3}\right) - \left(-\frac{1}{3\cdot 1^3}\right) = -\frac{1}{24} + \frac{1}{3} = \frac{7}{24}. \ \blacktriangle$$

We will now give a complete proof of the fundamental theorem of calculus. The basic idea is as follows: letting F be an antiderivative for f on $[a,b]$, we will show that the number $F(b) - F(a)$ lies between any lower and upper sums for f on $[a,b]$. Since f is assumed integrable, it has upper and lower sums arbitrarily close together, and the only number with this property is the integral of f (see page 217). Thus, we will have $F(b) - F(a) = \int_a^b f(x)\,dx$.

Proof of the Fundamental Theorem   For the lower sums, we must show that any step function g below f on (a,b) has integral at most $F(b) - F(a)$. So let $k_1, k_2, \ldots, k_n$ be the values of g on the partition intervals $(x_0, x_1), (x_1, x_2), \ldots, (x_{n-1}, x_n)$ (See Fig. 4.4.2). On

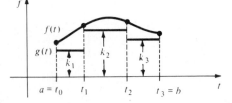

Figure 4.4.2. The integral of g is a lower sum for f on $[a,b]$.

(x_{i-1}, x_i), $k_i \leqslant f(x) = F'(x)$, so the difference quotient for F satisfies the inequality $k_i \leqslant [F(x_i) - F(x_{i-1})]/[x_i - x_{i-1}]$, by the first consequence of the mean value theorem (Section 3.6). Thus, $k_i \Delta x_i \leqslant F(x_i) - F(x_{i-1})$. Summing from $i = 1$ to n, we get

$$\sum_{i=1}^n k_i \Delta x_i \leqslant \big[F(x_1) - F(x_0)\big] + \big[F(x_2) - F(x_1)\big] + \cdots + \big[F(x_n) - F(x_{n-1})\big].$$

The left-hand side is just the integral of g on $[a,b]$, while the right-hand side is a telescoping sum which collapses to $F(x_n) - F(x_0)$; so we have proven that $\int_a^b g(x)\,dx \leqslant F(b) - F(a)$.

An identical argument works for upper sums: If h is a step function above f on (a,b), then $F(b) - F(a) \leqslant \int_a^b h(x)\,dx$ (see Exercise 49). Thus the proof of the fundamental theorem is complete. ∎

Here are two more examples illustrating the use of the fundamental theorem. Notice that any letter can be used as the variable of integration, just like the "dummy variable" in summation.

Example 6 Find $\displaystyle\int_0^4 (t^2 + 3t^{7/2})\,dt$.

Solution By the sum, constant multiple, and power rules for antiderivatives, an antiderivative for $t^2 + 3t^{7/2}$ is $(t^3/3) + 3 \cdot (2/9)t^{9/2}$. Thus,

$$\int_0^4 (t^2 + 3t^{7/2})\,dt = \left(\frac{t^3}{3} + \frac{2t^{9/2}}{3} \right)\Bigg|_0^4$$

$$= \frac{4^3}{3} + \frac{2 \cdot 2^9}{3} = \frac{1088}{3}. \; \blacktriangle$$

In the next example, some algebraic manipulations are needed before the integral is computed.

Example 7 Compute $\displaystyle\int_1^2 \frac{(s+5)^2}{s^4}\,ds$.

Solution The integrand may be broken apart:

$$\frac{(s+5)^2}{s^4} = \frac{s^2 + 10s + 25}{s^4} = \frac{1}{s^2} + \frac{10}{s^3} + \frac{25}{s^4}.$$

We can find an antiderivative term by term, by the power rule:

$$\int_1^2 \left(\frac{1}{s^2} + \frac{10}{s^3} + \frac{25}{s^4} \right)ds = \int_1^2 (s^{-2} + 10s^{-3} + 25s^{-4})\,ds$$

$$= \left(\frac{s^{-1}}{(-1)} + \frac{10s^{-2}}{(-2)} + \frac{25s^{-3}}{(-3)} \right)\Bigg|_1^2$$

$$= \left(-\frac{1}{s} - \frac{10}{2s^2} - \frac{25}{3s^3} \right)\Bigg|_1^2$$

$$= -\left(\frac{1}{s} + \frac{10}{2s^2} + \frac{25}{3s^3} \right)\Bigg|_1^2$$

$$= -\left(\left(\frac{1}{2} + \frac{5}{4} + \frac{25}{3 \cdot 8} \right) - \left(1 + 5 + \frac{25}{3} \right) \right)$$

$$= -\left(\frac{67}{24} - \frac{43}{3} \right) = \frac{277}{24} \approx 11.54. \; \blacktriangle$$

Next we use the fundamental theorem to solve area and distance–velocity problems. Let us first recall, from Sections 4.2 and 4.3, the situation for areas under graphs.

Area under a Graph

If $f(x) \geqslant 0$ for x in $[a,b]$, the area under the graph of f between $x = a$ and $x = b$ is

$$\int_a^b f(x)\,dx.$$

If f is negative at some points of $[a,b]$, then $\int_a^b f(x)\,dx$ is the signed area of the region between the graph of f, the x axis, and the lines $x = a$ and $x = b$.

Example 8   (a) Find the area of the region bounded by the x axis, the y axis, the line $x = 2$, and the parabola $y = x^2$. (b) Compute the area of the region shown in Fig. 4.4.3.

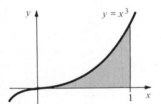

Figure 4.4.3. Compute this area.

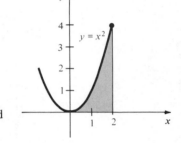

Figure 4.4.4. The shaded area equals $\int_0^2 x^2\,dx$.

Solution   (a) The region described is that under the graph of $f(x) = x^2$ on $[0,2]$ (Fig. 4.4.4). The area of the region is $\int_0^2 x^2\,dx = \frac{1}{3}x^3\big|_0^2 = \frac{8}{3}$.

(b) The region is that under the graph of $y = x^3$ from $x = 0$ to $x = 1$, so its area is $\int_0^1 x^3\,dx$. By the fundamental theorem,

$$\int_0^1 x^3\,dx = \frac{x^4}{4}\bigg|_0^1 = \frac{1}{4}.$$

Thus, the area is $\frac{1}{4}$. ▲

Example 9   (a) Interpret $\displaystyle\int_0^2 (x^2 - 1)\,dx$ in terms of areas and evaluate. (b) Find the shaded area in Figure 4.4.5.

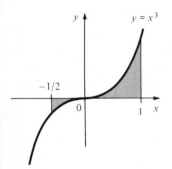

Figure 4.4.5. Find the area of this region.

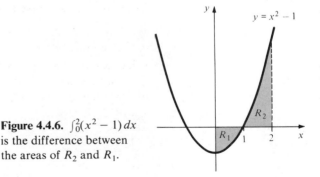

Figure 4.4.6. $\int_0^2 (x^2 - 1)\,dx$ is the difference between the areas of R_2 and R_1.

Solution   (a) Refer to Fig. 4.4.6. We know that the integral represents the signed area of the region between the graph of $y = x^2 - 1$ and the x axis. In other words, it is the area of R_2 minus the area of R_1. Evaluating,

$$\int_0^2 (x^2 - 1)\, dx = \left(\frac{x^3}{3} - x \right)\Big|_0^2 = \frac{8}{3} - 2 = \frac{2}{3}.$$

(b) For functions which are negative on part of an interval, we must recall, from Section 4.3, that the integral represents the *signed* area between the graph and the x axis. To get the ordinary area, we must integrate piece by piece.

The area from $x = 0$ to $x = 1$ is $\int_0^1 x^3\, dx$. The negative of the area from $x = -\frac{1}{2}$ to $x = 0$ is $\int_{-1/2}^0 x^3\, dx$. Thus the total area is

$$A = -\int_{-1/2}^0 x^3\, dx + \int_0^1 x^3\, dx$$

$$= -\frac{x^4}{4}\Big|_{-1/2}^0 + \frac{x^4}{4}\Big|_0^1 = \frac{(1/2)^4}{4} + \frac{1}{4}$$

$$= \frac{1}{4 \cdot 16} + \frac{1}{4} = \frac{17}{64}. \ \blacktriangle$$

Finally, in this section, we consider the use of the fundamental theorem to solve displacement problems. The following box summarizes the method, which was justified earlier in this section.

Displacements and Velocity

If a particle on the x axis has velocity $v = f(t)$ and position $x = F(t)$, then the displacement $F(b) - F(a)$ between the times $t = a$ and $t = b$ is obtained by integrating the velocity from $t = a$ to $t = b$:

$$\left(\begin{array}{c} \text{Displacement from} \\ \text{time } t = a \text{ to } t = b \end{array} \right) = \int_a^b (\text{velocity})\, dt.$$

Example 10   An object moving in a straight line has velocity $v = 5t^4 + 3t^2$ at time t. How far does the object travel between $t = 1$ and $t = 2$?

Solution   The displacement equals the total distance travelled in this case, since $v > 0$. Thus, the displacement is

$$\Delta d = \int_1^2 (5t^4 + 3t^2)\, dt = (t^5 + t^3)\Big|_1^2 = (32 + 8) - (1 + 1) = 38.$$

Thus, the object travels 38 units of length between $t = 1$ and $t = 2$. $\blacktriangle$

We have seen that the geometric interpretation of integrals of functions that can sometimes be negative requires the notion of signed area. Likewise, when velocities are negative, we have to be careful with signs. The integral is always the displacement; to get the actual distance travelled, we must change the sign of the integral over the periods when the velocity is negative. See Fig. 4.4.7 for a typical situation.

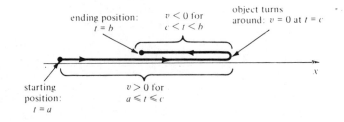

Figure 4.4.7. The total distance travelled is $\int_a^c v\, dt - \int_c^b v\, dt$; the displacement is $\int_a^b v\, dt$.

Example 11   An object on the x axis has velocity $v = 2t - t^2$ at time t. If it starts out at $x = -1$ at time t, where is it at time $t = 3$? How far has it travelled?

Solution   Let $x = f(t)$ be the position at time t. Then

$$f(3) - f(0) = \int_0^3 (2t - t^2)\, dt$$

$$= \left(t^2 - \frac{t^3}{3} \right)\Bigg|_0^3 = 9 - \frac{27}{3} = 0.$$

Since $f(0) = -1$, the object is again at $x = 0 + f(0) = -1$ at time $t = 3$.

The object turns around when v changes sign, namely, at those t where $2t - t^2 = 0$ or $t = 0, 2$. For $0 < t < 2$, $v > 0$, and for $2 < t < 3$, $v < 0$. The total distance travelled is therefore

$$\int_0^2 (2t - t^2)\, dt - \int_2^3 (2t - t^2)\, dt$$

$$= \left(t^2 - \frac{t^3}{3} \right)\Bigg|_0^2 - \left(t^2 - \frac{t^3}{3} \right)\Bigg|_2^3$$

$$= \left(4 - \frac{8}{3} \right) - \left(9 - \frac{27}{3} \right) + \left(4 - \frac{8}{3} \right)$$

$$= \frac{8}{3}. \ \blacktriangle$$

Exercises for Section 4.4

Using the fundamental theorem of calculus, compute the integrals in Exercises 1–4.

1. $\int_1^3 x^3\, dx.$

2. $\int_2^3 x^2\, dx.$

3. $\int_4^6 3x\, dx.$

4. $\int_1^8 (1 + \sqrt{x})\, dx.$

Compute the quantities in Exercises 5–8.

5. $x^{3/4}\big|_0^2.$

6. $(x^2 + 2\sqrt[3]{x}\,)\big|_0^8.$

7. $(3x^2 + 5)\big|_1^3.$

8. $(x^4 + x^2 + 2)\big|_{-2}^2.$

Evaluate the integrals in Exercises 9–24.

9. $\int_a^b s^{4/3}\, ds.$

10. $\int_{-1}^2 (t^4 + 8t)\, dt.$

11. $\int_1^2 4\pi r^{2/3}\, dr.$

12. $\int_{-1}^1 (t^4 + t^{917})\, dt.$

13. $\int_0^{10} \left(\frac{t^4}{100} - t^2 \right) dt.$

14. $\int_{-4}^0 (1 + x^2 - x^3)\, dx.$

15. $\int_{-1}^2 (1 + t^2)^2\, dt.$

16. $\int_1^2 \left(s^3 + \frac{1}{s^2} \right) ds.$

17. $\int_1^2 \frac{dt}{(t + 4)^3}.$

18. $\int_{\pi/2}^{\pi} (3 + z^2)\, dz.$

19. $\int_1^2 \frac{(1 + t^2)^2}{t^2}\, dt.$

20. $\int_1^2 \frac{t^2 + 8t + 1}{t^4}\, dt.$

21. $\int_1^2 \frac{(x^2 + 5)^2}{x^4}\, dx.$

22. $\int_{-2}^{-1} \frac{(x^2 + x)^2}{x}\, dx.$

23. $\int_2^3 \frac{u^3 - 1}{u - 1}\, du.$

24. $\int_2^4 \frac{u^4 - 1}{u - 1}\, du.$

Calculate the areas of the regions in Exercises 25–28 (Figure 4.4.8).

25.

26.

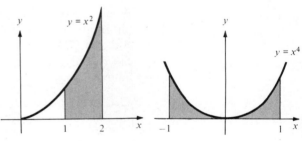

27.

28.

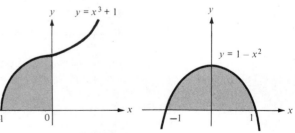

Figure 4.4.8. Regions for Exercises 25–28.

Interpret the integrals in Exercises 29 and 30 in terms of areas, sketch, and evaluate.

29. $\int_0^2 (x^3 - 1)\,dx$.

30. $\int_1^2 (x^2 - 3)\,dx$.

In Exercises 31–40, find the area of the region between the graph of each of the following functions and the x axis on the given interval and sketch.

31. x^3 on $[0, 2]$.
32. $1/x^2$ on $[1, 2]$.
33. $x^2 + 2x + 3$ on $[1, 2]$.
34. $x^3 + 3x + 2$ on $[0, 2]$.
35. $x^4 + 2$ on $[-1, 1]$.
36. $3x^4 - 2x + 2$ on $[-1, 1]$.
37. $x^4 + 3x^2 + 1$; $-2 \leqslant x \leqslant 1$.
38. $8x^6 + 3x^4 - 2$; $1 \leqslant x \leqslant 2$.
39. $(1/x^3) + x^2$; $1 \leqslant x \leqslant 3$.
40. $(3x + 5)/x^3$; $1 \leqslant x \leqslant 2$.

41. An object moving in a straight line has velocity $v = 6t^4 + 3t^2$ at time t. How far does the object travel between $t = 1$ and $t = 10$?

42. An object moving in a straight line has velocity $v = 2t^3 + t^4$ at time t. How far does the object travel between $t = 0$ and $t = 2$?

43. The velocity of an object on the x axis is $v = 4t - 2t^2$. If it is at $x = 1$ at $t = 0$, where is it at $t = 4$? How far has it travelled?

44. The velocity of an object on the x axis is $v = t^2 - 3t + 2$. If the object is at $x = -1$ at $t = 0$, where is it at $t = 2$? How far has it travelled?

45. The velocity of a stone dropped from a balloon is $32t$ feet per second, where t is the time in seconds after release. How far does the stone travel in the first 10 seconds?

46. How far does the stone in Exercise 45 travel in the second 10 seconds after its release? The third 10 seconds?

★47. An object is thrown upwards from the earth's surface with a velocity v_0. (a) How far has it travelled after it returns? (b) How far has it travelled when its velocity is $-\frac{1}{2}v_0$?

★48. Suppose that F is continuous on $[0, 2]$, that $F'(x) < 2$ for $0 \leqslant x \leqslant \frac{1}{3}$, and that $F'(x) < 1$ whenever $\frac{1}{3} \leqslant x \leqslant 2$. What can you say about the difference $F(2) - F(0)$?

★49. Prove that if $h(t)$ is a step function on $[a, b]$ such that $f(t) \leqslant h(t)$ for all t in the interval (a, b), then $F(b) - F(a) \leqslant \int_a^b h(t)\,dt$, where F is any antiderivative for f on $[a, b]$.

★50. Let $a_0, \ldots, a_n$ be a given set of numbers and $\delta_i = a_i - a_{i-1}$. Let $b_k = \sum_{i=1}^k \delta_i$, $d_i = b_i - b_{i-1}$. Express the b's in terms of the a's and the d's in terms of the δ's.

4.5 Definite and Indefinite Integrals

Integrals and sums have similar properties.

When we studied antiderivatives in Section 2.5, we used the notation $\int f(x)\,dx$ for an antiderivative of f, and we called it an *indefinite integral*. This notation and terminology are consistent with the fundamental theorem of calculus. We can rewrite the fundamental theorem in terms of the indefinite integral in the following way.

Definite and Indefinite Integrals

$$\int_a^b f(x)\,dx = \int f(x)\,dx \Big|_a^b$$

Notice that although the indefinite integral is a function involving an arbitrary constant, the expression

$$\left(\int f(x)\,dx \right) \Big|_a^b$$

represents a well-defined number, since the constant cancels when we subtract the value at a from the value at b.

An expression of the form $\int_a^b f(x)\,dx$ with the endpoints specified, which we have been calling simply "an integral," is sometimes called a *definite* integral to distinguish it from an indefinite integral.

Note that a definite integral is a number, while an indefinite integral is a *function* (determined up to an additive constant).

Remember that one may check an indefinite integral formula by differentiating.

Indefinite Integral Test

To check a given formula $\int f(x)\,dx = F(x) + C$, differentiate the right-hand side and see if you get the integrand $f(x)$.

Example 1   Check the formula $\int 3x^8\,dx = x^9/3 + C$.

Solution   We differentiate the right-hand side using the power rule:

$$\frac{d}{dx}\left(\frac{x^9}{3} + C\right) = \frac{9x^8}{3} = 3x^8;$$

so the formula checks. ▲

The next example involves an integral that cannot be readily found with the antidifferentiation rules.

Example 2   (a) Check the formula $\int x(1 + x)^6\,dx = \frac{1}{56}(7x - 1)(1 + x)^7 + C$. (Do not attempt to derive the formula.)

(b) Find $\int_0^2 x(1 + x)^6\,dx$.

Solution   (a) We differentiate the right-hand side using the product rule and power of a function rule:

$$\frac{d}{dx}\left[\frac{1}{56}(7x - 1)(1 + x)^7\right] = \frac{1}{56}\left[7(1 + x)^7 + (7x - 1)7(1 + x)^6\right]$$

$$= \frac{1}{56}(1 + x)^6\left[7(1 + x) + 7(7x - 1)\right]$$

$$= (1 + x)^6 x.$$

Thus the formula checks.

(b) By the fundamental theorem and the formula we just checked, we have

$$\int_0^2 x(1 + x)^6\,dx = \frac{1}{56}(7x - 1)(1 + x)^7\Big|_0^2$$

$$= \frac{1}{56}\left[13 \cdot 3^7 - (-1)\right]$$

$$= \frac{28{,}432}{56} = \frac{3554}{7} \approx 507.7 \ ▲$$

In the box on page 204 we listed five key properties of the summation process. In the following box we list the corresponding properties of the definite integral.

Properties of the Definite Integral

1. $\int_a^b [f(x) + g(x)]\,dx = \int_a^b f(x)\,dx + \int_a^b g(x)\,dx$ (sum rule).

2. $\int_a^b cf(x)\,dx = c\int_a^b f(x)\,dx$, c a constant (constant multiple rule).

3. If $a < b < c$, then $\int_a^c f(x)\,dx = \int_a^b f(x)\,dx + \int_b^c f(x)\,dx$.

4. If $f(x) = C$ is constant, then $\int_a^b f(x)\,dx = C(b - a)$.

5. If $f(x) \leqslant g(x)$ for all x satisfying $a \leqslant x \leqslant b$, then

$$\int_a^b f(x)\,dx \leqslant \int_a^b g(x)\,dx.$$

These properties hold for all functions f and g that have integrals. However, while it is technically a bit less general, it is much easier to deduce the properties from the antidifferentiation rules and the fundamental theorem of calculus, assuming not only that f and g have integrals, but that they have antiderivatives as well.

Example 3   Prove property 1 in the display above (assuming that f and g have antiderivatives).

Solution   Let F be an antiderivative for f and G be one for g. Then $F + G$ is an antiderivative for $f + g$ by the sum rule for antiderivatives. Thus,

$$\int_a^b \left[f(x) + g(x) \right] dx = \left[F(x) + G(x) \right]\Big|_a^b$$
$$= \left[F(b) + G(b) \right] - \left[F(a) + G(a) \right]$$
$$= \left[F(b) - F(a) \right] + \left[G(b) - G(a) \right]$$
$$= \int_a^b f(x)\,dx + \int_a^b g(x)\,dx. \quad \blacktriangle$$

Example 4   Prove property 5.

Solution   If $f(x) \leqslant g(x)$ on (a, b), then $(F - G)'(x) = F'(x) - G'(x) = f(x) - g(x) \leqslant 0$ for x in (a, b). Since a function with a negative derivative is decreasing, we get

$$\left[F(b) - G(b) \right] - \left[F(a) - G(a) \right] \leqslant 0,$$

and so $F(b) - F(a) \leqslant G(b) - G(a)$. By the fundamental theorem of calculus, the last inequality can be written

$$\int_a^b f(x)\,dx \leqslant \int_a^b g(x)\,dx$$

as required. $\blacktriangle$

Properties 2 and 3 can be proved in a way similar to property 1. Note that property 4 is obvious, since we know how to compute areas of rectangles.

Example 5   Explain property 3 in terms of (a) areas (assume that f is a positive function) and (b) distances and velocities.

Solution   (a) Since $\int_a^c f(x)\,dx$ is the area under the graph of f from $x = a$ to $x = c$, property 3 merely states that the sum of the areas of regions A and B in Fig. 4.5.1 is the total area.

Figure 4.5.1. Illustrating the rule $\int_a^c f(x)\,dx = \int_a^b f(x)\,dx + \int_b^c f(x)\,dx$.

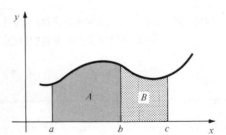

(b) Property 3 states that the displacement for a moving object between times a and c equals the sum of the displacements between a and b and between b and c. ▲

We have defined the integral $\int_a^b f(x)\,dx$ when a is less than b; however, the right-hand side of the equation

$$\int_a^b F'(x)\,dx = F(b) - F(a)$$

makes sense even when $a \geqslant b$. Can we define $\int_a^b f(x)\,dx$ for the case $a \geqslant b$ so that this equation will still be true? The answer is simple:

If $b < a$ and f is integrable on $[b, a]$, we define

$$\int_a^b f(x)\,dx \quad \text{to be} \quad -\int_b^a f(x)\,dx.$$

If $a = b$, we define $\int_a^b f(x)\,dx$ to be zero.

Notice that if F' is integrable on $[b, a]$, where $b < a$, then by the preceding definition and the fundamental theorem,

$$\int_a^b F'(x)\,dx = -\int_b^a F'(x)\,dx = -\left[F(a) - F(b) \right] = F(b) - F(a),$$

so the equation $\int_a^b F'(x)\,dx = F(b) - F(a)$ is still valid.

"Wrong-Way" Integrals

1. $\displaystyle\int_a^b f(x)\,dx = -\int_b^a f(x)\,dx;$

2. $\displaystyle\int_a^a f(x)\,dx = 0;$

3. $\displaystyle\int_a^b F'(x)\,dx = F(b) - F(a),$ for all a and b.

Example 6   Find $\displaystyle\int_6^2 x^3\,dx.$

Solution $\int_6^2 x^3\,dx = (x^4/4)\big|_6^2 = \frac{1}{4}(16 - 1296) = -320.$ (Although the function $f(x) = x^3$ is positive, the integral is negative. To explain this, we remark that as x goes from 6 to 2, "dx is negative.") ▲

We have seen that the fundamental theorem of calculus enables us to compute integrals by using antiderivatives. The relationship between integration and differentiation is completed by an alternative version of the fundamental theorem. Let us first state and prove it; its geometric meaning will be given shortly.

Fundamental Theorem of Calculus: Alternative Version

If f is continuous on $[a,b]$, then $\dfrac{d}{dx}\displaystyle\int_a^x f(s)\,ds = f(x)$.

We now justify the alternative version of the fundamental theorem. In Exercises 49–53, it is shown that f has an antiderivative F. Let us accept this fact here.

The fundamental theorem applied to f on the interval $[a, x]$ gives

$$\int_a^x f(s)\,ds = F(x) - F(a).$$

Differentiating both sides,

$$\frac{d}{dx}\int_a^x f(s)\,ds = \frac{d}{dx}\big[F(x) - F(a) \big]$$

$$= \frac{d}{dx} F(x) \qquad \text{(since } F(a) \text{ is constant)}$$

$$= f(x) \qquad \text{(since } F \text{ is an antiderivative of } f\text{).}$$

Thus the alternative version is proved.

Notice that in the statement of the theorem we have changed the (dummy) variable of integration to the letter "s" to avoid confusion with the endpoint "x."

Example 7   Verify the formula $\dfrac{d}{dx}\displaystyle\int_a^x f(s)\,ds = f(x)$ for $f(x) = x$.

Solution   The integral in question is

$$\int_a^x f(s)\,ds = \int_a^x s\,ds = \frac{s^2}{2}\bigg|_a^x = \frac{x^2}{2} - \frac{a^2}{2}\,.$$

Thus, $\dfrac{d}{dx}\left(\displaystyle\int_a^x f(s)\,ds\right) = \dfrac{d}{dx}\left(\dfrac{x^2}{2} - \dfrac{a^2}{2}\right) = x = f(x),$

so the formula holds. ▲

Example 8   Let $F(x) = \displaystyle\int_2^x \frac{1}{1 + s^2 + s^3}\,ds$. Find $F'(3)$.

Solution   Using the alternative version of the fundamental theorem, with $f(s) = 1/(1 + s^2 + s^3)$, we have $F'(3) = f(3) = 1/(1 + 3^2 + 3^3) = \frac{1}{37}$. Notice that we did not need to differentiate or integrate $1/(1 + s^2 + s^3)$ to get the answer. ▲

At the top of the next page, we summarize the two forms of the fundamental theorem.

Fundamental Theorem of Calculus

Usual Version: $\displaystyle\int_a^b F'(x)\,dx = F(b) - F(a).$

Integrating the derivative of F gives the change in F.

Alternative Version: $\displaystyle\frac{d}{dx}\int_a^x f(s)\,ds = f(x).$

Differentiating the integral of f with respect to the upper limit gives f.

The alternative form of the fundamental theorem of calculus has an illuminating interpretation and explanation in terms of areas. Suppose that $f(x)$ is non-negative on $[a, b]$. Imagine uncovering the graph of f by moving a screen to the right, as in Fig. 4.5.2. When the screen is at x, the exposed area is

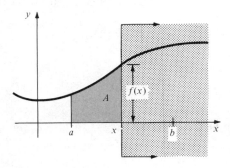

Figure 4.5.2. $dA/dx = f(x).$

$A = \int_a^x f(s)\,ds$. The alternative version of the fundamental theorem can be phrased as follows: *as the screen moves to the right, the rate of change of exposed area A with respect to x, dA/dx, equals $f(x)$.* This same conclusion can be seen graphically by investigating the difference quotient,

$$\frac{A(x + \Delta x) - A(x)}{\Delta x}.$$

The quantity $A(x + \Delta x) - A(x)$ is the area under the graph of f *between* x and $x + \Delta x$. For Δx small, this area is approximately the area of the rectangle

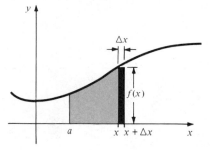

Figure 4.5.3. The geometry needed to explain why $dA/dx = f(x).$

with base Δx and height $f(x)$, as in Fig. 4.5.3. Therefore,

$$\frac{A(x + \Delta x) - A(x)}{\Delta x} \approx \frac{f(x)\,\Delta x}{\Delta x} = f(x),$$

and the approximation gets better as Δx becomes smaller. Thus

$$\frac{A(x + \Delta x) - A(x)}{\Delta x}$$

approaches $f(x)$ as $\Delta x \to 0$, which means that $dA/dx = f(x)$. If f is continuous, this argument is the basis for a rigorous proof of the alternative version of the fundamental theorem. See Exercises 49–53 for additional details.

Exercises for Section 4.5

In Exercises 1–4, check the integration formula by differentiating the right-hand side.

1. $\int 5x^4\, dx = x^5 + C$.

2. $\int \dfrac{1+t}{t^3}\, dt = -\dfrac{1}{2t^2} - \dfrac{1}{t} + C$.

3. $\int 5(t^9 + t^4)\, dt = \dfrac{t^{10}}{2} + t^5 + C$.

4. $\int \dfrac{x - x^3 + 1}{x^3}\, dx = -\dfrac{1}{x} - x - \dfrac{1}{2x^2} + C$.

5. (a) Check the following integral

$$\int \frac{3t^2}{(1 + t^3)^2}\, dt = \frac{t^3}{1 + t^3} + C.$$

   (b) Evaluate $\displaystyle\int_0^1 \dfrac{3t^2}{(1 + t^3)^2}\, dt$.

6. (a) Check the following integration formula:

$$\int \frac{x^3 + 2x + 1}{(1 - x)^5}\, dx = \frac{1}{x - 1} + \frac{3}{2(x - 1)^2}$$

$$+ \frac{5}{3(x - 1)^3} + \frac{1}{(x - 1)^4} + C$$

   (b) Evaluate $\displaystyle\int_2^3 [(s^3 + 2s + 1)/(1 - s)^5]\, ds$.

7. (a) Calculate the derivative of $\dfrac{x^3}{x^2 + 1}$.

   (b) Find $\displaystyle\int_0^1 \dfrac{(3x^2 + x^4)}{(1 + x^2)^2}\, dx$.

8. (a) Differentiate $\dfrac{x}{1 + x}$ and $-\dfrac{1}{1 + x}$.

   (b) Find $\displaystyle\int_3^2 \dfrac{1}{(1 + x)^2}\, dx$ in two ways.

Calculate the definite integrals in Exercises 9–18.

9. $\displaystyle\int_{-2}^3 (x^4 + 5x^2 + 2x + 1)\, dx$.

10. $\displaystyle\int_{-1}^1 (x^3 + 7)\, dx$.

11. $\displaystyle\int_{-2}^4 x^6\, dx$.

12. $\displaystyle\int_{-3}^{472} 0\, dt$.

13. $\displaystyle\int_1^2 \dfrac{x^2 + 2x + 2}{x^4}\, dx$.

14. $\displaystyle\int_1^8 \dfrac{1 + \theta^2}{\theta^4}\, d\theta$.

15. $\displaystyle\int_2^3 \dfrac{dt}{t^2}$.

16. $\displaystyle\int_2^{-2} t^4\, dt$.

17. $\displaystyle\int_1^2 (1 + 2t)^5\, dt$.

18. $\displaystyle\int_2^1 (1 - x)^6\, dx$.

19. Explain property 2 of integration in terms of (a) areas and (b) distances and velocities.

20. Explain property 5 of integration in terms of (a) areas and (b) distances and velocities.

If $\displaystyle\int_0^1 f(x)\, dx = 3$, $\displaystyle\int_1^2 f(x)\, dx = 4$, and $\displaystyle\int_2^3 f(x)\, dx = -8$, calculate the quantities in Exercises 21–24, using the properties of integration.

21. $\displaystyle\int_0^2 f(x)\, dx$.

22. $\displaystyle\int_0^1 3f(x)\, dx$.

23. $\displaystyle\int_0^3 8f(x)\, dx$.

24. $\displaystyle\int_1^3 10f(x)\, dx$.

Calculate the integrals in Exercises 25–28.

25. $\displaystyle\int_3^2 x\, dx$.

26. $\displaystyle\int_8^4 (x^2 - 1)\, dx$.

27. $\displaystyle\int_{10}^9 \dfrac{x + 1}{x^3}\, dx$.

28. $\displaystyle\int_{-3}^{-2} \dfrac{x^3 - 1}{x - 1}\, dx$.

Verify the formula $\dfrac{d}{dx} \displaystyle\int_a^x f(s)\, ds = f(x)$ for the functions in Exercises 29 and 30.

29. $f(x) = x^3 - 1$.

30. $f(x) = x^4 - x^2 + x$.

31. Let $F(t) = \displaystyle\int_3^t \dfrac{1}{\left[(4 - s)^2 + 8\right]^3}\, ds$. Find $F'(4)$.

32. Find $\dfrac{d}{dx} \displaystyle\int_0^x \dfrac{t^4}{1 + t^6}\, dt$.

Evaluate the derivatives in Exercises 33–36.

33. $\dfrac{d}{dt} \displaystyle\int_0^t \dfrac{3}{(x^4 + x^3 + 1)^6}\, dx$.

34. $\dfrac{d}{dt} \displaystyle\int_3^t \dfrac{1}{x^4 + x^6}\, dx$.

35. $\dfrac{d}{dt} \displaystyle\int_t^3 x^2(1 + x)^5\, dx$.

36. $\dfrac{d}{dt} \displaystyle\int_t^4 \dfrac{u^4}{(u^2 + 1)^3}\, du$.

37. Let $v(t)$ be the velocity of a moving object. In this context, interpret the formula

$$\frac{d}{dt} \int_a^t v(s)\, ds = v(t).$$

38. Interpret the alternative version of the fundamental theorem of calculus in the context of the solar energy example in the Supplement to Section 4.3.

39. Suppose that

$$f(t) = \begin{cases} t^2, & 0 \leq t \leq 1, \\ 1, & 1 \leq t < 5, \\ (t-6)^2, & 5 \leq t \leq 6. \end{cases}$$

(a) Draw a graph of f on the interval $[0, 6]$.

(b) Find $\int_0^6 f(t)\, dt$.

(c) Find $\int_0^6 f(x)\, dx$.

(d) Let $F(t) = \int_0^t f(s)\, ds$. Find the formula for $F(t)$ in $[0, 6]$ and draw a graph of F.

(e) Find $F'(t)$ for t in $(0, 6)$.

40. (a) Give a formula for a function f whose graph is the broken line segment $ABCD$ in Fig. 4.5.4.

(b) Find $\int_3^{10} f(t)\, dt$.

(c) Find the area of quadrilateral $ABCD$ by means of geometry and compare the result with the integral in part (b).

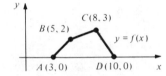

Figure 4.5.4. Find a formula for f.

41. Let f be continuous on the interval I and let a_1 and a_2 be in I. Define the functions:

$$F_1(t) = \int_{a_1}^t f(s)\, ds \quad \text{and} \quad F_2(t) = \int_{a_2}^t f(s)\, ds.$$

(a) Show that F_1 and F_2 differ by a constant.

(b) Express the constant $F_2 - F_1$ as an integral.

42. Develop a formula for $\int x(1 + x)^n\, dx$ for $n \neq -1$ or -2 by studying Example 2 [*Hint:* Guess the answer $(ax + b)(1 + x)^{n+1}$ and determine what a and b have to be.]

★43. (a) Combine the alternative version of the fundamental theorem of calculus with the chain rule to prove that

$$\frac{d}{dt} \int_a^{g(t)} f(s)\, ds = f(g(t)) \cdot g'(t).$$

(b) Interpret (a) in terms of Fig. 4.5.5.

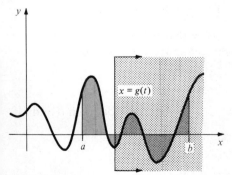

Figure 4.5.5. The rate of change of the exposed area with respect to t is $f(g(t)) \cdot g'(t)$.

★44. Compute $\dfrac{d}{dt} \displaystyle\int_0^{t^2} \dfrac{dx}{1 + x^2}$.

★45. Let $F(x) = \displaystyle\int_1^{x^2} \dfrac{dt}{t}$. What is $F'(x)$?

★46. Calculate $\dfrac{d}{ds} \displaystyle\int_0^{s^4 + s^2} \dfrac{dx}{1 + x^4}$.

★47. Find a formula for $\dfrac{d}{dt} \displaystyle\int_{h(t)}^{g(t)} f(x)\, dx$ and explain your formula in terms of areas.

★48. Compute $\dfrac{d}{ds} \displaystyle\int_{s^2}^{s^3} \dfrac{x^3}{1 + x^4}\, dx$.

Exercises 49–53 outline a proof of this fact: if f is continuous on $[a, b]$, then $F(t) = \displaystyle\int_a^t f(s)\, ds$ is an antiderivative of f.

★49. Prove property 3 for an integrable function f; that is, if f is integrable on $[a, b]$ and on $[b, c]$, then f is integrable on $[a, c]$ and

$$\int_a^c f(x)\, dx = \int_a^b f(x)\, dx + \int_b^c f(x)\, dx.$$

[*Hint:* Let I be the right-hand side. Show that every number less than I is a lower sum for f on $[a, c]$ and, likewise, every number greater than I is an upper sum. If $S < I$, show by a general fact about inequalities that you can write $S = S_1 + S_2$, where $S_1 < \int_a^b f(x)\, dx$ and $S_2 < \int_b^c f(x)\, dx$. Now piece together a lower sum corresponding to S_1 with one for S_2.]

★50. Prove property 5 for integrable functions f and g. [*Hint:* Every lower sum for f is also one for g.]

★51. Show that

$$\frac{F(t + h) - F(t)}{h} = \frac{1}{h} \int_t^{t+h} f(s)\, ds$$

using property 3 of the integral.

★52. Show that $(1/h) \int_t^{t+h} f(s)\, ds$ lies between the maximum and minimum values of f on the interval $[t, t + h]$ (you may assume $h > 0$; a similar argument is needed for $h < 0$). Conclude that

$$\frac{1}{h} \int_t^{t+h} f(s)\, ds = f(c)$$

for some c between t and $t + h$, by the intermediate value theorem. (This result is sometimes called the *mean value theorem for integrals*; we will treat it again in Section 9.3.)

★53. Use continuity of f and the results from Exercises 51 and 52 to show that $F' = f$.

★54. Exercises 49–53 outlined a complete proof of the alternative version of the fundamental theorem of calculus. Use the "alternative version" to prove the "main version" in Section 4.4.

4.6 Applications of the Integral

Areas between graphs can be calculated as integrals.

We have seen that area under the graph of a function can be expressed as an integral. After a word problem based on this fact, we will learn how to calculate areas *between* graphs in the plane. Other applications of integration concern recovering the total change in a quantity from its rate of change.

Example 1   A parabolic doorway with base 6 feet and height 8 feet is cut out of a wall. How many square feet of wall space are removed?

Solution   Place the coordinate system as shown in Fig. 4.6.1 and let the parabola have equation $y = ax^2 + c$. Since $y = 8$ when $x = 0$, $c = 8$. Also, $y = 0$ when $x = 3$, so $0 = a3^2 + 8$, so $a = -\frac{8}{9}$. Thus the parabola is $y = -\frac{8}{9}x^2 + 8$. The area under its graph is

$$\int_{-3}^{3}\left(-\frac{8}{9}x^2 + 8\right)dx = \left(-\frac{8}{27}x^3 + 8x\right)\Big|_{-3}^{3}$$

$$= \left(-\frac{8}{27}\cdot 27 + 8\cdot 3\right) - \left(\frac{8}{27}\cdot 27 - 8\cdot 3\right) = 32,$$

so 32 square feet have been cut out. ▲

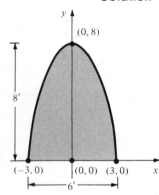

Figure 4.6.1. Find the area of the parabolic doorway.

Now we turn to the problem of finding the area between the graphs of two functions. If f and g are two functions defined on $[a, b]$, with $f(x) \leq g(x)$ for all x in $[a, b]$, we define *the region between the graphs* of f and g on $[a, b]$ to be the set of those points (x, y) such that $a \leq x \leq b$ and $f(x) \leq y \leq g(x)$.

Example 2   Sketch and find the area of the region between the graphs of x^2 and $x + 3$ on $[-1, 1]$.

Solution   The region is shaded in Fig. 4.6.2. It is not quite of the form we have been dealing with; however, we may note that if we add to it the region under the graph of x^2 on $[-1, 1]$, we obtain the region under the graph of $x + 3$ on $[-1, 1]$. Denoting the area of the shaded region by A, we have

$$\int_{-1}^{1} x^2\, dx + A = \int_{-1}^{1}(x + 3)\, dx$$

or

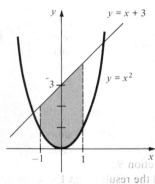

Figure 4.6.2. The area of the shaded region is the difference between two areas under graphs.

$$A = \int_{-1}^{1}(x + 3)\, dx - \int_{-1}^{1} x^2\, dx = \int_{-1}^{1}(x + 3 - x^2)\, dx.$$

Evaluating the integral yields $A = (\frac{1}{2}x^2 + 3x - \frac{1}{3}x^3)\big|_{-1}^{1} = 5\frac{1}{3}$. ▲

The method of Example 2 can be used to show that if $0 \leq f(x) \leq g(x)$ for x in $[a, b]$, then the area of the region between the graphs of f and g on $[a, b]$ is equal to $\int_{a}^{b} g(x)\, dx - \int_{a}^{b} f(x)\, dx = \int_{a}^{b}[g(x) - f(x)]\, dx$.

Example 3   Find the area between the graphs of $y = x^2$ and $y = x^3$ for x between 0 and 1.

Solution   Since $0 \leq x^3 \leq x^2$ on $[0, 1]$, by the principle just stated the area is

$$\int_{0}^{1}(x^2 - x^3)\, dx = \left(\frac{x^3}{3} - \frac{x^4}{4}\right)\Big|_{0}^{1} = \frac{1}{3} - \frac{1}{4} = \frac{1}{12}. \quad ▲$$

The same method works even if $f(x)$ can take negative values. It is only the *difference* between $f(x)$ and $g(x)$ which matters.

Area Between Graphs

If $f(x) \leqslant g(x)$ for all x in $[a,b]$, and f and g are integrable on $[a,b]$, then the area between the graphs of f and g on $[a,b]$ equals

$$\int_a^b \left[g(x) - f(x) \right] dx.$$

There is a heuristic argument for this formula for the area between graphs which gives a useful way of remembering the formula and deriving similar ones. We can think of the region between the graphs as being composed of infinitely many "infinitesimally wide" rectangles, of width dx, one for each x in $[a,b]$. (See Fig. 4.6.3.)

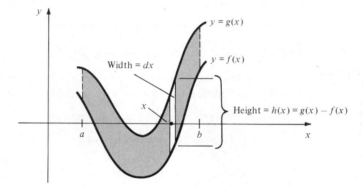

Figure 4.6.3. We may think of the shaded region as being composed of infinitely many rectangles, each of infinitesimal width.

The total area is then the "continuous sum" of the areas of these rectangles. The height of the rectangle over x is $h(x) = g(x) - f(x)$, the area of the rectangle is $[g(x) - f(x)]dx$, and the continuous sum of these areas is the integral $\int_a^b [g(x) - f(x)]dx$. This kind of infinitesimal argument was used frequently in the early days of calculus, when it was considered to be perfectly acceptable. Nowadays, we usually take the viewpoint of Archimedes, who used infinitesimals to discover results which he later proved by more rigorous, but much more tedious, arguments.

Example 4   Sketch and find the area of the region between the graphs of x and $x^2 + 1$ on $[-2, 2]$.

Solution   The region is shaded in Fig. 4.6.4. By the formula for the area between curves, the area is

$$\int_{-2}^2 \left[(x^2 + 1) - (x) \right] dx = \int_{-2}^2 (x^2 + 1 - x)\, dx = \left(\frac{x^3}{3} + x - \frac{x^2}{2} \right)\Big|_{-2}^2$$

$$= \left(\frac{8}{3} + 2 - \frac{4}{2} \right) - \left(-\frac{8}{3} - 2 - \frac{4}{2} \right) = \frac{28}{3}. \; \blacktriangle$$

If the graphs of f and g intersect, then the area of the region between them must be found by breaking the region up into smaller pieces and applying the preceding method to each piece.

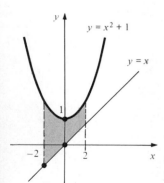

Figure 4.6.4. What is the area of the shaded region?

Area Between Intersecting Graphs

To find the area of the region between the graphs of f and g and between $x = a$ and $x = b$, first plot the graphs and locate points where $f(x) = g(x)$. Suppose, for example, that $f(x) \geqslant g(x)$ for $a \leqslant x \leqslant c$, $f(c) = g(c)$ and $f(x) \leqslant g(x)$ for $c \leqslant x \leqslant b$, as in Fig. 4.6.5. Then the area is

$$A = \int_a^c [f(x) - g(x)] \, dx + \int_c^b [g(x) - f(x)] \, dx.$$

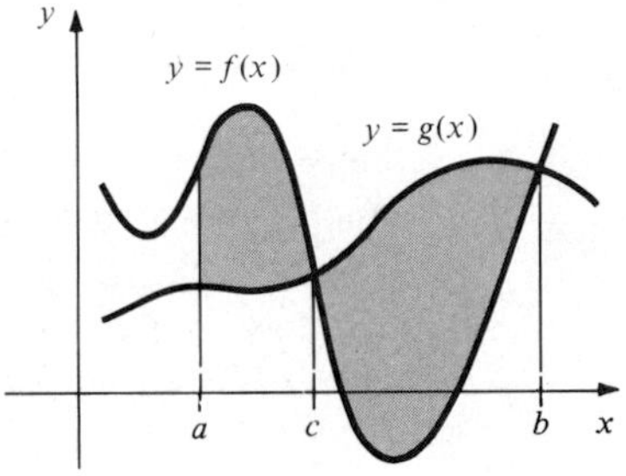

Figure 4.6.5. The area of the shaded region equals $\int_a^c [f(x) - g(x)] \, dx + \int_c^b [g(x) - f(x)] \, dx$.

Example 5 Find the area of the shaded region in Fig. 4.6.6.

Solution First we locate the intersection points by setting $x^2 = x$. This has the solutions $x = 0$ or 1. Between 0 and 1, $x^2 \leqslant x$, and between 1 and 2, $x^2 \geqslant x$, so the area is

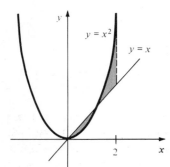

$$A = \int_0^1 (x - x^2) \, dx + \int_1^2 (x^2 - x) \, dx$$

$$= \left(\frac{x^2}{2} - \frac{x^3}{3} \right)\Big|_0^1 + \left(\frac{x^3}{3} - \frac{x^2}{2} \right)\Big|_1^2$$

$$= \left(\frac{1}{2} - \frac{1}{3} \right) + \left[\left(\frac{8}{3} - \frac{4}{2} \right) - \left(\frac{1}{3} - \frac{1}{2} \right) \right] = 1. \ \blacktriangle$$

Figure 4.6.6. Find the shaded area.

Example 6 Find the area between the graphs of $y = x^3$ and $y = 3x^2 - 2x$ between $x = 0$ and $x = 2$.

Solution The graphs are plotted in Fig. 4.6.7. They intersect when $x^3 = 3x^2 - 2x$, i.e., $x(x^2 - 3x + 2) = 0$, i.e., $x(x - 2)(x - 1) = 0$, which has solutions $x = 0$, 1, and 2, as in the figure. The area is thus

Figure 4.6.7. The graphs needed to find the area between $y = x^3$ and $y = 3x^2 - 2x$.

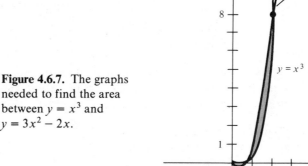

$$\int_0^1 \left[x^3 - (3x^2 - 2x) \right] dx + \int_1^2 \left[(3x^2 - 2x) - x^3 \right] dx$$

$$= \left[\frac{x^4}{4} - (x^3 - x^2) \right] \Big|_0^1 + \left[(x^3 - x^2) - \frac{x^4}{4} \right] \Big|_1^2$$

$$= \frac{1}{4} \left[(8 - 4) - \frac{16}{4} + \frac{1}{4} \right] = \frac{1}{2} . \ \blacktriangle$$

In the next problem, the intersection points of two graphs determine the limits of integration.

Example 7   The curves $x = y^2$ and $x = 1 + \frac{1}{2} y^2$ (neither of which is the graph of a function $y = f(x)$) divide the xy plane into five regions, only one of which is bounded. Sketch and find the area of this bounded region.

Solution   If we plot x as a function of y, we obtain the graphs and region shown in Fig. 4.6.8. We use our general rule for the area between graphs, which gives

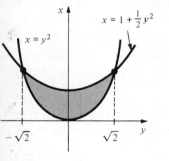

$$A = \int_{-\sqrt{2}}^{\sqrt{2}} \left(1 + \frac{1}{2} y^2 - y^2 \right) dy$$

$$= \int_{-\sqrt{2}}^{\sqrt{2}} \left(1 - \frac{1}{2} y^2 \right) dy = \left(y - \frac{1}{6} y^3 \right) \Big|_{-\sqrt{2}}^{\sqrt{2}}$$

$$= \sqrt{2} - \frac{1}{6} \left(\sqrt{2} \right)^3 - \left[-\sqrt{2} - \frac{1}{6} \left(-\sqrt{2} \right)^3 \right]$$

$$= 2\sqrt{2} - \frac{1}{3} \left(\sqrt{2} \right)^3 = \sqrt{2} \left(2 - \frac{2}{3} \right) = \frac{4}{3} \sqrt{2} = \frac{2^{5/2}}{3} .$$

Figure 4.6.8. The bounded region determined by $x = y^2$ and $x = 1 + \frac{1}{2} y^2$.

(Note that the roles of x and y have been reversed in this example.) $\blacktriangle$

Example 8   Find the area between the graphs $x = y^2 - 2$ and $y = x$.

Solution   This is a good illustration of the fact that sometimes it is wise to pause and think about various methods at our disposal rather than simply plunging ahead.

First of all, we sketch the graphs, as in Fig. 4.6.9. We can plot $x = y^2 - 2$ either by writing $y = \pm \sqrt{x + 2}$ and graphing these two square-root functions or, preferably, by regarding x as a function of y and drawing the corresponding parabola.

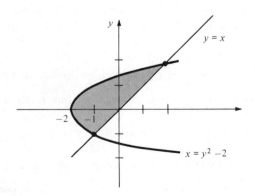

Figure 4.6.9. Find the shaded area.

The intersection points are obtained by setting $y^2 - 2 = y$, which gives $y = -1$ and 2.

Method 1 Write $y = \pm\sqrt{x+2}$ for the first graph and obtain the area in two pieces as

$$A = \int_{-2}^{-1}\left[\sqrt{x+2} - \left(-\sqrt{x+2}\right)\right]dx + \int_{-1}^{2}\left[\sqrt{x+2} - x\right]dx$$

$$= \frac{4}{3}(x+2)^{3/2}\Big|_{-2}^{-1} + \left(\frac{2}{3}(x+2)^{3/2} - \frac{x^2}{2}\right)\Big|_{-1}^{2}$$

$$= \frac{4}{3} - 0 + \left(\frac{2}{3}\cdot 8 - 2\right) - \left(\frac{2}{3} - \frac{1}{2}\right) = \frac{9}{2}\,.$$

Method 2 Regard x as a function of y and obtain the area as

$$\int_{-1}^{2}\left[y - (y^2 - 2)\right]dy = \left(\frac{y^2}{2} - \frac{y^3}{3} + 2y\right)\Big|_{-1}^{2}$$

$$= \left(2 - \frac{8}{3} + 4\right) - \left(\frac{1}{2} + \frac{1}{3} - 2\right) = \frac{9}{2}\,.$$

Method 2, while requiring the slight trick of regarding x as a function of y, is simpler. ▲

The velocity–displacement relationship holds for all rates of change. If a quantity Q depends on x and has a rate of change r, then by the fundamental theorem,

$$Q(b) - Q(a) = \int_{a}^{b} r\,dx.$$

Total Changes from Rates of Change

If the rate of change of Q with respect to x for $a \leqslant x \leqslant b$ is given by $r = f(x)$, then the change in Q is obtained by integrating:

$$\Delta Q = Q(b) - Q(a) = \int_{a}^{b} f(x)\,dx.$$

This relationship can be used in a variety of ways, depending on the interpretation of Q, r, and x. For example, we can view the box on page 230 as a special instance with Q the position, r the velocity, and x the time.

Example 9 Water is flowing into a tub at $3t^2 + 6t$ liters per minute at time t, between $t = 0$ and $t = 2$. How many liters enter the tub during this period?

Solution Let $Q(t)$ denote the number of liters at time t. Then $Q'(t) = 3t^2 + 6t$, so

$$Q(2) - Q(0) = \int_{0}^{2}(3t^2 + 6t)\,dt = (t^3 + 3t^2)\Big|_{0}^{2} = 20.$$

Thus, 20 liters enter the tub during the 3-minute interval. ▲

The final example comes from economics. We recall from the discussion on page 106 that the marginal revenue at production level x is $R'(x)$, where R is the revenue.

Example 10   The marginal revenue for a company at production level x is given by $15 - 0.1x$. If $R(x)$ denotes the revenue and $R(0) = 0$, find $R(100)$.

Solution   By the fundamental theorem, $R(100) - R(0) = \int_0^{100} R'(x)\,dx$. But $R(0) = 0$, and the marginal revenue $R'(x)$ is $15 - 0.1x$. Thus,

$$R(100) = \int_0^{100} (15 - 0.1x)\,dx$$

$$= \left(15x - \frac{0.1x^2}{2}\right)\Bigg|_0^{100}$$

$$= 1500 - 500 = 1000. \ \blacktriangle$$

Exercises for Section 4.6

1. A parabolic arch with base 8 meters and height 10 meters is erected. How much area does it enclose?

2. A parabolic arch with base 10 meters and height 12 meters is erected. How much area does it enclose?

3. A swimming pool has the shape of the region bounded by $y = x^2$ and $y = 2$. A swimming pool cover is estimated to cost $2.00 per square foot. If one unit along each of the x and y axes is 50 feet, then how much should the cover cost?

4. An artificial lake with two bays has the shape of the region above the curve $y = x^4 - x^2$ and below the line $y = 8$ (x and y are measured in kilometers). If the lake is 10 meters deep, how many cubic meters of water does it hold?

5. Find the area of the shaded region in Fig. 4.6.10.

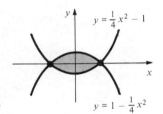

Figure 4.6.10. What is the area of the shaded region?

6. Find the area of the shaded region in Fig. 4.6.11.
Find the area between the graphs on the designated intervals in Exercises 7–10.

7. $y = (2/x^2) + x^4$ and $y = 1$ between $x = 1$ and $x = 2$.

8. $y = x^4$ and $y = x^3$ between $x = -1$ and $x = 0$.

9. $y = \sqrt{x}$ and $y = x$ between $x = 0$ and $x = 1$.

10. $y = \sqrt[3]{x}$ and $y = 1/x^2$ between $x = 8$ and $x = 27$.

In Exercises 11–16, find the area between the graphs of each pair of functions on the given interval:

11. x and x^4 on $[0, 1]$.

12. x^2 and $4x^4$ on $[2, 3]$.

13. $3x^2$ and $x^4 + 2$ on $[-\frac{1}{2}, \frac{1}{2}]$.

14. $x^4 + 1$ and $1/x^2$ on $[1, 2]$.

15. $3 + \dfrac{x^4 + x^2}{x^{59} + x^4}$ and $7 + \dfrac{x^4 + x^2}{x^{59} + x^4}$ on $[46917, 46919]$.

16. $\dfrac{4(x^6 - 1)}{x^6 + 1}$ and $\dfrac{(3x^6 - 1)(x^6 - 1)}{x^6(x^6 + 1)}$ on $[1, 2]$.

17. Find the area between the graphs of $y = x^3$ and $y = 5x^2 + 6x$ between $x = 0$ and $x = 3$.

18. Find the area between the graphs of $y = x^3 + 1$ and $y = x^2 - 1$ between $x = -1$ and $x = 1$.

19. The curves $y = x^3$ and $y = x$ divide the plane into six regions, only two of which are bounded. Find the areas of the bounded regions.

20. The lines $y = x$ and $y = 2x$ and the curve $y = 2/x^2$ together divide the plane into several regions, one of which is bounded. (a) How many regions are there? (b) Find the area of the bounded region.

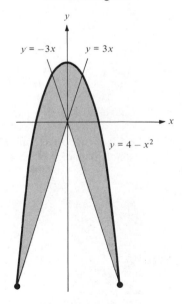

Figure 4.6.11. What is the area of the shaded region?

In Exercises 21–24, find the area between the given graphs.

21. $x = y^2 - 3$ and $x = 2y$
22. $x = y^2 + 8$ and $x = -6y$
23. $x = y^3$ and $y = 2x$
24. $x = y^4 - 2$ and $x = y^2$
25. Water is flowing out of a reservoir at $300t^2$ liters per second for t between 0 and 5. How many liters are released in this period?
26. Air is escaping from a balloon at $3t^2 + 2t$ cubic centimeters per second for t between 1 and 3. How much air escapes during this period?
27. Suppose that the marginal revenue of a company at production level x is given by $30 - 0.02x - 0.0001x^2$. If $R(0) = 0$, find $R(300)$.
28. (a) Find the total revenue $R(x)$ from selling x units of a product, if the marginal revenue is $36 - 0.01x + 0.00015x^2$ and $R(0) = 0$.
    (b) How much revenue is produced when the items from $x = 50$ to $x = 100$ are sold?
29. (a) Use calculus to find the area of the triangle whose vertices are $(0,0)$, (a,h), and $(b,0)$. (Assume $0 < a < b$ and $0 < h$.) Compare your result with a formula from geometry.
    (b) Repeat part (a) for the case $0 < b < a$.
30. In Example 2 of Section 4.5, it was shown that $\int x(1 + x)^6\, dx = \frac{1}{56}(7x - 1)(1 + x)^7 + C$. Use this result to find the area under the graph of $1 + x(1 + x)^6$ between $x = -1$ and $x = 1$.
31. Fill in the blank, referring to the Supplement to Section 4.3: Light meter is to energy-storage dial as _____ is to odometer.
32. Fill in the blank: Marginal revenue is to _____ as $f(x)$ is to $\int_a^x f(s)\, ds$.
33. A circus tent is equipped with four exhaust fans at one end, each capable of moving 5500 cubic feet of air per minute. The rectangular base of the tent is 80 feet by 180 feet. Each corner is supported by a 20-foot-high post and the roof is supported by a center beam which is 32 feet off the ground and runs down the center of the tent for its 180-foot length. Canvas drapes in a parabolic shape from the center beam to the sides 20 feet off the ground. (See Fig. 4.6.12.) Determine the elapsed time for a complete change of air in the tent enclosure.

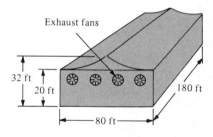

Figure 4.6.12. The circus tent in Exercise 33.

34. A small gold mine in northern Nevada was reopened in January 1979, producing 500,000 tons of ore in the first year. Let $A(t)$ be the number of tons of ore produced, in thousands, t years after 1979. Productivity $A'(t)$ is expected to decline by 20,000 tons per year until 1990.
    (a) Find a formula for $A'(t)$, assuming that the production decline is constant.
    (b) How much ore is mined, approximately, T years after 1979 during a time period Δt?
    (c) Find, by definite integration, the predicted number of tons of ore to be mined from 1981 through 1986.
35. Let $W(t)$ be the number of words learned after t minutes are spent memorizing a French vocabulary list. Typically, $W(0) = 0$ and $W'(t) = 4(t/100) - 3(t/100)^2$.
    (a) Apply the fundamental theorem of calculus to show that

$$W(t) = \int_0^t \left[4(x/100) - 3(x/100)^2\right] dx.$$

    (b) Evaluate the integral in part (a).
    (c) How many words are learned after 1 hour and 40 minutes of study?
36. The region under the graph $y = 1/x^2$ on $[1,4]$ is to be divided into two parts of equal area by a vertical line. Where should the line be drawn?
★37. Where would you draw a horizontal line to divide the region in the preceding exercise into two parts of equal area?
★38. Find the area in square centimeters, correct to within 1 square centimeter, of the region in Fig. 4.6.13.

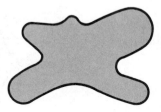

Figure 4.6.13. Find the area of the "blob."

Review Exercises for Chapter 4

Compute the sums in Exercises 1–8.

1. $\displaystyle\sum_{i=1}^{4} i^2$.

2. $\displaystyle\sum_{j=1}^{3} j^3$.

3. $\displaystyle\sum_{i=1}^{5} \frac{2^i}{i(i+1)}$.

4. $\displaystyle\sum_{j=4}^{8} \frac{j^2 - 10}{3j}$.

5. $\displaystyle\sum_{i=1}^{500} (3i + 7)$.

6. $\displaystyle\sum_{i=n}^{n+3} \frac{i^2 - 1}{i + 1}$ (n is a non-negative integer).

7. $\displaystyle\sum_{i=0}^{10} [(i + 1)^4 - i^4]$.

8. $\displaystyle\sum_{i=2}^{60} \left(\frac{1}{i} - \frac{1}{i - 1} \right)$.

9. Let f be defined on $[0, 1]$ by

$$f(x) = \begin{cases} 1, & 0 \leqslant x < \frac{1}{5}, \\ 2, & \frac{1}{5} \leqslant x < \frac{1}{4}, \\ 3, & \frac{1}{4} \leqslant x < \frac{1}{3}, \\ 4, & \frac{1}{3} \leqslant x < \frac{1}{2}, \\ 5, & \frac{1}{2} \leqslant x \leqslant 1. \end{cases}$$

Find $\displaystyle\int_0^1 f(x)\,dx$.

10. Let f be defined by

$$f(x) = \begin{cases} -1, & -1 \leqslant x < 0, \\ 2, & 0 \leqslant x < 1, \\ 3, & 1 \leqslant x \leqslant 2. \end{cases}$$

Find $\displaystyle\int_{-1}^{2} f(x)\,dx$.

11. Interpret the integral in Exercise 9 in terms of distances and velocities.

12. Interpret the integral in Exercise 10 in terms of distances and velocities.

Evaluate the definite integrals in Exercises 13–16.

13. $\displaystyle\int_3^5 (-2x^3 + x^2)\,dx$.

14. $\displaystyle\int_1^3 \frac{x^3 - 5}{x^2}\,dx$.

15. $\displaystyle\int_1^2 \frac{\frac{1}{3}s^2 - (s^4 + 1)}{2s^2}\,ds$.

16. $\displaystyle\int_1^2 \frac{x^2 + 3x + 2}{x + 1}\,dx$.

Find the area under the graphs of the functions between the indicated limits in Exercises 17–20.

17. $y = x^3 + x^2, \; 0 \leqslant x \leqslant 1$.

18. $y = \dfrac{x^2 + 2x + 1}{x^4}, \; 1 \leqslant x \leqslant 2$.

19. $y = (x + 3)^{4/3}, \; 0 \leqslant x \leqslant 2$.

20. $y = (x - 1)^{1/2}, \; 1 \leqslant x \leqslant 2$.

21. (a) Find upper and lower sums for

$$\int_0^1 \frac{4}{1 + x^2}\,dx$$

within 0.2 of one another. (b) Look at the average of these sums. Can you guess what the exact integral is?

22. Find upper and lower sums for $\displaystyle\int_2^3 \frac{1}{x}\,dx$ within $\frac{1}{10}$ of one another.

23. (a) Find $\dfrac{d}{dx} \left[\dfrac{1}{(1 + x^2)} \right]$.

(b) Find the area under the graph of the function $x/(1 + x^2)^2$, from $x = 0$ to $x = 1$.

24. Find $(d/dx)[x^3/(1 + x^3)]$. (b) Find the area under the graph of $x^2/(1 + x^3)^2$ from $x = 1$ to $x = 2$.

25. Find the area under the graph of $y = mx + b$ from $x = a_1$ to $x = a_2$ and verify your answer by using plane geometry. Assume that $mx + b \geqslant 0$ on $[a_1, a_2]$.

26. Find the area under the graph of the function $y = (1/x^2) + x + 1$ from $x = 1$ to $x = 2$.

27. Find the area under the graph of $y = x^2 + 1$ from -1 to 2 and sketch the region.

28. Find the area under the graph of

$$f(x) = \begin{cases} -x^3 & \text{if} \quad x \leqslant 0, \\ x^3 & \text{if} \quad x > 0 \end{cases}$$

from $x = -1$ to $x = 1$ and sketch.

29. (a) Verify the integration formula

$$\int \frac{x^2}{(x^3 + 6)^2}\,dx = \frac{1}{12}\left[\frac{x^3 + 2}{(x^3 + 6)} \right] + C.$$

(b) Find the area under the graph of $y = x^2/(x^3 + 6)^2$ between $x = 0$ and $x = 2$.

30. Find the area between the graphs of $y = x^3$ and $y = 5x^2 + 2x$ between $x = 0$ and $x = 2$. Sketch.

31. The curves $y = x^6 - 3$ and $y = -x^2 - 1$ divide the plane into five regions, one of which is bounded. Find its area.

32. Find the area of each of the numbered regions in Fig. 4.R.1.

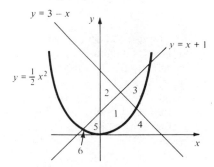

Figure 4.R.1. Find the area of the numbered regions.

33. Find the area of the region bounded by the graphs $x = y^2 - 6$ and $x = y$.

34. Find the area of the region bounded by the graphs $y = x^2 - 2$ and $y = 2 - x^2$.

35. An object is thrown at $t = 0$ from an airplane, and it has vertical velocity $v = -10 - 32t$ feet per second at time t. If the object is still falling after 10 seconds, what can you say about the altitude of the plane at $t = 0$?

36. Suppose the velocity of an object at position x is x^n, where n is some integer $\neq 1$. Find the time required to travel from $x = \frac{1}{1000}$ to $x = 1$.

37. (a) At time $t = 0$, a container has 1 liter of water in it. Water is poured in at the rate of $3t^2 - 2t + 3$ liters per minute (t = time in minutes). If the container has a leak which can drain 2 liters per minute, how much water is in the container at the end of 3 minutes?

    (b) What if the leak is 4 liters per minute?

    (c) What if the leak is 8 liters per minute? [*Hint*: What happens if the tank is empty for a while?]

38. Water is poured into a container at a rate of t liters per minute. At the same time, water is leaking out at the rate of t^2 liters per minute. Assume that the container is empty at $t = 0$.

    (a) When does the amount of water in the container reach its maximum?

    (b) When is the container empty again?

39. Suppose that a *supply curve* $p = S(x)$ and a *demand curve* $p = D(x)$ are graphed and that there is a unique point (a, b) at which supply equals demand (p = price/unit in dollars, x = number of units). The (signed) area enclosed by $x = 0$, $x = a$, $p = b$, and $p = D(x)$ is called the *consumer's surplus* or the *consumer's loss* depending on whether the sign is positive or negative. Similarly, the (signed) area enclosed by $x = 0$, $x = a$, $p = b$, and $p = S(x)$ is called the *producer's surplus* or the *producer's loss* depending on whether the sign is positive or negative.

    (a) Let $D(x) \geq b$. Explain why the *consumer's surplus* is $\int_0^a [D(x) - b] \, dx$.

    (b) Let $S(x) \leq b$. Explain why the *producer's surplus* is $\int_0^a [b - S(x)] \, dx$.

    (c) "If the price stabilizes at \$6 per unit, then some people are still willing to pay a higher price, but benefit by paying the lower price of \$6 per unit. The total of these benefits over $[0, a]$ is the consumer's surplus." Explain this in the language of integration.

    (d) Find the consumer's and producer's surplus for the supply curve $p = x^2/8$ and the demand curve $p = -(x/4) + 1$.

40. The demand for wood products in 1975 was about 12.6 billion cubic feet. By measuring order increases, it was determined that x years after 1975, the demand increased by $9x/1000$; that is, $D'(x) = 9x/1000$, where $D(x)$ is the demand x years after 1975, in billions of cubic feet.

    (a) Use the fundamental theorem of calculus to show that

$$D(x) = D(0) + \int_0^x (9t/1000) \, dt.$$

    (b) Find $D(x)$.

    (c) Find the demand for wood in 1982.

41. Suppose that an object on the x axis has velocity $v = t^2 - 4t - 5$. How far does it travel between $t = 0$ and $t = 6$?

42. Show that the actual distance travelled by a bus with velocity $v = f(t)$ is $\int_a^b |f(t)| \, dt$. What condition on $v = f(t)$ means that the bus made a round trip between $t = a$ and $t = b$?

43. A rock is dropped off a bridge over a gorge. The sound of the splash is heard 5.6 seconds after the rock was dropped. (Assume the rock falls with velocity $32t$ feet per second and sound travels at 1080 feet per second.)

    (a) Show by integration that the rock falls $16t^2$ feet after t seconds, and that the sound of the splash travels $1080t$ feet in t seconds.

    (b) The time T required for the rock to hit the water must satisfy $16T^2 = 1080(5.6 - T)$, because the rock and the sound wave travel equal distances. Find T.

    (c) Find the height of the bridge.

    (d) Find the number of seconds required for the sound of the splash to travel from the water to the bridge.

44. The current $I(t)$ and charge $Q(t)$ at time t (in amperes and coulombs, respectively) in a circuit are related by the equation $I(t) = Q'(t)$.

    (a) Given $Q(0) = 1$, use the fundamental theorem of calculus to justify the formula $Q(t) = 1 + \int_0^t I(r) \, dr$.

    (b) The voltage drop V (in volts) across a resistor of resistance R ohms is related to the current I (in amperes) by the formula $V = RI$. Suppose that in a simple circuit with a resistor made of nichrome wire, $V = 4.36$, $R = 1$, and $Q(0) = 1$. Find $Q(t)$.

    (c) Repeat (b) for a circuit with a 12-volt battery and 4-ohm resistance.

45. A ruptured sewer line causes lake contamination near a ski resort. The concentration $C(t)$ of bacteria (number per cubic centimeter) after t days is given by $C'(t) = 10^3(t - 7)$, $0 \leq t \leq 6$, after treatment of the lake at $t = 0$.

    (a) An inspector will be sent out after the bacteria concentration has dropped to half its original value $C(0)$. On which day should the inspector be sent if $C(0) = 40,000$?

    (b) What is the total change in the concentration from the fourth day to the sixth day?

46. A baseball is thrown vertically upwards from the ground with an initial upward velocity of 50 feet per second. How far has it travelled when it strikes the ground?

47. Let

$$g(y) = \begin{cases} y, & 0 \leqslant y < 1, \\ 2, & 1 \leqslant y < 2, \\ y, & 2 \leqslant y \leqslant 4. \end{cases}$$

Compute $\int_0^1 g(y)\,dy + \int_3^4 g(y)\,dy$.

48. Let

$$y(t) = \begin{cases} t, & 2 \leqslant t < 3, \\ -4, & 3 \leqslant t < 4, \\ 1, & 4 \leqslant t \leqslant 5. \end{cases}$$

Compute $\int_5^2 y(t)\,dt$.

49. Calculate $\dfrac{d}{dx}\int_0^x \dfrac{s^2}{1+s^3}\,ds$.

50. Calculate $\dfrac{d}{dx}\int_2^x \dfrac{t^3}{1-t^4}\,dt$.

51. Find the area between the graph of the function in Exercise 47 and the x axis.

★52. Find $\dfrac{d}{ds}\int_0^{s^2} \dfrac{1}{\sqrt[3]{x^2+1}}\,dx$. (*Hint*: See Exercise 43, Section 4.5.)

★53. Find $\dfrac{d}{dt}\int_0^{t^3+2} \dfrac{1}{y^2+1}\,dy$.

★54. Find $\int_a^b (3x^2+x)\,dx$ "by hand."

★55. Let f be defined on $[0, 1]$ by

$$f(t) = \begin{cases} 0, & x = 0, \\ 1/\sqrt{x}, & 0 < x \leqslant 1. \end{cases}$$

(a) Show that there are no upper sums for f on $[0, 1]$, and hence that f is not integrable.

(b) Show that every number less than 2 is a lower sum. [*Hint*: Use step functions which are zero on an interval $[0, \epsilon)$ and approximate f very closely on $[\epsilon, 1]$. Take ϵ small and use the integrability of f on $[\epsilon, 1]$.]

(c) Show that no number greater than or equal to 2 is a lower sum. [*Hint*: Show $\epsilon f(\epsilon) + \int_\epsilon^1 f(x)\,dx < 2$ for all ϵ in $(0, 1)$.]

(d) If you had to assign a value to $\int_0^1 f(x)\,dx$, what value would you assign?

★56. Modeling your discussion after the preceding exercise, find the upper or lower sums for each of the following functions on $[0, 1]$:

(a) $f(x) = \begin{cases} 0, & x = 0, \\ -1/\sqrt[3]{x}, & x > 0; \end{cases}$

(b) $f(x) = \begin{cases} 0, & x = 0, \\ \dfrac{1}{x^2}, & x > 0. \end{cases}$

Chapter 5

Trigonometric Functions

The derivative of $\sin x$ *is* $\cos x$ *and of* $\cos x$ *is* $-\sin x$; *everything else follows from this.*

Many problems involving angles, circles, and periodic motion lead to trigonometric functions. In this chapter, we study the calculus of these functions, and we apply our knowledge to solve new problems.

The chapter begins with a review of trigonometry. Well-prepared students may skim this material and move on quickly to the second section. Students who do not feel prepared or who failed Orientation Quiz C at the beginning of the book should study this review material carefully.

5.1 Polar Coordinates and Trigonometry

Trigonometric functions provide the link between polar and cartesian coordinates.

This section contains a review of trigonometry, with an emphasis on the topics which are most important for calculus. The derivatives of the trigonometric functions will be calculated in the next section.

The circumference C and area A of a circle of radius r are given by

$$C = 2\pi r, \qquad A = \pi r^2$$

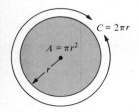

Figure 5.1.1. The circumference and area of a circle.

(see Fig. 5.1.1), where π is an irrational number whose value is approximately $3.14159 \ldots$.[1]

---

[1] For details on the fascinating history of π, see P. Beckman, *A History of π*, Golem Press, 1970. To establish deeper properties of π such as its irrationality (discovered by Lambert and Legendre around 1780), a careful and critical examination of the definition of π is needed. The first explicit expression for π was given by Viète (1540–1603) as

$$\frac{2}{\pi} = \sqrt{\frac{1}{2}} \cdot \sqrt{\frac{1}{2} + \frac{1}{2}\sqrt{\frac{1}{2}}} \cdot \sqrt{\frac{1}{2} + \frac{1}{2}\sqrt{\frac{1}{2} + \frac{1}{2}\sqrt{\frac{1}{2}}}} \cdots$$

which is obtained by inscribing regular polygons in a circle. Euler's famous expression $\pi/4 = 1 - \frac{1}{3} + \frac{1}{5} - \cdots$ is discussed in Example 3, Section 12.5. For an elementary proof of the irrationality of π, see M. Spivak, *Calculus*, Benjamin, 1967.

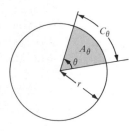

Figure 5.1.2. The length C_θ and area A_θ are proportional to θ.

If two rays are drawn from the center of the circle, both the length and area of the part of the circle between the rays are proportional to the angle between the rays. Thus, if we measure angles in degrees, the length C_θ and area A_θ between rays making an angle θ (see Fig. 5.1.2) are determined by the relations

$$\frac{C_\theta}{2\pi r} = \frac{\theta}{360}, \qquad \frac{A_\theta}{\pi r^2} = \frac{\theta}{360} \qquad (\theta \text{ in degrees}),$$

since a full circle corresponds to an angle of 360 degrees.

These formulas become simpler if we adopt the *radian* unit of measure, in which the total angular measure of a circle is defined to be 2π. Then our previous formulas become

$$\frac{C_\theta}{2\pi r} = \frac{\theta}{2\pi}, \qquad \frac{A_\theta}{\pi r^2} = \frac{\theta}{2\pi} \qquad (\theta \text{ in radians})$$

or simply

$$C_\theta = r\theta, \qquad A_\theta = \tfrac{1}{2} r^2 \theta.$$

The formulas of calculus are also simpler when angles are measured in radians rather than degrees. Unless explicit mention is made of degrees, all angles in this book will be expressed in radians. If you use a calculator to do computations with angles measured in radians, be sure that it is operating in the radian mode.

Example 1   An arc of length 10 meters on a circle of radius 4 meters subtends what angle at the center of the circle? How much area is enclosed in this part of the circle?

Solution   In the formula $C_\theta = r\theta$, we have $C_\theta = 10$ and $r = 4$, so $\theta = \frac{10}{4} = 2\frac{1}{2}$ (radians). The area enclosed is $A_\theta = \frac{1}{2} r^2 \theta = \frac{1}{2} \cdot 16 \cdot \frac{5}{2} = 20$ square meters. ▲

Conversions between degrees and radians are made by multiplying or dividing by the factor $360/2\pi = 180/\pi \approx 57.296$ degrees per radian.

Degrees and Radians

To convert from radians to degrees, multiply by $\dfrac{180°}{\pi} \approx 57°18' \approx 57.296°$.

To convert from degrees to radians, multiply by $\dfrac{\pi}{180°} \approx 0.01745$.

The following table gives some important angles in degrees and radians:

Degrees	0°	30°	45°	60°	90°	120°	135°	150°	180°	270°	360°
Radians	0	$\frac{\pi}{6}$	$\frac{\pi}{4}$	$\frac{\pi}{3}$	$\frac{\pi}{2}$	$\frac{2\pi}{3}$	$\frac{3\pi}{4}$	$\frac{5\pi}{6}$	π	$\frac{3\pi}{2}$	2π

The measures of right angles and straight angles are shown in Fig. 5.1.3.

Figure 5.1.3. A complete circle, a right angle, and a straight angle in degrees and radians.

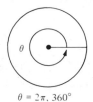

$\theta = 2\pi,\ 360°$

$\theta = \frac{\pi}{2},\ 90°$

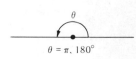

$\theta = \pi,\ 180°$

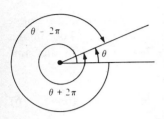

Figure 5.1.4. $\theta, \theta - 2\pi$, and $\theta + 2\pi$ measure the same geometric angle.

Negative numbers and numbers larger than 2π (or 360°) can also be used to represent angles. The convention is that θ and $\theta + 2\pi$ represent the same geometric angle; hence so do $\theta + 4\pi, \theta + 6\pi, \ldots$ as well as $\theta - 2\pi$, $\theta - 4\pi, \ldots$ (see Fig. 5.1.4). The angle $-\theta$ equals $2\pi - \theta$ and is thus the "mirror image" of θ (see Fig. 5.1.5). Note, also, that rays making angles of θ and $\theta + \pi$ with a given ray point in opposite directions along the same straight line (see Fig. 5.1.6).

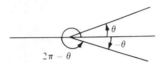

Figure 5.1.5. The angle $-\theta$, or $2\pi - \theta$, is the mirror image of θ.

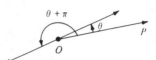

Figure 5.1.6. The rays making angles of θ and $\theta + \pi$ with OP point in opposite directions.

Example 2   (a) Convert to radians: 36°, 160°, 280°, $-300°$, 460°.
(b) Convert to degrees: $5\pi/18$, 2.6, 6.27, 0.2, -9.23.

Solution   (a) $36° \rightarrow 36 \times 0.01745 = 0.6282$ radian;
$160° \rightarrow 160 \times 0.01745 = 2.792$ radians;
$280° \rightarrow 280 \times 0.01745 = 4.886$ radians;
$-300° \rightarrow -300 \times 0.01745 = -5.235$ radians, or
$\qquad -300 \times \pi/180 = -5\pi/3$ radians;
$460° \rightarrow 460° - 360° = 100° \rightarrow 100 \times 0.01745 = 1.745$ radians.
(b) $5\pi/18 \rightarrow 5\pi/18 \times 180/\pi = 50°$;
$2.6 \rightarrow 2.6 \times 57.296 = 148.97°$;
$6.27 \rightarrow 6.27 \times 57.296 = 359.25°$;
$0.2 \rightarrow 0.2 \times 57.296 = 11.46°$.
$-9.23 \rightarrow -9.23 \times 180/\pi = -528.84° \rightarrow 720° - 528.84° = 191.16°$. ▲

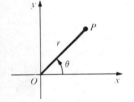

Figure 5.1.7. The polar coordinates (r, θ) of a point P.

Cartesian coordinates (x, y) represent points in the plane by their distances from two perpendicular lines. In the *polar coordinate* representation, a point P is associated with each pair (r, θ) of numbers in the following way.[2] First, a ray is drawn through the origin making an angle of θ with the positive x axis. Then one travels a distance r along this ray, if r is positive. (See Fig. 5.1.7.) If r is negative, one travels a distance $-r$ along the ray traced in the opposite

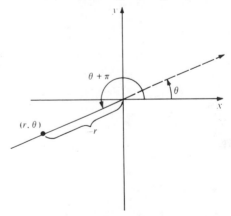

Figure 5.1.8. Plotting (r, θ) for negative r.

[2] Polar coordinates were first used successfully by Newton (1671) and Jacques Bernoulli (1691). The definitive treatment of polar coordinates in their modern form was given by Leonhard Euler in his 1748 textbook *Introductio in analysis infinitorium*. See C. B. Boyer. "The foremost textbook of modern times," American Mathematical Monthly **58** (1951): 223–226.

direction. One arrives at the point P; we call (r, θ) its *polar coordinates*. Notice that the resulting point is the same as the one with polar coordinates $(-r, \theta + \pi)$ (see Fig. 5.1.8) and that the pair $(r, \theta + 2\pi n)$ represents the same point as (r, θ), for any integer n.

Example 3   Plot the points P_1, P_2, P_3, and P_4 whose polar coordinates are $(5, \pi/6)$, $(-5, \pi/6)$, $(5, -\pi/6)$, and $(-5, -\pi/6)$, respectively.

Solution   (See Fig. 5.1.9.) The point $(-5, -\pi/6)$ is obtained by rotating $\pi/6 = 30°$ clockwise to give an angle of $-\pi/6$ and then moving 5 units backwards on

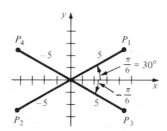

Figure 5.1.9. Some points in polar coordinates.

this line to the point P_4 shown. The other points are plotted in a similar way. ▲

Example 4   Describe the set of points P whose polar coordinates (r, θ) satisfy $0 \leqslant r \leqslant 2$ and $0 \leqslant \theta < \pi$.

Solution   Since $0 \leqslant r \leqslant 2$, we can range from the origin to 2 units from the origin. Our angle with the x axis varies from 0 to π, but not including π. Thus we are confined to the region in Fig. 5.1.10. The negative x axis is dashed since it is not included in the region. ▲

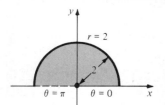

Figure 5.1.10. The region $0 \leqslant r \leqslant 2, 0 \leqslant \theta < \pi$.

If θ is a real number, we define $\cos \theta$ to be x and $\sin \theta$ to be y, where (x, y) are the cartesian coordinates of the point P on the circle of radius one whose polar coordinates are $(1, \theta)$. (See Fig. 5.1.11.) If an angle $\phi°$ is given in degrees, $\sin \phi°$ or $\cos \phi°$ means $\sin \theta$ or $\cos \theta$, where θ is the same angle measured in radians. Thus $\sin 45° = \sin(\pi/4)$, $\cos 60° = \cos(\pi/3)$, and so on.

The sine and cosine functions can also be defined in terms of ratios of sides of right triangles. (See Fig. 5.1.12.) By definition, $\cos \theta = |OA'|$, and by similar triangles,

$$\cos \theta = |OA'| = \left| \frac{OA'}{1} \right| = \frac{|OA'|}{|OB'|} = \frac{|OA|}{|OB|} = \frac{\text{side adjacent to } \theta}{\text{hypotenuse}}.$$

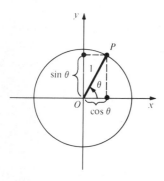

Figure 5.1.11. The cosine and sine of θ are the x and y coordinates of the point P.

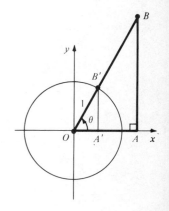

Figure 5.1.12. The triangles OAB and $OA'B'$ are similar; $\cos \theta = |OA|/|OB|$ and $\sin \theta = |AB|/|OB|$.

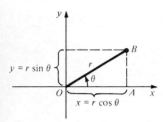

Figure 5.1.13. Converting polar to cartesian coordinates.

In the same way, we see that

$$\sin\theta = \frac{|AB|}{|OB|} = \frac{\text{side opposite to }\theta}{\text{hypotenuse}} .$$

It follows (see Fig. 5.1.13) that if the point B has cartesian coordinates (x, y) and polar coordinates (r, θ), then $\cos\theta = |OA|/|OB| = x/r$ and $\sin\theta = |AB|/|OB| = y/r$, so we obtain the following relations.

Cartesian and Polar Coordinates

$$x = r\cos\theta \quad \text{and} \quad y = r\sin\theta,$$

where (x, y) are cartesian coordinates and (r, θ) are polar coordinates.

Example 5 Convert from cartesian to polar coordinates: $(2, -4)$; and from polar to cartesian coordinates: $(6, -\pi/8)$.

Solution We plot $(2, -4)$ as in Fig. 5.1.14. Then $r = \sqrt{2^2 + (-4)^2} = \sqrt{20} = 2\sqrt{5}$ and $\cos\theta = 2/(2\sqrt{5}) = 1/\sqrt{5} \approx 0.447214$, so from tables or a calculator[3], $\theta = 1.107$; but we must take $\theta = -1.107$ (or 5.176) since we are in the fourth quadrant. Thus the polar coordinates of $(2, -4)$ are $(2\sqrt{5}, -1.107)$.

The cartesian coordinates of the point with polar coordinates $(6, -\pi/8)$ are

$$x = r\cos\theta = 6\cos(-\pi/8) = (6)(0.92388) = 5.5433$$

and

$$y = r\sin\theta = 6\sin(-\pi/8) = (6)(-0.38268) = -2.2961.$$

That is, $(5.5433, -2.2961)$. This point is also in the fourth quadrant as it should be. ▲

Figure 5.1.14. Find the polar coordinates of $(2, -4)$.

Example 6 (a) Show that $\sin\theta = \cos(\pi/2 - \theta)$ for $0 \leqslant \theta \leqslant \pi/2$. (b) Show that sin is an odd function: $\sin(-\theta) = -\sin\theta$ (assume that $0 \leqslant \theta \leqslant \pi/2$).

Solution (a) In Fig. 5.1.15, the angle OBA is $\pi/2 - \theta$ since the three angles must add up to π by plane geometry. Therefore $\sin\theta = (\text{opposite}/\text{hypotenuse}) = |AB|/|OB|$ and $\cos(\pi/2 - \theta) = (\text{adjacent}/\text{hypotenuse}) = |AB|/|OB| = \sin\theta$.
(b) Referring to Fig. 5.1.16, we see that if θ is switched to $-\theta$, this changes the sign of $y = \sin\theta$. Hence $\sin(-\theta) = -\sin\theta$. ▲

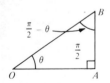

Figure 5.1.15. If $\theta = \angle BOA$, then $\angle OBA = \pi/2 - \theta$.

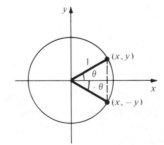

Figure 5.1.16. $\sin\theta$ is an odd function.

---

[3] Many calculators are equipped with a $\cos^{-1}$ (arc cos) function which computes the angle whose cosine is given. If your calculator does not have such a function, you can use the cosine function together with the method of bisection (see Example 7, Section 3.1). The inverse cosine (and other trigonometric) functions are discussed in Section 5.4.

The other trigonometric functions can be defined in terms of the sine and cosine:

Tangent: $\quad\tan\theta = \dfrac{\sin\theta}{\cos\theta} = \dfrac{|AB|}{|OA|} = \dfrac{\text{side opposite}}{\text{side adjacent}}.$

Cotangent: $\quad\cot\theta = \dfrac{\cos\theta}{\sin\theta} = \dfrac{1}{\tan\theta} = \dfrac{|OA|}{|AB|}.$

Secant: $\quad\sec\theta = \dfrac{1}{\cos\theta} = \dfrac{|OB|}{|OA|}.$

Cosecant: $\quad\csc\theta = \dfrac{1}{\sin\theta} = \dfrac{|OB|}{|AB|}.$

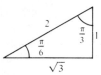

Some frequently used values of the trigonometric functions can be read off the right triangles shown in Fig. 5.1.17. For example, $\cos(\pi/4) = 1/\sqrt{2}$, $\sin(\pi/4) = 1/\sqrt{2}$, $\tan(\pi/4) = 1$, $\cos(\pi/6) = \sqrt{3}/2$, and $\tan(\pi/3) = \sqrt{3}$. (The proof that the $1,2,\sqrt{3}$ triangle has angles $\pi/3, \pi/6, \pi/2$ is an exercise in euclidean geometry; see Fig. 5.1.18.)

Figure 5.1.17. Two basic examples.

Figure 5.1.18. The angles of an equilateral triangle are all equal to $\pi/3$.

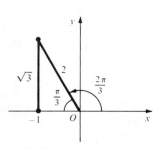

Figure 5.1.19. Illustrating the sine and cosine of $2\pi/3$.

Special care should be taken with functions of angles which are not between 0 and $\pi/2$—that is, angles not in the first quadrant—to ensure that their signs are correct. For instance, we notice in Fig. 5.1.19 that $\sin(2\pi/3) = \sqrt{3}/2$ and $\cos(2\pi/3) = -\frac{1}{2}$.

The following table gives some commonly used values of sin, cos, and tan:

$\theta°$	0°	30°	45°	60°	90°	120°	135°	150°	180°	270°	360°
θ	0	$\dfrac{\pi}{6}$	$\dfrac{\pi}{4}$	$\dfrac{\pi}{3}$	$\dfrac{\pi}{2}$	$\dfrac{2\pi}{3}$	$\dfrac{3\pi}{4}$	$\dfrac{5\pi}{6}$	π	$\dfrac{3\pi}{2}$	2π
$\cos\theta$	1	$\dfrac{\sqrt{3}}{2}$	$\dfrac{\sqrt{2}}{2}$	$\dfrac{1}{2}$	0	$-\dfrac{1}{2}$	$-\dfrac{\sqrt{2}}{2}$	$-\dfrac{\sqrt{3}}{2}$	-1	0	1
$\sin\theta$	0	$\dfrac{1}{2}$	$\dfrac{\sqrt{2}}{2}$	$\dfrac{\sqrt{3}}{2}$	1	$\dfrac{\sqrt{3}}{2}$	$\dfrac{\sqrt{2}}{2}$	$\dfrac{1}{2}$	0	-1	0
$\tan\theta$	0	$\dfrac{\sqrt{3}}{3}$	1	$\sqrt{3}$	$\pm\infty$	$-\sqrt{3}$	-1	$-\dfrac{\sqrt{3}}{3}$	0	$\pm\infty$	0

Over the centuries, large tables of values of the trigonometric functions have been complied. The first such table, compiled by Hipparchus and Ptolemy, appeared in Ptolemy's *Almagest*. Today these values are also on many pocket calculators. Since angles as well as some lengths can be directly measured (as in surveying), the trigonometric relations can then enable us to compute lengths which may be inaccessible (see Example 7).

▦ Calculator Discussion

You may be curious about how pocket calculators compute their values of $\sin\theta$ and $\cos\theta$. Some analytic expressions are available, such as

$$\sin\theta = \theta - \frac{\theta^3}{3\cdot2} + \frac{\theta^5}{5\cdot4\cdot3\cdot2} - \frac{\theta^7}{7\cdot6\cdot5\cdot4\cdot3\cdot2} + \cdots$$

(as will be proved in Section 12.5), but using these is inefficient and inaccurate. Instead a rational function of θ is fitted to many known values of $\sin\theta$ (or $\cos\theta$, $\tan\theta$, and so on) and this rational function is used to calculate approximate values at the remaining points. Thus when θ is entered and $\sin\theta$ pressed, a program in the calculator calculates the value of this rational function.

If you experiment with your calculator—for example, by calculating $\tan\theta$ for θ near $\pi/2$—you might discover some inaccuracies in this method.[4] ▲

Example 7   (a) In Fig. 5.1.20, find x. (b) A tree 50 meters away subtends an angle of 53° as seen by an observer. How tall is the tree?

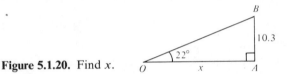

Figure 5.1.20. Find x.

Solution   (a) We find that $\tan22° = 10.3/x$, so $x = 10.3/\tan22°$. From tables or a calculator, $\tan22° \approx 0.404026$, so $x \approx 25.4934$. (b) Refer to Fig. 5.1.21. $|AB| = |OA|\tan53° = 50\tan53°$. Using tables or a calculator, this becomes $50(1.3270) = 66.35$ meters. ▲

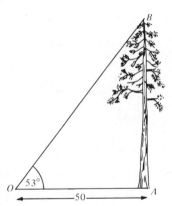

Figure 5.1.21. Trigonometry used to find the height of a tree.

From the definition of sin and cos, the point P with cartesian coordinates $x = \cos\theta$ and $y = \sin\theta$ lies on the unit circle $x^2 + y^2 = 1$. Therefore, for any value of θ,

$$\cos^2\theta + \sin^2\theta = 1. \tag{1}$$

This is an example of *trigonometric identity*—a relationship among the trigonometric functions which is valid for all θ.

---

[4] For more details on how calculators do those computations, see "Calculator Function Approximation" by C. W. Schelin, Am. Math. Monthly Vol. **90** (1983), 317–325.

Relationship (1) is, in essence, a statement of Pythagoras' theorem for a right triangle ($OA'B'$ in Fig. 5.1.12). For a general triangle, the correct relationship between the three sides is given by the *law of cosines*: with notation as in Fig. 5.1.22, we have

$$c^2 = a^2 + b^2 - 2ab\cos\theta. \tag{2}$$

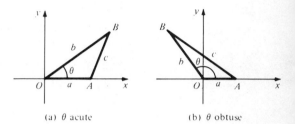

Figure 5.1.22. Proving the law of cosines.

(a) θ acute   (b) θ obtuse

To prove equation (2), note that the (x, y) coordinates of B are $x = b\cos\theta$ and $y = b\sin\theta$; those of A are $x = a, y = 0$. By the distance formula and equation (1),

$$c^2 = (b\cos\theta - a)^2 + (b\sin\theta)^2 = b^2\cos^2\theta - 2ab\cos\theta + a^2 + b^2\sin^2\theta$$

$$= b^2(\cos^2\theta + \sin^2\theta) + a^2 - 2ab\cos\theta = b^2 + a^2 - 2ab\cos\theta,$$

so equation (2) is proved.

In Fig. 5.1.22 we situated the triangles in a particular way, but this was just a device to prove equation (2); since any triangle can be moved into this special position, the formula holds in the general situation of Fig. 5.1.23.

Note that when $\theta = \pi/2$, $\cos\theta = 0$ and so equation (2) reduces to Pythagoras' theorem: $c^2 = a^2 + b^2$.

Figure 5.1.23. Data for the law of cosines: $c^2 = a^2 + b^2 - 2ab\cos\theta$.

Example 8   In Fig. 5.1.24, find x.

Solution   By the law of cosines

$$x^2 = (20.2)^2 + (13.4)^2 - 2(20.2)(13.4)\cos(10.3°)$$

$$= (408.04) + (179.56) - 532.64 = 54.96.$$

Taking square roots, we find $x \approx 7.41$. ▲

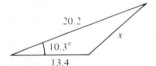

Figure 5.1.24. Find x.

Now consider the situation in Fig. 5.1.25. By the distance formula,

$$|PQ|^2 = (\cos\phi - \cos\theta)^2 + (\sin\phi - \sin\theta)^2$$

$$= \cos^2\phi - 2\cos\phi\cos\theta + \cos^2\theta + \sin^2\phi - 2\sin\phi\sin\theta + \sin^2\theta$$

$$= 2 - 2\cos\phi\cos\theta - 2\sin\phi\sin\theta.$$

On the other hand, by the law of cosines (2) applied to $\triangle OPQ$,

$$|PQ|^2 = 1^2 + 1^2 - 2\cos(\phi - \theta),$$

since $\phi - \theta$ is the angle at the vertex O. Comparing our two expressions for $|PQ|^2$ gives the identity

$$\cos(\phi - \theta) = \cos\phi\cos\theta + \sin\phi\sin\theta,$$

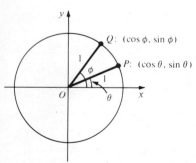

Figure 5.1.25. Geometry for the proof of the addition formulas.

which is valid for all ϕ and θ. If we replace θ by $-\theta$ and recall that $\cos(-\theta) = \cos\theta$ and $\sin(-\theta) = -\sin\theta$, this identity yields

$$\cos(\theta + \phi) = \cos\theta\cos\phi - \sin\theta\sin\phi. \tag{3}$$

Now if we write $\pi/2 - \phi$ for ϕ and recall that $\cos(\pi/2 - \phi) = \sin\phi$ and $\sin(\pi/2 - \phi) = \cos\phi$, the same identity gives

$$\sin(\theta + \phi) = \sin\theta\cos\phi + \cos\theta\sin\phi. \tag{4}$$

Identities (3) and (4), called the *addition formulas* for sine and cosine, will be essential for calculus. From these basic identities, we can also derive many others by algebraic manipulation.

For integral calculus, two of the most important consequences of (3) are the *double-angle formulas*. Setting $\theta = \phi$ in (3) gives

$$\cos 2\theta = \cos^2\theta - \sin^2\theta = (1 - \sin^2\theta) - \sin^2\theta = 1 - 2\sin^2\theta.$$

Thus

$$\sin^2\theta = \tfrac{1}{2}(1 - \cos 2\theta). \tag{5}$$

Similarly,

$$\cos 2\theta = \cos^2\theta - \sin^2\theta = 2\cos^2\theta - 1$$

gives

$$\cos^2\theta = \tfrac{1}{2}(\cos 2\theta + 1). \tag{6}$$

Example 9   (a) Prove the product formula: $\cos\theta\cos\phi = \tfrac{1}{2}[\cos(\theta - \phi) + \cos(\theta + \phi)]$.
(b) Prove that $1 + \tan^2\theta = \sec^2\theta$.

Solution   (a) Add the identity for $\cos(\theta - \phi)$ to that for $\cos(\theta + \phi)$:

$$\cos(\theta - \phi) + \cos(\theta + \phi) = \cos\phi\cos\theta + \sin\phi\sin\theta + \cos\theta\cos\phi - \sin\theta\sin\phi$$
$$= 2\cos\theta\cos\phi.$$

Dividing by 2 gives the product formula.
(b) Divide both sides of $\sin^2\theta + \cos^2\theta = 1$ by $\cos^2\theta$; then, using $\tan\theta = \sin\theta/\cos\theta$ and $1/\cos\theta = \sec\theta$, we get $\tan^2\theta + 1 = \sec^2\theta$ as required. ▲

Some of the most important trigonometric identities are listed on the inside front cover of the book for handy reference. They are all useful, but you can get by quite well by memorizing only (1) through (4) above and deriving the rest when you need them.

From the available values of the trigonometric functions, one can accurately draw their graphs. The calculus of these functions, studied in the next section, confirms that these graphs are correct, so there are no maxima, minima, or inflection points other than those in plain view in Fig. 5.1.26 on the following page.

Perhaps the most important fact about these functions is their *periodicity*: When a function f satisfies $f(\theta + \tau) = f(\theta)$ for all θ and a given positive τ, f is said to be *periodic* with *period* τ. The reciprocal $1/\tau$ is called the *frequency*. All the functions in Fig. 5.1.26 are periodic with period 2π; this enables us to draw the entire graph by repeating the segment over an interval of length 2π. The trigonometric functions also have $4\pi, 6\pi, 8\pi, \ldots$ as additional periods, but the least period of a periodic function is always unique. Note that the functions tan and cot have π as their least period, while 2π is the least period of the other four trigonometric functions.

Figure 5.1.26. Graphs of the trigonometric functions.

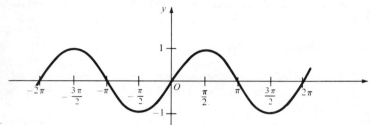

(a) $y = \sin \theta$

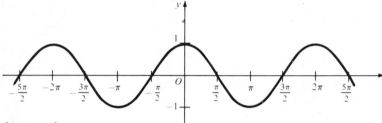

(b) $y = \cos \theta$

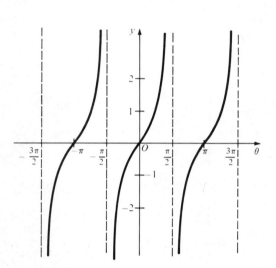

(c) $y = \tan \theta$

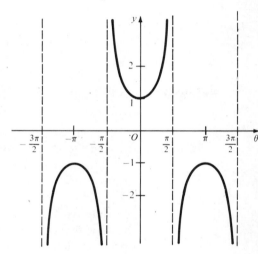

(d) $y = \sec \theta$

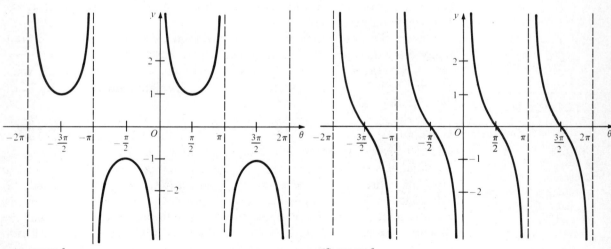

(e) $y = \csc \theta$ (f) $y = \cot \theta$

Example 10   (a) Sketch the graph $y = \cos 2x$. What is the (least) period of this function?
(b) Sketch the graph of $y = 3 \cos 5\theta$.

Solution   (a) We obtain $y = \cos 2x$ by taking the graph of $y = \cos x$ and compressing the graph horizontally by a factor of 2 (see Fig. 5.1.27). The function repeats every π units on the x axis, so it is periodic with (least) period π.

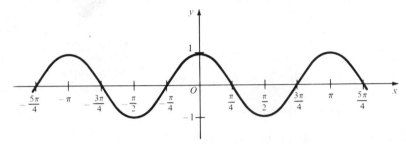

Figure 5.1.27. The graph of $y = \cos 2x$.

(b) We obtain $y = 3 \cos 5\theta$ by compressing the graph of $y = \cos \theta$ horizontally by a factor of 5 and stretching it vertically by a factor of 3 (see Fig. 5.1.28). ▲

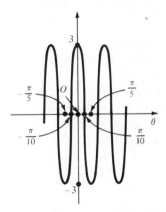

Figure 5.1.28. The graph of $y = 3 \cos 5\theta$.

Example 11   Where are the inflection points of $\tan \theta$? For which values of θ do you expect $\tan \theta$ to be a differentiable function?

Solution   Recall that an inflection point is a point where the second derivative changes sign—that is, a point between different types of concavity. On the graph of $\tan \theta$, notice that the graph is concave upward on $(0, \pi/2)$ and downward on $(-\pi/2, 0)$. Hence 0 is an inflection point, as are π, $-\pi$, and so forth. The general inflection point is $n\pi$, where $n = 0, \pm 1, \pm 2, \ldots$. (But $\pi/2$ is not an inflection point because $\tan \theta$ is not defined there.)

From the graph, we expect that $\tan \theta$ will be a differentiable function of θ except at $\pm \pi/2, \pm 3\pi/2, \ldots$. ▲

Exercises for Section 5.1

1. If an arc of a circle with radius 10 meters subtends an angle of 22°, how long is the arc? How much area is enclosed in this part of the circle?
2. An arc of radius 15 feet subtends an angle of 2.1 radians. How long is the arc? How much area is enclosed in this part of the circle?
3. An arc of radius 18 meters has length 5 meters. What angle does it subtend? How much area is enclosed in this part of the circle?
4. An arc of length 110 meters subtends an angle of

24°. What is the radius of the arc? How much area is enclosed in this part of the circle?
5. Convert to radians: 29°, 54°, 255°, 130°, 320°.
6. Convert to degrees: 5, $\pi/7$, 3.2, $2\pi/9$, $\frac{1}{2}$, 0.7.
7. Simplify so that $0 \leqslant \theta < 2\pi$ or $0 \leqslant \theta° < 360°$.
   (a) Radians: $7\pi/3$, $16\pi/5$, 15π.
   (b) Degrees: 520°, 1745°, 385°.
8. Simplify so that $0 \leqslant \theta < 2\pi$ or $0 \leqslant \theta° < 360°$.
   (a) Radians: $5\pi/2$, $48\pi/11$, $13\pi + 1$.
   (b) Degrees: 470°, 604°, 75°, 999°.

9. Plot the following points given in polar coordinates: $(3, \pi/2)$, $(5, -\pi/4)$, $(1, 2\pi/3)$, $(-3, \pi/2)$.

10. Plot the following points given in polar coordinates: $(6, 3\pi/2)$, $(-2, \pi/6)$, $(7, -2\pi/3)$, $(1, \pi/2)$, $(4, -\pi/6)$.

In Exercises 11–14, sketch the set of points whose polar coordinates (r, θ) satisfy the given conditions.

11. $-1 \leqslant r \leqslant 2$; $\pi/3 \leqslant \theta < \pi/2$.

12. $0 < r < 4$; $-\pi/6 < \theta < \pi/6$.

13. $2 \leqslant r < 3$; $-\pi/2 \leqslant \theta \leqslant \pi$.

14. $-2 \leqslant r \leqslant -1$; $-\pi/4 < \theta < 0$.

15. Find the polar coordinates of $(x, y) = (5, -2)$?

16. Find the cartesian coordinates of $(r, \theta) = (2, \pi/6)$.

17. Convert from cartesian to polar coordinates:
(a) $(1, 0)$, (b) $(3, 4)$, (c) $(\sqrt{3}, 1)$, (d) $(\sqrt{3}, -1)$, (e) $(-\sqrt{3}, 1)$.

18. Convert from polar to cartesian coordinates:
(a) $(0, \pi/8)$, (b) $(1, 0)$, (c) $(2, \pi/4)$, (d) $(8, 3\pi/2)$, (e) $(2, \pi)$.

19. Convert from cartesian to polar coordinates:
(a) $(1, -1)$, (b) $(0, 2)$, (c) $(\frac{1}{2}, 7)$, (d) $(-12, -5)$, (e) $(-3, 8)$, (f) $(\frac{3}{4}, \frac{3}{4})$.

20. Convert from cartesian to polar coordinates:
(a) $(-4, 4)$, (b) $(1, 15)$, (c) $(19, -3)$, (d) $(-5, -6)$, (e) $(0.3, 0.9)$, (f) $(-\frac{3}{2}, \frac{1}{2})$.

21. Convert from polar to cartesian coordinates:
(a) $(6, \pi/2)$, (b) $(-12, 3\pi/4)$, (c) $(4, -\pi)$, (d) $(2, 13\pi/2)$, (e) $(8, -2\pi/3)$, (f) $(-1, 2)$.

22. Convert from polar to cartesian coordinates:
(a) $(-1, -1)$, (b) $(1, \pi)$, (c) $(10, 2.7)$, (d) $(5, 7\pi/2)$, (e) $(8, 7\pi)$, (f) $(4, -3\pi)$.

23. Show that $\tan\theta = \cot(\pi/2 - \theta)$ assuming that $0 \leqslant \theta \leqslant \pi/2$.

24. Show that $\sec\theta = \csc(\pi/2 - \theta)$ for $0 \leqslant \theta \leqslant \pi/2$.

25. Show that $\cos\theta = \cos(-\theta)$ for $0 \leqslant \theta \leqslant \pi$.

26. Show that $\tan\theta = -\tan(-\theta)$ for $0 \leqslant \theta \leqslant \pi$, $\theta \neq \pi/2$.

Refer to Fig. 5.1.29 for Exercises 27–30.

27. Find a. 28. Find b.

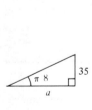

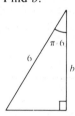

29. Find c. 30. Find d.

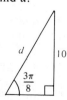

Figure 5.1.29. Find a, b, c, d.

31. An airplane flying at 5000 feet has an angle of elevation of 25° from observer A. Observer B sees that airplane directly overhead. How far apart are A and B?

32. A leaning tower tilts at 9° from the vertical directly away from an observer who is 500 meters away from its base. If the observer sees the top of the tower at an angle of elevation of 22°, how high is the tower?

33. A mountain 3000 meters away subtends an angle of 17° at an observer. How tall is the mountain?

34. A pedestrian 100 meters from the outdoor elevator at the Fairhill Hotel at noon sees the elevator at an angle of 10°. The elevator, steadily rising, makes an angle of 20° after 30 seconds has elapsed. How fast is the elevator rising? When will it make an angle of 30°?

Refer to Fig. 5.1.30 for Exercises 35–38.

35. Find p. 36. Find q.

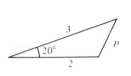

37. Find r. 38. Find s.

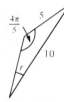

Figure 5.1.30. Find $p, q, r,$ and s.

39. Prove that $\cos^2\dfrac{\theta}{2} = \dfrac{1 + \cos\theta}{2}$.

40. Prove that $\tan\dfrac{\theta}{2} = \dfrac{\sin\theta}{1 + \cos\theta}$.

41. Prove that $\sin\theta\sin\phi = \dfrac{\cos(\theta - \phi) - \cos(\theta + \phi)}{2}$.

42. Express $\sin(3\theta)$ in terms of $\sin\theta$ and $\cos\theta$.

Simplify the expressions in Exercises 43–50.

43. $\sin\left(\theta + \dfrac{\pi}{2}\right)$

44. $\sin\left(\dfrac{3\pi}{2} + \theta\right)$

45. $\cos\left(\dfrac{3\pi}{2} - \theta\right)$

46. $\tan(\theta + \frac{7}{2}\pi)$

47. $\sec(6\pi + \theta)$

48. $\sin(\theta - \frac{9}{2}\pi)$

49. $\cos\left(\theta + \dfrac{\pi}{2}\right)\sin\left(\phi - \dfrac{3\pi}{2}\right)$.

50. $\dfrac{\sin(\theta + \frac{5}{2}\pi)}{\cos(\frac{\pi}{2} - \theta)}$.

Compute the quantities in Exercises 51–54 by using trigonometric identities, without using tables or a calculator.

51. $\cos 7\frac{1}{2}°$

52. $\tan 22\frac{1}{2}°$

53. $\sec \dfrac{\pi}{12}$

54. $\sin \dfrac{\pi}{12}$

Derive the identities in Exercises 55–60, making use of the table of identities on the inside front cover of the book.

55. $\sec\theta + \tan\theta = \dfrac{1 + \sin\theta}{1 - 2\sin^2(\theta/2)}$.

56. $8\cos\theta + 8\cos 2\theta = -9 + 16(\cos\theta + \frac{1}{4})^2$.

57. $\sec^2 \dfrac{\theta}{2} = \dfrac{2\sec\theta}{\sec\theta + 1}$.

58. $\csc^2 \dfrac{\theta}{2} = \dfrac{2\sec\theta}{\sec\theta - 1}$.

59. $\csc\theta\csc\phi = \dfrac{2\sec(\theta + \phi)\sec(\theta - \phi)}{\sec(\theta + \phi) - \sec(\theta - \phi)}$.

60. $\tan\theta\tan\phi = \dfrac{\sec(\theta + \phi) - \sec(\theta - \phi)}{\sec(\theta + \phi) + \sec(\theta - \phi)}$.

Sketch the graph of the functions in Exercises 61–68.

61. $2\cos 3\theta$

62. $\cos\left(3\theta + \dfrac{\pi}{6}\right)$

63. $\tan \dfrac{3\theta}{2}$

64. $\tan \dfrac{\theta}{2}$

65. $4\sin 2x\cos 2x$

66. $\sin x + \cos x$

67. $\sin 3\theta + 1$

68. $\csc 2\theta$

69. Locate the inflection points of $\cot\theta$ by inspecting the graph.

70. Locate the maximum and minimum points of $\sin\theta$ by inspecting the graph.

71. For what θ do you expect $\sec\theta$ and $\cot\theta$ to be differentiable?

72. Where is $\sec\theta$ concave upward?

73. Light travels at velocity v_1 in a certain medium, enters a second medium at angle of incidence θ_1 (measured from the normal to the surface), and refracts at angle θ_2 while travelling a different velocity v_2 (in the second medium). According to Snell's law, $v_1/v_2 = \sin\theta_1/\sin\theta_2$.
  (a) Light enters at 60° and refracts at 30°. The first medium is air ($v_1 = 3 \times 10^{10}$ centimeters per second). Find the velocity in the second medium.
  (b) Show that if $v_1 = v_2$, then the light travels in a straight line.
  (c) The speed halves in passing from one medium to another, the angle of incidence being 45°. Calculate the angle of refraction.

74. Scientists and engineers often use the approximations $\sin\theta = \theta$ and $\cos\theta = 1$, valid for θ near zero. Experiment with your hand calculator to determine a region of validity for θ that guarantees eight-place accuracy for these approximations.

Exercises 75–78 concern the *law of sines*.

75. Using the notation in Fig. 5.1.31, prove that

$$\frac{\sin\alpha}{a} = \frac{\sin\beta}{b} = \frac{\sin\gamma}{c}.$$

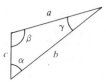

Figure 5.1.31. Law of sines.

76. Find a in Fig. 5.1.32.

Figure 5.1.32. Find a.

77. Show that the common value of $(\sin\alpha)/a$, $(\sin\beta)/b$, and $(\sin\gamma)/c$ is the reciprocal of twice the radius of the circumscribed circle.

78. (a) A parallelogram is formed with acute angle θ and sides l, L. Find a formula for its area. (b) A parallelogram is formed with an acute angle θ, a base of length L, and a diagonal opposite θ of length d. Find a formula for its area.

79. Show that $A\cos\omega t + B\sin\omega t = \alpha\cos(\omega t - \theta)$, where (α, θ), are the polar coordinates of (A, B).

80. Use Exercise 79 to write $f(t) = \cos t + \sqrt{3}\sin t$ as $\alpha\cos(t - \theta)$ for some α and θ. Use this to graph f.

81. Light of wavelength λ is diffracted through a single slit of width a. This light then passes through a lens and falls on a screen. A point P is on the screen, making an angle θ with the lens axis. The intensity I at the point P on the screen is

$$I = I_0\left[\frac{\sin([\pi a\sin\theta]/\lambda)}{[\pi a\sin\theta]/\lambda}\right]^2, \qquad 0 < \theta < \pi,$$

where I_0 is the intensity when P is on the lens axis.
  (a) Show that the intensity is zero for $\sin\theta = \lambda/a$.
  (b) Find all values of θ for which $\theta > 0$ and $I = 0$.
  (c) Verify that I is approximately I_0 when $[\pi a\sin\theta]/\lambda$ is close enough to zero. (Use $\sin\theta \approx \theta$; see Exercise 74.)
  (d) In practice, $\lambda = 5 \times 10^{-5}$ centimeters, and $a = 10^{-2}$ centimeters. Check by means of a calculator or table that $\sin\theta = \lambda/a$ is approximately $\theta = \lambda/a$.

82. The current I in a circuit is given by the formula $I(t) = 20\sin(311t) + 40\cos(311t)$. Let $r = (20^2 + 40^2)^{1/2}$ and define the angle θ by $\cos\theta = 20/r$, $\sin\theta = 40/r$.
  (a) Verify by use of the sum formula for the sine function that $I(t) = r\sin(311t + \theta)$.
  (b) Show that the *peak current* is r (the maximum value of I).
  (c) Find the *period* and *frequency*.
  (d) Determine the *phase shift* (in radians), that is, the value of t which makes $311t + \theta = 0$.

83. The instantaneous power input to an AC circuit is $p = vi$, where v is the instantaneous potential difference between the circuit terminals and i is the instantaneous current. If the circuit is a pure resistor, then $v = V\sin(\omega t)$ and $i = I\sin(\omega t)$.
    (a) Verify by means of trigonometric identities that

$$p = \frac{VI}{2} - \frac{VI}{2}\cos(2\omega t).$$

    (b) Draw a graph of p as a function of t.

84. Two points are located on one bank of the Colorado River, 500 meters apart. A point on the opposite bank makes angles of 88° and 80° with the line joining the two points. Find the lengths of two cables to be stretched across the river connecting the points.

5.2 Differentiation of the Trigonometric Functions

Differentiation rules for sine and cosine follow from arguments using limits and the addition formulas.

In this section, we will derive differentiation formulas for the trigonometric functions. In the course of doing so, we will use many of the basic properties of limits and derivatives introduced in the first two chapters.

The unit circle $x^2 + y^2 = 1$ can be described by the parametric equations $x = \cos\theta$, $y = \sin\theta$. As θ increases, the point $(x, y) = (\cos\theta, \sin\theta)$ moves along the circle in a counterclockwise direction (see Fig. 5.2.1).

The length of arc on the circle between the point $(1, 0)$ (corresponding to $\theta = 0$) and the point $(\cos\theta, \sin\theta)$ equals the angle θ subtended by the arc. If we think of θ as time, the point $(\cos\theta, \sin\theta)$ travels a distance θ in time θ, so it is moving with unit speed around the circle. At $\theta = 0$, the tangent line to the circle is vertical, so the velocity of the point is 1 in the vertical direction; thus, we expect that

$$\left.\frac{dx}{d\theta}\right|_{\theta=0} = 0 \quad \text{and} \quad \left.\frac{dy}{d\theta}\right|_{\theta=0} = 1.$$

That is,

$$\left.\frac{d\cos\theta}{d\theta}\right|_{\theta=0} = 0, \tag{1}$$

$$\left.\frac{d\sin\theta}{d\theta}\right|_{\theta=0} = 1. \tag{2}$$

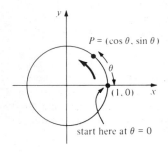

Figure 5.2.1. The point P moves at unit speed around the circle.

We will now derive (1) and (2) using limits.

According to the definition of the derivative, formulas (1) and (2) amount to the following statements about limits:

$$\lim_{\Delta\theta \to 0}\frac{\cos\Delta\theta - 1}{\Delta\theta} = 0 \tag{3}$$

and

$$\lim_{\Delta\theta \to 0}\frac{\sin\Delta\theta}{\Delta\theta} = 1. \tag{4}$$

To prove (3) and (4), we use the geometry in Fig. 5.2.2 and shall denote $\Delta\theta$ by the letter ϕ for simplicity of notation. When $0 < \phi < \pi/2$, we have

$$\text{area triangle } OCB = \tfrac{1}{2}|OC| \cdot |AB| = \tfrac{1}{2}\sin\phi,$$

$$\text{area triangle } OCB < \text{area sector } OCB = \tfrac{1}{2}\phi,$$

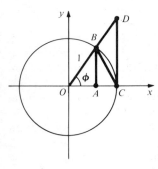

Figure 5.2.2. Geometry used to determine cos′0 and sin′0.

and

$$\text{area sector } OCB < \text{area triangle } OCD = \tfrac{1}{2}|OC| \cdot |CD|$$

$$= \tfrac{1}{2}\tan\phi = \tfrac{1}{2}\frac{\sin\phi}{\cos\phi}.$$

Thus,

$$\sin\phi < \phi \tag{5}$$

and

$$\phi < \frac{\sin\phi}{\cos\phi}. \tag{6}$$

For $-\pi/2 < \phi < 0$, $\sin\phi$ is negative, and from (5) we have $-\sin\phi = \sin(-\phi) < -\phi$, so $|\sin\phi| < |\phi|$ for all $\phi \neq 0$. Thus, if ϕ approaches zero, so must $\sin\phi$:

$$\lim_{\phi\to 0}\sin\phi = 0. \tag{7}$$

Now, $\lim_{\phi\to 0}\cos\phi = \lim_{\phi\to 0}\sqrt{1 - \sin^2\phi} = \sqrt{1 - (\lim_{\phi\to 0}\sin\phi)^2} = 1$. (By continuity of the square root function, we can take the limit under the root sign.) Thus we have

$$\lim_{\phi\to 0}\cos\phi = 1. \tag{8}$$

Next, we use (5) and (6) to get

$$\cos\phi < \frac{\sin\phi}{\phi} < 1 \tag{9}$$

for $0 < \phi < \pi/2$. However, the expressions in (9) are unchanged if ϕ is replaced by $-\phi$, so (9) holds for $-\pi/2 < \phi < 0$ as well. Now (9) and (8) imply

$$\lim_{\phi\to 0}\frac{\sin\phi}{\phi} = 1,$$

since $\sin\phi/\phi$ is squeezed between 1 and a function which approaches 1. Thus, the limit statement (4) is proved.

To prove (3), we again use the Pythagorean identity:

$$\sin^2\phi = 1 - \cos^2\phi = (1 - \cos\phi)(1 + \cos\phi)$$

which implies $(1 - \cos\phi)/\phi = \sin^2\phi/\phi(1 + \cos\phi)$; using the product and quotient rules for limits,

$$\lim_{\phi\to 0}\frac{1 - \cos\phi}{\phi} = \lim_{\phi\to 0}\frac{\sin\phi}{\phi}\frac{\lim_{\phi\to 0}\sin\phi}{1 + \lim_{\phi\to 0}\cos\phi} = 1 \cdot \frac{0}{1 + 1} = 0.$$

▦ Calculator Discussion

We can confirm (3) and (4) by some numerical experiments. For instance, on our HP-15C calculator, we compute

$$\frac{1 - \cos(\Delta\theta)}{\Delta\theta} \quad \text{and} \quad \frac{\sin(\Delta\theta)}{\Delta\theta}$$

for $\Delta\theta = 0.02$ and 0.001 to be

$$\frac{1 - \cos\Delta\theta}{\Delta\theta} = 0.009999665 \qquad \text{for} \quad \Delta\theta = 0.02,$$

$$\frac{1 - \cos\Delta\theta}{\Delta\theta} = 0.000500000 \qquad \text{for} \quad \Delta\theta = 0.001$$

and

$$\frac{\sin \Delta\theta}{\Delta\theta} = 0.999933335 \qquad \text{for} \quad \Delta\theta = 0.02,$$

$$\frac{\sin \Delta\theta}{\Delta\theta} = 0.999999833 \qquad \text{for} \quad \Delta\theta = 0.001.$$

Your answer may differ because of calculator inaccuracies. However, these numbers confirm that $(1 - \cos\Delta\theta)/\Delta\theta$ is near zero for $\Delta\theta$ small and that $(\sin\Delta\theta)/\Delta\theta$ is near 1 for $\Delta\theta$ small. ▲

Now we are ready to compute the derivatives of $\sin\theta$ and $\cos\theta$ at all values of θ. According to the definition of the derivative,

$$\frac{d}{d\theta} \sin\theta = \lim_{\Delta\theta\to0} \left[\frac{\sin(\theta + \Delta\theta) - \sin\theta}{\Delta\theta} \right].$$

From the addition formula for sin, the right-hand side equals

$$\lim_{\Delta\theta\to0} \left[\frac{\sin\theta\cos\Delta\theta + \cos\theta\sin\Delta\theta - \sin\theta}{\Delta\theta} \right]$$

$$= \lim_{\Delta\theta\to0} \left[\frac{\sin\theta(\cos(\Delta\theta) - 1)}{\Delta\theta} + \frac{\cos\theta\sin(\Delta\theta)}{\Delta\theta} \right]$$

$$= \sin\theta \lim_{\Delta\theta\to0} \left[\frac{\cos(\Delta\theta) - 1}{\Delta\theta} \right] + \cos\theta \lim_{\Delta\theta\to0} \left[\frac{\sin(\Delta\theta)}{\Delta\theta} \right]$$

by the sum rule and constant multiple rule for limits. Substituting from (3) and (4) gives $(\sin\theta) \cdot 0 + (\cos\theta) \cdot 1 = \cos\theta$. Thus

$$\frac{d}{d\theta} \sin\theta = \cos\theta.$$

We compute the derivative of $\cos\theta$ in a similar way:

$$\frac{d}{d\theta} \cos\theta = \lim_{\Delta\theta\to0} \left[\frac{\cos(\theta + \Delta\theta) - \cos\theta}{\Delta\theta} \right]$$

$$= \lim_{\Delta\theta\to0} \left[\frac{\cos\theta\cos(\Delta\theta) - \sin\theta\sin(\Delta\theta) - \cos\theta}{\Delta\theta} \right]$$

$$= \lim_{\Delta\theta\to0} \left[\cos\theta \left(\frac{\cos(\Delta\theta) - 1}{\Delta\theta} \right) - \frac{\sin\theta\sin(\Delta\theta)}{\Delta\theta} \right]$$

$$= -\sin\theta.$$

Derivative of Sine and Cosine

$$\frac{d}{d\theta} \sin\theta = \cos\theta \quad \text{and} \quad \frac{d}{d\theta} \cos\theta = -\sin\theta$$

In words, the derivative of the sine function is the cosine and the derivative of the cosine is *minus* the sine. These formulas are worth memorizing. Study Fig. 5.2.3 to check that they are consistent with the graphs of sine and cosine. For example, notice that on the interval $(0, \pi/2)$, $\sin\theta$ is increasing and its derivative $\cos\theta$ is positive.

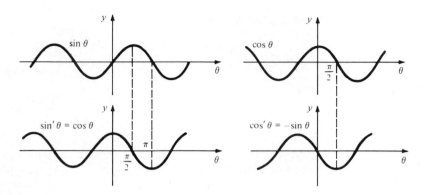

Figure 5.2.3. Graphs of sin, cos, and their derivatives.

Example 1   Differentiate

    (a) $(\sin\theta)(\cos\theta)$,   (b) $\sin^2\theta$

    (c) $\sin 5\theta$,   (d) $\dfrac{\sin 3\theta}{\cos\theta + \theta^4}$.

Solution   (a) By the product rule,

$$\frac{d}{d\theta}(\sin\theta)(\cos\theta) = \left(\frac{d}{d\theta}\sin\theta\right)\cos\theta + \sin\theta\left(\frac{d}{d\theta}\cos\theta\right)$$

$$= \cos\theta\cos\theta - \sin\theta\sin\theta = \cos^2\theta - \sin^2\theta.$$

(b) By the power of a function rule,

$$\frac{d}{d\theta}\sin^2\theta = \frac{d}{d\theta}(\sin\theta)^2 = 2\sin\theta\frac{d}{d\theta}\sin\theta = 2\sin\theta\cos\theta.$$

(c) By the chain rule,

$$\frac{d}{d\theta}\sin 5\theta = \frac{d}{du}\sin u\frac{du}{d\theta}\qquad(\text{where } u = 5\theta)$$

$$= (\cos u)(5) = 5\cos 5\theta.$$

(d) By the quotient rule and chain rule,

$$\frac{d}{d\theta}\frac{\sin 3\theta}{\cos\theta + \theta^4} = \frac{(\cos\theta + \theta^4)(d/d\theta)\sin 3\theta - \sin 3\theta(d/d\theta)(\cos\theta + \theta^4)}{(\cos\theta + \theta^4)^2}$$

$$= \frac{(\cos\theta + \theta^4)3\cos 3\theta - \sin 3\theta(-\sin\theta + 4\theta^3)}{(\cos\theta + \theta^4)^2}$$

$$= \frac{3\cos\theta\cos 3\theta + 3\theta^4\cos 3\theta + \sin\theta\sin 3\theta - 4\theta^3\sin 3\theta}{(\cos\theta + \theta^4)^2}.\ \blacktriangle$$

Example 2   Differentiate (a) $\cos\theta\sin^2\theta$ and (b) $(\sin 3x)/(1 + \cos^2 x)$.

Solution   (a) By the product rule and the power rule,

$$\frac{d}{d\theta}(\cos\theta\sin^2\theta) = \left(\frac{d}{d\theta}\cos\theta\right)\sin^2\theta + \cos\theta\left(\frac{d}{d\theta}\sin^2\theta\right)$$

$$= (-\sin\theta)\sin^2\theta + \cos\theta\cdot 2\sin\theta\cos\theta$$

$$= 2\cos^2\theta\sin\theta - \sin^3\theta.$$

(b) Here the independent variable is called x rather than θ. By the chain rule,

$$\frac{d}{dx}\sin 3x = 3\cos 3x,$$

so, by the quotient rule,

$$\frac{d}{dx}\frac{\sin 3x}{1+\cos^2 x} = \frac{(1+\cos^2 x)\cdot 3\cos 3x - \sin 3x \cdot 2\cos x(-\sin x)}{(1+\cos^2 x)^2}$$

$$= \frac{3\cos 3x(1+\cos^2 x) + 2\cos x\sin x\cdot\sin 3x}{(1+\cos^2 x)^2}\cdot\blacktriangle$$

Now that we know how to differentiate the sine and cosine functions, we can differentiate the remaining trigonometric functions by using the rules of calculus. For example, consider $\tan\theta = \sin\theta/\cos\theta$. The quotient rule gives:

$$\frac{d}{d\theta}\tan\theta = \frac{\cos\theta(d/d\theta)\sin\theta - \sin\theta(d/d\theta)\cos\theta}{\cos^2\theta}$$

$$= \frac{\cos\theta\cdot\cos\theta + \sin\theta\cdot\sin\theta}{\cos^2\theta} = \frac{1}{\cos^2\theta} = \sec^2\theta.$$

In a similar way, we see that

$$\frac{d}{d\theta}\cot\theta = -\csc^2\theta.$$

Writing $\csc\theta = 1/\sin\theta$, we get $\csc'\theta = (-\sin'\theta)/(\sin^2\theta) = (-\cos\theta)/(\sin^2\theta) = -\cot\theta\csc\theta$ and, similarly, $\sec'\theta = \tan\theta\sec\theta$.

The results we have obtained are summarized in the following box.

Differentiation of Trigonometric Functions

Function	Derivative	Leibniz notation
$\sin\theta$	$\cos\theta$	$\dfrac{d(\sin\theta)}{d\theta} = \cos\theta$
$\cos\theta$	$-\sin\theta$	$\dfrac{d(\cos\theta)}{d\theta} = -\sin\theta$
$\tan\theta$	$\sec^2\theta$	$\dfrac{d(\tan\theta)}{d\theta} = \sec^2\theta$
$\cot\theta$	$-\csc^2\theta$	$\dfrac{d(\cot\theta)}{d\theta} = -\csc^2\theta$
$\sec\theta$	$\tan\theta\sec\theta$	$\dfrac{d(\sec\theta)}{d\theta} = \tan\theta\sec\theta$
$\csc\theta$	$-\cot\theta\csc\theta$	$\dfrac{d(\csc\theta)}{d\theta} = -\cot\theta\csc\theta$

Example 3   Differentiate $\csc x\tan 2x$.

Solution   Using the product rule and chain rule,

$$\frac{d}{dx}\csc x\tan 2x = \left(\frac{d}{dx}\csc x\right)(\tan 2x) + \csc x\left(\frac{d}{dx}\tan 2x\right)$$

$$= -\cot x\cdot\csc x\cdot\tan 2x + \csc x\cdot 2\cdot\sec^2 2x$$

$$= 2\csc x\sec^2 2x - \cot x\csc x\tan 2x.\ \blacktriangle$$

Example 4   Differentiate: (a) $(\tan 3x)/(1 + \sin^2 x)$, (b) $1 - \csc^2 5x$, (c) $\sin(\sqrt{3\theta^2 + 1}\,)$.

Solution   (a) By the quotient and chain rules,

$$\frac{d}{dx}\,\frac{\tan 3x}{1 + \sin^2 x} = \frac{(3\sec^2 3x)(1 + \sin^2 x) - (\tan 3x)(2\sin x\cos x)}{(1 + \sin^2 x)^2}.$$

(b) $\dfrac{d}{dx}(1 - \csc^2 5x) = -2\csc 5x\,\dfrac{d}{dx}(\csc 5x) = 10\csc^2 5x\cot 5x.$

(c) By the chain rule,

$$\frac{d}{d\theta}\sin(\sqrt{3\theta^2 + 1}\,) = \cos(\sqrt{3\theta^2 + 1}\,)\,\frac{d}{d\theta}\sqrt{3\theta^2 + 1}$$

$$= \cos(\sqrt{3\theta^2 + 1}\,)\cdot\frac{1}{2}\,\frac{3\cdot 2\theta}{\sqrt{3\theta^2 + 1}}$$

$$= \frac{3\theta}{\sqrt{3\theta^2 + 1}}\cos\sqrt{3\theta^2 + 1}\,.\ \blacktriangle$$

Example 5   Differentiate: (a) $\tan(\cos\sqrt{x}\,)$, (b) $\csc\sqrt{x}$.

Solution   (a) By the chain rule,

$$\frac{d}{dx}\tan(\cos\sqrt{x}\,) = \frac{d}{du}\tan u\,\frac{du}{dx} \qquad (u = \cos\sqrt{x}\,)$$

$$= \sec^2 u\cdot\frac{d}{dx}\cos\sqrt{x}$$

$$= \sec^2(\cos\sqrt{x}\,)(-\sin\sqrt{x}\,)\frac{d}{dx}\sqrt{x}$$

$$= \frac{-\sec^2(\cos\sqrt{x}\,)\sin\sqrt{x}}{2\sqrt{x}}.$$

(b) By the chain rule,

$$\frac{d}{dx}\csc\sqrt{x} = -\frac{\csc\sqrt{x}\cot\sqrt{x}}{2\sqrt{x}}.\ \blacktriangle$$

By reversing the formulas for derivatives of trigonometric functions and multiplying through by -1 where necessary, we obtain the following indefinite integrals (antiderivatives).

Antiderivatives of Some Trigonometric Functions

$$\int\cos\theta\,d\theta = \sin\theta + C \qquad\qquad \int\sin\theta\,d\theta = -\cos\theta + C$$

$$\int\sec^2\theta\,d\theta = \tan\theta + C \qquad\qquad \int\csc^2\theta\,d\theta = -\cot\theta + C$$

$$\int\tan\theta\sec\theta\,d\theta = \sec\theta + C \qquad\qquad \int\cot\theta\csc\theta\,d\theta = -\csc\theta + C$$

For instance, to check that

$$\int \sec^2\theta \, d\theta = \tan\theta + C,$$

we simply recall from the preceding display that $(d/d\theta)(\tan\theta) = \sec^2\theta$.

Example 6 Find $\int \sec\theta(\sec\theta + 3\tan\theta) \, d\theta$.

Solution Multiplying out, we have

$$\int (\sec^2\theta + 3\sec\theta\tan\theta) \, d\theta = \int \sec^2\theta \, d\theta + 3\int \sec\theta\tan\theta \, d\theta$$

$$= \tan\theta + 3\sec\theta + C. \; \blacktriangle$$

Example 7 Find $\int \sin 4u \, du$.

Solution If we guess $-\cos 4u$ as the antiderivative, we find $(d/du)(-\cos 4u) = 4\sin 4u$, which is four times too big, so

$$\int \sin 4u \, du = -\frac{1}{4}\cos 4u + C. \; \blacktriangle$$

Example 8 Find the following antiderivatives: (a) $\int 2\cos 4s \, ds$, (b) $\int (1 + \sec^2\theta) \, d\theta$, (c) $\int \tan 2x \sec 2x \, dx$, (d) $\int (\sin x + \sqrt{x}) \, dx$.

Solution (a) Since $(d/ds)\sin 4s = 4\cos 4s$, $(d/ds)\frac{1}{2}\sin 4s = \frac{1}{2}4\cos 4s = 2\cos 4s$. Thus $\int 2\cos 4s \, ds = \frac{1}{2}\sin 4s + C$.
(b) By the sum rule for integrals, $\int (1 + \sec^2\theta) \, d\theta = \theta + \tan\theta + C$.
(c) Since $(d/dx)\sec(2x) = (\sec 2x \tan 2x) \cdot 2$, we have

$$\int \tan 2x \sec 2x \, dx = \frac{1}{2}\sec(2x) + C.$$

(d) $\int (\sin x + \sqrt{x}) \, dx = -\cos x + (x^{3/2}/\frac{3}{2}) + C = -\cos x + \frac{2}{3}x^{3/2} + C. \; \blacktriangle$

We can use these methods for indefinite integrals to calculate some definite integrals as well.

Example 9 Calculate: (a) $\int_0^1 (2\sin x + x^3) \, dx$, (b) $\int_{\pi/4}^{\pi/2} \cos 2x \, dx$.

Solution (a) By the sum and power rules,

$$\int (2\sin x + x^3) \, dx = 2\int \sin x \, dx + \int x^3 \, dx = -2\cos x + \frac{x^4}{4} + C.$$

Thus, by the fundamental theorem of calculus,

$$\int_0^1 (2\sin x + x^3) \, dx = \left(-2\cos x + \frac{x^4}{4}\right)\Big|_0^1 = -2(\cos 1 - \cos 0) + \frac{1}{4}$$

$$= \frac{9}{4} - 2\cos 1.$$

(b) As an Example 7, we find that an antiderivative for $\cos 2x$ is $(1/2)\sin 2x$. Therefore, by the fundamental theorem,

$$\int_{\pi/4}^{\pi/2} \cos 2x \, dx = \frac{1}{2}\sin 2x\Big|_{\pi/4}^{\pi/2} = \frac{1}{2}(\sin\pi - \sin\pi/2) = -\frac{1}{2}. \; \blacktriangle$$

Our list of antiderivatives still leaves much to be desired. Where, for instance, is $\int \tan \theta \, d\theta$? The absence of this and other entries in the previous display is related to the missing antiderivative for $1/x$ (see Exercise 65). The gap will be filled in the next chapter.

Exercises for Section 5.2

Differentiate the functions of θ in Exercises 1–12.

1. $\cos \theta + \sin \theta$
2. $8 \cos \theta - 10 \sin \theta$
3. $5 \cos 3\theta + 10 \sin 2\theta$
4. $8 \sin 10\theta - 10 \cos 8\theta$
5. $(\cos \theta)(\sin \theta + \theta)$
6. $\cos^2\theta - 3 \sin^3\theta$
7. $\cos^3 3\theta$
8. $\sin^4 5\theta$
9. $\dfrac{\cos \theta}{\cos \theta - 1}$
10. $\dfrac{\sin \theta}{\cos \theta - 1}$
11. $\dfrac{\cos \theta + \sin \theta}{\sin \theta + 1}$
12. $\dfrac{\theta + \cos \theta}{\cos \theta \sin \theta}$

Differentiate the functions of x in Exercises 13–24.

13. $(\cos x)^3$
14. $\sin(20x^2)$
15. $(\sqrt{x} + \cos x)^4$
16. $\dfrac{\cos\sqrt{x}}{1 + \sqrt{x}}$
17. $\sin(x + \sqrt{x})$
18. $\csc(x^2 + \sqrt{x})$
19. $\dfrac{x}{\cos x + \sin(x^2)}$
20. $\dfrac{\cos x}{\tan x + \sqrt{x}}$
21. $\tan x + 2 \cos x$
22. $\sec x + 8 \csc x$
23. $\sec 3x$
24. $\tan 10x$

Differentiate each of the functions in Exercises 25–36.

25. $f(x) = \sqrt{x} + \cos 3x$
26. $f(x) = [(\sin 2x)^2 + x^2]$
27. $f(x) = \sqrt{\cos x}$
28. $f(x) = \sin^2 x$
29. $f(t) = (4t^3 + 1)\sin\sqrt{t}$
30. $f(t) = \csc t \cdot \sec^2 3t$
31. $f(x) = \sin(\sqrt{1 - x^3}) + \tan\left(\dfrac{x}{x^4 + 1}\right)$.
32. $f(\theta) = \tan\left(\theta + \dfrac{1}{\theta}\right)$.
33. $f(\theta) = \left[\csc\left(\dfrac{\theta}{\sqrt{\theta^2 + 1}}\right) + 1\right]^{3/2}$.
34. $f(r) = \dfrac{r^2 + \sqrt{1 - r^2}}{r \sin r}$.
35. $f(v) = (\tan\sqrt{v^2 + 1})\left(\sec\left[\dfrac{1}{(v^2 + 1)}\right]\right)$.
36. $f(s) = \dfrac{\sin(s\sqrt[3]{1 + s^2})}{\cos s}$.

Find the antiderivatives in Exercises 37–44.

37. $\int (x^3 + \sin x) \, dx$
38. $\int (\cos 3x + 5x^{3/2}) \, dx$
39. $\int (x^4 + \sec 2x \tan 2x) \, dx$
40. $\int (\sin 2x + \sqrt{x}) \, dx$
41. $\int \sin\left(\dfrac{u}{2}\right) du$
42. $\int \cos\left(\dfrac{3p}{5}\right) dv$

43. $\int \cos(\theta^2)\theta \, d\theta$. (*Hint*: Compute the derivative of $\sin(\theta^2)$.)
44. $\int \sin(\phi^3)\phi^2 \, d\phi$.
45. Find $\int \cos \theta \sin \theta \, d\theta$ by using a trigonometric identity.
46. Find $\int \cos^2\theta \, d\theta$ by using (a) a half-angle formula; (b) a double-angle formula.
47. Find $\int \sin^2\theta \, d\theta + \int \cos^2\theta \, d\theta$.
48. Find $\int \sin^2\theta \, d\theta$ by using a trigonometric identity.

Evaluate the definite integrals in Exercises 49–56

49. $\int_0^{\pi/2} \sin\left(\dfrac{\theta}{4}\right) d\theta$
50. $\int_0^{2\pi} \sin\left(\dfrac{2\theta}{3}\right) d\theta$
51. $\int_{-7}^{7} (\sin t + \sin^3 t) \, dt$. (*Hint*: No calculation is necessary.)
52. $\int_{-5}^{5} (2\cos t + \sin^9 t) \, dt$
53. $\int_0^1 \cos(3\pi t) \, dt$
54. $\int_{1/4}^{1/2} \sin\left(\dfrac{\pi}{3} s\right) ds$
55. $\int_0^{\pi} \cos^2\theta \, d\theta$ (see Exercise 46).
56. $\int_0^{10\pi} \sin^2\theta \, d\theta$ (see Exercise 48).
57. Prove the following inequalities by using trigonometric identities and the inequalities established at the beginning of this section:
    (a) $\phi < \dfrac{\sin 2\phi}{1 + \cos 2\phi}$ for $0 < \phi < \pi/2$;
    (b) $\sec\phi > \phi$ for $0 < \phi < \dfrac{\pi}{2}$.
58. Find $\lim_{\phi \to 0} \dfrac{\tan 2\phi}{3\phi}$.
59. Find $\lim_{\Delta\theta \to 0} \dfrac{\sin a\Delta\theta}{\Delta\theta}$ where a is any constant.
60. Find $\lim_{\phi \to 0} \dfrac{\sin a\phi}{\sin b\phi}$, where a and b are constants.
61. Show that $f(x) = \sin x$ and $f(x) = \cos x$ satisfy $f''(x) + f(x) = 0$.
62. Find a function $f(x)$ which satisfies $f''(x) + 4f(x) = 0$.
63. Show that $f(x) = \tan x$ satisfies the equation $f'(x) = 1 + [f(x)]^2$ and $f(x) = \cot x$ satisfies $f'(x) = -(1 + [f(x)]^2)$.
64. Show that $f(x) = \sec x$ satisfies $f'' + f - 2f^3 = 0$.

65. Suppose that $f'(x) = 1/x$. (We do not yet have such a function at our disposal, but we will see in Chapter 6 that there is one.) Show that $\int \tan \theta \, d\theta = -f(\cos \theta) + C$.

66. Show that $\int \tan \theta \, d\theta = f(\sec \theta) + C$, where f is a function such that $f'(x) = 1/x$. (The apparent conflict with Exercise 65 will be resolved in the next chapter.)

67. Show that $(d/d\theta)\cos \theta = -\sin \theta$ can be derived from $(d/d\theta)\sin \theta = \cos \theta$ by using the identity $\cos \theta = \sin(\pi/2 - \theta)$ and the chain rule.

68. (a) Evaluate $\lim_{\Delta\theta \to 0}(\tan \Delta\theta)/\Delta\theta$ using the methods of the beginning of this section.
    (b) Use part (a) and the addition formula for the tangent (see the inside front cover) to prove that $\tan'\theta = \sec^2\theta$, $-\pi/2 < \theta < \pi/2$, that is, prove that
$$\lim_{\Delta\theta \to 0} \frac{\tan(\theta + \Delta\theta) - \tan \theta}{\Delta\theta} = \sec^2\theta.$$

★69. Suppose that $\phi(x)$ is a function "appearing from the blue" with the property that
$$\frac{d\phi}{dx} = \frac{1}{\cos x}.$$
Calculate: (a) $\dfrac{d}{dx}(\phi(3x)\cos x)$, (b) $\displaystyle\int_0^1 \frac{1}{\cos x}\, dx$,

(c) $\dfrac{d^2}{dx^2}(\phi(2x)\sin 2x)$.

★70. Let ψ be a function such that
$$\frac{d\psi}{dx} = \phi,$$
where ϕ is described in Exercise 69. Prove that
$$\frac{d}{dx}\left(\psi(x)\sin x + \frac{d}{dx}\left(\psi(x)\cos x - \frac{x^2}{2}\right)\right)$$
$$= -\phi(x)\sin x$$

★71. Give a geometric "proof" that $\sin' = \cos$ and $\cos' = -\sin$ by these steps:
    (a) Consider the parametric curve $(\cos \theta, \sin \theta)$ as in the beginning of this section. Argue that $(\cos'\theta)^2 + (\sin'\theta)^2 = 1$, since the point moves at unit speed around the circle (see Exercise 34, Section 2.4).
    (b) Use the relation $\cos^2\theta + \sin^2\theta = 1$ to show that $\sin \theta \sin'\theta + \cos \theta \cos'\theta = 0$.
    (c) Conclude from (a) and (b) that $(\sin'\theta)^2 = \cos^2\theta$ and $(\cos'\theta)^2 = \sin^2\theta$.
    (d) Give a geometric argument to get the correct signs in the square roots of the relations in (c).

5.3 Inverse Functions

The derivative of an inverse function is the reciprocal of its derivative.

Sometimes two variable quantities are related in such a way that either one may be considered as a function of the other. The relationship between the quantities may be expressed by either of two functions, which are called *inverses* of one another. In this section, we will learn when a given function has an inverse, and we will see a useful relationship between the derivative of a function and the derivative of its inverse.

A simple example of a function with an inverse is the linear function $y = f(x) = mx + b$ with $m \neq 0$. We can solve for x in terms of y to get $x = (1/m)y - b/m$. Considering the expression $(1/m)y - b/m$ as a function $g(y)$ of y, we find that $y = f(x)$ whenever $x = g(y)$, and vice versa.

The graphs of f and g are very simply related. We interchange the role of the x and y axes by flipping the graph over the diagonal line $y = x$ (which bisects the right angle between the axes) or by viewing the graph through the back of the page held so that the x axis is vertical. (See Fig. 5.3.1 for a specific example.) Notice that the (constant) slopes of the two graphs are reciprocals of one another.

Whenever two functions, f and g, have the property that $y = f(x)$ whenever $x = g(y)$, and vice versa, we say that f and g are *inverses* of one another; the graphs of f and g are then related by flipping the x and y axes, as in Fig. 5.3.1. If we are given a formula for $y = f(x)$, we can try to find the inverse function by solving for x in terms of y.

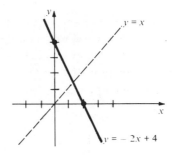

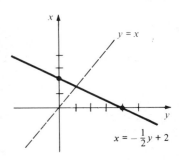

Figure 5.3.1. The functions $f(x) = -2x + 4$ and $g(y) = -\frac{1}{2}y + 2$ are inverses to one another.

Example 1 Find an inverse function for $f(x) = x^3$. Graph f and its inverse.

Solution Solving the relation $y = x^3$ for x in terms of y, we find $x = \sqrt[3]{y}$, so the cuberoot function $g(y) = \sqrt[3]{y}$ is the inverse function to the cubing function $f(x) = x^3$. The graphs of f and g are shown in parts (a) and (b) of Fig. 5.3.2. In part (c), we have illustrated the important fact that the names of variables

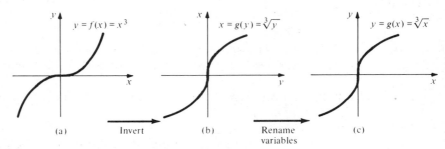

Figure 5.3.2. The function $y = x^3$ and its inverse $x = \sqrt[3]{y}$.

used with a function are arbitrary; since we like to have y as a function of x, we may write $g(x) = \sqrt[3]{x}$, and so the graph $y = \sqrt[3]{x}$ is another acceptable picture of the cuberoot function. ▲

Not every function has an inverse. For instance, if $y = f(x) = x^2$, we may solve for x in terms of y to get $x = \pm\sqrt{y}$, but this does not give x as a function defined for all y, for two reasons:

(1) if $y < 0$, the square root $\sqrt{y}$ is not defined;
(2) if $y > 0$, the two choices of positive or negative sign give two different values for x.

We can also see the difficulty geometrically. If we interchange the axes by flipping the parabola $y = x^2$ (Fig. 5.3.3), the resulting "horizontal parabola" is

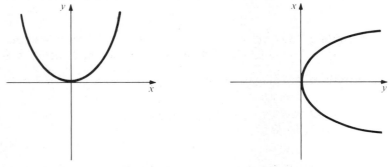

Figure 5.3.3. Flipping the parabola $y = f(x) = x^2$ produces a curve which is not the graph of a function.

not the graph of a function defined on the real numbers, since it intersects some vertical lines twice and others not at all.

This situation is similar to one we encountered in Section 2.3, where a curve like $x^2 + y^2 = 1$ defined y as an implicit function of x only if we looked at *part* of the curve. In the present case, we can obtain an invertible function if we restrict $f(x) = x^2$ to the domain $[0, \infty)$. (The choice $(-\infty, 0]$ would do as well.) Then the inverse function $g(y) = \sqrt{y}$ is well defined with domain $[0, \infty)$ (see Fig. 5.3.4). The domain $(-\infty, 0]$ for f would have led to the other square root sign $-\sqrt{y}$ for g.

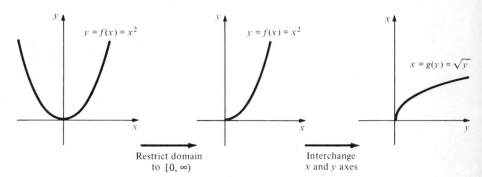

Figure 5.3.4. The function $f(x) = x^2$ and its inverse $g(y) = \sqrt{y}$.

The inverse to a function f, when it exists, is sometimes denoted by f^{-1} and read "f inverse". A function with an inverse is said to be *invertible*.

Warning   Notice from this example that the inverse f^{-1} is not in general the same as $1/f$.

We summarize our work to this point in the following box.

Inverse Functions

The inverse function to a function f is a function g for which $g(y) = x$ when $y = f(x)$, and vice versa.

The inverse function to f is denoted by f^{-1}.

To find a formula for f^{-1}, try to solve the equation $y = f(x)$ for x in terms of y. If the solution is unique, set $f^{-1}(y) = x$.

The graph of f^{-1} is obtained from that of f by flipping the figure to interchange the horizontal and vertical axes.

It may be necessary to restrict the domain of f before there is an inverse function.

Example 2   Let $f(x) = x^2 + 2x + 3$. Restrict f to a suitable interval so that it has an inverse. Find the inverse function and sketch its graph.

Solution   We may solve the equation $y = x^2 + 2x + 3$ for x in terms of y by the quadratic formula:

$$x^2 + 2x + 3 - y = 0$$

gives $x = [-2 \pm \sqrt{4 - 4(3 - y)}\,]/2 = -1 \pm \sqrt{y - 2}$. If we choose $y \geqslant 2$ and the $+$ sign, we get $x = -1 + \sqrt{y - 2}$ for the inverse function. The restriction $y \geqslant 2$ corresponds to $x \geqslant -1$ (see Fig. 5.3.5). (The answer $x = -1 - \sqrt{y - 2}$ for $y \geqslant 2$ and $x \leqslant -1$ is also acceptable—this is represented by the dashed portion of the graph. ▲

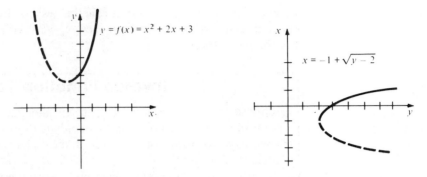

Figure 5.3.5. Restricting the domain of $f(x)$ to $[-1, \infty)$ gives a function with an inverse defined on $[2, \infty)$.

Example 3   Sketch the graph of the inverse function for each function in Fig. 5.3.6.

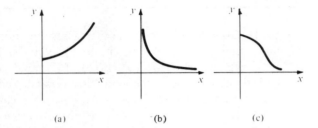

Figure 5.3.6. Sketch the graph of the inverse.

(a)   (b)   (c)

Solution   The graphs which we obtain by viewing the graphs in Fig. 5.3.6 from the reverse side of the page, are shown in Fig. 5.3.7. ▲

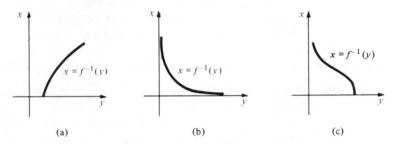

Figure 5.3.7. Graphs of the inverse functions (compare Fig. 5.3.6).

(a)   (b)   (c)

There is a simple geometric test for invertibility; a function is invertible if each horizontal line meets the graph in at most one point.

Example 4   Determine whether or not each function in Fig. 5.3.8 is invertible on its domain.

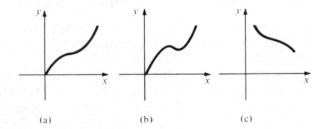

Figure 5.3.8. Is there an inverse?

(a)   (b)   (c)

Solution   Applying the test just mentioned, we find that the functions (a) and (c) are invertible while (b) is not. ▲

A function may be invertible even though we cannot find an explicit formula for the inverse function. This fact gives us a way of obtaining "new functions." The following is a useful calculus test for finding intervals on which a function is invertible. In the next section, we shall use it to obtain inverses for the trigonometric functions.

Inverse Function Test

Suppose that f is continuous on $[a,b]$ and that f is increasing at each point of (a,b). (For instance, this holds if $f'(x) > 0$ for each x in (a,b).) Then f is invertible on $[a,b]$, and the inverse f^{-1} is defined on the interval $[f(a), f(b)]$.

     If f is decreasing rather than increasing at each point of (a,b), then f is still invertible; in this case, the domain of f^{-1} is $[f(b), f(a)]$.

To justify this test, we apply the increasing function test from Section 3.2 to conclude that f is increasing on $[a,b]$; that is, if $a \leqslant x_1 < x_2 \leqslant b$, then $f(x_1) < f(x_2)$. In particular, $f(a) < f(b)$. If y is any number in $(f(a), f(b))$, then by the intermediate value theorem (first version, Section 3.1), there is an x in (a,b) such that $f(x) = y$. If $y = f(a)$ or $f(b)$, we can choose $x = a$ or $x = b$. Since f is increasing on $[a,b]$, for any y in $[f(a), f(b)]$ there can only be one x such that $y = f(x)$. Thus, by definition, f is invertible on $[a,b]$ and the domain of f^{-1} is the range $[f(a), f(b)]$ of values of f on $[a,b]$. The proof of the second assertion in the inverse function test is similar.

     We can allow open or infinite intervals in the inverse function test. For instance, if f is continuous and increasing on $[a, \infty)$, and if $\lim_{x \to \infty} f(x) = \infty$, then f has an inverse defined on $[f(a), \infty)$.

Example 5   Verify that $f(x) = x^2 + x$ has an inverse if f is defined on $[-\frac{1}{2}, \infty)$.

Solution   Since f is differentiable on $(-\infty, \infty)$, it is continuous on $(-\infty, \infty)$ and hence on $[-\frac{1}{2}, \infty)$. But $f'(x) = 2x + 1 > 0$ for $x > -\frac{1}{2}$. Thus f is increasing. Also, $\lim_{x \to \infty} f(x) = \infty$. Hence the inverse function test guarantees that f has an inverse defined on $[-\frac{1}{2}, \infty)$. ▲

Example 6   Let $f(x) = x^5 + x$.

    (a) Show that f has an inverse on $[-2, 2]$. What is the domain of this inverse?
    (b) Show that f has an inverse on $(-\infty, \infty)$.
    (c) What is $f^{-1}(2)$?
    ▦(d) Numerically calculate $f^{-1}(3)$ to two decimal places of accuracy.

Solution   (a) $f'(x) = 5x^4 + 1 > 0$, so by the inverse function test, f is invertible on $[-2, 2]$. The domain of the inverse is $[f(-2), f(2)]$, which is $[-34, 34]$.
    (b) Since $f'(x) > 0$ for all x in $(-\infty, \infty)$, f is increasing on $(-\infty, \infty)$. Now f takes arbitrarily large positive and negative values as x varies over $(-\infty, \infty)$; it takes all values in between by the intermediate value theorem, so the domain of f^{-1} is $(-\infty, \infty)$. There is no simple formula for $f^{-1}(y)$, the solution of $x^5 + x = y$, but we can calculate $f^{-1}(y)$ for any specific values of y to any desired degree of accuracy. (This is really no worse than the situation for $\sqrt{x}$. If the inverse function to $x^5 + x$ had as many applications as the square-root function, we would learn about it in high school, tables would be readily available for it, calculators would calculate it at the touch of a key, and there would be a standard notation like $\mathscr{S}y$ for the solution of $x^5 + x = y$, just as $\sqrt[5]{y}$ is the standard notation for the solution of $x^5 = y$.)

(c) Since $f(1) = 1^5 + 1 = 2$, $f^{-1}(2)$ must equal 1.

(d) To calculate $f^{-1}(3)$—that is, to seek an x such that $x^5 + x = 3$—we use the method of bisection described in Example 7, Section 3.1. Since $f(1) = 2 < 3$ and $f(2) = 34 > 3$, x must lie between 1 and 2. We can squeeze toward the correct answer by calculating:

$$f(1.5) = 9.09375 \quad \text{so} \quad 1 < x < 1.5,$$

$$f(1.25) = 4.30176 \quad \text{so} \quad 1 < x < 1.25,$$

$$f(1.1) = 2.71051 \quad \text{so} \quad 1.1 < x < 1.25,$$

$$f(1.15) = 3.16135 \quad \text{so} \quad 1.1 < x < 1.15,$$

$$f(1.14) = 3.06541 \quad \text{so} \quad 1.1 < x < 1.14,$$

$$f(1.13) = 2.97244 \quad \text{so} \quad 1.13 < x < 1.14,$$

$$f(1.135) = 3.01856 \quad \text{so} \quad 1.13 < x < 1.135.$$

Thus, to two decimal places, $x \approx 1.13$. (About 10 minutes of further experimentation gave $f(1.132997566) = 3.000000002$ and $f(1.132997565) = 2.999999991$. What does this tell you about $f^{-1}(3)$?) ▲

▦ Calculator Discussion

Recall (see Section R.6) that a function f may be thought of as an operation key on a calculator. The inverse function should be another key, which we can label f^{-1}. According to the definition, if we feed in any x, then push f to get $y = f(x)$, then push f^{-1}, we get back $x = f^{-1}(y)$, the number we started with. Likewise if we feed in a number y and first push f^{-1} and then f, we get y back again. By Fig. 5.3.4, $y = x^2$ and $x = \sqrt{y}$ (for $x \geq 0$, $y \geq 0$) are inverse functions. Try it out numerically, by pushing $x = 3.0248759$, then the x^2 key, then the $\sqrt{x}$ key. Try it also in the reverse order. (The answer may not come out exactly right because of roundoff errors.) ▲

As the preceding calculator discussion suggests, there is a close relation between inverse functions and composition of functions as discussed in connection with the chain rule in Section 2.2. If f and g are inverse functions, then $g(y)$ is that number x for which $f(x) = y$, so $f(g(y)) = y$; i.e., $f \circ g$ is the *identity function* which takes each y to itself: $(f \circ g)(y) = y$. Similarly, $g(f(x)) = x$ for all x, so $g \circ f$ is the identity function as well.

If we assume that f^{-1} and f are differentiable, we can apply the chain rule to the equation

$$f^{-1}(f(x)) = x$$

to obtain $(f^{-1})'(f(x)) \cdot f'(x) = 1$, which gives

$$(f^{-1})'(f(x)) = \frac{1}{f'(x)}.$$

Writing y for $f(x)$, so that $x = f^{-1}(y)$, we obtain the formula

$$(f^{-1})'(y) = \frac{1}{f'(f^{-1}(y))}.$$

Since the expression $(f^{-1})'$ is awkward, we sometimes revert to the notation $g(y)$ for the inverse function and write

$$g'(y) = \frac{1}{f'(x)}.$$

Notice that although dy/dx is not an ordinary fraction, the rule

$$\frac{dx}{dy} = \frac{1}{dy/dx}$$

is valid as long as $dy/dx \neq 0$. (Maybe the "reciprocal" notation f^{-1} is not so bad after all!)

Inverse Function Rule

To differentiate the inverse function $g = f^{-1}$ at y, take the reciprocal of the derivative of the given function at $x = f^{-1}(y)$:

$$g'(y) = \frac{1}{f'(g(y))} \qquad \text{if} \quad g = f^{-1},$$

$$\frac{dx}{dy} = \frac{1}{dy/dx}.$$

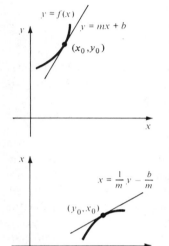

Figure 5.3.9. The inverse of the tangent line is the tangent line of the inverse.

Notice that our earlier observation that the slopes of inverse linear functions are reciprocals is just a special case of the inverse function rule. Figure 5.3.9 illustrates how the general rule is related to this special case.

Assuming that the inverse f^{-1} is continuous (see Exercise 41), we can prove that f^{-1} is differentiable whenever $f' \neq 0$ in the following way. Recall that $f'(x_0) = dy/dx = \lim_{\Delta x \to 0}(\Delta y/\Delta x)$, where Δx and Δy denote changes in x and y. On the other hand,

$$g'(y_0) = \frac{dx}{dy} = \lim_{\Delta y \to 0} \frac{\Delta x}{\Delta y} = \frac{1}{\lim_{\Delta y \to 0}(\Delta y/\Delta x)},$$

by the reciprocal rule for limits. But $\Delta x \to 0$ when $\Delta y \to 0$, since g is continuous, so

$$g'(y_0) = \frac{1}{\lim_{\Delta y \to 0}(\Delta y/\Delta x)} = \frac{1}{dy/dx}.$$

Example 7   Use the inverse function rule to compute the derivative of $\sqrt{x}$. Evaluate the derivative at $x = 2$.

Solution   Let us write $g(y) = \sqrt{y}$. This is the inverse function of $f(x) = x^2$. Since $f'(x) = 2x$,

$$g'(y) = \frac{1}{f'(g(y))} = \frac{1}{2g(y)} = \frac{1}{2\sqrt{y}},$$

so $(d/dy)(\sqrt{y}) = 1/(2\sqrt{y})$. We may substitute any letter for y in this result, including x, so we get the formula

$$\frac{d}{dx}\sqrt{x} = \frac{1}{2\sqrt{x}}.$$

When $x = 2$, the derivative is $1/(2\sqrt{2})$. ▲

Example 7 reproduces the rule for differentiating $x^{1/2}$ that we learned in Section 2.3. In fact, one can similarly use inverse functions to obtain an alternative proof of the rule for differentiating fractional powers: $(d/dx)x^{p/q} = (p/q)x^{(p/q)-1}$.

Example 8 Find $(f^{-1})'(2)$, where f is the function in Example 6.

Solution We know that $f^{-1}(2) = 1$, so $(f^{-1})'(2) = 1/f'(1)$. But $f'(x) = 5x^4 + 1$, so $f'(1) = 6$, and hence $(f^{-1})'(2) = \frac{1}{6}$. ▲

The important point in Example 8 is that we differentiated f^{-1} without having an explicit formula for it. We will exploit this idea in the next section.

Supplement to Section 5.3
Inverse Functions and Yogurt

If a yogurt culture is added to a quart of boiled milk and set aside for 4 hours, then the sourness of the resulting yogurt depends upon the temperature at which the mixture was kept. By performing a series of experiments, we can plot the graph of a function $S = f(T)$, where T is temperature and S is the sourness measured by the amount of lactic acid in grams in the completed yogurt. (See Fig. 5.3.10.) If T is too low, the culture is dormant; if T is too high, the culture is killed.

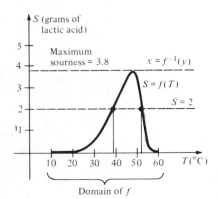

Figure 5.3.10. The sourness of yogurt as a function of fermentation temperature.

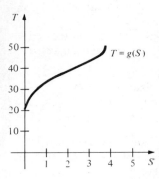

Figure 5.3.11. The graph of T as a function of S.

In making yogurt to suit one's taste, one might desire a certain degree of sourness and wish to know what temperature to use. (Remember that we are fixing all other variables, including the time of fermentation.) To find the temperature which gives $S = 2$, for instance, one may draw the horizontal line $S = 2$, see if it intersects the graph of f, and read off the value of T (Fig. 5.3.10).

From the graph, we see that there are two possible values of T: 38°C and 52°C. Similarly, there are two possible temperatures to achieve any value of S strictly between zero and the maximum value 3.8 of f. If, however, we restrict the allowable temperatures to the interval $[20, 47]$, then we will get a unique value of T for each S in $[0, 3.8]$. The new function $T = g(S)$, which assigns to each the sourness value the proper temperature for producing it, is the inverse function to f (Fig. 5.3.11).

Exercises for Section 5.3

Find the inverse for each of the functions in Exercises 1–6 on the given interval.

1. $f(x) = 2x + 5$ on $[-4, 4]$.
2. $f(x) = -\frac{1}{3}x + 2$ on $(-\infty, \infty)$.
3. $f(x) = x^5$ on $(-\infty, \infty)$.
4. $f(x) = x^8$ on $(0, 1]$.
5. $h(t) = \dfrac{t - 10}{t + 3}$ on $[-1, 1]$.
6. $a(s) = \dfrac{2s + 5}{-s + 1}$ on $[-\frac{1}{2}, \frac{1}{2}]$.
7. Find the inverse function for the function $f(x) = (ax + b)/(cx + d)$ on its domain. What must you assume about a, b, c, and d?

8. Find an inverse function g for $f(x) = x^2 + 2x + 1$ on some interval containing zero. What is $g(9)$? What is $g(x)$?

9. Sketch the graph of the inverse of each of the functions in Fig. 5.3.12.

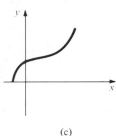

(a)

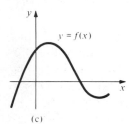

(b)

(c)

Figure 5.3.12. Sketch the graph of the inverse of each of these functions.

10. Determine whether each function in Fig. 5.3.13 has an inverse. Sketch the inverse if there is one.

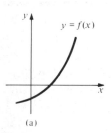

$y = f(x)$

(a)

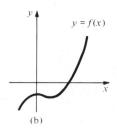

$y = f(x)$

(b)

$y = f(x)$

(c)

Figure 5.3.13. Which functions have inverses?

11. Draw a graph of $f(x) = (3x + 1)/(2x - 2)$ and a graph of its inverse function.

12. Draw a graph of $f(x) = (x - 1)/(x + 1)$ and a graph of its inverse function.

13. Find the largest possible intervals on which $f(x) = 1/(x^2 - 1)$ is invertible. Sketch the graphs of the inverse functions.

14. Sketch a graph of $f(x) = x/(1 + x^2)$ and find an interval on which f is invertible.

15. Let $f(x) = x^3 - 4x^2 + 1$.
   (a) Find an interval containing 1 on which f is invertible. Denote the inverse by g.
   (b) Compute $g(-7)$ and $g'(-7)$.
   (c) What is the domain of g?
   ▦(d) Compute $g(-5)$ and $g'(-5)$.

16. Let $f(x) = x^5 + x$. (a) Find $f^{-1}(246)$.
   ▦(b) Find $f^{-1}(4)$, correct to at least two decimal places.

17. Show that $f(x) = \frac{1}{3}x^3 - x$ is not invertible on any open interval containing 1.

18. Find intervals on which $f(x) = x^5 - x$ is invertible.

19. Show that if n is odd, $f(x) = x^n$ is invertible on $(-\infty, \infty)$. What is the domain of the inverse function?

20. Discuss the invertibility of $f(x) = x^n$ for n even.

21. Show that $f(x) = -x^3 - 2x + 1$ is invertible on $[-1, 2]$. What is the domain of the inverse?

22. (a) Show that $f(x) = x^3 - 2x + 1$ is invertible on $[2, 4]$. What is the domain of the inverse function?
   (b) Find the largest possible intervals on which f is invertible.

23. Verify the inverse function rule for $y = x^3$.

24. Verify the inverse function rule for the function $y = (ax + b)/(cx + d)$ by finding dy/dx and dx/dy directly. (See Exercise 7.)

25. If $f(x) = x^3 + 2x + 1$, show that f has an inverse on $[0, 2]$. Find the derivative of the inverse function at $y = 4$.

26. Find $g'(0)$, where g is the inverse function to $f(x) = x^9 + x^5 + x$.

27. Let $y = x^3 + 2$. Find dx/dy when $y = 3$.

28. If $f(x) = x^5 + x$, find the derivative of the inverse function when $y = 34$.

For each function f in Exercises 29–32, find the derivative of the inverse function g at the points indicated.

29. $f(x) = 3x + 5$; find $g'(2)$, $g'(\frac{3}{4})$.

30. $f(x) = x^5 + x^3 + 2x$; find $g'(0)$, $g'(4)$.

31. $f(x) = \frac{1}{12}x^3 - x$ on $[-1, 1]$; find $g'(0)$, $g'(\frac{11}{12})$.

32. $f(x) = \sqrt{x - 3}$ on $[3, \infty)$; find $g'(4)$, $g'(8)$.

▦33. Enter the number 2.6 on your calculator, then push the x^2 key followed by the $\sqrt{x}$ key. Is there any roundoff error? Try the $\sqrt{x}$ key, then the x^2 key. Also try a sequence such as $x^2, \sqrt{x}, x^2,$ $\sqrt{x}, x^2, \sqrt{x}$. Do you get the original number? Try these experiments with different starting numbers.

34. If we think of a French–English dictionary as defining a function from the set of French words to the set of English words (does it really?), how is the inverse function defined? Discuss.

35. Suppose that $f(x)$ is the number of pounds of beans you can buy for x dollars. Let $g(y)$ be the inverse function. What does $g(y)$ represent?

36. Let $f_1(x) = x$, $f_2(x) = 1/x$, $f_3(x) = 1 - x$, $f_4(x) = 1/(1 - x)$, $f_5(x) = (x - 1)/x$, and $f_6(x) = x/(x - 1)$.
    (a) Show that the composition of any two functions in this list is again in the list. Complete the "composition table" below. For example $(f_2 \circ f_3)(x) = f_2(1 - x) = 1/(1 - x) = f_4(x)$.

$\circ$	f_1	f_2	f_3	f_4	f_5	f_6
f_1			f_3			
f_2			f_4			
f_3		f_5				
f_4				f_5		
f_5						
f_6						

   (b) Show that the inverse of any function in the preceding list is again in the list. Which of the functions equal their own inverses?
37. (a) Find the domain of
$$f(x) = \sqrt{(2x + 5)/(3x + 7)}\,.$$
   (b) By solving for the x in the equation $y = \sqrt{(2x + 5)/(3x + 7)}$, find a formula for the inverse function $g(y)$.

   (c) Using the inverse function rule, find a formula for $f'(x)$.
   (d) Check your answer by using the chain rule.
38. Suppose that f is concave upward and increasing on $[a, b]$.
   (a) By drawing a graph, guess whether f^{-1} is concave upward or downward on the interval $[f(a), f(b)]$.
   (b) What if f is concave upward and decreasing on $[a, b]$?
39. Show that if the inverse function to f on S is g, with domain T, then the inverse function to g on T is f, with domain S. Thus, the inverse of the inverse function is the original function, that is, $(f^{-1})^{-1} = f$.
★40. Under what conditions on a, b, c, and d is the function $f(x) = (ax + b)/(cx + d)$ equal to its own inverse function?
★41. Suppose that $f'(x_0) > 0$ and $f(x_0) = y_0$. If g is the inverse of f with $g(y_0) = x_0$, show that g is continuous at y_0 by filling in the details of the following argument:
   (a) For Δx sufficiently small,
$$\tfrac{3}{2}f'(x_0) > \Delta y/\Delta x > \tfrac{1}{2}f'(x_0).$$
   (b) As $\Delta y \to 0$, $\Delta x \to 0$ as well.
   (c) Let $\Delta y = f(x_0 + \Delta x) - f(x_0)$. Then $\Delta x = g(y_0 + \Delta y) - g(y_0)$.

5.4 The Inverse Trigonometric Functions

The derivatives of the inverse trigonometric functions have algebraic formulas.

In the previous section, we discussed the general concept of inverse function and developed a formula for differentiating the inverse. Now we will apply this formula to study the inverses of the sine, cosine, and the other trigonometric functions.

We begin with the function $y = \sin x$, using the inverse function test to locate an interval on which $\sin x$ has an inverse. Since $\sin' x = \cos x > 0$ on $(-\pi/2, \pi/2)$, $\sin x$ is increasing on this interval, so $\sin x$ has an inverse on the interval $[-\pi/2, \pi/2]$. The inverse is denoted $\sin^{-1} y$.[4] We obtain the graph of $\sin^{-1} y$ by interchanging the x and y coordinates. (See Fig. 5.4.1.)

The values of $\sin^{-1} y$ may be obtained from a table for $\sin x$. (Many pocket calculators can evaluate the inverse trigonometric functions as well as the trigonometric functions.)

The Inverse Sine Function

$x = \sin^{-1} y$ means that $\sin x = y$ and $-\pi/2 \leqslant x \leqslant \pi/2$. The number $\sin^{-1} y$ is expressed in radians unless a degree sign is explicitly shown.

[4] Although the notation $\sin^2 y$ is commonly used to mean $(\sin y)^2$, $\sin^{-1} y$ does not mean $(\sin y)^{-1} = 1/\sin y$. Sometimes the notation arcsin y is used for the inverse sine function to avoid confusion.

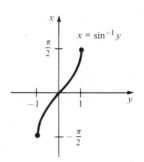

Figure 5.4.1. The graph of $\sin x$ on $[-\pi/2, \pi/2]$ together with its inverse.

Example 1 Calculate $\sin^{-1}1$, $\sin^{-1}0$, $\sin^{-1}(-1)$, $\sin^{-1}(-\frac{1}{2})$, and $\sin^{-1}(0.342)$.

Solution Since $\sin(\pi/2) = 1$, $\sin^{-1}1 = \pi/2$. Similarly, $\sin^{-1}0 = 0$, $\sin^{-1}(-1) = -\pi/2$. Also $\sin(-\pi/6) = -\frac{1}{2}$, so $\sin^{-1}(-\frac{1}{2}) = -\pi/6$. Using a calculator, or tables, we find $\sin^{-1}(0.342) = 0.349$ (or $20°$). ▲

We could have used any other interval on which $\sin x$ has an inverse, such as $[\pi/2, 3\pi/2]$, to define an inverse sine function; had we done so, the function obtained would have been different. The choice $[-\pi/2, \pi/2]$ is standard and is usually the most convenient.

Example 2 (a) Calculate $\sin^{-1}(\frac{1}{2})$, $\sin^{-1}(-\sqrt{3}/2)$, and $\sin^{-1}(2)$. (b) Simplify $\tan(\sin^{-1}y)$.

Solution (a) Since $\sin(\pi/6) = \frac{1}{2}$, $\sin^{-1}(\frac{1}{2}) = \pi/6$. Similarly, $\sin^{-1}(-\sqrt{3}/2) = -\pi/3$. Finally, $\sin^{-1}(2)$ is not defined since $\sin x$ always lies between -1 and 1.
(b) From Fig. 5.4.2 we see that $\theta = \sin^{-1}y$ (that is, $\sin\theta = |AB|/|OB|$

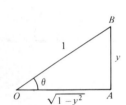

$= y$) and $\tan\theta = y/\sqrt{1-y^2}$, so $\tan(\sin^{-1}y) = y/\sqrt{1-y^2}$. ▲

Figure 5.4.2. $\tan(\sin^{-1}y)$ $= y/\sqrt{1-y^2}$.

Let us now calculate the derivative of $\sin^{-1}y$. By the formula for the derivative of an inverse function from page 278,

$$\frac{d}{dy}\sin^{-1}y = \frac{1}{(d/dx)\sin x} = \frac{1}{\cos x},$$

where $y = \sin x$. However, $\cos^2 x + \sin^2 x = 1$, so $\cos x = \sqrt{1-y^2}$. (The negative root does not occur since $\cos x$ is positive on $(-\pi/2, \pi/2)$.)
Thus,

$$\frac{d}{dy}\sin^{-1}y = \frac{1}{\sqrt{1-y^2}} = (1-y^2)^{-1/2}, \qquad -1 < y < 1. \tag{1}$$

Notice that the derivative of $\sin^{-1}y$ is not defined at $y = \pm 1$ but is "infinite" there. This is consistent with the appearance of the graph in Fig. 5.4.1.

Example 3 (a) Differentiate $h(y) = \sin^{-1}(3y^2)$. (b) Differentiate $f(x) = x\sin^{-1}2x$.
(c) Calculate $(d/dx)(\sin^{-1}2x)^{3/2}$.

Solution (a) From (1) and the chain rule, with $u = 3y^2$,

$$h'(y) = (1-u^2)^{-1/2}\frac{du}{dy} = 6y(1-9y^4)^{-1/2}.$$

(b) Here we are using x for the variable name. Of course we can use any letter we please. By the product and chain rules, and equation (1),

$$f'(x) = \left(\frac{dx}{dx}\right)(\sin^{-1}2x) + x\frac{d}{dx}(\sin^{-1}2x)$$

$$= \sin^{-1}2x + 2x(1-4x^2)^{-1/2}.$$

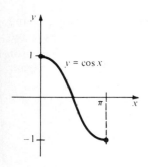

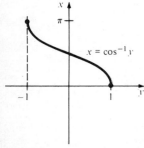

Figure 5.4.3. The graph of cos and its inverse.

(c) By the power rule, chain rule, and (1),

$$\frac{d}{dx}(\sin^{-1}2x)^{3/2} = \frac{3}{2}(\sin^{-1}2x)^{1/2}\frac{d}{dx}\sin^{-1}2x$$

$$= \frac{3}{2}(\sin^{-1}2x)^{1/2}\cdot 2\cdot\frac{1}{\sqrt{1-(2x)^2}}$$

$$= 3\left(\frac{\sin^{-1}2x}{1-4x^2}\right)^{1/2}.\ \blacktriangle$$

It is interesting that while $\sin^{-1}y$ is defined in terms of trigonometric functions, its derivative is an algebraic function, even though the derivatives of the trigonometric functions themselves are still trigonometric.

The rest of the inverse trigonometric functions can be introduced in the same way as $\sin^{-1}y$. The derivative of $\cos x$, $-\sin x$, is negative on $(0,\pi)$, so $\cos x$ on $(0,\pi)$ has an inverse $\cos^{-1}y$. Thus for $-1 \le y \le 1$, $\cos^{-1}y$ is that number (expressed in radians) in $[0,\pi]$ whose cosine is y. The graph of $\cos^{-1}y$ is shown in Fig. 5.4.3.

The derivative of $\cos^{-1}y$ can be calculated in the same manner as we calculated $(d/dy)\sin^{-1}y$:

$$\frac{d}{dy}\cos^{-1}y = \frac{1}{(d/dx)\cos x} = \frac{1}{-\sin x}$$

$$= \frac{-1}{\sqrt{1-y^2}}. \qquad (2)$$

Example 4   Differentiate $\tan(\cos^{-1}x)$.

Solution   By the chain rule and equation (2) (with x in place of y),

$$\frac{d}{dx}\tan(\cos^{-1}x) = \sec^2(\cos^{-1}x)\cdot\left(\frac{-1}{\sqrt{1-x^2}}\right).$$

From Fig. 5.4.4 we see that $\sec(\cos^{-1}x) = 1/x$, so

$$\frac{d}{dx}\left[\tan(\cos^{-1}x)\right] = \frac{-1}{x^2\sqrt{1-x^2}}.$$

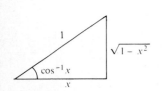

Figure 5.4.4. $\sec(\cos^{-1}x)$ $= 1/x$.

Another method is to use Fig. 5.4.4 directly to obtain

$$\tan(\cos^{-1}x) = \frac{\sqrt{1-x^2}}{x}.$$

Differentiating by the quotient and chain rules,

$$\frac{d}{dx}\frac{\sqrt{1-x^2}}{x} = \frac{x\cdot(-2x)\cdot\frac{1}{2}(1-x^2)^{-1/2} - (1-x^2)^{1/2}}{x^2}$$

$$= -\frac{1}{x^2}\left(\frac{x^2}{\sqrt{1-x^2}} + \sqrt{1-x^2}\right)$$

$$= \frac{-1}{x^2\sqrt{1-x^2}},$$

which agrees with our previous answer. ▲

Next, we construct the inverse tangent. Since $\tan' x = \sec^2 x$, $\tan x$ is increasing at every point of its domain. It is continuous on $(-\pi/2, \pi/2)$ and has range $(-\infty, \infty)$, so $\tan^{-1} y$ is defined on this domain; see Fig. 5.4.5. Thus, for $-\infty < y < \infty$, $\tan^{-1} y$ is the unique number in $(-\pi/2, \pi/2)$ whose tangent is y.

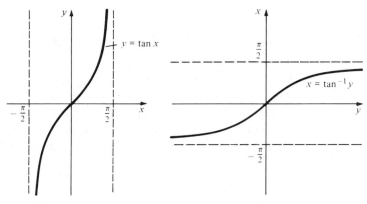

Figure 5.4.5. $\tan x$ and its inverse.

The derivative of $\tan^{-1} y$ can be calculated as in (1) and (2):

$$\frac{d}{dy} \tan^{-1} y = \frac{1}{(d/dx)\tan x} = \frac{1}{\sec^2 x} = \frac{1}{1 + \tan^2 x} = \frac{1}{1 + y^2}.$$

Thus $\dfrac{d}{dy} \tan^{-1} y = \dfrac{1}{1 + y^2}$. (3)

Example 5 Differentiate $f(x) = (\tan^{-1}\sqrt{x})/(\cos^{-1}x)$. Find the domain of f and f'.

Solution By the quotient rule, the chain rule, and (3),

$$\frac{d}{dx} \frac{\tan^{-1}\sqrt{x}}{\cos^{-1}x}$$

$$= \left\{ \cos^{-1}x \cdot \left[\frac{1}{1 + \left(\sqrt{x}\right)^2} \cdot \frac{1}{2\sqrt{x}} \right] - \tan^{-1}\sqrt{x} \cdot \left(-\frac{1}{\sqrt{1 - x^2}} \right) \right\} (\cos^{-1}x)$$

$$= \frac{\sqrt{1 - x^2}\,\cos^{-1}x + 2\sqrt{x}\,(1 + x)\tan^{-1}\sqrt{x}}{2\sqrt{x}\,(1 + x)\sqrt{1 - x^2}\,(\cos^{-1}x)^2}.$$

The domain of f consists of those x for which $x \geqslant 0$ (so that $\sqrt{x}$ is defined) and $-1 < x < 1$ (so that $\cos^{-1}x$ is defined and not zero)—that is, the domain of f is $[0, 1)$. For f' to be defined, the denominator in the derivative must be nonzero. This requires x to belong to the interval $(0, 1)$. Thus, the domain of f' is $(0, 1)$. ▲

The remaining inverse trigonometric functions can be treated in the same way. Their graphs are shown in Fig. 5.4.6 and their properties are summarized in the box on the next page.

Remembering formulas such as those on the next page is an unpleasant chore for most students (and professional mathematicians as well). Many people prefer to memorize only a few basic formulas and to derive the rest as needed. It is also useful to develop a short mental checklist: Is the sign right? Is the sign consistent with the appearance of the graph? Is the derivative undefined at the proper points?

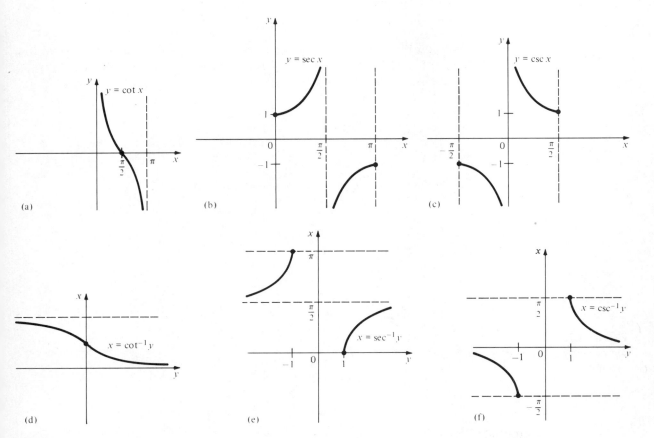

Figure 5.4.6. cot, sec, csc, and their inverses.

Inverse Trigonometric Functions

Function	Domain on which function has an inverse	Derivative of function	Inverse	Domain of inverse	Derivative of inverse	
$\sin x$	$\left[-\dfrac{\pi}{2},\dfrac{\pi}{2}\right]$	$\cos x$	$\sin^{-1}y$	$[-1,1]$	$\dfrac{1}{\sqrt{1-y^2}}$,	$-1<y<1$
$\cos x$	$[0,\pi]$	$-\sin x$	$\cos^{-1}y$	$[-1,1]$	$-\dfrac{1}{\sqrt{1-y^2}}$,	$-1<y<1$
$\tan x$	$\left(-\dfrac{\pi}{2},\dfrac{\pi}{2}\right)$	$\sec^2 x$	$\tan^{-1}y$	$(-\infty,\infty)$	$\dfrac{1}{1+y^2}$,	$-\infty<y<\infty$
$\cot x$	$(0,\pi)$	$-\csc^2 x$	$\cot^{-1}y$	$(-\infty,\infty)$	$-\dfrac{1}{1+y^2}$,	$-\infty<y<\infty$
$\sec x$	$\left[0,\dfrac{\pi}{2}\right)$ and $\left(\dfrac{\pi}{2},\pi\right]$	$\tan x\sec x$	$\sec^{-1}y$	$(-\infty,-1]$ and $[1,\infty)$	$\dfrac{1}{\sqrt{y^2(y^2-1)}}$,	$-\infty<y<-1,$ $1<y<\infty$
$\csc x$	$\left[-\dfrac{\pi}{2},0\right)$, and $\left(0,\dfrac{\pi}{2}\right]$	$-\cot x\csc x$	$\csc^{-1}y$	$(-\infty,-1]$ and $[1,\infty)$	$-\dfrac{1}{\sqrt{y^2(y^2-1)}}$,	$-\infty<y<-1,$ $1<y<\infty$

Example 6 Calculate $\cos^{-1}(\frac{1}{2})$, $\tan^{-1}(1)$, and $\csc^{-1}(2/\sqrt{3})$.

Solution We find that $\cos^{-1}(-\frac{1}{2}) = 2\pi/3$ since $\cos(2\pi/3) = -\frac{1}{2}$, as is seen from Fig. 5.4.7. Similarly, $\tan^{-1}(1) = \pi/4$ since $\tan(\pi/4) = 1$. Finally, $\csc^{-1}(2/\sqrt{3}) = \pi/3$ since $\csc(\pi/3) = 2/\sqrt{3}$. ▲

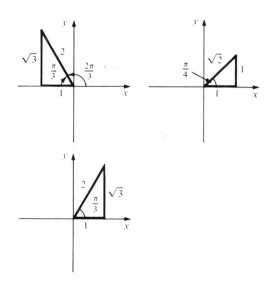

Figure 5.4.7. Evaluating some inverse trigonometric functions.

Example 7 Differentiate: (a) $\sec^{-1}(y^2)$, $y > 0$; (b) $\cot^{-1}[(x^3 + 1)/(x^3 - 1)]$.

Solution (a) By the chain rule,

$$\frac{d}{dy} \sec^{-1}(y^2) = \frac{1}{\sqrt{y^4(y^4 - 1)}} \cdot 2y = \frac{2}{y\sqrt{y^4 - 1}}.$$

(b) By the chain rule,

$$\frac{d}{dx} \cot^{-1}\left(\frac{x^3 + 1}{x^3 - 1} \right) = \frac{-1}{1 + \left[(x^3 + 1)/(x^3 - 1) \right]^2} \frac{d}{dx}\left(\frac{x^3 + 1}{x^3 - 1} \right)$$

$$= \frac{-(x^3 - 1)^2}{(x^3 - 1)^2 + (x^3 + 1)^2} \left[\frac{(x^3 - 1) \cdot 3x^2 - (x^3 + 1)3x^2}{(x^3 - 1)^2} \right.$$

$$= \frac{6x^2}{(x^3 - 1)^2 + (x^3 + 1)^2} = \frac{3x^2}{x^6 + 1}. \quad ▲$$

Example 8 Differentiate $f(x) = (\csc^{-1}3x)^2$. Find the domain of f and f'.

Solution By the chain rule,

$$\frac{d}{dx} (\csc^{-1}3x)^2 = 2(\csc^{-1}3x) \frac{d}{dx} \csc^{-1}3x$$

$$= 2\csc^{-1}3x \cdot 3 \cdot \frac{-1}{\sqrt{(3x)^2[(3x)^2 - 1]}}$$

$$= \frac{-2\csc^{-1}3x}{\sqrt{x^2(9x^2 - 1)}} = \frac{-2\csc^{-1}3x}{|x|\sqrt{9x^2 - 1}}.$$

For $f(x)$ to be defined, $3x$ should lie in $[1, \infty)$ or $(-\infty - 1]$; that is, x should lie in $[\frac{1}{3}, \infty)$ or $(-\infty, -\frac{1}{3}]$. The domain of f' is $(\frac{1}{3}, \infty)$, together with $(-\infty, -\frac{1}{3})$. ▲

Example 9 Explain why the derivative of every inverse cofunction in the preceding box is the negative of that of the inverse function.

Solution Let $f(x)$ be one of the functions $\sin x$, $\tan x$, or $\sec x$, and let $g(x)$ be the corresponding cofunction $\cos x$, $\cot x$ or $\csc x$. Then we know that

$$f\left(\frac{\pi}{2} - x\right) = g(x).$$

If we let y denote $g(x)$, then we get

$$\frac{\pi}{2} - x = f^{-1}(y) \quad \text{and} \quad x = g^{-1}(y),$$

so

$$\frac{\pi}{2} - f^{-1}(y) = g^{-1}(y).$$

It follows by differentation in y that

$$-\frac{d}{dy} f^{-1}(y) = \frac{d}{dy} g^{-1}(y).$$

Hence, the derivatives of $f^{-1}(y)$ and $g^{-1}(y)$ are negatives, which is the general reason why this same phenomenon occurred three times in the box. ▲

The differentiation formulas for the inverse trigonometric functions may be read backwards to yield some interesting antidifferentiation formulas. For example, since $(d/dx)\tan^{-1}x = 1/(1 + x^2)$, we get

$$\int \frac{dx}{1 + x^2} = \tan^{-1}x + C. \tag{4}$$

Formulas like this will play an important role in the techniques of integration.

Example 10 Find $\int \dfrac{x^2}{1 + x^2}\, dx$ [*Hint:* Divide first.]

Solution Using long division, $x^2/(1 + x^2) = 1 - 1/(1 + x^2)$. Thus, by (4),

$$\int \frac{x^2}{1 + x^2}\, dx = \int 1\, dx - \int \frac{1}{1 + x^2}\, dx$$
$$= x - \tan^{-1}x + C. \;\blacktriangle$$

Exercises for Section 5.4

Calculate the quantities in Exercises 1–10.

1. $\sin^{-1}\left(\dfrac{1}{\sqrt{3}}\right)$

2. $\sin^{-1}\left(\dfrac{2}{\sqrt{3}}\right)$

3. $\sin^{-1}\left(-\dfrac{\pi}{4}\right)$

4. $\sin^{-1}(0.4)$

5. $\cos^{-1}(1)$

6. $\cos^{-1}(0.3)$

7. $\tan^{-1}(\sqrt{3})$

8. $\cot^{-1}(2.3)$

9. $\sec^{-1}\left(\dfrac{2}{\sqrt{3}}\right)$

10. $\csc^{-1}(-5)$

Differentiate the functions in Exercises 11–28.

11. $\sin^{-1}(8x)$

12. $\sin^{-1}(\sqrt{1 - x^2})$

13. $x^2\sin^{-1}x$

14. $(x^2 - 1)\sin^{-1}(x^2)$

15. $(\sin^{-1}x)^2$

16. $\cos^{-1}\left(\dfrac{t}{t + 1}\right)$

17. $\tan^{-1}\left(\dfrac{2x^5 + x}{1 - x^2}\right)$

18. $\cot^{-1}(1 - y^2)$

19. $\sec^{-1}\left(y - \dfrac{1}{y^2}\right)$

20. $\left(\dfrac{1}{x}\right)\csc^{-1}\left(\dfrac{1}{x}\right)$

21. $\dfrac{\sin^{-1}3x}{x^2 + 2}$

22. $\dfrac{\sin^{-1}(2t)}{3t^2 - t + 1}$

23. $\sin^{-1}\left(\dfrac{t^7 + t^4 + 1}{t^5 + 2t}\right)$.

24. $\dfrac{u^2}{4} + 2\sin^{-1}\left[\dfrac{1}{(u^2+1)^{3/2}}\right]$.

25. $\dfrac{\cos^{-1}x}{1 - \sin^{-1}x}$.

26. $[\sin^{-1}(3x) + \cos^{-1}(5x)][\tan^{-1}(3x)]$.

27. $(x^2\cos^{-1}x + \tan x)^{3/2}$.

28. $(x^3\sin^{-1}x + \cot x)^{5/3}$.

Find the antiderivatives in Exercises 29–36.

29. $\displaystyle\int\left(\dfrac{3}{1+x^2} + x\right)dx$.

30. $\displaystyle\int\dfrac{1}{\sqrt{1-y^2}}\,dy$

31. $\displaystyle\int\dfrac{4}{\sqrt{1-x^2}}\,dx$

32. $\displaystyle\int\dfrac{1}{\sqrt{4-x^2}}\,dx$

33. $\displaystyle\int\left(\dfrac{3}{1+4x^2}\right)dx$

34. $\displaystyle\int\left(\dfrac{1}{\sqrt{1-4x^2}}\right)dx$

35. $\displaystyle\int\left(\dfrac{2}{\sqrt{y^2(y^2-1)}}\right)dy$

36. $\displaystyle\int\left(\dfrac{3}{\sqrt{4u^2(u^2-1)}}\right)du$

37. Prove: $\tan(\sin^{-1}x) = x/\sqrt{1-x^2}$.

38. Prove $\csc^{-1}(1/x) = \sin^{-1}x = \pi/2 - \cos^{-1}x$.

39. Is $\cot^{-1}y = 1/(\tan^{-1}y)$? Explain.

40. Is the following correct:

$(d/dx)\cos^{-1}x = (-1)(\cos^{-2}x)[(d/dx)\cos x]$?

Explain.

Calculate the quantities in Exercises 41–46.

41. $\dfrac{d^2}{d\theta^2}\cos^{-1}\theta$

42. $\dfrac{d}{d\theta}\cot^{-1}(\theta^2 + 1)$

43. $\dfrac{d}{dx}[(\sin^{-1}2x)^2 + x^2]$

44. $\dfrac{d}{dx}\left[\sin^{-1}\left(\dfrac{1}{2}\sec\dfrac{2x}{3}\right)\right]$

45. The rate of change of $\cos^{-1}(8s^2 + 2)$ with respect to s at $s = 0$.

46. The rate of change of $h(t) = \sin^{-1}(3t + \frac{1}{2})$ with respect to t at $t = 0$.

47. What are the maxima, minima, and inflection points of $f(x) = \sin^{-1}x$?

48. Prove that $y = \tan^{-1}x$ has an inflection point at $x = 0$.

49. Derive the formula $(d/dy)\cot^{-1}y = -\dfrac{1}{1+y^2}$,

$-\infty < y < \infty$.

50. Derive the formula

$$\dfrac{d}{dy}\csc^{-1}y = -\dfrac{1}{\sqrt{y^2(y^2-1)}}\,,$$

$-\infty < y < -1,\ 1 < y < \infty$.

51. (a) What is the domain of $\cos^{-1}(x^2 - 3)$? Differentiate. (b) Sketch the graph of $\cos^{-1}(x^2 - 3)$.

52. What is the equation of the line tangent to the graph of $\cos^{-1}(x^2)$ at $x = 0$?

53. Let x and y be related by the equation

$$\dfrac{\sin(x+y)}{xy} = 1,$$

and assume that $\cos(x + y) > 0$.
(a) Find dy/dx.
(b) If $x = t/(1 - t^2)$, find dy/dt.
(c) If $y = \sin^{-1}t$, find dx/dt.
(d) If $x = t^3 + 2t - 1$, find dy/dt.

54. Find a function $f(x)$ which is differentiable and increasing for all x, yet $f(x) < \pi/2$ for all x.

Calculate the definite integrals in Exercises 55–58.

55. $\displaystyle\int_1^2\dfrac{dx}{1+x^2}$.

56. $\displaystyle\int_0^1\dfrac{dx}{1+(2x)^2}$.

57. $\displaystyle\int_{1/2}^{\sqrt{2}/2}\left(\dfrac{x^2+1}{x^2} + \dfrac{1}{\sqrt{1-x^2}} + \cos x\right)dx$.

58. $\displaystyle\int_2^{10}\dfrac{dy}{\sqrt{y^2(y^2-1)}}$.

★59. Suppose that $\sin^{-1}$ had been defined by inverting $\sin x$ on $[\pi/2, 3\pi/2]$ instead of on $[0, \pi]$. What would the derivative of $\sin^{-1}$ have been?

★60. It is possible to approach the trigonometric functions without using geometry by *defining* the function $a(x)$ to be $\int_0^x du/\sqrt{1-u^2}$ and the letting $\sin$ be the inverse function of a. Using this definition prove that:
(a) $(\sin\theta)^2 + (\sin'\theta)^2$ is constant and equal to 1;
(b) $\sin''\theta = -\sin\theta$.

5.5 Graphing and Word Problems

Many interesting word problems involve trigonometric functions.

The graphing and word problems in Chapter 3 were limited since, at that point, we could differentiate only algebraic functions. Now that we have more functions at our disposal, we can solve a wider variety of problems.

Example 1   A searchlight 10 kilometers from a straight coast makes one revolution every 30 seconds. How fast is the spot of light moving along a wall on the coast at the point P in Fig. 5.5.1?

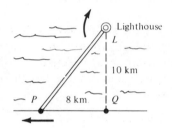

Figure 5.5.1. How fast is spot P moving if the beam LP revolves once every 30 seconds?

Solution   Let θ denote the angle PLQ, and let x denote the distance $|PQ|$. We know that $d\theta/dt = 2\pi/30$, since θ changes by 2π in 30 seconds. Now $x = 10\tan\theta$, so $dx/d\theta = 10\sec^2\theta$, and the velocity of the spot is $dx/dt = (dx/d\theta)(d\theta/dt)$ $= 10 \cdot \sec^2\theta \cdot 2\pi/30 = (2\pi/3)\sec^2\theta$. At $x = 8$, $\sec\theta = |PL|/|LQ| = \sqrt{8^2 + 10^2}/10$, so the velocity is $dx/dt = (2\pi/3) \times (8^2 + 10^2)/10^2 = (82\pi/75)$ ≈ 3.4 kilometers per second (this is very fast!). ▲

Example 2   Find the point on the x axis for which the sum of the distances from $(0, 1)$ and (p, q) is a minimum. (Assume that p and q are positive.)

Solution   Let $(x, 0)$ be a point on the x axis. The distance from $(0, 1)$ is $\sqrt{1 + x^2}$ and the distance from (p, q) is $\sqrt{(x - p)^2 + q^2}$. The sum of the distances is

$$S = \sqrt{1 + x^2} + \sqrt{(x - p)^2 + q^2}.$$

To minimize, we find:

$$\frac{dS}{dx} = \frac{1}{2}(1 + x^2)^{-1/2}(2x) + \frac{1}{2}\left[(x - p)^2 + q^2\right]^{-1/2}2(x - p)$$

$$= \frac{x}{\sqrt{1 + x^2}} + \frac{x - p}{\sqrt{(x - p)^2 + q^2}}.$$

Setting this equal to zero gives

$$\frac{x}{\sqrt{1 + x^2}} = \frac{p - x}{\sqrt{(x - p)^2 + q^2}}.$$

Instead of solving for x, we will interpret the preceding equation geometrically. Referring to Fig. 5.5.2, we find that $\sin\theta_1 = x/\sqrt{1 + x^2}$ and $\sin\theta_2 = (p - x)/\sqrt{(x - p)^2 + q^2}$; our equation says that these are equal, so $\theta_1 = \theta_2$. Thus $(x, 0)$ is located at the point for which the lines from $(x, 0)$ to

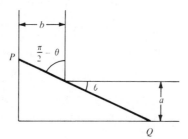

Figure 5.5.2. The shortest path from $(0, 1)$ to (p, q) via the x axis has $\theta_1 = \theta_2$.

$(0, 1)$ and (p, q) make equal angles with a line parallel to the y axis. This result is sometimes called the *law of reflection*. ▲

Example 3 Two hallways, meeting at right angles, have widths a and b. Find the length of the longest pole which will go around the corner; the pole must be in a horizontal position. (In Exercise 51, Section 3.5, you were asked to do the problem by minimizing the square of the length; redo the problem here by minimizing the length itself.)

Solution Refer to Fig. 5.5.3. The length of PQ is

$$f(\theta) = \frac{a}{\sin\theta} + \frac{b}{\cos\theta}.$$

Figure 5.5.3. The pole in the corner.

The minimum of $f(\theta)$, $0 \leqslant \theta \leqslant \pi/2$, will give the length of the longest pole which will fit around the corner. The derivative is

$$f'(\theta) = -\frac{a\cos\theta}{\sin^2\theta} + \frac{b\sin\theta}{\cos^2\theta},$$

which is zero when $a\cos^3\theta = b\sin^3\theta$; that is, when $\tan^3\theta = a/b$; hence $\theta = \tan^{-1}(\sqrt[3]{a/b})$. Since f is large positive near 0 and $\pi/2$, and there are no other critical points, this is a global minimum. (You can also use the second derivative test.) Thus, the answer is

$$\frac{a}{\sin\theta} + \frac{b}{\cos\theta},$$

where $\theta = \tan^{-1}(\sqrt[3]{a/b})$. Using $\sin(\tan^{-1}\alpha) = \alpha/\sqrt{1 + \alpha^2}$ and $\cos(\tan^{-1}\alpha) = 1/\sqrt{1 + \alpha^2}$ (Fig. 5.5.4), one can express the answer, after some simplification, as $(a^{2/3} + b^{2/3})^{3/2}$.

One way to check the answer (which the authors actually used to catch an error) is to note its "dimension." The result must have the dimension of a length. Thus an answer like $a^{1/3}(a^{2/3} + b^{2/3})^{3/2}$, which has dimension of (length)$^{1/3}$ × length, cannot be correct. ▲

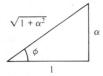

Figure 5.5.4. $\phi = \tan^{-1}\alpha$.

Example 4   (This problem was written on a train.) One normally chooses the window seat on a train to have the best view. Imagine the situation in Fig. 5.5.5 and see if this is really the best choice. (Ignore the extra advantage of the window seat which enables you to lean forward to see a special view.)

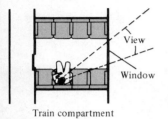

Train compartment

Figure 5.5.5. Where should you sit to get the widest view?

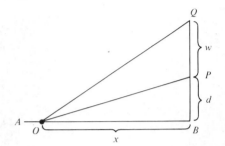

Figure 5.5.6. Which value of x maximizes $\angle POQ$?

Solution   It is convenient to replace the diagram by a more abstract one (Fig. 5.5.6). We assume that the passenger's eye is located at a point O on the line AB when he is sitting upright and that he wishes to maximize the angle $\angle POQ$ subtended by the window PQ. Denote by x the distance from O to B, which can be varied. Let the width of the window be w and the distance from AB to the window be d. Then we have

$$\angle POQ = \angle BOQ - \angle POB,$$

$$\tan(\angle BOQ) = \frac{w+d}{x},$$

and

$$\tan(\angle POB) = \frac{d}{x}.$$

So we wish to maximize

$$f(x) = \angle POQ = \tan^{-1}\left(\frac{w+d}{x}\right) - \tan^{-1}\left(\frac{d}{x}\right)$$

Differentiating, we have

$$f'(x) = \frac{1}{1+\left[(w+d)/x\right]^2} \cdot \left(-\frac{w+d}{x^2}\right) - \frac{1}{1+(d/x)^2} \cdot \left(-\frac{d}{x^2}\right)$$

$$= -\frac{w+d}{x^2+(w+d)^2} + \frac{d}{x^2+d^2}.$$

Setting $f'(x) = 0$ yields the equation

$$(x^2+d^2)(w+d) = \left[x^2+(w+d)^2\right]d,$$

$$wx^2 = (w+d)^2 d - d^2(w+d),$$

$$x^2 = (w+d) \cdot d.$$

The solution, therefore, is $x = \sqrt{(w+d)d}$.

For example (all distances measured in feet), if $d = 1$ and $w = 5$, we should take $x = \sqrt{6} \approx 2.45$. Thus, it is probably better to take the second seat from the window, rather than the window seat.

There is a geometric interpretation for the solution of this problem. We may rewrite the solution as

$$x^2 = d^2 + wd \quad \text{or} \quad x^2 = \left(d + \tfrac{1}{2}w\right)^2 - \left(\tfrac{1}{2}w\right)^2.$$

This second formula leads to the following construction (which you may be able to carry out mentally before choosing your seat). Draw a line RP through P and parallel to AB. Now construct a circle with center at the midpoint M of PQ and with radius MB. Let Z be the point where the circle intersects RP. Then $x = ZP$. (See Fig. 5.5.7.) ▲

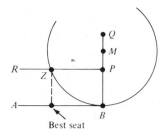

Figure 5.5.7. Geometric construction for the best seat.

Let us turn from word problems to some problems in graphing, using the methods of Chapter 3.

Example 5   Discuss maxima, minima, concavity, and points of inflection for $f(x) = \sin^2 x$. Sketch its graph.

Solution   If $f(x) = \sin^2 x$, $f'(x) = 2 \sin x \cos x$ and $f''(x) = 2(\cos^2 x - \sin^2 x)$. The first derivative vanishes when either $\sin x = 0$ or $\cos x = 0$, at which points f'' is positive and negative, yielding minima and maxima. Thus, the minima of f are at $0, \pm \pi, \pm 2\pi, \ldots$, where $f = 0$, and the maxima are at $\pm \pi/2$, $\pm 3\pi/2, \ldots$, where $f = 1$.

The function $f(x)$ is concave upward when $f''(x) > 0$ (that is, $\cos^2 x > \sin^2 x$) and downward when $f''(x) < 0$ (that is, $\cos^2 x < \sin^2 x$). Also, $\cos x = \pm \sin x$ exactly if $x = \pm \pi/4$, $\pm \pi/4 \pm \pi$, $\pm \pi/4 \pm 2\pi$, and so on (see the graphs of sine and cosine). These are then inflections points separating regions where $\sin^2 x$ is concave up and concave down. The graph is shown in Fig. 5.5.8.

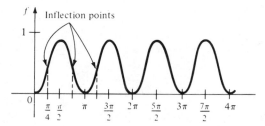

Figure 5.5.8. The graph of $\sin^2 x$.

Now that we have done all this work, we observe that the graph could also have been found from the half-angle formula $\sin^2 \theta = \frac{1}{2}(1 - \cos 2\theta)$. ▲

Example 6   Sketch the graph of the function $f(x) = \cos x + \cos 2x$.

Solution   The function is defined on $(-\infty, \infty)$; there are no asymptotes. We find that $f(-x) = \cos(-x) + \cos(-2x) = \cos x + \cos 2x = f(x)$, so f is even. Furthermore, $f(x + 2\pi) = f(x)$, so the graph repeats itself every 2π units (like that of $\cos x$). It follows that we need only look for features of the graph on $[0, \pi]$, because we can obtain $[-\pi, 0]$ by reflection across the y axis and the rest of the graph by repetition of the part over $[-\pi, \pi]$.

We have

$$f'(x) = -\sin x - 2 \sin 2x \quad \text{and} \quad f''(x) = -\cos x - 4 \cos 2x.$$

To find roots of the first and second derivatives, it is best to factor using the formulas

$$\sin 2x = 2 \sin x \cos x \quad \text{and} \quad \cos 2x = 2 \cos^2 x - 1.$$

We obtain

$$f'(x) = -\sin x - 4 \sin x \cos x = -\sin x (1 + 4 \cos x).$$

and

$$f''(x) = -\cos x - 4(2 \cos^2 x - 1) = -8 \cos^2 x - \cos x + 4.$$

The critical points occur when $\sin x = 0$ or $1 + 4 \cos x = 0$—that is, when $x = 0$, π, or $\cos^{-1}(-\frac{1}{4}) \approx 1.82$ (radians).

We have

$$f(0) = 2, \quad f(\pi) = 0, \quad f\left[\cos^{-1}(-\tfrac{1}{4})\right] = -1.125;$$

$$f''(0) = -5, \quad f''(\pi) = -3, \quad f''\left[\cos^{-1}(-\tfrac{1}{4})\right] = 3.75.$$

Hence 0 and π are local maximum points, and $\cos^{-1}(-\frac{1}{4})$ is a local minimum point, by the second derivative test.

To find the points of inflection, we first find the roots of $f''(x) = 0$; that is,

$$-8 \cos^2 x - \cos x + 4 = 0.$$

This is a quadratic equation in which $\cos x$ is the unknown, so

$$\cos x = \frac{-1 \pm \sqrt{129}}{16} \approx 0.647 \quad \text{and} \quad -0.772.$$

Thus our candidates for points of inflection are

$$x_1 = \cos^{-1}(0.647) \approx 0.87 \quad \text{and} \quad x_2 = \cos^{-1}(-0.772) \approx 2.45.$$

We can see from the previously calculated values for $f''(x)$ that f'' does change sign at these points, so they are inflection points. We calculate $f(x)$ and $f'(x)$ at the inflection points:

$$f(x_1) \approx 0.48, \quad f'(x_1) \approx -2.74;$$

$$f(x_2) \approx -0.58, \quad f'(x_2) \approx 1.33.$$

Finally, the zeros of f may be found by writing

$$f(x) = \cos x + \cos 2x$$

$$= \cos x + 2 \cos^2 x - 1$$

$$= 2(\cos x + 1)(\cos x - \tfrac{1}{2}).$$

Thus, $f(x) = 0$ at $x = \pi$ and $x = \cos^{-1}(\frac{1}{2}) \approx 1.047$. The graph on $[0, \pi]$ obtained from this information is shown in Fig. 5.5.9. Reflecting across the y axis

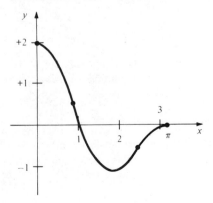

Figure 5.5.9. The graph of $\cos x + \cos 2x$ on $[0, \pi]$.

and then repeating the pattern, we obtain the graph shown in Fig. 5.5.10. Such graphs, with oscillations of varying amplitudes, are typical when sine and cosine functions with different frequencies are added. ▲

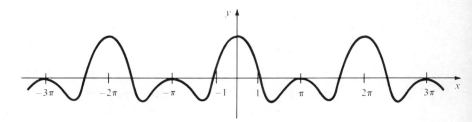

Figure 5.5.10. The full graph of $\cos x + \cos 2x$.

Exercises for Section 5.5

1. The height of an object thrown straight down from an initial altitude of 1000 feet is given by $h(t) = 1000 - 40t - 16t^2$. The object is being tracked by a searchlight 200 feet from where the object will hit. How fast is the angle of elevation of the searchlight changing after 4 seconds?

2. A searchlight 100 meters from a road is tracking a car moving at 100 kilometers per hour. At what rate (in degrees per second) is the searchlight turning when the car is 141 meters away?

3. A child is whirling a stone on a string 0.5 meter long in a vertical circle at 5 revolutions per second. The sun is shining directly overhead. What is the velocity of the stone's shadow when the stone is at the 10 o'clock position?

4. A bicycle is moving 10 feet per second. It has wheels of radius 16 inches and a reflector attached to the front spokes 12 inches from the center. If the reflector is at its lowest point at $t = 0$, how fast is the reflector accelerating vertically at $t = 5$ seconds?

5. Two weights A and B together on the ground are joined by a 20-meter wire. The wire passes over a pulley 10 meters above the ground. Weight A is slid along the ground at 2 meters per second. How fast is the distance between the weights changing after 3 seconds? (See Fig. 5.5.11.)

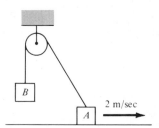

Figure 5.5.11. How fast are A and B separating?

6. Consider the two posts in Fig. 5.5.12. The light atop post A moves vertically up and down the

post according to $h(t) = 55 + 5\sin t$ (t in seconds, height in meters). How fast is the length of the shadow of the 2-meter statue changing at $t = 20$ seconds?

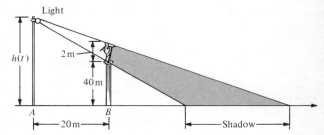

Figure 5.5.12. How fast is the shadow's length changing when the light oscillates up and down?

7. Consider the situation sketched in Fig. 5.5.13. At what position on the road is the angle θ maximized?

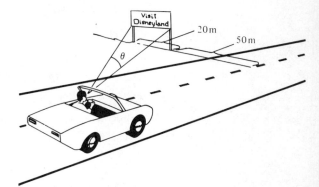

Figure 5.5.13. Maximize θ.

8. Two trains, each 50 meters long, are moving away from the intersection point of perpendicular tracks at the same speed. Where are the trains when train A subtends the the largest angle as seen from the front of train B?

9. Particle A is moving in the plane according to $x = 3 \sin 3t$ and $y = 3 \cos 3t$ and particle B is moving according to $x = 3 \cos 2t$ and $y = 3 \sin 2t$. Find the maximum distance between A and B.

10. Which points on the parametric curve $x = \cos t$, $y = 4 \sin t$ are closest to $(0, 1)$?

11. A slot racer travels at constant speed around a circular track, doing each lap in 3.1 seconds. The track is 3 feet in diameter.
   (a) The position (x, y) of the racer can be written as $x = r \cos(\omega t)$, $y = r \sin(\omega t)$. Find the values of r and ω.
   (b) The *speed* of the racer is the elapsed time divided into the distance traveled. Find its value and check that it is equal to $\sqrt{(dx/dt)^2 + (dy/dt)^2}$.

12. The motion of a projectile (neglecting air friction and the curvature of the earth) is governed by the equations
$$x = v_0 t \cos \alpha \qquad y = v_0 t \sin \alpha - 4.9 t^2,$$
where v_0 is the initial velocity and α is the initial angle of elevation. Distances are measured in meters and t is the time from launch. (See Fig. 5.5.14.)

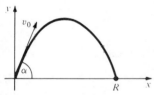

Figure 5.5.14. Path of a projectile near the surface of the earth.

   (a) Find the maximum height of the projectile and the distance R from launch to fall as a function of α.
   (b) Show that R is maximized when $\alpha = \pi/4$.

13. Molasses is smeared over the upper half ($y > 0$) of the (x, y) plane. A bug crawls at 1 centimeter per minute where there is molasses and 3 centimeters per minute where there is none. Suppose that the bug is to travel from some point in the upper half-plane ($y > 0$) to some point in the lower half-plane ($y < 0$). The fastest route then consists of a broken line segment with a break on the x axis. Find a relation between the sines of the angles made by the two parts of the segments with the y axis.

14. Drywall sheets weighing 6000 lbs are moved across a level floor. The method is to attach a chain to the skids under the drywall stack, then pull it with a truck. The angle θ made by the chain and the floor is related to the force F along the chain by $F = 6000k/(k \sin \theta + \cos \theta)$, where the number k is the coefficient of friction.
   (a) Compute $dF/d\theta$.
   (b) Find θ for which $dF/d\theta = 0$. (This is the angle that requires the least force.)

15. Discuss the maxima, minima, concavity, and points of inflection for $y = \sin 2x - 1$.

16. Find the maxima, minima, concavity, and inflection points of $f(x) = \cos^2 3x$.

17. Where is $f(x) = x \sin x + 2 \cos x$ concave up? Concave down?

18. Where is $g(\theta) = \sin^3 \theta$ concave up? Concave down?

Sketch the graphs of the functions in Exercises 19–26.

19. $y = \cos^2 x$ 20. $y = 1 + \sin 2x$
21. $y = x + \cos x$ 22. $y = x \sin x$
23. $y = 2 \cos x + \cos 2x$ 24. $y = \cos 2x + \cos 4x$
25. $y = x^{2/3} \cos x$ 26. $y = x^{1/2} \sin x$

27. Do Example 2 without using calculus. [*Hint*: Replace (p, q) by $(p, -q)$.]

28. The displacement $x(t)$ from equilibrium of a mass m undergoing harmonic motion on a spring of Hooke's constant k is known to satisfy the equation $mx''(t) + kx(t) = 0$. Check that $x(t) = A \cos(\omega t + \theta)$ is a solution of this equation, where $\omega = \sqrt{k/m}$; A, θ and ω are constants.

29. Determine the equations of the tangent and normal lines to the curve $y = \cos^{-1} 2x + \cos^{-1} x$ at $(0, \pi)$.

30. Find the equation of the tangent line to the parametric curve $x = t^2$, $y = \cos^{-1} t$ when $t = \frac{1}{2}$.

31. Sketch the graph of the function $(\sin x)/(1 + x^2)$ for $0 \leqslant x < 2\pi$. (You may need to use a calculator to locate the critical points.)

★32. Let $f(x) = \sin^{-1}[2x/(x^2 + 1)]$.
   (a) Show that $f(x)$ is defined for x in $(-\infty, \infty)$.
   (b) Compute $f'(x)$. Where is it defined?
   (c) Show that the maxima and minima occur at points where f is not differentiable.
   (d) Sketch the graph of f.

★33. Show that the function $(\sin x)/x$ has infinitely many local maxima and minima, and that they become approximately evenly spaced as $x \to \infty$.

★34. Given two points outside a circle, find the shortest path between them which touches the circle. (*Hint*: First assume that the points are equidistant from the center of the circle, and put the figure in a standard position.)

5.6 Graphing in Polar Coordinates

Periodic functions are graphed as closed curves in polar coordinates.

The graph in polar coordinates of a function f consists of all those points in the plane whose polar coordinates (r, θ) satisfy the relation $r = f(\theta)$. Such graphs are especially useful when f is built up from trigonometric functions, since the entire graph is drawn when we let θ vary from 0 to 2π.

Various properties of f may appear as symmetries of the graph. For instance, if $f(\theta) = f(-\theta)$, then its graph is symmetric in the x axis; if $f(\pi - \theta) = f(\theta)$, it is symmetric in the y axis; and if $f(\theta) = f(\pi + \theta)$, it is symmetric in the origin (see Fig. 5.6.1).

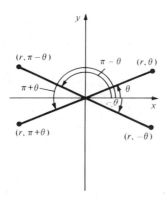

Figure 5.6.1. Symmetry in x axis: $f(\theta) = f(-\theta)$; symmetry in y axis: $f(\pi - \theta) = f(\theta)$; symmetry in origin: $f(\theta) = f(\pi + \theta)$.

Example 1 Plot the graph of $r = \cos 2\theta$ in the xy plane and discuss its symmetry.

Solution We know from the *cartesian* graph $y = \cos 2x$ (Fig. 5.1.27) that as θ increases from 0 to $\pi/4$, $r = \cos 2\theta$ decreases from 1 to 0. As θ continues from $\pi/4$ to $\pi/2$, $r = \cos 2\theta$ becomes negative and decreases to -1. Thus (r, θ) traces out the path in Fig. 5.6.2.

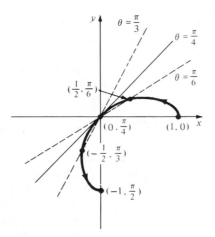

Figure 5.6.2. The graph $r = \cos 2\theta$ for $0 \leqslant \theta \leqslant \pi/2$.

We can complete the path as θ goes through all values between 0 and 2π, sweeping out the four petals in Fig. 5.6.3, or else we may use symmetry in the x and y axes. In fact, $f(-\theta) = \cos(-2\theta) = \cos 2\theta = f(\theta)$, so we have symmetry in the x axis. Also,

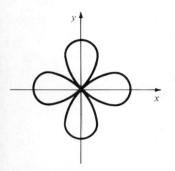

$$f(\pi - \theta) = \cos\left[2(\pi - \theta)\right]$$
$$= \cos 2\pi \cos 2\theta + \sin 2\pi \sin 2\theta$$
$$= \cos 2\theta = f(\theta),$$

which gives symmetry in the y axis. Finally,

$$f\left(\theta + \frac{\pi}{2}\right) = \cos 2\left(\theta + \frac{\pi}{2}\right) = \cos(2\theta + \pi)$$
$$= \cos 2\theta \cos \pi - \sin 2\theta \sin \pi$$
$$= -\cos 2\theta = -f(\theta).$$

Figure 5.6.3. The full graph $r = \cos 2\theta$, the "four-petaled rose."

Thus, the graph is unchanged when reflected in the x axis and the y axis. When we rotate by 90°, $r = f(\theta)$ reflects through the origin; that is, r changes to $-r$. ▲

Example 2   Sketch the graph of $r = f(\theta) = \cos 3\theta$.

Solution   The graph is symmetric in the x axis and, moreover, $f(\theta + \pi/3) = -f(\theta)$. This means that we need only sketch the graph for $0 \leqslant \theta \leqslant \pi/3$ and obtain the rest by reflection and rotations. Thus, we expect a three- or six-petaled rose. As θ varies from 0 to $\pi/3$, 3θ increases from 0 to π, and $\cos 3\theta$ decreases from 1 to 0. Hence, we get the graph in Fig. 5.6.4.

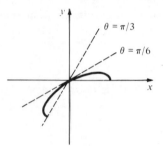

Figure 5.6.4. Beginning the graph of $r = \cos 3\theta$.

Reflect across the x axis to complete the petal and then rotate by $\pi/3$ and reflect through the origin (see Fig. 5.6.5). ▲

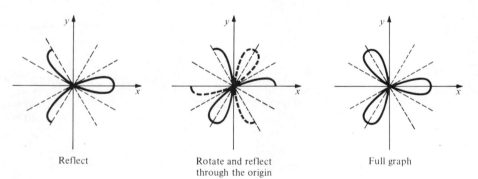

Figure 5.6.5. The graphing of $r = \cos 3\theta$.

Reflect   Rotate and reflect through the origin   Full graph

Example 3   If $f(\theta + \pi/2) = f(\theta)$, what does this tell you about the graph of $r = f(\theta)$?

Solution   This means that the graph will have the same appearance if it is rotated by 90°, since replacing θ by $\theta + \pi/2$ means that we rotate through an angle $\pi/2$. ▲

Figure 5.6.6 shows two other graphs in polar coordinates with striking symmetry. (The curve in (b) is discussed in Example 7 below.)

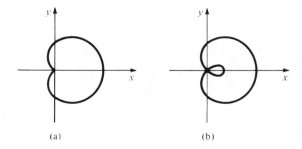

Figure 5.6.6.
(a) $r = 1 + \cos\theta$
("cardioid");
(b) $r = 1 + 2\cos\theta$
("limaçon").

(a)   (b)

Example 4   Convert the relation $r = 1 + 2\cos\theta$ to cartesian coordinates.

Solution   We substitute $r = \sqrt{x^2 + y^2}$ and $\cos\theta = x/r = x/\sqrt{x^2 + y^2}$ to get

$$\sqrt{x^2 + y^2} = 1 + \frac{2x}{\sqrt{x^2 + y^2}} \; .$$

That is, $x^2 + y^2 - \sqrt{x^2 + y^2} - 2x = 0$. ▲

Calculus can help us to draw graphs in polar coordinates by telling us the slope of tangent lines (see Fig. 5.6.7).

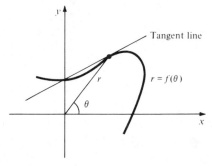

Figure 5.6.7. What is the slope of the tangent line?

This slope at a point (r, θ) is not $f'(\theta)$, since $f'(\theta)$ is the rate of change of r with respect to θ, while the slope is the rate of change of y with respect to x. To calculate dy/dx, we write

$$x = r\cos\theta = f(\theta)\cos\theta \quad \text{and} \quad y = r\sin\theta = f(\theta)\sin\theta.$$

This is a parametric curve with θ as the parameter. According to the formula $dy/dx = (dy/dt)/(dx/dt)$ from Section 2.4, with t replaced by θ,

$$\frac{dy}{dx} = \frac{dy/d\theta}{dx/d\theta} = \frac{f'(\theta)\sin\theta + f(\theta)\cos\theta}{f'(\theta)\cos\theta - f(\theta)\sin\theta} \; .$$

Dividing numerator and denominator by $\cos\theta$ gives

$$\frac{dy}{dx} = \frac{f'(\theta)\tan\theta + f(\theta)}{f'(\theta) - f(\theta)\tan\theta} \; . \tag{1}$$

Tangents to Graphs in Polar Coordinates

The slope of the tangent to the graph of $r = f(\theta)$ at (r, θ) is

$$\frac{(\tan\theta)\, dr/d\theta + r}{dr/d\theta - r\tan\theta} \; .$$

Example 5   (a) Find the slope of the line tangent to the graph of $r = 3\cos^2 2\theta$ at $\theta = \pi/6$.
(b) Find the slope of the line tangent to the graph of $r = \cos 3\theta$ at $(r,\theta) = (-1, \pi/3)$.

Solution   (a) Here, $f(\theta) = 3\cos^2 2\theta$, so $dr/d\theta = f'(\theta) = -12\cos 2\theta \sin 2\theta$ (by the chain rule). Now $f(\pi/6) = 3\cos^2(\pi/3) = \frac{3}{4}$ and $f'(\pi/6) = -12\cos(\pi/3)\sin(\pi/3)$
$= -12 \cdot \frac{1}{2} \cdot \sqrt{3}/2 = -3\sqrt{3}$. Thus formula (1) gives

$$\frac{dy}{dx} = \frac{-3\sqrt{3}\,(1/\sqrt{3}) + 3/4}{-3\sqrt{3} - (3/4)(1/\sqrt{3})} = \frac{3\sqrt{3}}{13},$$

so the slope of the tangent line is $3\sqrt{3}/13$.
(b) Here, $f(\theta) = \cos 3\theta$, so the slope is, by formula (1),

$$\frac{f'(\theta)\tan\theta + f(\theta)}{f'(\theta) - f(\theta)\tan\theta} = \frac{-3\sin 3\theta \tan\theta + \cos 3\theta}{-3\sin 3\theta - \cos 3\theta \tan\theta}$$

$$= \frac{1 - 3\tan 3\theta \tan\theta}{-\tan\theta - 3\tan 3\theta}.$$

Hence, at $\theta = \pi/3$, the slope is $1/-1.732 \approx -0.577$. ▲

Calculus can aid us in other ways. A local maximum of $f(\theta)$ will be a point on the graph where the distance from the origin is a local maximum, as in Fig. 5.6.8. The methods of Chapter 3 can be used to locate these local maxima (as well as the local minima).

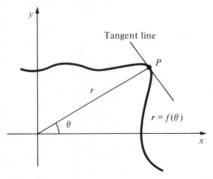

Figure 5.6.8. The point P corresponds to a local maximum point of $f(\theta)$.

Example 6   Calculate the slope of the line tangent to $r = f(\theta)$ at (r,θ) if f has a local maximum there. Interpret geometrically.

Solution   At a local maximum, $f'(\theta) = 0$. Plugging this into formula (1) gives

$$\frac{dy}{dx} = \frac{f(\theta)}{-f(\theta)\tan\theta}$$

$$= -\frac{1}{\tan\theta}.$$

Since dy/dx is the negative reciprocal of $\tan\theta$, the tangent line is perpendicular to the line from the origin to (r,θ). (See Fig. 5.6.8.) ▲

Example 7   Find the maxima and minima of $f(\theta) = 1 + 2\cos\theta$. Sketch the graph of $r = 1 + 2\cos\theta$ in the xy plane.

Solution   Here $dr/d\theta = -2\sin\theta$, which vanishes if $\theta = 0, \pi$. Also, $d^2r/d\theta^2 = -2\cos\theta$, which is -2 at $\theta = 0$, $+2$ at $\theta = \pi$. Hence $r = 3$, $\theta = 0$ is a local maximum and $r = -1$, $\theta = \pi$ is a local minimum. The tangent lines are vertical there. The curve passes through $r = 0$ when $\theta = \pm 2\pi/3$. The curve is symmetric in the x axis and can be plotted as in Fig. 5.6.9. ▲

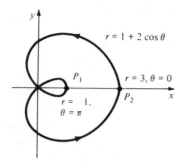

Figure 5.6.9. The maxima and minima of $1 + 2\cos\theta$ correspond to the points P_1 and P_2.

Exercises for Section 5.6

In Exercises 1–10, sketch the graph of the given function in polar coordinates. Also convert the given equation to cartesian coordinates.

1. $r = \cos\theta$

2. $r = 2\sin\theta$

3. $r = 1 - \sin\theta$

4. $r = \sin\left(\dfrac{\theta}{2}\right) + 1$

5. $r = 3$

6. $\sin\theta = 1$

7. $r^2 + 2r\cos\theta + 1 = 0$

8. $r^2\sin 2\theta = \frac{1}{2}$

9. $r = \sin 3\theta$

10. $r = \cos\theta - \sin\theta$

11. Describe the graph of the equation $r = $ constant.
12. What is the equation of a line through the origin?
13. If $f(\pi/2 - \theta) = f(\theta)$, what does this tell you about the graph of $r = f(\theta)$?
14. If $f(\theta + \pi) = -f(\theta)$, what does this tell you about the graph of $r = f(\theta)$?

Convert the relations in Exercises 15–22 to polar coordinates.

15. $x^2 + y^2 = 1$

16. $xy = 1$

17. $x^2 + xy + y^2 = 1$

18. $y = x^5 + x^3 + 2$

19. $y = \dfrac{1}{1 - x^2}$

20. $y = \dfrac{x}{1 + x}$

21. $y = x + 1$

22. $y = \dfrac{1}{x} + x^2$

In Exercises 23–32, find the slope of the tangent line at the indicated point.

23. $r = \tan\theta$; $\theta = \dfrac{2\pi}{3}$.

24. $r = \tan\theta$; $\theta = \dfrac{\pi}{4}$.

25. $r = 2\sin 5\theta$; $\theta = \dfrac{\pi}{2}$.

26. $r = 1 + 2\sin 2\theta$; $\theta = 0$.

27. $r = \cos 4\theta$; $\theta = \dfrac{\pi}{3}$.

28. $r = 2 - \sin\theta$; $\theta = \dfrac{\pi}{6}$.

29. $r = 3\sin\theta + \cos(\theta^2)$; $\theta = 0$.

30. $r = \sin 3\theta\cos 2\theta$; $\theta = 0$.

31. $r = \theta^2 + 1$; $\theta = 5$.

32. $r = \sec\theta + 2\theta^3$; $\theta = \dfrac{\pi}{6}$.

Find the maximum and minimum values of r for the functions in Exercises 33–38. Sketch the graphs in the xy plane.

33. $r = \cos 4\theta$

34. $r = 2\sin 5\theta$

35. $r = \tan\theta$

36. $r = 2 - \sin\theta$

37. $r = 1 + 2\sin 2\theta$

38. $r = \theta^2 + 1$

39. Find the maximum and minimum values of $r = \sin 3\theta\cos 2\theta$. Sketch its graph in the xy plane.

★40. Sketch the graph of $r = \sin^3 3\theta$.

Supplement to Chapter 5
Length of days

We outline an application of calculus to a phenomenon which requires no specialized equipment or knowledge for its observation—the setting of the sun.

Using spherical trigonometry or vector methods, one can derive a formula relating the following variables:

A = angle of elevation of the sun above the horizon;

l = latitude of a place on the earth's surface;

α = inclination of the earth's axis (23.5° or 0.41 radian);

T = time of year, measured in days from the first day of summer in the northern hemisphere (June 21);

t = time of day, measured in hours from noon.[5]

The formula reads:

$$\sin A = \cos l \sqrt{1 - \sin^2\alpha \cos^2\left(\frac{2\pi T}{365}\right)} \, \cos\left(\frac{2\pi t}{24}\right) + \sin l \sin\alpha \cos\left(\frac{2\pi T}{365}\right).$$

(1)

We will derive (1) in the supplement to Chapter 14. For now, we will simply assume the formula and find some of its consequences.[6]

At the time S of sunset, $A = 0$. That is,

$$\cos\left(\frac{2\pi S}{24}\right) = -\tan l \frac{\sin\alpha \cos(2\pi T/365)}{\sqrt{1 - \sin^2\alpha \cos^2(2\pi T/365)}}.$$

(2)

Solving for S, and remembering that $S \geqslant 0$ since sunset occurs after noon, we get

$$S = \frac{12}{\pi} \cos^{-1}\left[-\tan l \frac{\sin\alpha \cos(2\pi T/365)}{\sqrt{1 - \sin^2\alpha \cos^2(2\pi T/365)}} \right].$$

(3)

For example, let us compute when the sun sets on July 1 at 39° latitude. We have $l = 39°$, $\alpha = 23.5°$, and $T = 11$. Substituting these values in (3), we find $\tan l = 0.8098$, $2\pi T/365 = 0.1894$ (that is, 10.85°), $\cos(2\pi T/365) = 0.9821$, and $\sin\alpha = 0.3987$. Therefore, $S = (12/\pi)\cos^{-1}(-0.3447) = (12/\pi)(1.9227) = 7.344$. Thus $S = 7.344$ (hours after noon); that is, the sun sets at 7:20:38 (if noon is at 12:00).

For a fixed point on the earth, S may be considered a function of T. Differentiating (3) and simplifying, we find

$$\frac{dS}{dT} = -\frac{24}{365} \tan l \left\{ \frac{\sin\alpha \sin(2\pi T/365)}{\left[1 - \sin^2\alpha \cos^2(2\pi T/365)\right]\sqrt{1 - \sin^2\alpha \cos^2(2\pi T/365)\sec^2 l}} \right\}.$$

(4)

The critical points of S occur when $2\pi T/365 = 0$, π, or 2π; that is, $T = 0$, $365/2$, or $T = 365$—the first day of summer and the first day of winter. For the northern hemisphere, $\tan l$ is positive. By the first derivative test, $T = 0$ (or 365) is a local maximum and $T = 365/2$ a minimum. Thus we get the graph shown in Fig. 5.S.1.

---

[5] By noon we mean the moment at which the sun is highest in the sky. To find out when noon occurs in your area, look in a newspaper for the times of sunrise and sunset, and take the midpoint of these times. It will probably not be 12:00, but it should change only very slowly from day to day (except when daylight savings time comes or goes).

[6] If $\pi/2 - \alpha < |l| < \pi/2$ (inside the polar circles), there will be some values of t for which the right-hand side of formula (1) does not lie in the interval $[-1, 1]$. On the days corresponding to these values of t, the sun will never set ("midnight sun").

If $l = \pm \pi/2$, then $\tan l = \infty$, and the right-hand side does not make sense at all. This reflects the fact that, at the poles, it is either light all day or dark all day, depending upon the season.

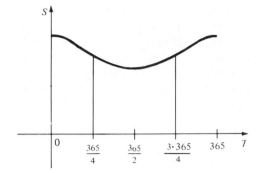

Figure 5.S.1. Sunset time as a function of the date. The sun sets late in summer, early in winter.

Using the first-order approximation, we can also determine how much the sunset time changes from one day to the next. If we set $\Delta T = 1$, we have

$$\Delta S \approx \frac{dS}{dT} \Delta T = \frac{dS}{dT},\qquad (5)$$

and dS/dT is given by (4). Thus, if the number of days after June 21 is inserted, along with the latitude l, formula (4) gives an approximation for the number of minutes later (or earlier) the sun will set the following evening. Note that the *difference* between sunset times on two days is the same whether we measure time in minutes from noon or by the clock ("standard time").

Formula (3) also tells us how long the days are as a function of latitude and day of the year. Plotting the formula on a computer (taking careful account of the polar regions) gives Fig. 5.S.2.[7] Graphs such as the one shown in Fig. 5.S.1 result if l is fixed and only T varies.

Figure 5.S.2. Day length as a function of latitude and day of the year.

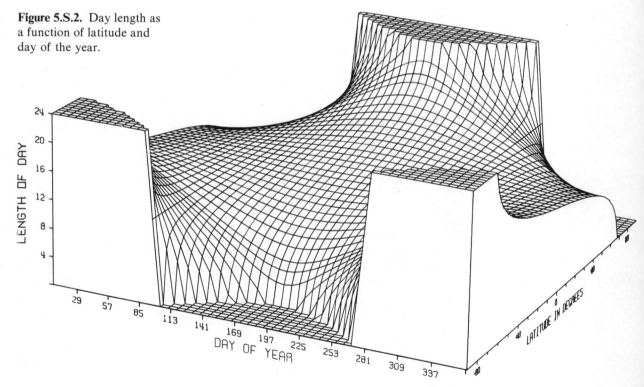

Example Use (4) and (5) to compute how many minutes earlier the sun will set on July 2 at latitude 39° than on July 1.

Solution Substituting $T = 11$ into (4), we obtain $\Delta S \approx -0.0055$; that is, the sun should set 0.0055 hour, or 20 seconds, earlier on July 2 than on July 1. Computing the time of sunset on July 2 by formula (3), with $T = 12$, we obtain $S = 7.338$, or

[7]Note that, in this figure, the date is measured from December 21st rather than June 21st.

7:20:17, which is 21 seconds earlier than the time computed by (4) and (5) for sunset on July 2. The error in the first-order approximation to ΔS is thus about 1 part in 20, or 5%. ▲

Differentiating formula (4), we find that the extreme values of dS/dT occur when $2\pi T/365 = \pi/2$ or $3\pi/2$. (These are the inflection points in Fig. 5.6.1.) When $2\pi T/365 = \pi/2$ (the first day of fall), dS/dT is equal to $-(24/365)\tan l \sin \alpha$; at this time, the days are getting shorter most rapidly.[8] When $2\pi T/365 = 3\pi/2$ (the first day of spring), the days are lengthening most rapidly, with $dS/dT = (24/365)\tan l \sin \alpha$.

It is interesting to note how this maximal rate, $(24/365)\tan l \sin \alpha$, depends on latitude. Near the equator, $\tan l$ is very small, so the rate is near zero, corresponding to the fact that seasons don't make much difference near the equator. Near the poles, $\tan l$ is very large, so the rate is enormous. This large rate corresponds to the sudden switch from nearly 6 months of sunlight to nearly 6 months of darkness. At the poles, the rate is "infinite." (Of course, in reality the change isn't quite sudden because of the sun's diameter, the fact that the earth isn't a perfect sphere, refraction by the atmosphere, and so forth.)

The reader who wishes to explore these topics further should read the supplements to Section 9.5 and Chapter 14 and try the following exercises.

Exercises for the Supplement to Chapter 5

1. According to the *Los Angeles Times* for July 12, 1975, the sun set at 8:06 P.M. The latitude of Los Angeles is 33.57° North. Guess what time the July 13 paper gave for sunset? What about July 14?

2. Determine your latitude (approximately) by measuring the times of sunrise and sunset.

3. Calculate d^2S/dT^2 to confirm that the inflection points of S occur at the first days of spring and fall.

4. Derive formula (4) by differentiating (3). (It may help you to use the chain rule with $u = \sin \alpha \cos(2\pi T/365)$ as the intermediate variable.)

5. At latitude l on the earth, on what day of the year is the day 13 hours long? Sketch the relation between T and l in this case.

6. Near springtime in the temperate zone (near 45°), show that the sunsets are getting later at a rate of about 1.6 minutes a day (or 11 minutes a week).

7. Planet VCH revolves about its sun once every 590 VCH "days." Each VCH "day" = 19 earth hours = 1140 earth minutes. Planet VCH's axis is inclined at 31°. What time is sunset, at a latitude of 12°, 16 VCH days after the first day of summer? (Assume that each VCH "day" is divided into 1440 "minutes.")

8. For which values of T does the sun never set if $l > 90° - \alpha$ (that is, near the North Pole)? Discuss.

9. (a) When the sun rises at the equator on June 21, how fast is its angle of elevation changing?
   (b) Using the linear approximation, estimate how long it takes for the sun to rise 5° above the horizon.

10. Let L be the latitude at which the sun has the highest noon elevation on day T. (a) Find a formula for L in terms of T. (b) Graph L as a function of T.

---

[8] One of the authors was stimulated to do these calculations by the observation that he was most aware of the shortening of the days at the beginning of the school year. This calculation provides one explanation for the observation; perhaps the reader can think of others.

Review Exercises for Chapter 5

Perform the conversions in Exercises 1–6.

1. $66°$ to radians.
2. $\pi/10$ radians to degrees.
3. $(4, \pi/2)$ from polar to cartesian coordinates.
4. $(3, 6)$ from cartesian to polar coordinates.
5. The equation $y = x^2$ from cartesian to polar coordinates.
6. The equation $r^2 = \cos 2\theta$ from polar to cartesian coordinates.

7. Prove that
$$\tan(\theta + \phi) = \frac{\tan\theta + \tan\phi}{1 - \tan\theta\tan\phi}.$$

8. Prove that
$$\tan\theta = \frac{\sin(\theta + \phi) + \sin(\theta - \phi)}{\cos(\theta + \phi) + \cos(\theta - \phi)}.$$

Refer to Fig. 5.R.1 for Exercises 9–12.

9. Find a.
10. Find b.

11. Find c.
12. Find d.

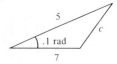
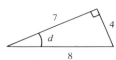

Figure 5.R.1. Find a, b, c, and d.

13. Side BC of the equilateral triangle ABC is trisected by points P and Q. What is the angle between AP and AQ?
14. A 100-meter building and a 200-meter building stand 500 meters apart. Where between the buildings should an observer stand so that the taller building subtends twice the angle of the shorter one?

Differentiate the functions in Exercises 15–34.

15. $y = -3\sin 2x$.
16. $y = 8\tan 10x$.
17. $y = x + x\sin 3x$.
18. $y = x^2\cos x^2$.
19. $f(\theta) = \theta^2 + \dfrac{\theta}{\sin\theta}$.
20. $g(x) = \sec x + \left[\dfrac{1}{x\cos(x + 1)}\right]$.
21. $h(y) = y^3 + 2y\tan(y^3) + 1$.
22. $x(\theta) = \left[\sin\left(\dfrac{\theta}{2}\right)\right]^{4/7} + \theta^9 + 11\sqrt{\theta} + 4$.
23. $y(x) = \cos(x^8 - 7x^4 - 10)$.
24. $f(y) = (2y^3 - 3\csc\sqrt{y})^{1/3}$.

25. $f(x) = \sec^{-1}[(x + \sin x)^2]$.
26. $f(x) = \cot^{-1}(20 - \sqrt[4]{x})$.
27. $r(\theta) = 6\cos^3(\theta^2 + 1) + 1$.
28. $r(\theta) = \dfrac{7\sin a\theta}{\sin b\theta + \cos c\theta}$; a, b, c constants.
29. $\sin^{-1}(\sqrt{x})$.
30. $\tan^{-1}(\sqrt{1 + x^2})$.
31. $\tan(\sin\sqrt{x})$.
32. $\tan(\cos x + \csc\sqrt{x})$.
33. $\sin^{-1}(\sqrt{x} + \cos 3x)$.
34. $\dfrac{\sin\sqrt{x}}{\cos^{-1}(\sqrt{x} + 1)}$.

35. Let $y = x^2 + \sin(2x + 1)$ and $x = t^3 + 1$. Find dy/dx and dy/dt.
36. Let $g = 1/r^2 + (r^2 + 4)^{1/3}$ and $r = \sin 2\theta$. Find dg/dr and $dg/d\theta$.
37. Let $h = x\sin^{-1}(x + 1)$ and $x = y - y^3$. Find dh/dx and dh/dy.
38. Let $f = x^{3/5} + \sqrt{2x^4 + x^2 - 6}$ and $x = y + \sin y$. Find df/dx and df/dy.
39. Let $f = \tan^{-1}(2x^3)$ and $x = a + bt$, where a and b are constants. Find df/dx and df/dt.
40. Let $y = \sin^{-1}(u^2)$ and $u = \dfrac{\cos x + 1}{x^2 + 1}$. Find dy/du and dy/dx.

Find the antiderivatives in Exercises 41–50.

41. $\displaystyle\int \sin 3x\, dx$.
42. $\displaystyle\int (x^{3/2} + \cos 2x)\, dx$.
43. $\displaystyle\int (4\cos 4x - 4\sin 4x)\, dx$.
44. $\displaystyle\int (\cos x + \sin 2x + \cos 3x)\, dx$.
45. $\displaystyle\int (3x^2\sin x^3 + 2x)\, dx$.
46. $\displaystyle\int \left[\left(\dfrac{4}{x^2}\right)\sin\left(1 + \dfrac{1}{x}\right)\right] dx$.
47. $\displaystyle\int \sin(u + 1)\, du$.
48. $\displaystyle\int (x^3 + 3\sec^2 x)\, dx$.
49. $\displaystyle\int \left[\dfrac{1}{(4 + y^2)}\right] dy$.
50. $\displaystyle\int \left[\dfrac{1}{(s^2 + 1)} - 2\cos 5s\right] ds$.

Find the definite integrals in Exercises 51–54.

51. $\displaystyle\int_0^\pi \left(\sin\dfrac{\theta}{2} + \sin 2\theta\right) d\theta$.
52. $\displaystyle\int_{-\pi}^{3\pi} \cos^2 u\, du$ [*Hint*: Use a trigonometric identity.]
53. $\displaystyle\int_{-1}^1 \left[\dfrac{1}{\sqrt{4 - x^2}}\right] dx$.
54. $\displaystyle\int_2^3 \left[\dfrac{4}{(5 + 6x^2)}\right] dx$.

55. (a) Verify that
$$\int \sin^{-1}x \, dx = x \sin^{-1}x + \sqrt{1 - x^2} + C.$$

(b) Differentiate $f(x) = \cos^{-1}x + \sin^{-1}x$ to conclude that f is constant. What is the constant?

(c) Find $\int \cos^{-1}x \, dx$.

(d) Find $\int \sin^{-1}3x \, dx$.

56. (a) If F and G are antiderivatives for f and g, show that $F(x)G(x) + C$ is the antiderivative of $f(x)G(x) + F(x)g(x)$.

(b) Find the antiderivative of $x \sin x - \cos x$.

(c) Find the antiderivative of $x \sin(x + 3) - \cos(x + 3)$.

57. Let $f(x) = x^3 - 3x + 7$.
(a) Find an interval containing zero on which f is invertible.

(b) Denote the inverse by g. What is the domain of g?

(c) Calculate $g'(7)$.

58. (a) Show that the function $f(x) = \sin x + x$ has an inverse g defined on the whole real line.

(b) Find $g'(0)$.

(c) Find $g'(2\pi)$.

(d) Find $g'\left(1 + \dfrac{\pi}{2}\right)$.

59. Let f be a function such that $f'(x) = 1/x$. (We will find such a function in the next chapter.) Show that the inverse function to f is equal to its own derivative.

60. Find a formula for the inverse function to $y = \sin^2 x$ on $[0, \pi/2]$. Where is this function differentiable?

61. A balloon is released from the ground 10 meters from the base of a 30-meter lamp post. The balloon rises steadily at 2 meters per second. How fast is the shadow of the balloon moving away from the base of the lamp after 4 seconds?

62. Three runners are going around a track which is an equilateral triangle with sides 50 meters long. If the runners are equally spread and all running counterclockwise at 20 kilometers per hour, at what rate is the distance between a pair of them changing when they are:
(a) leaving the vertices?

(b) arriving at the vertices?

(c) at the midpoints of the sides?

63. A pocket watch is swung counterclockwise on the end of its chain in a vertical circle; it undergoes circular motion, but not uniform, and the tension T in the chain is given by $T = m[(v^2/R) + g \cos \theta]$, where θ is the angle from the downward direction. Suppose the length is $R = 0.5$ meter, the watch mass is $m = 0.1$ kilograms, and the tangential velocity is $v = f(\theta)$; $g = 9.8$ meters per second per second.

(a) At what points on the circular path do you expect $dv/d\theta = 0$?

(b) Compute $dT/d\theta$ when $dv/d\theta = 0$.

(c) If the speed v is low enough at the highest point on the path, then the chain will become slack. Find the critical speed v_c below which the chain becomes slack.

64. A wheel of unit radius rolls along the x axis uniformly, rotating one half-turn per second. A point P on the circumference at time t (in seconds) has coordinates (x, y) given by $x = \pi t - \sin \pi t$, $y = 1 - \cos \pi t$.

(a) Find the velocities $dx/dt, dy/dt$ and the accelerations $d^2x/dt^2, d^2y/dt^2$.

(b) Find the *speed* $\sqrt{(dx/dt)^2 + (dy/dt)^2}$ when y is a maximum.

65. Refer to Fig. 5.R.2. A girl at point G on the riverbank wishes to reach point B on the opposite side of the river as quickly as possible. She starts off in a rowboat which she can row at 4 kilometers per hour, and she can run at 16 kilometers per hour. What path should she take? (Ignore any current in the river.)

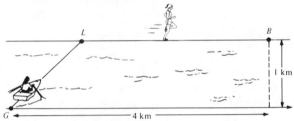

Figure 5.R.2. Find the best path from G to B.

66. The angle of deviation δ of a light ray entering a prism of Snell's index n and apex angle A is given by
$$\delta = \sin^{-1}(n \sin \rho) + \sin^{-1}(n \sin(A - \rho)) - A$$
where ρ depends on the angle of incidence ϕ of the light ray. (See Fig. 5.R.3.)

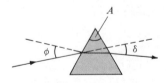

Figure 5.R.3. Light passing through a prism.

(a) Find $d\delta/d\rho$.

(b) Show that $d\delta/d\rho = 0$ occurs for $\rho = A/2$. This is a minimum value for δ.

(c) By Snell's law, $n = \sin \phi/\sin \rho$. Verify that $n = \sin[(A + \delta)/2]/\sin(A/2)$ when $\rho = A/2$.

67. A man is driving on the freeway at 50 miles per hour. He sees the sign in Fig. 5.R.4. How far from the sign is θ maximum? How fast is θ changing at this time?

Figure 5.R.4. At what distance is θ biggest?

68. Two lighthouses on a straight coastline are 10 kilometers apart. A ship sees the two lights, and the lines of sight make an angle of 120° with one another. How far from the shore could the ship possibly be?

69. Where is $f(x) = 1 + 2 \sin x \cos x$ concave upward? Concave downward? Find its inflection points and sketch its graph.

70. Sketch the graph of $y = x + \sin x$ on $[-2\pi, 2\pi]$.

71. Prove that $f(x) = x - 1 - \cos x$ is increasing on $[0, \infty)$. What inequality can you deduce?

72. Suppose that the graph of $r = f(\theta)$ is symmetric in the line $x = y$. What does this imply about f?

Sketch the graphs in the xy plane of the functions in Exercises 73–76 given in polar coordinates.

73. $r = \cos 6\theta$.

74. $r = 1 + 3 \cos \theta$.

75. $r = \sin \theta + \cos \theta$.

76. $r = \theta^2 \cos^2 \theta$. (Use your calculator to locate the zeros of $dr/d\theta$.)

In Exercises 77 and 78, find a formula for the tangent line to the graph at the indicated point.

77. $r = \cos 4\theta$; $\theta = \pi/4$.

78. $r = \dfrac{1}{1 - \sin^2\theta}$; $\theta = \pi/2$.

79. (a) Using a calculator, try to determine whether $\tan^{-1}[\tan(\pi + 10^{-20})]$ is in the interval $(0, 1)$.
(b) Do part (a) without using a calculator.
(c) Do some other calculator experiments with trigonometric functions. How else can you "fool" your calculator (or vice versa!)?

80. On a calculator, put any angle in radians on the display and successively press the buttons "sin" and "cos," alternatively, until you see the numbers 0.76816 and 0.69481 appear on the display.
(a) Try to explain this phenomenon from the graphs of $\sin x$ and $\cos x$, using composition of functions.
(b) Can you guess the solutions x, y of the equations $\sin(\cos x) = x$, $\cos(\sin y) = y$?
★(c) Using the mean value and intermediate value theorems, show that the equations in (b) have exactly one solution.

★81. If f is differentiable with a differentiable inverse, and $g(x) = f^{-1}(\sqrt{x})$, what is $g'(x)$?

★82. Consider water waves impinging on a breakwater which has two gaps as in Fig. 5.R.5. With the notation in the figure, analyze the maximum and minimum points for wave amplitude along the shore. The two wave forms emanating from P and Q can be described at any point R as $\alpha \cos(k\rho - \omega t)$, where ρ is the distance from the source P or Q; k, ω, α are constant. The net wave is described by their sum; the amplitudes do *not* add; ignore complications such as reflections of waves off the beach.[9]

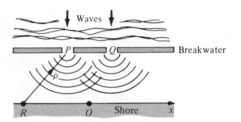

Figure 5.R.5. Find the wave pattern on the shore.

★83. Find a formula for the second derivative of an inverse function.

★84. Prove that the function
$$f(x) = \begin{cases} x^2 \sin \dfrac{1}{x} & x \neq 0, \\ 0 & x = 0 \end{cases}$$
is differentiable for all x, but that $f'(x)$ is not continuous at zero.

★85. Show that the function
$$f(x) = \begin{cases} x^4 \sin \dfrac{1}{x} & x \neq 0, \\ 0 & x = 0 \end{cases}$$
is twice differentiable but that the second derivative is not continuous.

---

[9] We recommend the book *Waves and Beaches* by W. Bascom (Anchor Books, 1965; revised, 1980) as a fascinating study of the mathematics, physics, engineering, and aesthetics of water waves.

Chapter 6

Exponentials and Logarithms

The inverse of $y = e^x$ is $x = \ln y$.

In our work so far, we have studied integer powers (b^n) and rational powers ($b^{m/n}$) as functions of a variable base, i.e., $y = x^m$ or $y = x^{m/n}$. In this chapter, we will study powers as functions of a variable exponent, i.e., $y = b^x$. To do this, we must first *define* b^x when x is not a rational number. This we do in Section 6.1; the rest of the chapter is devoted to the differential and integral calculus of the exponential functions $y = b^x$ and their inverses, the logarithms. The special value $b = e = 2.7182818285\ldots$ leads to especially simple formulas.

6.1 Exponential Functions

Any real number can be used as an exponent if the base is positive.

In Section R.3, we reviewed the properties of the powers b^r, where r was first a positive integer and then a negative number or a fraction. The calculus of the *power function* $g(x) = x^r$ has been studied in Section 2.3. We can also consider b as fixed and r as variable. This gives the function $f(x) = b^x$, whose domain consists of all rational numbers. The following example shows how such *exponential functions* occur naturally and suggests why we would like to have them defined for all real x.

Example 1 The mass of a bacterial colony doubles after every hour. By what factor does the mass grow after: (a) 5 hours; (b) 20 minutes; (c) $2\frac{1}{2}$ hours; (d) x hours, if x is rational?

Solution (a) In 5 hours, the colony doubles five times, so it grows by a factor of $2 \cdot 2 \cdot 2 \cdot 2 \cdot 2 = 2^5 = 32$.

(b) If the colony grows by a factor of k in $\frac{1}{3}$ hour, it grows by a factor $k \cdot k \cdot k = k^3$ in 1 hour. Thus $k^3 = 2$, so $k = 2^{1/3} = \sqrt[3]{2} \approx 1.26$.

(c) In $\frac{1}{2}$ hour, the colony grows by a factor of $2^{1/2}$, so it grows by a factor of $(2^{1/2})^5 = 2^{5/2} \approx 5.66$ in $2\frac{1}{2}$ hours.

(d) Reasoning as in parts (a), (b), and (c) leads to the conclusion that the mass of the colony grows by a factor of 2^x in x hours. ▲

Time is not limited to rational values; we should be able to ask how much the colony in Example 1 grows after $\sqrt{3}$ hours or π hours. Since the colony is increasing in size, we are led to the following mathematical problem: Find a function f defined for all real x such that f is increasing, and $f(x) = 2^x$ for all rational x.

Computing some values of 2^x and plotting, we obtain the graph shown in Fig. 6.1.1. By doing more computations, we can fill in more points between those in Fig. 6.1.1, and the graph looks more and more like a smooth curve. It is therefore plausible that a smooth curve can be drawn through all these points (Fig. 6.1.2).

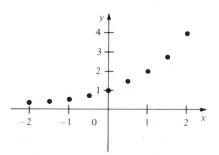

Figure 6.1.1. Some points on the graph $y = 2^x$ for rational x.

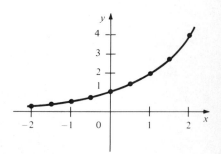

Figure 6.1.2. Interpolating a smooth curve between the points in the previous graph.

The proof that one can really fill in the graph of b^x for irrational x to produce a continuous function defined for all real x is rather technical, so we shall omit it (it is found in more theoretical texts such as the authors' *Calculus Unlimited*). It can also be shown that the laws of exponents, as given in Section R.3 for rational powers, carry over to all real x.

Real Powers

(a) For any $b > 0$, $f(x) = b^x$ is a continuous function.
(b) Let b, c, x, and y be real numbers with $b > 0$ and $c > 0$. Then:
  1. $b^{x+y} = b^x b^y$.
  2. $b^{xy} = (b^x)^y$.
  3. $(bc)^x = b^x c^x$.
  4. f is increasing if $b > 1$, constant if $b = 1$, and decreasing if $0 < b < 1$.

Property 4 says that $b^x < b^y$ whenever $b > 1$ and $x < y$; i.e., larger powers of $b > 1$ give larger numbers. If $b < 1$ and $x < y$, then $b^x > b^y$. For example, for $b = 2$, $2^3 < 2^4$, but for $b = \frac{1}{2}$, $(\frac{1}{2})^3 > (\frac{1}{2})^4$. If you examine property 4 for rational powers given in Section R.3, you will see that it corresponds to property 4 here.

Example 2   Simplify (a) $(\sqrt{3^{\pi}})(3^{-\pi/4})$ and (b) $(2^{\sqrt{3}} + 2^{-\sqrt{3}})(2^{\sqrt{3}} - 2^{-\sqrt{3}})$.

Solution   (a) $\sqrt{3^{\pi}}\,3^{-\pi/4} = (3^{\pi})^{1/2}3^{-\pi/4} = 3^{\pi/2 - \pi/4} = 3^{\pi/4}$.

(b) $(2^{\sqrt{3}} + 2^{-\sqrt{3}})(2^{\sqrt{3}} - 2^{-\sqrt{3}}) = (2^{\sqrt{3}})^2 - (2^{-\sqrt{3}})^2 = 2^{2\sqrt{3}} - 2^{-2\sqrt{3}}$. ▲

Sometimes the notation $\exp_b x$ is used for b^x; exp stands for "exponential." One reason for this is typographical: an expression like $\exp_b(x^2/2 + 3x)$ is easier on the eyes and on the typesetter than $b^{(x^2/2 + 3x)}$. Another reason is mathematical: when we write $\exp_b x$, we indicate that we are thinking of b^x as a *function of x*.

Example 3   Which is larger, $2^{\sqrt{5}}$ or $4^{\sqrt{2}}$? (Do not use a calculator.)

Solution   We may write $4^{\sqrt{2}}$ as $(2^2)^{\sqrt{2}} = 2^{2\sqrt{2}} = 2^{\sqrt{8}}$. Since $\sqrt{8} > \sqrt{5}$, it follows from property 4 in the previous box that $4^{\sqrt{2}} = 2^{\sqrt{8}}$ is larger than $2^{\sqrt{5}}$. ▲

▦ Calculator Discussion

When we compute $2^{\sqrt{3}}$ on a calculator, we are implicitly using the continuity of $f(x) = 2^x$. The calculator in fact computes a rational power of 2—namely, $2^{1.732050808}$, where 1.732050808 is a decimal approximation to $\sqrt{3}$. Continuity of $f(x)$ means precisely that if the decimal approximation to x is good, then the answer is a good approximation to $f(x)$. The fact that f is increasing gives more information. For example, since

$$\frac{1732}{1000} < \sqrt{3} < \frac{17{,}321}{10{,}000},$$

we can be sure that

$$2^{1732/1000} = 3.32188 \cdots < 2^{\sqrt{3}} < 2^{17{,}321/10{,}000} = 3.32211 \cdots,$$

so $2^{\sqrt{3}} = 3.322$ is correct to three decimal places. ▲

Example 4   (a) Sketch the graphs of $\exp_2$, $\exp_{3/2}$, $\exp_1$, $\exp_{2/3}$, and $\exp_{1/2}$. (b) How are the graphs of $\exp_{1/2}$ and $\exp_2$ related?

Solution   (a) $1^x = 1$ for all x_1, so $\exp_1$ is the constant function with graph $y = 1$. The functions $\exp_2$ and $\exp_{3/2}$ are increasing, with $\exp_2 x > \exp_{3/2} x$ for $x > 0$ and $\exp_2 x < \exp_{3/2} x$ for $x < 0$ (by property 4).

Likewise, $\exp_{2/3} x > \exp_{1/2} x$ for $x > 0$ and $\exp_{2/3}$ and $\exp_{1/2}$ are decreasing. Using these facts and a few plotted points, we sketch the graphs in Fig. 6.1.3.

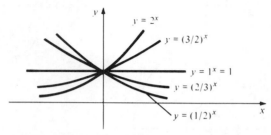

Figure 6.1.3. $y = \exp_b x$ for $b = \frac{1}{2}, \frac{2}{3}, 1, \frac{3}{2}$, and 2.

(b) Using the properties of exponentiation, $\exp_{1/2} x = (\frac{1}{2})^x = 2^{-x} = \exp_2(-x)$, so the graph $y = \exp_{1/2} x$ is obtained by reflecting $y = \exp_2 x$ in the y axis; $y = \exp_{2/3} x$ and $y = \exp_{3/2} x$ are similarly related. ▲

Example 5 Match the graphs and functions in Figure 6.1.4.

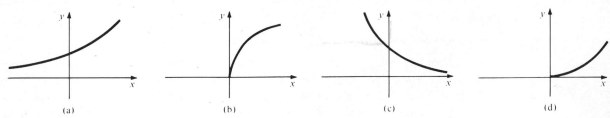

(a)　　　　　　　　(b)　　　　　　　　(c)　　　　　　　　(d)

Figure 6.1.4. Match the graphs and functions: (A) $y = x^{\sqrt{3}}$; (B) $y = x^{1/\sqrt{3}}$; (C) $y = (\sqrt{3})^x$; (D) $y = (1/\sqrt{3})^x$.

Solution Only functions (A) and (B) have graphs going through the origin; $x^{\sqrt{3}} < x$ for $x < 1$, so (A) matches (d) and (B) matches (b). The function $y = (\sqrt{3})^x$ is increasing, so (C) matches (a) and (D) matches (c). ▲

Example 6 Match the graphs and functions in Fig. 6.1.5.

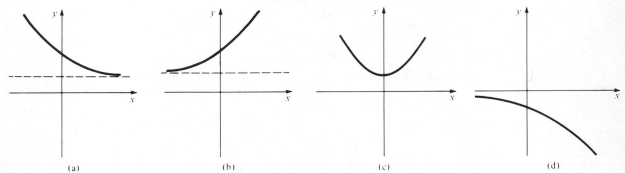

(a)　　　　　　　　(b)　　　　　　　　(c)　　　　　　　　(d)

Figure 6.1.5. Match the graphs and functions: (A) $y = -2^x$; (B) $y = x^2 + 1$; (C) $y = 2^{-x} + 1$; (D) $y = 2^x + 1$.

Solution (a) must be the graph of $y = (\frac{1}{2})^x$ shifted up one unit, so it matches (C).
(b) is the graph of 2^x shifted up one unit, so it matches (D).
(c) is a parabola, so it matches (B).
(d) is $y = 2^x$ reflected in the x axis, so it matches (A). ▲

Example 7 A curve whose equation is polar coordinates has the form $r = b^\theta$ for some b is called an *exponential spiral*. Sketch the exponential spiral for $b = 1.1$.

Solution We observe that $r \to \infty$ as $\theta \to \infty$ and that $r \to 0$ as $\theta \to -\infty$. To graph the spiral, we note that r increases with θ; we then plot several points (using a calculator) and connect them with a smooth curve. (See Fig. 6.1.6.) Every turn of the spiral is $(1.1)^{2\pi} \approx 1.82$ times as big as the previous one. ▲

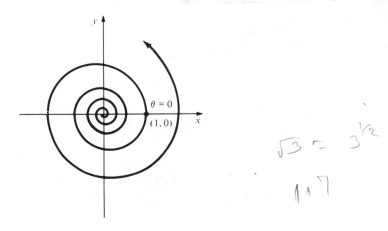

Figure 6.1.6. The exponential spiral $r = (1.1)^\theta$.

Exercises for Section 6.1

1. The mass of a certain bacterial culture triples every 2 hours. By what factor does the mass grow after (a) 4 hours; (b) 6 hours, (c) 7 hours, (d) x hours?

2. The amount of a radioactive substance in an ore sample halves every 5 years. How much is left after (a) 10 years, (b) 30 years, (c) 45 years, (d) x years?

▣3. The amount of money in a bank account increases by 8% after being deposited for 1 year. How much is there in the account after (a) 2 years (b) 10 years, (c) x years?

▣4. A lender supplies an amount P to a borrower at an annual interest rate of r. After t years with interest compounded n times a year, the borrower will owe the lender the amount $A = P[1 + (r/n)]^{nt}$ (compound interest). Suppose $P = 100$, $r = 0.06$, $t = 2$ years. Find the amount owed for interest compounded: (a) monthly, (b) weekly, (c) daily, (d) twice daily. Draw a conclusion.

Simplify the expressions in Exercises 5–12.

5. $(2^{\sqrt{2}})^{\sqrt{2}}$

6. $(2^{1/\sqrt{2}})^2$

7. $\dfrac{3^{\sqrt{3}} + 3^{2\sqrt{3}}}{3^{\sqrt{3}}}$

8. $\dfrac{\pi^{-\sqrt{2}} - \pi^{\sqrt{2}}}{\pi^{\sqrt{2}} + 1}$

9. $\dfrac{5^{\pi/2} \cdot 10^{\pi}}{15^{-\pi}}$

10. $\dfrac{8^{\sqrt{3}} 2^{-\sqrt{12}}}{4^{2\sqrt{3}}}$

11. $(3^{\pi} - 2^{(3^{1/2})})(3^{-\pi} + 2^{-(3^{1/2})})$

12. $\dfrac{(\sqrt{3})^{\pi} - (\sqrt{2})^{\sqrt{5}}}{\sqrt[4]{3^{\pi}} + 2^{\sqrt{5}/4}}$

In Exercises 13–16, decide which number is larger without using a calculator.

13. $3^{\sqrt{2}}$ or $9^{1/\sqrt{3}}$

15. $2^{\sqrt{3}}$ or $3^{\sqrt{2}}$

14. $8^{\sqrt{\pi}}$ or $2^{3\pi}$

16. $10^{\sqrt{8}}$ or $8^{\sqrt{10}}$

Sketch the graphs of the functions in Exercises 17–24.

17. $f(x) = \exp_3(x)$ 18. $f(x) = \exp_{1/3}(x)$

19. $f(x) = \exp_{4/3}(x)$ 20. $f(x) = \exp_{3/4}(x)$

21. $y = 2^{(x^2)}$ 22. $y = (2^x)^2$

23. $y = 2^{\sqrt{x}}$ 24. $y = 2^{1/x}$

25. How are the graphs of $\exp_3 x$ and $\exp_{1/3}(x)$ related?

26. How are the graphs of $\exp_{4/3}(x)$ and $\exp_{3/4}(x)$ related?

27. How are the graphs of $\exp_3(x)$ and $\exp_3(-x)$ related?

28. How are the graphs of $\exp_5(x)$ and $\exp_5(-x)$ related?

29. Match the graphs and functions in Fig. 6.1.7.

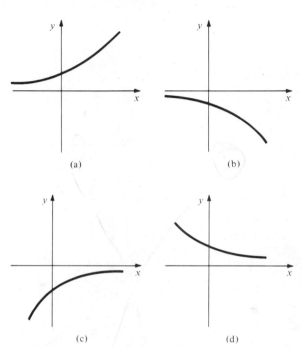

(a)   (b)

(c)   (d)

Figure 6.1.7. Match the graphs and functions:
(A) $y = -3^x$;   (B) $y = 3^{-x}$;
(C) $y = -3^{-x}$;   (D) $y = 3^x$.

30. Match the graphs and functions in Fig. 6.1.8.

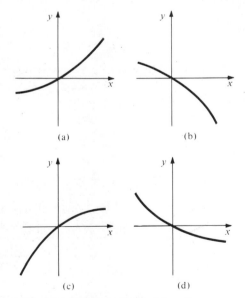

(a)   (b)

(c)   (d)

Figure 6.1.8. Match the graphs and functions:
(A) $y = -2^{-x} + 1$;   (B) $y = 2^x - 1$;
(C) $y = -2^x + 1$;   (D) $y = 2^{-x} - 1$.

31. Match the graphs and functions in Fig. 6.1.9.

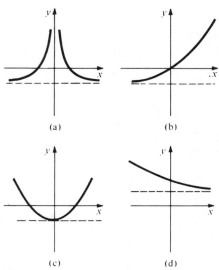

(a)   (b)

(c)   (d)

Figure 6.1.9. Match the graphs and functions:
(A) $y = x^2 - 1$;   (B) $y = 2^x - 1$;
(C) $y = 2^{-x} + 1$;   (D) $y = x^{-2} - 1$.

32. Match the graphs and functions in Fig. 6.1.10.

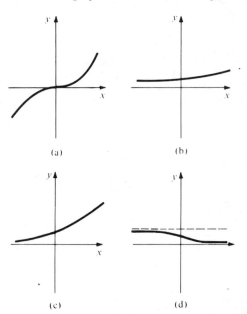

(a)   (b)

(c)   (d)

Figure 6.1.10. Match the graphs and functions:
(A) $y = x^3$ (B) $y = \sqrt{3^x}$
(C) $y = (2^x + 1)^{-1}$ (D) $y = 10^{x/100}$

33. Graph the exponential spiral $r = (1.2)^\theta$.
34. Graph the exponential spiral $r = (1/1.1)^\theta$.
35. Graph the exponential spiral $r = (1.1)^{2\theta}$.
36. Graph the exponential spiral $r^2 = (1.1)^\theta$.
37. Graph $y = 3^{x+2}$ by "shifting" the graph of $y = 3^x$ by 2 units to the left. Graph $y = 9(3^x)$ by "stretching" the graph of $y = 3^x$ by a factor of 9 in the direction of the y axis. Compare the two results. In general, how does shifting the graph of $y = 3^x$ by k units to the left compare with stretching the graph by a factor of 3^k in the direction of the y axis?
38. Carefully graph the following functions on one set of axes: (a) $f(x) = 2^x$, (b) $g(x) = x^2 + 1$, (c) $h(x) = x + 1$. Can you see why $f'(1)$ should be between 1 and 2?
39. From the graph of $f(x) = 2^x$, make a reasonable sketch of what the function $f'(x)$ might look like.
40. Answer the question in Exercise 39 for $f(x) = 2^{-x}$.
41. Compute the ratio of the area under the graph of $y = 3^x$ between $x = 0$ and $x = 2$ to that between $x = 2$ and $x = 4$ (see Exercise 37).
42. Compare the areas under the graph of $y = 3^x$ between $x = 1$ and $x = 2$ and between $x = 2$ and $x = 3$ (see Exercise 37).

Solve for x in Exercises 43–46.

43. $10^x = 0.001$
44. $5^x = 1$
45. $2^x = 0$
46. $x - 2\sqrt{x} - 3 = 0$ (*Hint*: factor)

6.2 Logarithms

The function $\log_b$ *is the inverse of* $\exp_b$.

If $b > 1$, the function $\exp_b x = b^x$ is positive, increasing, and continuous. As $x \to \infty$, $\exp_b x$ becomes arbitrarily large, while as $x \to -\infty$, $\exp_b x$ decreases to zero. (See Review Exercise 85 for an outline of a proof of these facts.) Thus the range of $\exp_b$ is $(0, \infty)$. It follows from the inverse function test in Section 5.3 that $\exp_b$ has a unique inverse function with domain $(0, \infty)$ and range $(-\infty, \infty)$. This function is called $\log_b$. By the definition of an inverse function, $\log_b y$ *is that number* x *such that* $b^x = y$. The number b is called the *base* of the logarithm.

Example 1   Find $\log_3 9$, $\log_{10} 10^a$, and $\log_9 3$.

Solution   Let $x = \log_3 9$. Then $3^x = 9$. Since $3^2 = 9$, x must be 2. Similarly, $\log_{10} 10^a$ is a, and $\log_9 3 = \frac{1}{2}$ since $9^{1/2} = 3$. ▲

The graph of $\log_b x$ for $b > 1$ is sketched in Fig. 6.2.1 and is obtained by flipping over the graph of $\exp_b x$ along the diagonal. As usual with inverse functions, the label y in $\log_b y$ is only temporary and merely stresses the fact that $\log_b y$ is the inverse of $y = \exp_b x$. From now on we will usually use the variable name x and write $\log_b x$. In Fig. 6.2.1, the negative y axis is a vertical asymptote for $y = \log_b x$.

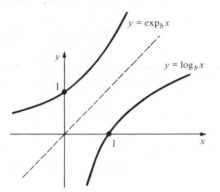

Figure 6.2.1. The graphs of $y = \exp_b x$ and $y = \log_b x$ for $b > 1$.

Example 2   Sketch the graphs of $\log_2 x$ and $\log_{1/2} x$.

Solution   This is done by flipping the graphs of 2^x and $(\frac{1}{2})^x$, as shown in Fig. 6.2.2. The graphs of $\log_2 x$ and $\log_{1/2} x$ are reflections of one another in the x axis. ▲

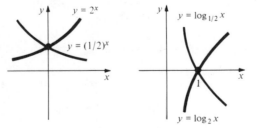

Figure 6.2.2. Exponential and logarithm functions with base $= 2 > 1$ and base $= \frac{1}{2} < 1$.

Notice that for $b > 1$, $\log_b x$ is increasing. If $b < 1$, $\exp_b x$ is decreasing and so is $\log_b x$. However, while $\exp_b x$ is always positive, $\log_b x$ can be either positive or negative. Since $\exp_b 0 = 1$, we can conclude that $\log_b 1 = 0$; since $\exp_b 1 = b$, $\log_b b = 1$. These properties are summarized in the following box.

Properties of $\log_b x$

Definition: $\log_b x$ is that number y such that $b^y = x$; i.e., $b^{\log_b x} = x$.

1. $\log_b x$ is defined for $x > 0$ and $b > 0$ (but $\log_b x$ can be positive or negative).
2. $\log_b 1 = 0$.
3. If $b < 1$, $\log_b x$ is a decreasing function of x; if $b > 1$, $\log_b x$ is increasing.

Example 3   Match the graphs and functions in Fig. 6.2.3.

Figure 6.2.3. Match the graphs and functions:
(A) $y = 3^x$;
(B) $y = \log_3 x$;
(C) $y = \log_{1/3} x$;
(D) $y = (\frac{1}{3})^x$.

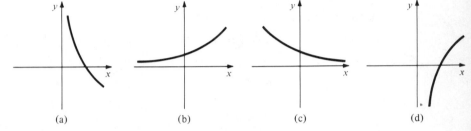

(a)   (b)   (c)   (d)

Solution   The functions (A) and (B) are increasing, but only (A) is defined for all x, so (A) matches (b) and (B) matches (d). Of (C) and (D), only (D) is defined for all x, so (C) matches (a) and (D) matches (c). ▲

From the laws of exponents given in Section 6.1, we can read off the corresponding laws for $\log_b x$.

Laws of Logarithms

1. $\log_b(xy) = \log_b x + \log_b y$ and $\log_b(x/y) = \log_b x - \log_b y$.
2. $\log_b(x^y) = y \log_b x$.
3. $\log_b x = (\log_b c)(\log_c x)$.

To prove Law 1, for instance, we remember that $\log_b(xy)$ is that number u such that $\exp_b u = xy$. So we must check that

$$\exp_b(\log_b x + \log_b y) = xy.$$

By the rule $\exp_b(v + w) = \exp_b v \cdot \exp_b w$, the left hand side is given by $\exp_b(\log_b x)\exp_b(\log_b y) = xy$, as is the right side. The other laws are proved in the same way (see Exercises 31 and 32.)

Example 4   Simplify $\log_{10}[(100^{3.2})\sqrt{10}\,]$ without using a calculator.

Solution
$$\log_{10}\left(100^{3.2}\sqrt{10}\,\right) = \log_{10}100^{3.2} + \log_{10}\sqrt{10} \qquad \text{(Law 1)}$$
$$= 3.2 \log_{10}100 + \tfrac{1}{2}\log_{10}10 \qquad \text{(Law 2)}$$
$$= 3.2 \log_{10}10^2 + \tfrac{1}{2}$$
$$= 6.4 + \tfrac{1}{2} = 6.9. \; ▲$$

Example 5   What is the relationship between $\log_b c$ and $\log_c b$?

Solution   Substituting b for x in Law 3, we get

$$\log_b b = (\log_b c)(\log_c b).$$

But $\log_b b = 1$, so $\log_b c = 1/\log_c b$. ▲

Example 6   Solve for x: (a) $\log_x 5 = 0$, (b) $\log_2(x^2) = 4$, (c) $2\log_3 x + \log_3 4 = 2$.

Solution   (a) $\log_x 5 = 0$ means $x^0 = 5$. Since any number to the zero power is 1, there is no solution for x.
(b) $\log_2(x^2) = 4$ means $2^4 = x^2$. This is the same as $16 = x^2$. Hence, $x = \pm 4$.
(c) Solving for $\log_3 x$, we get

$$\log_3 x = 1 - \tfrac{1}{2}\log_3 4 = 1 - \log_3 2.$$

Thus,

$$x = 3^{\log_3 x} = 3^{1 - \log_3 2} = 3 \cdot 3^{-\log_3 2} = \tfrac{3}{2}. \quad ▲$$

We conclude this section with a word problem involving exponentials and logarithms.

▥ Example 7   The number N of people who contract influenza t days after a group of 1000 people are put in contact with a single person with influenza can be modeled by $N = 1000/(1 + 999 \cdot 10^{-0.17t})$.

(a) How many people contract influenza after 20 days?
(b) Will everyone eventually contract the disease?
(c) In how many days will 600 people contract the disease?

Solution   (a) According to the given model, we substitute $t = 20$ into the fomrula for N to give

$$N = \frac{1000}{1 + 999 \cdot 10^{-0.17 \cdot 20}} = \frac{1000}{1 + 999 \cdot 10^{-3.4}} = \frac{1000}{1.398} \approx 715.$$

Thus 715 people will contract the disease after 20 days. (The calculation was done on a calculator.)
(b) "Eventually" is interpreted to mean "for t very large." For t large, $-0.17t$ will be a large negative number and so $10^{-0.17t}$ will be nearly zero (equivalently $10^{-0.17t} = 1/10^{0.17t}$ and $10^{0.17t}$ will be very large if t is very large). Thus the denominator in N will be nearly 1 and so N itself is nearly 1000. For instance, it is eventually larger than 999.9999. Thus, according to the model, all of the 1000 will eventually contract the disease.
(c) We must find the t for which $N = 600$:

$$600 = \frac{1000}{1 + 999 \cdot 10^{-0.17t}}, \quad \text{so} \quad (600)(1 + 999 \cdot 10^{-0.17t}) = 1000.$$

Thus $1 + 999 \cdot 10^{-0.17t} = \tfrac{10}{6} = \tfrac{5}{3}$. Solving for $10^{-0.17t}$, $10^{-0.17t} = (2/3)/999$. Therefore, $-0.17t = \log_{10}((2/3)/999) \approx -3.176$ (from our calculator) and so $t = 3.176/0.17 \approx 18.68$ days. ▲

Exercises for Section 6.2

Compute the logarithms in Exercises 1–10.

1. $\log_2 4$
2. $\log_3 81$
3. $\log_{10} 0.01$
4. $\log_{10}(10^{-8})$
5. $\log_{10}(0.001)$
6. $\log_{10}(1000)$
7. $\log_3 3$
8. $\log_5 125$
9. $\log_{1/2} 2$
10. $\log_{1/3} 9$

Sketch the graphs of the functions in Exercises 11–14.

11. $y = \log_{10} x$
12. $y = \log_{1/10} x$
13. $y = 8 \log_2 x$
14. $y = \log_{1/2}(x + 1)$

15. Match the graphs and functions in Fig. 6.2.4.

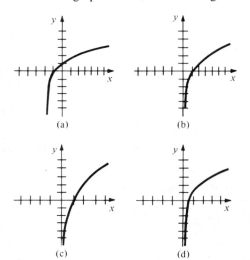

(a)   (b)

(c)   (d)

Figure 6.2.4. Match the graphs and functions:
(A) $y = \log_2 x$;   (B) $y = 2 \log_2 x$;
(C) $y = \log_2(x + 2)$;   (D) $y = \log_2(2x)$.

16. Match the graphs and functions in Fig. 6.2.5.

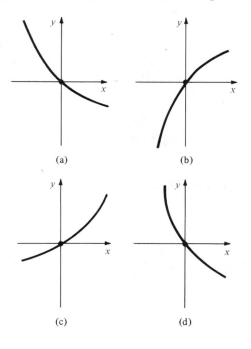

(a)   (b)

(c)   (d)

Simplify the expressions in Exercises 17–24 without using a calculator.

17. $\log_2(2^8/8^2)$
18. $\log_2(3^5 \cdot 4^{-6} \cdot 9^{-5.2})$
19. $2^{\log_2 4}$
20. $2^{\log_2 b}$
21. $\log_2(2^b)$
22. $\log_b[b^2 \cdot (2b)^3 \cdot (8b)^{-2/3}]$
23. $\log_b(b^{2x}/2b)$
24. $(\log_b c^2)(\log_c b^2)$

Given that $\log_7 2 \approx 0.356$, $\log_7 3 \approx 0.565$, and $\log_7 5 \approx 0.827$, calculate the quantities in Exercises 25–28 without using a calculator.

25. $\log_7(7.5)$
26. $\log_7 6$
27. $\log_7(3.333 \cdots)$
28. $\log_7(1.5)$

29. Suppose that $\log_b 10 = 2.5$. Use a $\log_{10}$ table or a calculator to find an approximate value for b.
30. Which is larger, $\log_{11} 2$ or $\log_{10} 2$?. How about $\log_{1/2} 2$ or $\log_{1/4} 2$? Do not use a calculator.

Use the definition of $\log_b x$ to prove the identities in Exercises 31 and 32.

31. $\log_b(x^y) = y \log_b x$.
32. $\log_b x = (\log_b c)(\log_c x)$.
33. Verify the formula $\log_{a^n} x = (1/n)\log_a x$. What restrictions must you make on a?
34. Prove that $\log_{b^n}(x^m) = (m/n)\log_b x$.

Write the expressions in Exercises 35–38 as sums of (rational) multiples of $\log_b A$, $\log_b B$, and $\log_b C$.

35. $\log_b(A^2 B/C)$.
36. $\log_b(\sqrt{AB^3}/C^4 B^2)$.
37. $2\log_b(A\sqrt{1 + B}/C^{1/3}B) - \log_b[(B + 1)/AC]$.
38. $\log_{b^2}(A^{-1}B^3) - \log_{b^{-1}}(C^{-1}B^2)$. [*Hint:* Use Exercise 34.]

Solve for x in Exercises 39–46.

39. $\log_x 9 = 2$
40. $\log_x 27 = 3$
41. $\log_3 x = 2$
42. $\log_3 x = 3$
43. $\log_2 x = \log_2 5 + 3 \log_2 3$
44. $\log_{25}(x + 1) = \log_5 x$
45. $\log_x(1 - x) = 2$
46. $\log_x(2x - 1) = 2$

Figure 6.2.5. Match the graphs with:
(A) $y = \log_2(x + 1)$;
(B) $y = \log_{1/2}(x + 1)$;
(C) $y = (\frac{1}{2})^x - 1$;
(D) $y = 2^x - 1$.

47. A biologist measures culture growth and gets the following data: After 1 day of growth the count is 1750 cells. After 2 days it is 3065 cells. After 4 days it is 9380 cells. Finish filling out the following table by using a table or calculator:

x = number of days of growth	1	2	4
n = number of cells	1750	3065	9380
$y = \log_{10}n$			

(a) Verify that the data fit a curve of the form $n = Mb^x$ by examining the linear equation $y = (\log_{10}b)x + \log_{10}M$ (with respect to the y and x values in the table). Using the slope and y intercept to evaluate M and b. If the biologist counts the culture on the fifth day, predict how many cells will be found.

(b) Suppose that you had originally known that the data would satisfy a relation of the form $n = Mb^x$. Solve for M and b without using logarithms.

48. Color analyzers are constructed from photomultipliers and various electronic parts to give a scale reading of light intensities falling on a light probe. These scales read relative densities directly, and the scale reading S can be given by $S = k\log_{10}(I/I')$ where I is a reference intensity, I' is the new intensity, and k is a positive constant.

(a) Show that the scale reads zero when $I = I'$.

(b) Assume the needle is vertical on the scale when $I = I'$. Find the sign of S when $I' = 2I$ and $I' = I/2$.

(c) In most photographic applications, the range of usable values of I' is given by $I/8 \leqslant I' \leqslant 8I$. What is the scale range?

49. The *opacity* of a photographic negative is the ratio I_0/I, where I_0 is the reference light intensity and I the intensity transmitted through the negative. The *density* of a negative is the quantity $D = \log_{10}(I_0/I)$. Find the density for opacities of 2, 4, 8, 10, 100, 1000.

50. The loudness, in decibels (dB), of a sound of intensity I is $L = 10\log_{10}(I/I_0)$, where I_0 is the threshold intensity for human hearing.

(a) Conversations have intensity $(1,000,000)I_0$. Find the dB level.

(b) An increase of 10 dB doubles the loudness of a particular sound. What is the effect of this increase on the intensity I?

(c) A jet airliner on takeoff has sound intensity $10^{12}I_0$. Levels above 90 dB are considered dangerous to the ears. Is this level dangerous?

51. The Richter scale for earthquake magnitude uses the formula $R = \log_{10}(I/I_0)$, where I_0 is a minimum reference intensity and I is the earthquake intensity.

(a) Compare the Richter scale magnitudes of the 1906 earthquake in San Francisco, $I = 10^{8.25}I_0$, and the 1971 earthquake in Los Angeles, $I = 10^{6.7}I_0$.

(b) Show that the difference between the Richter scale magnitude of the earthquakes depends only on the *ratio* of the intensities.

52. The pH value of a substance is determined by the concentration of $[H^+]$ of the hydrogen ions in the substance in moles per liter, via the formula $pH = -\log_{10}[H^+]$. The pH of distilled water is 7; acids have pH < 7; bases have pH > 7.

(a) Tomatoes have $[H^+] = (6.3)\cdot 10^{-5}$. Are tomatoes acidic?

(b) Milk has $[H^+] = 4\cdot 10^{-7}$. Is milk acidic?

(c) Find the hydrogen ion concentration of a skin cleanser of rated pH value 5.5.

53. The graph of $y = \log_b x$ contains the point $(3, \frac{1}{3})$. What is b?

54. Graph and compare the following functions: $f(x) = 2\log_2 x$; $g(x) = \log_2(x^2)$; $h(x) = 2\log_2|x|$. Which (if any) are the same?

★55. Give the domain of the following functions. Which (if any) are the same?

(a) $f(x) = \log_{10}\left[\dfrac{(1 - x^2)^4}{\sqrt{(x + 5)/(x^2 + 1)}}\right]$

(b) $g(x) = 4\log_{10}(1 - x) + 4\log_{10}(1 + x) + \frac{1}{2}\log_{10}(x^2 + 1) - \frac{1}{2}\log_{10}(x + 5)$

(c) $h(x) = 4\log_{10}|1 - x| + 4\log_{10}|1 + x| + \frac{1}{2}\log_{10}(x^2 + 1) - \frac{1}{2}\log_{10}(x + 5)$

★56. Give the domain and range of the following functions:

(a) $f(x) = \log_{10}(x^2 - 2x - 3)$,

(b) $g(x) = \log_2[(2x + 1)/2]$,

(c) $h(x) = \log_{10}(1 - x^2)$.

★57. Let $f(x) = \log_2(x - 1)$. Find a formula for the inverse function g of f. What is its domain and range?

★58. Is the logarithm to base 2 of an irrational number ever rational? If so, find an example.

6.3 Differentiation of the Exponential and Logarithm Functions

When a special number e is used as the base, the differentiation rules for the exponential and logarithm functions become particularly simple.

Since we have now defined b^x for all real x, we can attempt to differentiate with respect to x. The result is that $\exp_b$ reproduces itself up to a constant multiple when differentiated. Choosing b properly, we can make the constant equal to 1. The derivative of the corresponding logarithm function turns out to be simply $1/x$.

Consider the function $f(x) = \exp_b(x) = b^x$ defined in Section 6.1. If we assume that f is differentiable at zero, we can calculate $f'(x)$ for all x as follows:

$$\frac{f(x + \Delta x) - f(x)}{\Delta x} = \frac{b^{x+\Delta x} - b^x}{\Delta x} = \frac{b^x b^{\Delta x} - b^x}{\Delta x} = b^x \left(\frac{b^{\Delta x} - 1}{\Delta x} \right);$$

thus,

$$f'(x) = \lim_{\Delta x \to 0} \frac{f(x + \Delta x) - f(x)}{\Delta x} = \lim_{\Delta x \to 0} b^x \left(\frac{b^{\Delta x} - 1}{\Delta x} \right)$$

$$= b^x \lim_{\Delta x \to 0} \frac{f(\Delta x) - f(0)}{\Delta x} = b^x f'(0).$$

One can show that $f'(0)$ really does exist,[1] so it follows by the preceding argument that f is differentiable everywhere.

Derivative of b^x

If $b > 0$, then $\exp_b(x) = b^x$ is differentiable and

$$\exp_b'(x) = \exp_b'(0)\exp_b(x).$$

That is,

$$\frac{d}{dx} b^x = kb^x,$$

where $k = \exp_b'(0)$ is a number depending on b.

Notice that when we differentiate an exponential function, we reproduce it, multiplied by a constant k. If $b \neq 1$, then $k \neq 0$, for otherwise $\exp_b'(x)$ would be zero for all x, and $\exp_b$ would be constant.

Example 1 Let $f(x) = 3^x$. How much faster is f increasing at $x = 5$ than at $x = 0$?

Solution By the preceding display,

$$f'(5) = f'(0)f(5) = f'(0) \cdot 3^5.$$

Thus at $x = 5$, f is increasing $3^5 = 243$ times as fast as at $x = 0$. ▲

To differentiate effectively, we still need to find $\exp_b'(0)$ and see how it depends upon b. It would be nice to be able to adjust b so that $\exp_b'(0) = 1$, for then we would have simply $\exp_b'(x) = \exp_b(x)$. To find such a b, we

---

[1] For the proof of this fact, see Chapter 10 of *Calculus Unlimited* by the authors.

numerically compute the derivative of b^x for $b = 1, 2, 3, 4, 5$ at $x = 0$. These derivatives are obtained by computing $(b^{\Delta x} - b^0)/\Delta x = (b^{\Delta x} - 1)/\Delta x$ for various small values of Δx. The results are as follows:

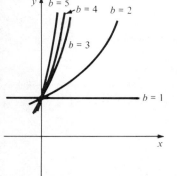

			Δx			Derivative at 0 Equals Approximately
$\dfrac{b^{\Delta x} - 1}{\Delta x}$ b	1	0.1	0.01	0.001	0.0001	
1	0	0	0	0	0	0
2	1	0.72	0.70	0.69	0.69	0.69
3	2	1.16	1.10	1.10	1.10	1.10
4	3	1.49	1.40	1.39	1.39	1.39
5	4	1.75	1.62	1.61	1.61	1.61

Figure 6.3.1. $y = b^x$ for $b = 1, 2, 3, 4, 5$.

The graphs of b^x for these values of b are shown in Fig. 6.3.1. The slopes of the graphs at $x = 0$ are given by the corresponding derivatives computed in the above table. We see that $b = 2$ gives a slope less than 1, while $b = 3$ gives a slope larger than 1. Since it is plausible that the slopes increase steadily with b, it is also plausible that there is a unique number somewhere between 2 and 3 that will give a slope exactly equal to 1. The number is called e, and further numerical experimentation shows that the value of e is approximately 2.718. (Exercise 92 shows another way to find e, and a formula for e in terms of limits is given on page 330.)

The Number e

The number e is chosen so that $\exp'_e(0) = 1$, that is, so that

$$\frac{d}{dx} e^x = e^x.$$

Logarithms to the base e are called *natural logarithms*. We denote $\log_e x$ by $\ln x$. (The notation $\log x$ is generally used in calculus books for the common logarithm $\log_{10} x$.) Since $e^1 = e$, we have the formula $\ln e = 1$.

Natural Logarithms and e

$\ln x$ means $\log_e x$ (natural logarithm).
$\log x$ means $\log_{10} x$ (common logarithm).
$\exp x$ means e^x.
$\ln(\exp x) = x$; $\exp(\ln x) = x$.
$\ln e = 1$, $\ln 1 = 0$.

Most scientific calculators have buttons for evaluating e^x and $\ln x$, but one can sometimes get answers faster and more accurately by hand, as the next example illustrates.

Example 2 Simplify $\ln[e^{205}/(e^{100})^2]$.

Solution By the laws of logarithms,

$$\ln\left[\frac{e^{205}}{\left(e^{100}\right)^2}\right] = \ln(e^{205}) - 2\ln(e^{100}) = 205 - 2 \cdot 100 = 5. \; \blacktriangle$$

We can now complete the differentiation formula for the general exponential function $\exp_b x$. Since $b = e^{\ln b}$ we have $b^x = e^{x \ln b}$. Using the chain rule we find, since $\ln b$ is a constant,

$$\frac{d}{dx} b^x = \frac{d}{dx} e^{x \ln b} = e^{x \ln b} \frac{d}{dx}(x \ln b) = e^{x \ln b} \ln b = b^x \ln b.$$

Thus, the unknown factor $\exp_b'(0)$ turns out to be just the natural logarithm of the base b.

Differentiation of the Exponential

$$\frac{d}{dx} e^x = e^x,$$

$$\frac{d}{dx} b^x = (\ln b) b^x.$$

Example 3   Differentiate: (a) $f(x) = e^{3x}$; (b) $g(x) = 3^x$.

Solution   (a) Let $u = 3x$ so $e^{3x} = e^u$ and use the chain rule:

$$\frac{d}{dx} e^u = \frac{d}{du}(e^u)\frac{du}{dx} = e^u \cdot 3 = 3e^{3x}.$$

(b) $\dfrac{d}{dx} 3^x = 3^x \ln 3,$

taking $b = 3$ in the preceding box. This expression cannot be simplified further; one can find the value $\ln 3 \approx 1.0986$ in a table or with a calculator. (Compare the third line of the table on p. 319.) ▲

Example 4   Differentiate the following functions: (a) xe^{3x}, (b) $\exp(x^2 + 2x)$, (c) x^2, (d) $e^{\sqrt{x}}$, (e) $e^{\sin x}$, (f) $2^{\sin x}$.

Solution   (a) $\dfrac{d}{dx}(xe^{3x}) = \dfrac{dx}{dx} e^{3x} + x \dfrac{d}{dx} e^{3x} = e^{3x} + x \cdot 3e^{3x} = (1 + 3x)e^{3x};$

(b) $\dfrac{d}{dx} \exp(x^2 + 2x) = \exp(x^2 + 2x) \dfrac{d}{dx}(x^2 + 2x)$

$$= \left[\exp(x^2 + 2x)\right](2x + 2);$$

(c) $\dfrac{d}{dx} x^2 = 2x;$

(d) $\dfrac{d}{dx} e^{\sqrt{x}} = e^{\sqrt{x}} \dfrac{d}{dx} \sqrt{x} = \dfrac{1}{2} x^{-1/2} e^{\sqrt{x}};$

(e) $\dfrac{d}{dx} e^{\sin x} = e^{\sin x} \dfrac{d}{dx} \sin x = e^{\sin x} \cos x;$

(f) $\dfrac{d}{dx} 2^{\sin x} = \dfrac{d}{du}(2^u)\dfrac{du}{dx} \qquad$ (with $u = \sin x$)

$$= \ln 2 \cdot 2^u \cdot \cos x$$

$$= \ln 2 \cdot 2^{\sin x} \cdot \cos x. \quad ▲$$

We can differentiate the logarithm function by using the inverse function rule of Section 5.3. If $y = \ln x$, then $x = e^y$ and

$$\frac{dy}{dx} = \frac{1}{dx/dy} = \frac{1}{e^y} = \frac{1}{x}.$$

Hence

$$\frac{d}{dx} \ln x = \frac{1}{x} \, .$$

For other bases, we use the same process; setting $y = \log_b x$ and $x = b^y$:

$$\frac{d}{dx} \log_b x = \frac{1}{(d/dy)b^y} = \frac{1}{\ln b \cdot b^y} = \frac{1}{\ln b \cdot x} \, .$$

That is,

$$\frac{d}{dx} \log_b x = \frac{1}{(\ln b)x} \, .$$

The last formula may also be proved by using Law 3 of logarithms given in Section 6.2:

$$\ln x = \log_e x = \log_b x \cdot \ln b,$$

so

$$\frac{d}{dx} \log_b x = \frac{d}{dx} \left(\frac{1}{\ln b} \ln x \right)$$

$$= \frac{1}{\ln b} \frac{d}{dx} \ln x = \frac{1}{(\ln b)x} \, .$$

Our discussion so far can be summarized as follows:

Derivative of the Logarithm

$$\frac{d}{dx} \ln x = \frac{1}{x} \, , \qquad x > 0;$$

$$\frac{d}{dx} \log_b x = \frac{1}{(\ln b)x} \, , \qquad x > 0.$$

Example 5   Differentiate: (a) $\ln(3x)$, (b) $xe^x \ln x$, (c) $8 \log_3 8x$.

Solution   (a) Setting $u = 3x$ and using the chain rule:

$$\frac{d}{dx} \ln 3x = \frac{d}{du}(\ln u) \cdot \frac{du}{dx} = \frac{1}{3x} \cdot 3 = \frac{1}{x} \, .$$

Alternatively, $\ln 3x = \ln 3 + \ln x$, so the derivative with respect to x is $1/x$.
(b) By the product rule:

$$\frac{d}{dx}(xe^x \ln x) = x \frac{d}{dx}(e^x \ln x) + e^x \ln x = xe^x \ln x + e^x + e^x \ln x.$$

(c) From the formula $(d/dx)\log_b x = 1/[(\ln b)x]$ with $b = 3$,

$$\frac{d}{dx} 8 \log_3 8x = 8 \frac{d}{dx} \log_3 8x = 8 \left(\frac{d}{du} \log_3 u \right) \frac{du}{dx} \qquad [u = 8x]$$

$$= 8 \cdot \frac{1}{(\ln 3) \cdot u} \cdot 8 = \frac{64}{(\ln 3)8x} = \frac{8}{(\ln 3)x} \, . \; \blacktriangle$$

Example 6   Differentiate: (a) $\ln(10x^2 + 1)$; (b) $\sin(\ln x^3)\exp(x^4)$.

Solution   (a) By the chain rule with $u = 10x^2 + 1$, we get

$$\frac{d}{dx}\ln(10x^2 + 1) = \frac{d}{du}\ln u \frac{du}{dx} = \frac{1}{u}\cdot 20x = \frac{20x}{10x^2 + 1}\ .$$

(b) By the product rule and chain rule,

$$\frac{d}{dx}\left(\sin(\ln x^3)\exp(x^4)\right)$$

$$= \left(\frac{d}{dx}\sin(\ln x^3)\right)\exp(x^4) + \sin(\ln x^3)\frac{d}{dx}\exp(x^4)$$

$$= \cos(\ln x^3)\cdot\frac{3x^2}{x^3}\cdot\exp(x^4) + \sin(\ln x^3)\cdot 4x^3\exp(x^4)$$

$$= \exp(x^4)\left(\frac{3}{x}\cos(\ln x^3) + 4x^3\sin(\ln x^3)\right).\ \blacktriangle$$

Previously, we knew the formula $(d/dx)x^n = nx^{n-1}$ for rational n. Now we are in a position to prove it for all n, rational or irrational, and $x > 0$. Indeed, write $x^n = e^{(\ln x)n}$ and differentiate using the chain rule and the laws of exponents:

$$\frac{d}{dx}x^n = \frac{d}{dx}e^{(\ln x)\cdot n} = \frac{n}{x}\cdot e^{(\ln x)n} = \frac{n}{x}x^n = nx^{n-1}.$$

For example, $(d/dx)x^\pi = \pi x^{\pi-1}$.

In order to differentiate complex expressions involving powers, it is sometimes convenient to begin by taking logarithms.

Example 7   Differentiate the functions (a) $y = x^x$ and (b) $y = x^x\cdot\sqrt{x}$.

Solution   (a) We take natural logarithms,

$$\ln y = \ln(x^x) = x\ln x.$$

Next, we differentiate using the chain rule, remembering that y is a function of x:

$$\frac{1}{y}\frac{dy}{dx} = x\cdot\frac{1}{x} + \ln x = 1 + \ln x.$$

Hence

$$\frac{dy}{dx} = y(1 + \ln x) = x^x(1 + \ln x).$$

Alternatively, we could have written $x^x = e^{x\ln x}$. Thus, by the chain rule,

$$\frac{d}{dx}x^x = e^{x\ln x}(\ln x + 1) = x^x(\ln x + 1).$$

(b) $y = x^x\cdot\sqrt{x} = x^{x+1/2}$, so $\ln y = (x + \frac{1}{2})\ln x$. Thus

$$\frac{1}{y}\frac{dy}{dx} = \left(x + \frac{1}{2}\right)\frac{1}{x} + \ln x,$$

and so

$$\frac{dy}{dx} = x^{x+1/2}\left[\left(x + \frac{1}{2}\right)\frac{1}{x} + \ln x\right].\ \blacktriangle$$

This method of differentiating functions by first taking logarithms and then differentiating is called *logarithmic differentiation*.

Example 8   Use logarithmic differentiation to calculate dy/dx, where

$$y = (2x + 3)^{3/2}/\sqrt{x^2 + 1} \ .$$

Solution $\ln y = \ln[(2x + 3)^{3/2}/(x^2 + 1)^{1/2}] = \frac{3}{2}\ln(2x + 3) - \frac{1}{2}\ln(x^2 + 1)$, so

$$\frac{1}{y}\frac{dy}{dx} = \frac{3}{2} \cdot \frac{2}{2x + 3} - \frac{1}{2} \cdot \frac{2x}{x^2 + 1}$$

$$= \frac{3}{2x + 3} - \frac{x}{x^2 + 1} = \frac{(x^2 - 3x + 3)}{(2x + 3)(x^2 + 1)} \ ,$$

and hence

$$\frac{dy}{dx} = \frac{(2x + 3)^{3/2}}{(x^2 + 1)^{1/2}} \cdot \frac{(x^2 - 3x + 3)}{(2x + 3)(x^2 + 1)}$$

$$= \frac{(x^2 - 3x + 3)(2x + 3)^{1/2}}{(x^2 + 1)^{3/2}} \cdot \blacktriangle$$

Since the derivative of $\ln x$ is $1/x$, $\ln x$ is an antiderivative of $1/x$; that is

$$\int \frac{1}{x} dx = \ln x + C \qquad \text{for} \quad x > 0.$$

This integration rule fills an important gap in our earlier formula

$$\int x^n dx = \frac{x^{n+1}}{n + 1} + C$$

from Section 2.5, which was valid only for $n \neq -1$.

Antidifferentiation Formulas for exp and log

1. $\int e^x dx = e^x + C$;

2. $\int b^x dx = \dfrac{b^x}{\ln b} + C \qquad (b > 0)$;

3. $\int \dfrac{1}{x} dx = \ln|x| + C \qquad (x \neq 0).$

To prove integration formula 3 in the preceding box, consider separately the cases $x > 0$ and $x < 0$. For $x > 0$, it is the inverse of our basic formula for differentiating the logarithm. For $x < 0$, $(d/dx)(\ln|x|) = (d/dx)[\ln(-x)] = [1/(-x)] \cdot [-1] = 1/x$, so $\ln|x|$ is an antiderivative for $1/x$, for $x \neq 0$.

Example 9   Find the indefinite integrals: (a) $\int e^{ax} dx$; (b) $\int \left(\dfrac{1}{3x + 2}\right) dx$.

Solution   (a) $(d/dx)e^{ax} = ae^{ax}$, by the chain rule, so $\int e^{ax} dx = (1/a)e^{ax} + C$.
(b) Differentiate $\ln|3x + 2|$ by the chain rule, setting $u = 3x + 2$. We get $(d/dx)\ln|u| = (d/du)\ln|u| \cdot du/dx = (1/u) \cdot 3 = 3/(3x + 2)$; hence

$$\int \frac{1}{(3x + 2)} dx = \frac{1}{3} \ln|3x + 2| + C. \ \blacktriangle$$

Example 10 Integrate

$$\text{(a)} \quad \int_0^1 \frac{1}{x+1} \, dx \qquad \text{(b)} \quad \int_1^2 \frac{x^3 + 3x + 2}{x} \, dx.$$

Solution (a) Since $(d/dx)\ln x = 1/x$, for $x > 0$, the chain rule gives

$$\frac{d}{dx} \ln(x+1) = \frac{1}{x+1} .$$

(Here $x + 1 > 0$, so we can omit the absolute value signs.) Thus

$$\int_0^1 \frac{1}{x+1} \, dx = \ln(x+1) \Big|_0^1 = \ln 2 - \ln 1 = \ln 2.$$

(b) $\displaystyle \int_1^2 \frac{x^3 + 3x + 2}{x} \, dx = \int_1^2 \left(x^2 + 3 + \frac{2}{x} \right) dx = \left(\frac{x^3}{3} + 3x + 2 \ln x \right) \Big|_1^2$

$$= \frac{7}{3} + 3 + 2 \ln 2 = \frac{16}{3} + 2 \ln 2. \; \blacktriangle$$

Example 11 Verify the formula $\displaystyle \int e^x \cos x \, dx = \frac{1}{2} e^x (\sin x + \cos x) + C.$

Solution We must check that the right-hand side is an antiderivative of the integrand. We compute, using the product rule:

$$\frac{d}{dx} \frac{1}{2} e^x (\sin x + \cos x)$$

$$= \frac{1}{2} \left(\frac{de^x}{dx} \right)(\sin x + \cos x) + \frac{1}{2} e^x \left(\frac{d}{dx} (\sin x + \cos \times) \right)$$

$$= \frac{1}{2} e^x (\sin x + \cos x) + \frac{1}{2} e^x (\cos x - \sin x) = e^x \cos x$$

Thus the formula is verified. $\blacktriangle$

Exercises for Section 6.3

1. How much faster is $f(x) = 2^x$ increasing at $x = 3$ than at $x = 0$?
2. How much faster is $f(x) = 4^x$ increasing at $x = 2$ than at $x = 0$?
3. How much faster is $f(x) = (\frac{1}{2})^x$ increasing at $x = \frac{1}{2}$ than at $x = 0$?
4. How much faster is $f(x) = (\frac{1}{4})^x$ increasing at $x = -3$ than at $x = 0$?

Simplify the expressions in Exercises 5–10.

5. $\ln(e^{x+1}) + \ln(e^2)$
6. $\ln(e^{\sin x}) - \ln(e^{\cos x})$
7. $e^{\ln x + \ln x^2}$
8. $e^{\ln \sin x} e^{-\ln \cos x}$
9. $e^{4x}[\ln(e^{3x-1}) - \ln(e^{1-x})]$
10. $e^{x \ln 3 + \ln 2^x}$

Differentiate the functions in Exercises 11–32.

11. e^{x^2+1}
12. $(e^{3x^3+x})(1 - e^x)$
13. $e^{1-x^2} + x^3$
14. $e^{2x} - \cos(x + e^{2x})$
15. $2^x + x$
16. $3^x + x^{-x}$
17. $3^x - 2^{x-1}$
18. $\tan(3^{2x})$
19. $\ln 10x$
20. $\ln x^2$
21. $\frac{\ln x}{x}$
22. $(\ln x)^3$
23. $\ln(\sin x)$
24. $\ln(\tan x)$
25. $\ln(2x + 1)$
26. $\ln(x^2 - 3x)$
27. $(\sin x)\ln x$
28. $(x^2 - 2x)\ln(2x + 1)$
29. $\frac{\ln(\tan 3x)}{1 + \ln x^2}$
30. $x^{\sqrt{2}} + (\ln \cos x)^{\sqrt{3}}$
31. $\log_5 x$
32. $\log_7(2x)$

Use logarithmic differentiation to differentiate the functions in Exercises 33–40.

33. $y = (\sin x)^x$.
34. $y = x^{\sin x}$.
35. $y = (\sin x)^{\cos x}$.
36. $y = (x^3 + 1)^{x^2 - 2}$.
37. $y = (x - 2)^{2/3}(4x + 3)^{8/7}$.
38. $y = (x + 2)^{5/8}(8x + 9)^{10/13}$.
39. $y = x^{(x^x)}$.
40. $y = x^{3x}$.

Differentiate the functions in Exercises 41–62.

41. $e^{x \sin x}$
42. x^e
43. $\ln(x^{-5} + x)$

44. $6\ln(x^3 - xe^x) + e^x\ln x$
45. $14^{x^2 - 8\sin x}$
46. $\log_2[\sin(x^2)]$
47. $\ln(x + \ln x)$
48. $e^x\sin(\ln x + 1)$
49. $\cos(x^{\sin x})$
50. $\sin(x^{\cos x})$
51. $x^{(x^2)}$
52. x^{e^x}
53. $(1/x)^{\tan x^2}$
54. $\ln(x^{\sec x^2})$
55. $\sin(x^4 + 1) \cdot \log_8(14x - \sin x)$
56. $\log_{5/3}(\cos 2x)$
57. $3x^{\sqrt{x}}$
58. $3x^{x/2}$
59. $\sin(x^x)$
60. $\ln(x^{x+1})$
61. $(\sin x)^{(\cos x)^x}$
62. 2^{2^x}

Find the indefinite integrals in Exercises 63–76.

63. $\int e^{2x}\,dx$
64. $\int (x^2 + e^x)\,dx$
65. $\int (\cos x + e^{4x})\,dx$
66. $\int 4e^{-2x}\,dx$
67. $\int \left(s^2 + \dfrac{2}{s}\right)ds$
68. $\int \left(s^2 + s + 1 + \dfrac{1}{s} + \dfrac{1}{s^2}\right)ds$
69. $\int \left(\dfrac{x^2 + 1}{2x}\right)dx$
70. $\int \left(e^{4x} - \dfrac{2}{x}\right)dx$
71. $\int \left(\dfrac{x}{x - 1}\right)dx$ [*Hint*: Divide.]
72. $\int \left(\dfrac{x}{x + 3}\right)dx$
73. $\int 3^x\,dx$
74. $\int x^3\,dx$
75. $\int \left(\dfrac{x^2 + 2x + 2}{x - 8}\right)dx$ [*Hint*: Divide.]
76. $\int \left(\dfrac{y - 1}{y^2 - 1}\right)dy$

Find the definite integrals in Exercises 77–84.

77. $\int_0^1 (x^2 + 3e^x)\,dx$
78. $\int_1^2 e^{-x}\,dx$
79. $\int_2^3 (x^3 + e^{2x})\,dx$
80. $\int_{50}^{100} (4/x)\,dx$
81. $\int_0^1 2^x\,dx$

82. $\int_{-1}^1 3^x\,dx$
83. $\int_0^1 \dfrac{dx}{x + 2}$
84. $\int_0^1 \dfrac{x}{x^2 + 2}\,dx$ [*Hint*: differentiate $\ln(x^2 + 2)$.]
85. (a) Differentiate $x\ln x$. (b) Find $\int \ln x\,dx$.
86. (a) By differentiating $\ln(\cos x)$, find $\int \tan x\,dx$.
    (b) Find $\int \cot x\,dx$.
87. (a) Verify the integration formula
$$\int e^{ax}\sin bx\,dx = \dfrac{e^{ax}(a\sin bx - b\cos bx)}{a^2 + b^2} + C.$$
    (b) Find a similar formula for
$$\int e^{ax}\cos bx\,dx.$$
88. Verify the following:
    (a) $\displaystyle\int (x^n e^x + nx^{n-1}e^x)\,dx = x^n e^x + C.$
    (b) $\displaystyle\int x^2 e^x\,dx = x^2 e^x - 2xe^x + 2e^x + C.$
89. Verify the following integration formulas:
    (a) $\displaystyle\int \dfrac{1}{\sqrt{1 + x^2}}\,dx = \ln\left(x + \sqrt{1 + x^2}\right) + C;$
    (b) $\displaystyle\int \dfrac{1}{x\sqrt{1 - x^2}}\,dx = -\ln\left|\dfrac{1 + \sqrt{1 - x^2}}{x}\right| + C.$
90. Use Exercises 88 and 89 to evaluate
    (a) $\displaystyle\int_0^1 \dfrac{dx}{\sqrt{1 + x^2}}$

    and

    (b) $\displaystyle\int_0^1 x^2 e^x\,dx.$
91. Express the derivatives of the following in terms of $f(x)$, $g(x)$, $f'(x)$, and $g'(x)$:
    (a) $f(x) \cdot e^x + g(x)$;
    (b) $e^{f(x) + x^2}$;
    (c) $f(x) \cdot e^{g(x)}$;
    (d) $f(e^x + g(x))$;
    (e) $f(x)^{g(x)}$.
★92. This exercise shows how to adjust b to make $\exp_b'(x) = \exp_b(x)$. We start with the base 10 of common logarithms and find another base b for which $\exp_b'(0) = 1$ as follows. By the definition of the logarithm, $b = 10^{\log_{10}b}$.
    (a) Deduce that $\exp_b(x) = \exp_{10}(x\log_{10}b)$.
    (b) Differentiate (a) to show that $\exp_b'(0) = \exp_{10}'(0) \cdot \log_{10}b$. To have $\exp_b'(0) = 1$, we should pick b in such a way that $\log_{10}b = 1/\exp_{10}'(0)$.
    (c) Deduce that $e = \exp_{10}[1/\exp_{10}'(0)]$ satisfies the condition $\exp_b'(0) = 1$ and so $\exp_e'(x) = \exp_e(x)$.
    (d) Show that for any b, $\exp_b[1/\exp_b'(0)] = e$.

★93. By calculating $(b^{\Delta x} - 1)/\Delta x$ for small Δx and various values of b, as at the beginning of this section, estimate e to within 0.01.

★94. Suppose that you *defined* $\ln x$ to be $\int_1^x dt/t$.
 (a) Use the fundamental theorem to show that $(d/dx)\ln x = 1/x$.
 (b) Define e^x to be the inverse function of $\ln x$ and show $(d/dx)e^x = e^x$.

★95. (a) Use the definition of $\ln x$ in Exercise 94 to show that $\ln xy = \ln x + \ln y$ by showing that for a given fixed x_0,

$$\frac{d}{dx}(\ln(xx_0) - \ln x - \ln x_0) = 0.$$

 (b) Deduce from (a) that $e^{x+y} = e^x e^y$.
 (c) Prove $e^{x+y} = e^x e^y$ by assuming only that $(d/dx)e^x = e^x$ and $e^0 = 1$.

★96. (a) Compute $\int_0^x e^{-t} dt$ for $x = 1$, 10, and 100.
 (b) How would you define $\int_0^\infty e^{-t} dt$? What number would this integral be?

 (c) Interpret the integral in (b) as an area.

★97. (a) Compute $\int_\epsilon^2 \ln x\, dx$ (see Exercise 85) for $\epsilon = 1$, 0.1, and 0.01.
 (b) How would you define $\int_0^2 \ln x\, dx$? Compute it by evaluating $\lim_{x\to 0}(x \ln x)$ numerically.
 (c) Why doesn't this integral exist in the ordinary sense?

★98. What do you see if you rotate an exponential spiral about the origin at a uniform rate? Compare with the spiral $r = \theta$.

★99. Differentiate $y = f(x)g(x)$ by writing the logarithm of y as a sum of logarithms. Show that you recover the product rule.

★100. Differentiate $y = f(x)/g(x)$ logarithmically to recover the quotient rule.

★101. Find a formula for the derivative of $f_1(x)^{n_1} f_2(x)^{n_2} \cdots f_k(x)^{n_k}$ using logarithmic differentiation.

6.4 Graphing and Word Problems

Money grows exponentially when interest is compounded continuously.

Now we turn to applications of the exponential and logarithm functions in graphing and word problems. Additional applications involving growth and decay are given in Chapter 8.

We begin by studying some useful facts about limits of exponential and logarithm functions.

We shall first show that

$$\lim_{x\to\infty} \frac{\ln x}{x} = 0. \tag{1}$$

Intuitively, this means that for large x, x is much larger than $\ln x$. Indeed this is plausible from their graphs (Fig. 6.4.1).

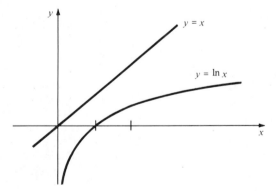

Figure 6.4.1. x is much larger than $\ln x$ for large x.

To prove (1), note that for any fixed integer n,

$$\frac{d}{dx}(x - n\ln x) = 1 - \frac{n}{x} \to 1 \qquad \text{as} \quad x \to \infty.$$

Thus

$$x - n \ln x \to \infty \qquad \text{as} \quad x \to \infty$$

(since its slope is nearly 1 for large x). In particular,

$$x - n \ln x > 0 \qquad \text{for large } x,$$

$$\frac{\ln x}{x} < \frac{1}{n} \qquad \text{for large } x.$$

This shows that $\ln x / x$ becomes arbitrarily small for x large, so (1) holds.

Now let $y = x^a$ for $a > 0$. Then $y \to \infty$ as $x \to \infty$, and so

$$\frac{\ln x}{x^a} = \frac{\ln(y^{1/a})}{y} = \frac{1}{a} \frac{\ln y}{y} \to 0$$

as $y \to \infty$ by (1), so we get

$$\lim_{x \to \infty} \frac{\ln x}{x^a} = 0 \quad \text{or} \quad \lim_{x \to \infty} \frac{x^a}{\ln x} = \infty. \tag{2}$$

Thus, not only does x become much larger than $\ln x$, but so does any positive power of x. For example, with $a = \frac{1}{2}$,

$$\lim_{x \to \infty} \frac{\ln x}{\sqrt{x}} = 0.$$

▣ Calculator Discussion

The validity of (1) and (2) for various a can also be readily checked by performing numerical calculations on a calculator. For example, on our calculator we got the following data:

x	1	2	3	5	50	500	50,000	10^6	10^{10}	10^{20}	10^{30}
$\ln x$	0	.69	1.09	1.609	3.912	6.214	10.8	13.8	23.02	46.05	69.08
$x^{0.1}$	1	1.07	1.11	1.17	1.47	1.86	2.95	3.98	10	100	1000

It takes $x^{0.1}$ a while to overtake $\ln x$, but eventually it does. ▲

If we let $y = 1/x$, then $y \to \infty$ as $x \to 0$, so

$$x^a \ln x = y^{-a} \ln\left(\frac{1}{y}\right) = -\frac{\ln y}{y^a} \to 0$$

by (2). Thus for $a > 0$, we have the limit

$$\lim_{x \to 0} x^a \ln x = 0. \tag{3}$$

This means that $\ln x$ approaches $-\infty$ more slowly than x^a approaches zero as $x \to 0$.

Finally, write

$$\frac{e^x}{x^n} = e^{(x - n \ln x)}.$$

As we have seen in deriving (1), $x - n \ln x \to \infty$ as $x \to \infty$. Thus,

$$\lim_{x \to \infty} \frac{e^x}{x^n} = \infty. \tag{4}$$

This says that the exponential function grows more rapidly than any power of x.

Alternative proofs of (1)–(4) are given in Review Exercixes 86–91 at the end of this chapter; simple proofs also follow from l'Hôpital's rule given in Chapter 11.

Limiting Behavior of exp and log

1. $\lim\limits_{x \to \infty} \dfrac{e^x}{x^n} = \infty$ for any n.

e^x grows more rapidly as $x \to \infty$ than any power of x (no matter how large).

2. $\lim\limits_{x \to \infty} \dfrac{\ln x}{x^a} = 0$ for any $a > 0$.

$\ln x$ grows more slowly as $x \to \infty$ than any positive power of x (no matter how small).

3. $\lim\limits_{x \to 0} x^a \ln x = \lim\limits_{x \to 0} \dfrac{\ln x}{x^{-a}} = 0$ for any $a > 0$.

$\ln x$ is dominated as $x \to 0$ by any negative power of x.

Example 1 Sketch the graph $y = x^2 e^{-x}$.

Solution We begin by noting that y is positive except at $x = 0$. Thus $x = 0$ is a minimum. There are no obvious symmetries. The positive x axis is an asymptote, since $x^2 e^{-x} = x^2/e^x = 1/(e^x/x^2)$, and $e^x/x^2 \to \infty$ as $x \to \infty$ by item 1 in the previous box. For $x \to -\infty$, both x^2 and e^{-x} become large, so $\lim_{x \to -\infty} x^2 e^{-x} = \infty$.

The critical points are obtained by setting $dy/dx = 0$; here $dy/dx = 2xe^{-x} - x^2 e^{-x} = (2x - x^2)e^{-x}$. Thus $dy/dx = 0$ when $x = 0$ and $x = 2$; y is decreasing on $(-\infty, 0)$, increasing on $(0, 2)$, and decreasing on $(2, \infty)$. The second derivative is

$$\frac{d^2y}{dx^2} = (2 - 2x)e^{-x} - (2x - x^2)e^{-x} = (2 - 4x + x^2)e^{-x},$$

which is positive at $x = 0$ and negative at $x = 2$. Thus 0 is a minimum and 2 is a maximum. There are inflection points where $d^2y/dx^2 = 0$; i.e., at $x = 2 \pm \sqrt{2}$. This information, together with the plot of a few points, enables us to sketch the graph in Fig. 6.4.2. ▲

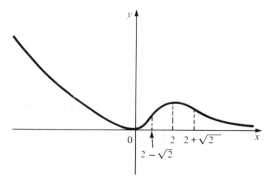

Figure 6.4.2. $y = x^2 e^{-x}$.

Example 2 Sketch the graph of $y = x \ln x$.

Solution The function is defined for $x > 0$. As $x \to 0$, $x \ln x \to 0$ by item 3 in the preceding box, so the graph approaches the origin. As $x \to \infty$, $x \ln x \to \infty$. The function changes sign from negative to positive at $x = 1$. $dy/dx = \ln x + 1$, which is zero when $x = 1/e$ and changes sign from negative to positive there, so $x = 1/e$ is a local minimum point. We also note that dy/dx approaches $-\infty$ as $x \to 0$, so the graph becomes "vertical" as it approaches the origin.

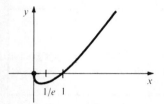

Figure 6.4.3. $y = x \ln x$.

Finally, we see that $d^2y/dx^2 = 1/x$ is positive for all $x > 0$, so the graph is everywhere concave upward. The graph is sketched in Fig. 6.4.3. ▲

Next we turn to an application of logarithmic differentiation. The expression $(d/dx)\ln f(x) = f'(x)/f(x)$ is called the *logarithmic derivative* of f. The quantity $f'(x)/f(x)$ is also called the *relative rate of change* of f, since it measures the rate of change of f per unit of f itself. This idea is explored in the following application.

Example 3 A certain company's profits are given by $P = 5000 \exp(0.3t - 0.001t^2)$ dollars where t is the time in years from January 1, 1980. By what percent per year were the profits increasing on July 1, 1981?

Solution We compute the relative rate of change of P by using logarithmic differentiation.

$$\frac{1}{P} \frac{dP}{dt} = \frac{d}{dt} (\ln P) = \frac{d}{dt} (\ln 5000 + 0.3t - 0.001t^2) = 0.3 - 0.002t.$$

Substituting $t = 1.5$ corresponding to July 1, 1981, we get

$$\frac{1}{P} \frac{dP}{dt} = 0.3 - (0.002)(1.5) = 0.2970.$$

Therefore on July 1, 1981, the company's profits are increasing at a rate of 29.7% per year. ▲

We shall be examining compound interest shortly, but before doing so, we shall need some further information about the number e.

In our previous discussion, the number e was obtained in an implicit way. Using limits, we can derive the more explicit expression

$$e = \lim_{h \to 0} (1 + h)^{1/h}. \tag{5}$$

To prove this, we write

$$(1 + h)^{1/h} = \exp\left[\ln(1 + h)^{1/h}\right]$$

$$= \exp\left[\frac{1}{h} \ln(1 + h)\right]$$

$$= \exp\left[\frac{1}{h} (\ln(1 + h) - \ln 1)\right] \quad \text{(since } \ln 1 = 0\text{)};$$

but

$$\lim_{h \to 0} \left[\frac{1}{h} (\ln(1 + h) - \ln 1)\right] = \lim_{\Delta x \to 0} \left(\frac{\ln(1 + \Delta x) - \ln 1}{\Delta x}\right) = \frac{d}{dx} \ln x \Big|_{x=1} = 1.$$

Substituting this in our expression for $(1 + h)^{1/h}$ and using continuity of the exponential function gives

$$\lim_{h \to 0} (1 + h)^{1/h} = \exp\left[\lim_{h \to 0} \frac{1}{h} (\ln(1 + h) - \ln 1)\right] = \exp(1) = e.$$

which proves (5).

One way to get approximations for e is by letting $h = \pm(1/n)$, where n is a large integer. We get

$$e = \lim_{n \to \infty} \left(1 + \frac{1}{n}\right)^n = \lim_{n \to \infty} \left(1 - \frac{1}{n}\right)^{-n}. \tag{6}$$

Notice that the numbers $(1 + 1/n)^n$ and $(1 - 1/n)^{-n}$ are all rational, so e is the limit of a sequence of rational numbers. It is known that e itself is irrational.[2]

---

[2] A proof is given in Review Exercise 128, Chapter 12.

---

e **as a Limit**

$$e = \lim_{h \to 0} (1 + h)^{1/h} = \lim_{n \to \infty} \left(1 + \frac{1}{n}\right)^n = \lim_{n \to \infty} \left(1 - \frac{1}{n}\right)^{-n} \qquad (6)$$

🔲 Calculator Discussion

Let us calculate $(1 - 1/n)^{-n}$ and $(1 + 1/n)^n$ for various values of n. By (6), the numbers should approach $e = 2.71828\ldots$ in both cases, as n becomes large.

One can achieve a fair degree of accuracy before round off errors make the operations meaningless. For example, on our calculator we obtained the following table. (Powers of 2 were chosen for n to permit computing the nth powers by repeated squaring.)

n	$\left(1 - \dfrac{1}{n}\right)^{-n} = \left(\dfrac{n}{n-1}\right)^n$	$\left(1 + \dfrac{1}{n}\right)^n = \left(\dfrac{n+1}{n}\right)^n$	
2	4.000000	2.250000	
4	3.160494	2.441406	
8	2.910285	2.565785	
16	2.808404	2.637928	
32	2.762009	2.676990	
64	2.739827	2.697345	
128	2.728977	2.707739	
256	2.723610	2.712992	
512	2.720941	2.715632	
1024	2.719610	2.716957	
2048	2.718947	2.717617	
4096	2.718611	2.717954	
8192	2.718443	2.718109	
16,384	2.718370	2.718192	
32,768	2.718367	2.718278	
65,536	2.718299	2.718299	
131,072	2.718131	2.718131	
262,144	2.718492	2.718492	Calculator
524,288	2.717782	2.717782	roundoff
1,048,576	2.719209	2.719209	errors
16,777,216	2.736372	2.736372	significant
536,870,912	2.926309	2.926309	beyond
2,147,483,648	1.00000	1.00000	↓ here

▲

Example 4 Express $\ln b$ as a limit by using the formula $\ln b = \exp_b'(0)$.

Solution By the definition of the derivative as a limit,

$$\ln b = \exp_b'(0) = \lim_{\Delta x \to 0} \frac{\exp_b(\Delta x) - \exp_b(0)}{\Delta x} = \lim_{\Delta x \to 0} \frac{b^{\Delta x} - 1}{\Delta x},$$

$$\text{or} \quad \ln b = \lim_{h \to 0} \left(\frac{b^h - 1}{h}\right). \ \blacktriangle$$

The limit formula (6) for e has the following generalization: for any real number a,

$$e^a = \lim_{n \to \infty} \left(1 + \frac{a}{n}\right)^n = \lim_{n \to \infty} \left(1 - \frac{a}{n}\right)^{-n}. \tag{7}$$

To prove (7), write $h = a/n$. Then $h \to 0$ as $n \to \infty$, so

$$\left(1 + \frac{a}{n}\right)^n = (1 + h)^{a/h} = \left((1 + h)^{1/h}\right)^a$$

which tends to e^a by (5). The second half of (7) follows by taking $h = -a/n$.

Formula (7) has an interpretation in terms of *compound interest*. If a bank offers $r\%$ interest on deposits, compounded n times per year, then any invested amount will grow by a factor of $1 + r/100n$ during each compounding period and hence by a factor of $(1 + r/100n)^n$ over a year. For instance, a deposit of $\$1000$ at 6% interest will become, at the end of the year,

$$1000(1 + 6/400)^4 = \$1061.36 \qquad \text{with quarterly compounding,}$$

$$1000(1 + 6/36500)^{365} = \$1061.8314 \qquad \text{with daily compounding,}$$

and

$$1000\left[1 + 6/(24 \cdot 36500)\right]^{24 \cdot 365} = \$1061.8362 \quad \text{with hourly compounding.}$$

Two lessons seem to come out of this calculation: the final balance is an increasing function of the number of compounding periods, but there may be an upper limit to how much interest could be earned at a given rate, even if the compounding period were to be decreased to the tiniest fraction of a second.

In fact, applying formula (7) with $a = r/100$ gives us

$$\lim_{n \to \infty} \left(1 + \frac{r}{100n}\right)^n = e^{r/100},$$

and so $\$1000$ invested at 6% interest can never grow in a year to more than $1000e^{0.06} = \$1061.8366$, no matter how frequent the compounding. (Strictly speaking, this assertion depends on the fact that $(1 + r/100n)^n$ is really an increasing function of n. This is intuitively clear from the compound interest interpretation; a proof is outlined in Exercise 37.)

In general, if P_0 dollars are invested at $r\%$ interest, compounded n times a year, then the account balance after a year will be $P_0(1 + r/100n)^n$, and the limit of this as $n \to \infty$ is $P_0 e^{r/100}$. This limiting case is often referred to as *continuously compounded interest*. The actual fraction by which the funds increase during each one year period with continuous compound interest is $(P_0 e^{r/100} - P_0)/P_0 = e^{r/100} - 1$.

Compound Interest

1. If an initial principal P_0 is invested at r percent interest compounded n times per year, then the balance after one year is $P_0(1 + r/100n)^n$.
2. If $n \to \infty$, so that the limit of continuous compounding is reached, then the balance after one year is $P_0 e^{r/100}$.
3. The annual percentage increase on funds invested at $r\%$ per year compounded continuously is

$$100(e^{r/100} - 1)\%. \tag{8}$$

Example 5   What is the yearly percent increase on a savings account with 5% interest compounded continuously?

Solution   By formula (8), the fraction by which funds increase is $e^{r/100} - 1$. Substituting $r = 5$ gives $e^{r/100} - 1 = e^{0.05} - 1 = 0.0513 = 5.13\%$. ▲

Continuous compounding of interest is an example of *exponential growth*, a topic that will be treated in detail in Section 8.1.

Reasoning as above, we find that if P_0 dollars is invested at r percent compounded annually and is left for t years, the amount accumulated is $P_0(1 + r/100)^t$. If it is compounded n times a year, the amount becomes $P_0(1 + r/n \cdot 100)^{nt}$, and if it is compounded continuously, it is $P_0 e^{rt/100}$.

Example 6   Manhattan Island was purchased by the Dutch in 1626 for the equivalent of $24. Assuming an interest rate of 6%, how much would the $24 have grown to be by 1984 if (a) compounded annually? (b) Compounded continuously?

Solution   (a) $24 \cdot (1.06)^{1984 - 1626} = 24 \cdot (1.06)^{358} \approx \$27.5$ billion.
(b) $24 \cdot e^{(0.06)(1984 - 1626)} = 24 e^{(0.06)(358)} \approx \$51.2$ billion. ▲

Exercises for Section 6.4

Sketch the graphs of the functions in Exercises $1 - 8$.

1. $y = e^{-x}\sin x$.
2. $y = (1 + x^2)e^{-x}$.
3. $y = xe^{-x}$.
4. $y = e^x/(1 + x^2)$.
5. $y = \log_x 2$. [*Hint:* $y = \ln 2/\ln x$.]
6. $y = (\ln x)/x$.
7. $y = x^\pi \ (x > 0)$.
8. $y = e^{-1/x^2}$.

9. By what percentage are the profits of the company in Example 3 increasing on January 1, 1982?
10. A certain company's profits are given by $P = 50{,}000 \exp(0.1t - 0.002t^2 + 0.00001t^3)$, where t is the time in years from July 1, 1975. By what percentage are the profits growing on January 1, 1980?
11. For $0 \leqslant t < 1000$, the height of a redwood tree in feet, t years after being planted, is given by $h = 300(1 - \exp[-t/(1000 - t)])$. By what percent per year is the height increasing when the tree is 500 years old?
12. A company truck and trailer has salvage value $y = 120{,}000 e^{-0.1x}$ dollars after x years of use. (a) Find the rate of depreciation in dollars per year after five years. (b) By what percentage is the value decreasing after three years?
13. Express $3^{\sqrt{2}}$ as a limit.
14. Express e^{a+1} as a limit.
15. Express $3 \ln b$ as a limit.
16. Express $\ln(\frac{1}{2})$ as a limit.
17. A bank offers 8% per year compounded continuously and advertises an actual yield of 8.33%. Verify that this is correct.

18. (a) What rate of interest compounded annually is equivalent to 7% compound continuously?
    (b) How much money would you need to invest at 7% to see the difference between continuous compounding and compounding by the minute over a year?
19. The amount A for principal P compounded continuously for t years at an annual interest rate of r is $A = Pe^{rt}$. Find the amount after three years for $100 principal compounded continuously at 6%.
20. (a) How long does it take to double your money at 6% compounded anually?
    (b) How long does it take to double your money at 6% compounded continuously?

Find the equation of the tangent line to the graph of the given functions at the indicated points in Exercises 21–26.

21. $y = xe^{2x}$ at $x = 1$.
22. $y = x^2 e^{x/2}$ at $x = 2$.
23. $y = \cos(\pi e^x/4)$ at $x = 0$.
24. $y = \sin(\ln x)$ at $x = 1$.
25. $y = \ln(x^2 + 1)$ at $x = 1$.
26. $y = x^{\ln x}$ at $x = 1$.

27. (a) Show that the first-order approximation to b^x, for x near zero, is $1 + x \ln b$.
    (b) Compare $2^{0.01}$ with $1 + 0.01 \ln 2$; compare $2^{0.0001}$ with $1 + 0.0001 \ln 2$. (Use a calculator or tables.)
    (c) By writing $e = (e^{1/n})^n$ and using the first-order approximation for $e^{1/n}$, obtain an approximation for e.
28. Show that $\ln b = \lim_{n \to \infty} n(\sqrt[n]{b} - 1)$.
29. Find the minimum value of $y = x^x$ for x in $(0, \infty)$.

30. Let f be a function satisfying $f'(t_0) = 0$ and $f(t_0) \neq 0$. Show that the relative rate of change of $P = \exp[f(t)]$ is zero at $t = t_0$.

31. One form of the Weber–Fechner law of mathematical psychology is $dS/dR = c/R$, where $S = $ perceived sensation, $R = $ stimulus strength. The law says, for example, that adding a fixed amount to the stimulus is less perceptible as the total stimulus is greater.
    (a) Show that $S = c \cdot \ln(R/R_0)$ satisfies the Weber–Fechner law and that $S(R_0) = 0$. What is the meaning of R_0?
    (b) The loudness L in decibels is given by $L = 10 \log_{10}(I/I_0)$, where I_0 is the least audible intensity. Find the value of the constant c in the Weber–Fechner law of loudness.

32. The rate of damping of waves in a plasma is proportional to $r^3 e^{-r^2/2}$, where r is the ratio of wave velocity to "thermal" velocity of electrons in the plasma. Find the value of r for which the damping rate is maximized.

33. The atmospheric pressure p at x feet above sea level is approximately given by $p = 2116 e^{-0.0000318x}$. Compute the decrease in outside pressure expected in one second by a balloon at 2000 feet which is rising at 10 feet per second. [*Hint*: Use $dp/dt = (dp/dx)(dx/dt)$.]

34. The pressure P in the aorta during the diastole phase—period of relaxation—can be modeled by the equation
$$\frac{dP}{dt} + \frac{C}{W} P = 0, \qquad P(0) = P_0.$$
    The numbers C and W are positive constants.
    (a) Verify that $P = P_0 e^{-Ct/W}$ is a solution.
    (b) Find $\ln(P_0/P)$ after 1 second.

35. The pressure P in the aorta during systole can be given by
$$P = \left(P_0 + \frac{CAW^2B}{C^2 + W^2B^2} \right) e^{-Ct/W}$$
$$+ \frac{CAW}{C^2 + W^2B^2} [C \sin Bt + (-WB)\cos Bt].$$

Show that $P(0) = P_0$ and $dP/dt + (C/W)P = CA \sin Bt$.

36. A rich uncle makes an endowment to his brother's firstborn son of $10,000, due on the child's twenty-first birthday. How much money should be put into a 9% continuous interest account to secure the endowment? [*Hint*: Use the formula $P = P_0 e^{kt}$, solving for P_0.]

★37. Let $a > 0$. Show that $[1 + (a/n)]^n$ is an increasing function of n by following this outline:
    (a) Suppose that $f(1) = 0$ and $f'(x)$ is positive and decreasing on $[1, \infty)$. Then show that $g(x) = xf(1 + (1/x))$ is increasing on $[1, \infty)$. [*Hint*: Compute $g'(x)$ and use the mean value theorem to show that it is positive.]
    (b) Apply the result of (a) to $f(x) = \ln(x)$.
    (c) Apply the result of (b) to
$$\frac{1}{a} \ln\left[\left(1 + \frac{a}{n}\right)^n \right].$$

★38. Let $r = b^\theta$ be an exponential spiral.
    (a) Show that the angle ϕ between the tangent line at any point of the spiral and the line from that point to the origin is the same for all points of the spiral. (Use the formula for the tangent line in polar coordinates given in Section 5.6.) Express ϕ in terms of b.
    (b) The tangent lines to a certain spiral make an angle of 45° with the lines to the origin. By what factor does the spiral grow after one turn about the origin?

★39. Determine the number of (real) solutions of the equation $x^3 - 4x + \frac{1}{2} = 2^x$ by a graphical method.

★40. (a) For any positive integer n, show by using calculus that $(1/n)e^x - x \geq 0$ for large x.
    (b) Use (a) to show that $\lim_{x \to \infty} (e^x/x) = \infty$.

Review Exercises for Chapter 6

Simpify the expressions in Exercises 1–8.

1. $(x^\pi + x^{-\pi})(x^\pi - x^{-\pi})$
2. $[(x^{-3/2})^2]^{1/4}$
3. $\log_2(8^3)$
4. $\log_3(9^2)$
5. $\ln(e^3) + \frac{1}{2}\ln(e^{-5})$
6. $\dfrac{3e^{-\ln 4}}{\ln e^4}$
7. $\ln \exp(-36)$
8. $\exp(\ln(\exp 3 + \exp 4) + \ln(8))$

Differentiate the functions in Exercises 9–38.

9. e^{x^3}
10. $(e^x)^3$
11. $e^x \cos x$
12. $\cos(e^x)$
13. $e^{\cos 2x}$
14. $e^{\cos x}$
15. $x^2 e^{10x}$
16. $xe^{(x+2)^3}$
17. e^{6x}
18. $xe^x - e^x$
19. $\dfrac{e^{\cos x}}{\cos(\sin x)}$
20. $\dfrac{x^2 + 2x}{1 + e^{\cos x}}$
21. $\dfrac{\sin(e^x)}{e^x + x^2}$
22. $\tan(\sin(e^x))$
23. $\cos\sqrt{1 + e^x}$
24. $e^x \cos(x^{3/2})$

25. $e^{\cos x + x}$

26. $\exp((\sin x) - x^2)$

27. $\dfrac{e^{-x^2}}{1 + x^2}$

28. $\cos(e^{x^2 + 2})$

29. $x \ln(x + 3)$

30. $x \ln x$

31. $\ln(\cos x)$

32. $\ln(\sqrt{x}\,)$

33. $\log_3(5x)$

34. $\log_2(3x)$

35. $\cos^{-1}(x + e^{-x})$

36. $\sin^{-1}(e^x - 1)$

37. $\dfrac{1}{(\ln t)^2 + 3}$

38. $\sin\left[(\ln t)^3 + \dfrac{\pi}{6}\right]$

Compute the integrals in Exercises 39–48.

39. $\displaystyle\int e^{3x}\, dx$

40. $\displaystyle\int (e^{6x} + e^{-6x})\, dx$

41. $\displaystyle\int \left(\cos x + \dfrac{1}{3x}\right) dx$

42. $\displaystyle\int \dfrac{1}{x + 2}\, dx$

43. $\displaystyle\int \left(\dfrac{x + 1}{x}\right) dx$

44. $\displaystyle\int \dfrac{x^2 + x + 2}{x}\, dx$

45. $\displaystyle\int_1^2 \dfrac{x + x^2 \sin \pi x + 1}{x^2}\, dx$

46. $\displaystyle\int_1^2 \left(\dfrac{1}{x} + \dfrac{1}{x^2} + \dfrac{1}{x^3}\right) dx$

47. $\displaystyle\int_1^2 (x - \cos x - e^x)\, dx$

48. $\displaystyle\int_0^5 e^{-5x}\, dx$

Use logarithmic differentiation to differentiate the functions in Exercises 49–52.

49. $(\ln x)^x$

50. $(\ln x)^{\exp x}$

51. $\dfrac{(x + 3)^{7/2}(x + 8)^{5/3}}{(x^2 + 1)^{6/11}}$

52. $(3x + 2)^{1/2}(8x^2 - 6)^{3/4}(\sin x - 3)^{6/17}$

Sketch the graphs of the functions in Exercises 53–56.

53. $y = \dfrac{e^x}{1 + e^x}$.

54. $y = \sin(\ln x)$.

55. $y = \dfrac{e^{-x}}{1 + x}$.

56. $y = \dfrac{1}{(\ln t)^2 + 1}$.

Find dy/dx in Exercises 57–60.

57. $e^{x \cdot y} = x + y$

58. $x^y + y = 3$.

59. $e^{-x} + e^{-y} = 2$.

60. $e^x + e^y = 1$.

61. Find the equation of the tangent line to the graph of $y = (x + 1)e^{(3x^2 + 4x)}$ at $(0, 1)$.

62. Find the tangent line at $(0, \ln 3)$ to the graph of the curve defined implicitly by the equation $e^y - 3 + \ln(x + 1)\cos y = 0$.

Differentiate the functions in Exercises 63–66 and write the corresponding integral formulas.

63. $\dfrac{1}{36} \sin(6x) - \dfrac{x}{6} \cos(6x)$

64. $2x^2 \ln x - x^2$

65. $\dfrac{e^x}{x + 1}$

66. $\frac{2}{5} e^{2x} \cos x + \frac{1}{5} e^{2x} \sin x$

Find the limits in Exercises 67–70.

67. $\lim_{n \to \infty} \left(1 + \dfrac{8}{n}\right)^n$

68. $\lim_{n \to \infty} \left(1 - \dfrac{3}{2n}\right)^{-2n}$

69. $\lim_{n \to \infty} \left(1 + \dfrac{10}{n}\right)^n$

70. $\lim_{n \to \infty} \left(1 - \dfrac{6}{n}\right)^{2n}$

71. The radius of a bacterial colony is growing with a percentage rate of 20% per hour. (a) If the colony maintains a disk shape, what is the growth rate of its area? (b) What if the colony is square instead of round, and the lengths of the sides are increasing at a percentage rate of 20% per hour?

72. If a quantity $f(t)$ is increasing by 10% a year and quantity $g(t)$ is decreasing by 5%, how is the product $f(t)g(t)$ changing?

73. What annual interest rate gives an effective 8% rate after continuous compounding?

74. How much difference does continuous versus daily compounding make on a one-million dollar investment at 10% per year?

75. The velocity of a particle moving on the line is given by $v(t) = 37 + 10e^{-0.07t}$ meters per second. (a) If the particle is at $x = 0$ at $t = 0$, how far has it travelled after 10 seconds? (b) How important is the term $e^{-0.07t}$ in the first 10 seconds of motion? In the second 10 seconds?

76. If \$1000 is to double in 10 years, at what rate of interest must it be invested if interest is compounded (a) continuously, and (b) quarterly?

77. If a deposit of A_0 dollars is made t times and is compounded n times during each deposit interval at an interest rate of i, then

$$A = A_0 \left\{ \dfrac{(1 + i/n)^{nt} - 1}{(1 + i/n)^n - 1} \right\}$$

is the amount after t intervals of deposit. (Deposits occur at the *end* of each deposit interval.)

(a) Justify the formula. [*Hint*:

$$x^{t-1} + \cdots + x + 1 = (x^t - 1)/(x - 1).]$$

(b) A person deposits \$400 every three months, to be compounded quarterly at 7% per annum. How much is in the bank after 6 years?

78. The *transmission density* of a test area in a color slide is $D = \log_{10}(I/I_0)$, where I_0 is a reference intensity and I is the intensity of light transmitted through the slide. Rewrite this equation in terms of the natural logarithm.

79. The salvage value of a tugboat is given by $y = 260,000\,e^{-0.15x}$ dollars after x years of use. What is the expected depreciation during the fifth year?

80. Find the marginal revenue of a commodity with demand curve $p = (1 + e^{-0.05x})10^3$ dollars per unit for x units produced. [Revenue equals (number of units)(price per unit) $= xp$; the marginal revenue is the derivative of the revenue with respect to x.]

81. A population model which takes birth and death rates into account is the *logistic model* for the population P: $dP/dt = P(a - bP)$. The constants a, b, where $a > 0$ and $b \neq 0$, are the *vital constants*.
    (a) Let $P(0) = P_0$. Check by differentiation that $P(t) = a/[b + ([a/P_0] - b)e^{-at}]$ is a solution of the logistic equation.
    (b) Show that the population size approaches a/b as t tends to ∞.

82. We have seen that the exponential function $\exp(x)$ satisfies $\exp(x) > 0$, $\exp(0) = 1$, and $\exp'(x) = \exp(x)$. Let $f(x)$ be a function such that $0 \leqslant f'(x) \leqslant f(x)$ and $f(0) = 0$. Prove that $f(x) = 0$ for all x. [*Hint*: Consider $g(x) = f(x)/\exp(x)$.]

83. Show that for any $x \neq 0$ there is a number c between zero and x such that $e^x = 1 + e^c x$. Deduce that $e^x > 1 + x$.

84. Fig. 6.R.1 shows population data from each U.S. census from 1790 to 1970.

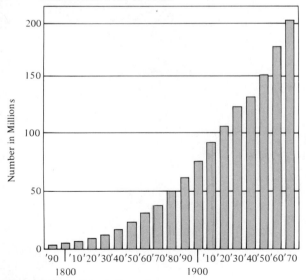

(a) Fit the data to an exponential curve $y = Ae^{\gamma t}$, where t is the time in years from 1900; i.e., find A and γ numerically.
(b) Use (a) to "predict" the 1980 census. How close was your prediction? (The actual 1980 census figure was 226,500,000.)
(c) Use (a) to predict when the U.S. population will be 400 million.

★85. (a) Show that if $b > 1$ and n is a positive integer, then
$$b^n \geqslant 1 + n(b - 1)$$
and
$$b^{-n} \leqslant \frac{1}{1 + n(b - 1)}.$$
[*Hint*: Write $b^n = [1 + (b - 1)]^n$ and expand.]
(b) Deduce from these inequalities that $\lim_{x \to \infty} b^x = \infty$ and $\lim_{x \to -\infty} b^x = 0$.

Exercises 86–91 form a unit.

★86. Show that $e^x \geqslant x^n/n!$ for all integers $n \geqslant 0$ and all real $x \geqslant 1$, as follows: (recall that $n! = 1 \cdot 2 \cdot 3 \cdots (n - 1) \cdot n$ and $0! = 1$).
(a) Let n be fixed, let x be variable, and let $f_n(x) = e^x - x^n/n!$. Show that $f_n'(x) = f_{n-1}(x)$.
(b) Show that $f_0(x) > 0$ for $x > 1$, and conclude that $f_1(x)$ is increasing on $[1, \infty)$.
(c) Conclude that $f_1(x) > 0$ for $x \geqslant 1$.
(d) Repeat the argument to show that $f_2(x) > 0$ for $x \geqslant 1$.
(e) Finish the proof.

★87. Show that $e^x \geqslant x^n$ for $x \geqslant (n + 1)!$

★88. Show that $\lim_{x \to \infty}(e^x/x^n) = \infty$.

★89. (a) By taking logarithms in the inequality derived in Review Exercise 87, show that $x/\ln x \geqslant n$ when $x \geqslant (n + 1)!$
(b) Conclude that $\lim_{x \to \infty} \ln \dfrac{\ln x}{x} = 0$.

★90. Use the result of Review Exercise 89 to show that $\lim_{x \to \infty}(\ln x/x^a) = 0$ for any $a > 0$. (*Hint*: Let $y = x^a$.)

★91. Use the result of Review Exercise 90 to show that $\lim_{x \to 0}(\ln x/x^{-a}) = 0$ for any $a > 0$.

★92. Look at Exercise 43 in Section 3.1. Explain why it takes, on the average, 3.32 bisections for every decimal place of accuracy.

★93. Prove that $3^{x-2} > 2x^2$ if $x \geqslant 7$. Can you improve this result?

Figure 6.R.1. Population of the United States. Total number of persons in each census: 1790–1970.

Answers to Orientation Quizzes

Answers to Quiz A on p. 13

1. $-\dfrac{4}{3}$ **2.** $x > -\dfrac{2}{3}$ **3.** $x < -\dfrac{1}{3}$ and $x > 1$ **4.** $x = -2 \pm \dfrac{\sqrt{38}}{2}$

5.

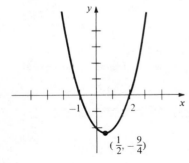

6. $g(2) = \dfrac{4}{3}$. The domain of g is all x such that $x \neq 0, \dfrac{1}{2}$.   **7.** $x > -10$ **8.** At the points $(1, 1)$ and $(2, 4)$

9.

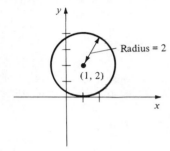

10. 2

Answers to Quiz B on p. 14

1. $\dfrac{11}{4} = 2\dfrac{3}{4}$ **2.** $x(x + 3)$ **3.** 10   **4.** 4 kilograms   **5.** $(2x - 1)/2x$ (or $1 - 1/2x$)   **6.** x^7

7. $-6, -4, 0, \dfrac{1}{2}, 8$ **8.** $-\dfrac{3}{7}$ **9.** $\dfrac{1}{3}$ **10.** $x + 4$

Answers to Quiz C on p. 14

1. $(-2, -3)$ **2.** $\dfrac{8}{7}$ **3.** $150°$ **4.** $\dfrac{4}{5}$ **5.** $\sqrt{5}$ **6.** 8π centimeters   **7.** 14 meters

8. 18π cubic centimeters   **9.** $\dfrac{1}{2}$ **10.** 2

Chapter R Answers

R.1 Basic Algebra: Real Numbers and Inequalities

1. Rational
3. Rational
5. $ab - 5b - 3c$
7. $a^2b - b^3 + b^2c$
9. $a^3 - 3a^2b + 3ab^2 - b^3$
11. $b^4 + 4b^3c + 6b^2c^2 + 4bc^3 + c^4$
13. $(x + 3)(x + 2)$
15. $(x - 6)(x + 1)$
17. $3(x - 4)(x + 2)$
19. $(x - 1)(x + 1)$
21. 2
23. 1
25. Multiply out $(x - 1)(x^2 + x + 1)$
27. $x(x + 2)(x - 1)$
29. $-4, -1$
31. $-1, 1/2$
33. $(5 \pm \sqrt{26.2})/2$
35. No real solution
37. $(1 \pm \sqrt{33})/4$
39. $9/2$
41. $\sqrt{7}/2$
43. $-\sqrt{3}, -5/3, -\sqrt{2}, -7/5, 0, 9/5, 23/8, 3, 22/7$
45. $a > c$
47. $b^2 < c$
49. $-2 > b$
51. (a) $x < 4$
   (b) $x \leqslant 13/3$
   (c) $x > 3$

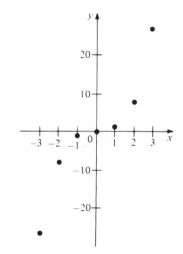

(a)

(b) $4\frac{1}{3}$

(c)

53. (a)
$$ax^2 + bx + c = a\left[\left(x + \frac{b}{2a}\right)^2 + \left(\frac{c}{a} - \frac{b^2}{4a^2}\right)\right] = 0$$

has solution $x + b/2a = \pm\sqrt{\left(\frac{c}{a} - \frac{b^2}{4a^2}\right)}$. Now

simplify.
   (b) $b^2 = 4ac$ exactly when the discriminant vanishes.

R.2 Intervals and Absolute Values

1. (a) True (b) False (c) True
   (d) False (e) True
3. If $a < b$, then $a = 3a - 2a < b$, so $3a < 2a + b$
   and $a < (2a + b)/3$.
5. $[3, \infty)$
7. $(-\infty, 2)$
9. $(-\infty, -3)$ or $(1, \infty)$
11. $(-\infty, 0]$ or $[1, \infty)$
13. 2
15. 8
17. 15
19. 15

21. ± 8
23. $|x + 5/2| > 5/2$
25. $|x - 1/2| > 3/2$
27. $(3, 4]$
29. $(-5, 5)$
31. $(-1, 1/3)$
33. $|x| < 3$
35. $|x| < 6$
37. $|x - 2| \leqslant 10$
39. Let $x = 1$, $y = 1$, and $z = -1$.
41. Take the cases $x \geqslant 0$ and $x < 0$ separately.
43. No

R.3 Laws of Exponents

1. 1
3. $64(4 \cdot 3^{10} + 1)$
5. $1/(2^9 \cdot 3^{12})$
7. 4^4
9. 3
11. 3
13. $1/8$
15. 36
17. $1 + x$
19. $x + 2x^3 + 3x^6$
21. Use $\sqrt[a]{x} = x^{1/a}$
23. $(ab)^{3/2}$
25. 4, 8, 32
27. $(x^{1/2} - 2y^{1/2})(x^{1/2} + y^{1/2})$
29. $(x^{1/2} - 4)(x^{1/2} + 2)$
31. Both choices are consistent.
33. Proceed as in the proof that $b^{p+q} = b^p b^q$.

R.4 Straight Lines

1.

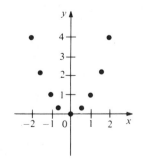

3.

5. 2

7. $\sqrt{314}$

9. 43718

11. $\sqrt{2}$

13. $2\sqrt{x^2 + 25}$

15. $2a\sqrt{2}$

17. 3

19. $-7/2$

21. $y = 2x - 1$

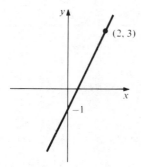

23. $y = 7$

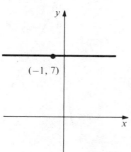

25. $2y = x + 9$

27. $y = 4$

29. $m = -1/2$, intercept is -2

31. $m = 0$, intercept is $17/4$

33. $m = -11/7$, intercept is $13/7$

35. $m = 0$, intercept is 17

37. (a) $m = -4/5$

   (b) $4y = 5x - 1$

39. $y = 5x + 14$

41. $y = -x + 6$

43. $(1, -3)$, $(1 + 2\sqrt{6}, 3)$ or $(1 - 2\sqrt{6}, 3)$

R.5 Circles and Parabolas

1. $(x - 1)^2 + (y - 1)^2 = 9$

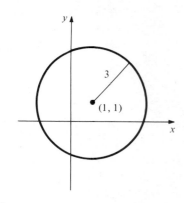

3. $x^2 + (y - 5)^2 = 25$

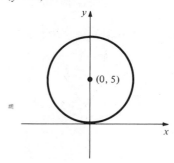

5. $(x + 1)^2 + (y - 4)^2 = 10$

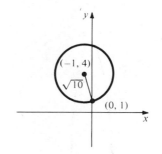

7. $r = \sqrt{2}$, center is $(1, -1/2)$

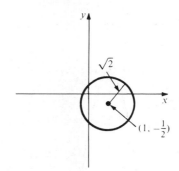

9. $r = 3$, center is $(4, -2)$

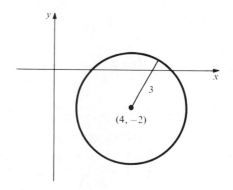

11. $y = -x^2 + 2x + 1$

13. $y = -x^2/5 + 2x$

15.

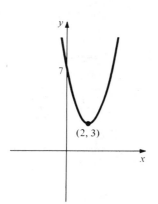

17.

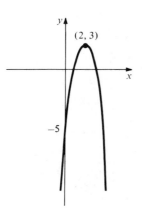

19.

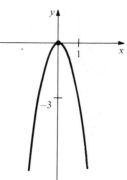

21.

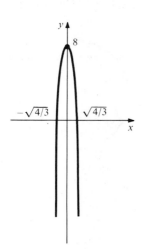

23.

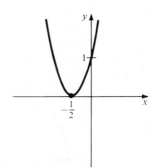

25. $(6/7, 37/7)$

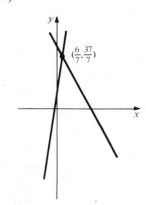

27. $\left(\dfrac{-3 + 2\sqrt{11}}{5} , \dfrac{53 - 12\sqrt{11}}{5} \right),$

$\left(\dfrac{-3 - 2\sqrt{11}}{5} , \dfrac{53 + 12\sqrt{11}}{5} \right)$

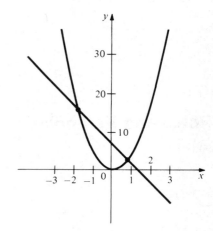

29. No intersection.

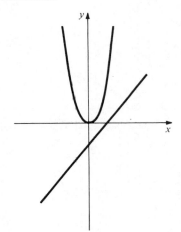

31. 0, 1, 2, or infinitely many intersections.

33. $\left(\sqrt{(-9 + \sqrt{273})/32} , (-9 + \sqrt{273})/8 \right)$
 and $\left(-\sqrt{(-9 + \sqrt{273})/32} , (-9 + \sqrt{273})/8 \right)$

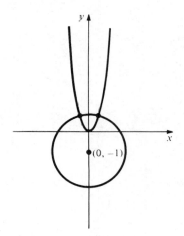

35. $(-3/2, 5/4)$

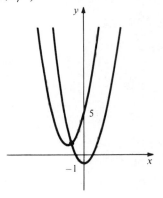

37. $x \in ((1 - \sqrt{37})/18, (1 + \sqrt{37})/18)$

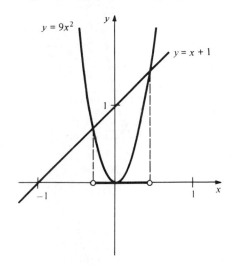

39. $x \in ((-1 - \sqrt{5})/4, (-1 + \sqrt{5})/4)$

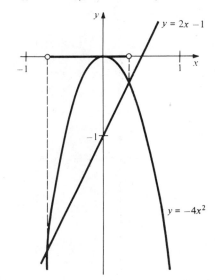

R.6 Functions and Graphs

1. 7; 3   **3.** -2; 0   **5.** 4; 0
7. Domain is $x \neq 1$; $f(10) = 100/9$.
9. Domain is $-1 \leqslant x \leqslant 1$; $f(10)$ does not exist.
11. Domain is $x \neq -2, 3$; $f(10) = 13/21$.
13.

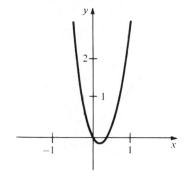

15.

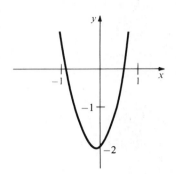

17.

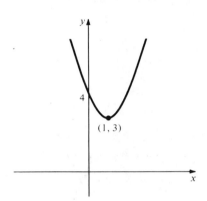

19.

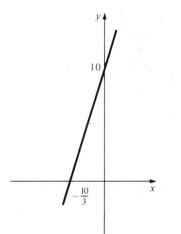

21.

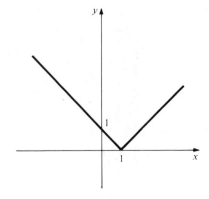

23.

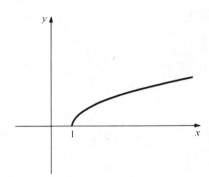

25.

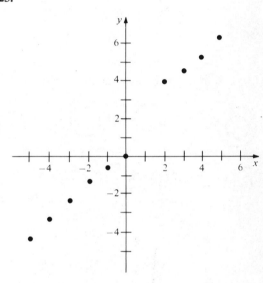

27. (a)

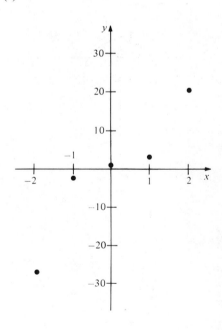

27. (b)

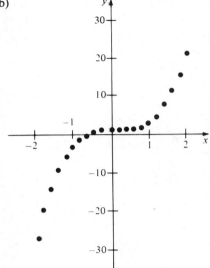

65.

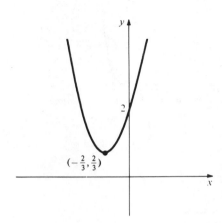

29. a, c
31. Yes; domain is $x \neq 0$.
33. Yes; domain is $x \leqslant -1$ and $x \geqslant 1$.
35. Yes; domain is all real numbers.

Review Exercises for Chapter R

1. $-2/3$
3. $(1 \pm \sqrt{5})/2$
5. $\frac{1}{2}$
7. 3
9. $x > -1/4$
11. $-\sqrt{5} < x < \sqrt{5}$
13. $x > 3/2$
15. $-\sqrt{7/3} < x < \sqrt{7/3}$
17. $(-1, 1)$
19. $(0, 2)$
21. $(-\infty, 1 - \sqrt{2}\,]$ and $[1 + \sqrt{2}, \infty)$
23. $(5, 6)$
25. $(-2, 3)$
27. $[5, 6)$
29. $(-\infty, 13/2]$ and $(22/3, \infty)$
31. $(-\infty, -5]$, $(1, 5/2)$ and $[3, \infty)$
33. 13
35. 8
37. 1
39. 4
41. $\sqrt[4]{xy}$
43. $(\sqrt{x-1} - \sqrt{x-8})/7$
45. $\sqrt{10}$
47. $5\sqrt{2}$
49. $13y - 8x + 17 = 0$
51. $x + y = 1$
53. $4y + 12x - 61 = 0$
55. $2y - 7x - 34 = 0$
57. $8y - 5x - 3 = 0$
59. $y - 4 = 0$
61. $x^2 - 24x + y^2 - 10y + 105 = 0$
63. $x^2 - 14x + y^2 + 2y + 41 = 0$

67.

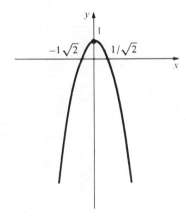

69. $(-\sqrt{2}, -\sqrt{2})$, $(\sqrt{2}, \sqrt{2})$
71. No intersection.
73.

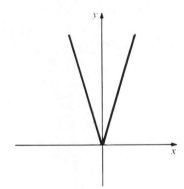

75.

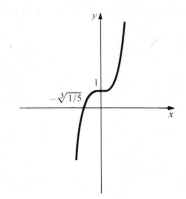

77.

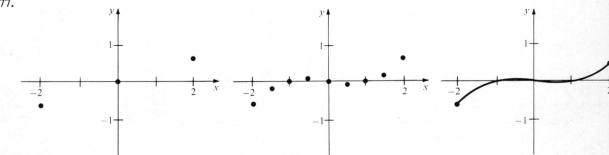

79.

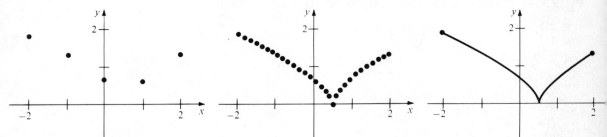

81. a, c

83. (a) $\dfrac{1}{1 + 1/(k + l)}$ has the smaller denominator

(b) use $\dfrac{1}{1 + 1/l} = \dfrac{l}{1 + l}$.

Chapter 1 Answers

1.1 Introduction to the Derivative

1. (a) $\Delta y = 3.75$; $\Delta y/\Delta x = 7.5$
   (b) $\Delta y = 0.0701$; $\Delta y/\Delta x = 7.01$
   (c) $\Delta y = 1.11$; $\Delta y/\Delta x = 11.1$
   (d) $\Delta y = 0.1101$; $\Delta y/\Delta x = 11.01$

3. (a) $\Delta y = 7.25$; $\Delta y/\Delta x = 14.5$
   (b) $\Delta y = 0.1401$; $\Delta y/\Delta x = 14.01$
   (c) $\Delta y = 1.81$; $\Delta y/\Delta x = 18.1$
   (d) $\Delta y = 0.1801$; $\Delta y/\Delta x = 18.01$

5. 7 meters/second

7. 13 meters/second

9. (a) $2x_0 + 3$ (b) $x_0 = 7/2$

11. (a) $2x_0 + 10$ (b) $x_0 = 0$

13. Slope is 2.

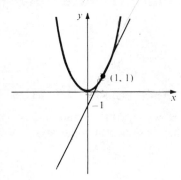

15. Slope is -3.

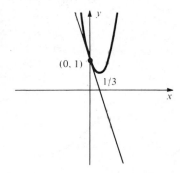

17. a

19. $2a$

21. 3

23. -11

25. $(8, -62)$

27. $(-2, 7)$

29. $2x + 3$

31. $2x$

33. $-8t + 3$

35. $-2s$

37. $t = 1$

39. 29.4 meters/second.

41. 6

43. 0

45. -5

47. -1

49. $8x + 3$

51. $-2x$

53. $-4x + 5$

55. $f'(x) = 0$ at $x = -3/4$; $f'(x) > 0$ for $x > -3/4$; $f'(x) < 0$ for $x < -3/4$

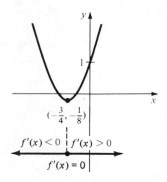

57. $A'(x) = 2x$

59. $(1, 0)$

61. $y = 10x - 1$

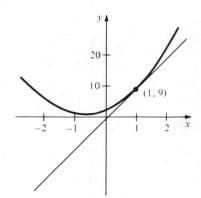

63. $y = 14x - 49$ and $y = 2x - 1$

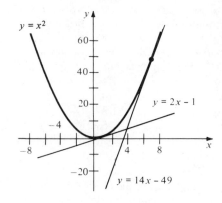

65. The equations of the lines are $y = x_0^2$ and $y = x_0^2 - (1/2x_0)(x - x_0)$

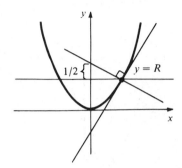

67. $f'(0) = 0$; $f'(\sqrt{2}) = -1$

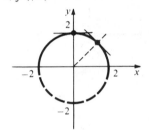

69. (a) 24.5, 14.7, 4.9, -4.9, -14.7, -24.5 meters/sec.
(b) 2.5 seconds
(c) 2.5 seconds
(d) 5 seconds

1.2 Limits

1. Limit is 2.

3. 2; does not exist

5. 20

7. 0

9. 0

11. 0

13. $\sqrt{3}/6$

15. 1

17. 1/2

19. 3

21. 2

23. -8

25. $-\sqrt{3} - 1$

27. No limit exists.

29. No limit exists.

31. 1/2

33. 2/3

35. 1

37. $1 = f(0)$; 2, $f(1)$ does not exist; no limit, $f(2) = 2$; 1, $f(3)$ does not exist; no limit, $f(4) = 2$.

39. 2, 4

41. $-\infty$

43. No limit exists.

45. 1

47. $+\infty$

49. 1/2

51. 5

53. 75

55. 9

57. No limit exists.

59. $f(1) = 5$

61. (a) $f(T) = g(T) = 0$
(b) A thin sheet of ice of area A could remain just before complete melting. Hence, $\lim_{t \to T} g(t) = 0 = g(T)$, but $\lim_{t \to T} f(t) = A \ne f(T)$.
(c) $\lim_{t \to T} [f(t) \cdot g(t)] = 0$
(d) $\lim_{t \to T} f(t) \cdot \lim_{t \to T} g(t) = 0$ as well.

63. By the quotient rule,
$$\lim_{x \to x_0} \left[\frac{1}{f(x)} \right] = 1/\lim_{x \to x_0} f(x).$$
As with the argument that $\lim_{x \to \infty} \dfrac{1}{x} = 0$, we can conclude that $\lim_{x \to x_0} [1/f(x)] = 0$.

65. (a) Write $f_1(x) + f_2(x) + f_3(x) + f_4(x) = [f_1(x) + f_2(x) + f_3(x)] + f_4(x)$.
(b) Write $f_1(x) + f_2(x) + \cdots + f_{17}(x) = [f_1(x) + \cdots + f_{16}(x)] + f_{17}(x)$.
(c) Use the method of (b) with n in place of 16 and $n + 1$ in place of 17.
(d) Choose $m = 1$ and use the basic sum rule together with (c).

1.3 The Derivative as a Limit and the Leibniz Notation

1. $2x + 1$

3. $15x^2$

5. $-3/x^2$

7. $2x - 3/x^2$

9. $1/\sqrt{x}$

11. $4x - 1/2\sqrt{x} - 1/x^2$

13. The difference quotient $\dfrac{|\Delta x|}{\Delta x}$ has no limit at $x_0 = 0$. However, $f(0) = 1$ and $\lim_{x \to 0}(1 + |x|) = 1$ imply that $f(x)$ is continuous at $x_0 = 0$.

15. $2x - 1$

17. $9x^2 + 1$

19. $4/3$

21. 15

23. -3

25. 10

27. 8

29. $-2/x^3$

31. 1/2

33. One answer is
$$f(x) = \begin{cases} 1 - x^2 & \text{for} \quad -1 < x < 1 \\ 0 & \text{for} \quad x \le -1 \text{ or } x \ge 1 \end{cases}$$
Another is given in this figure:

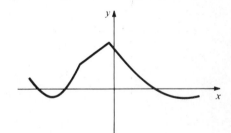

35. (a) 1, 0
(b) $\dfrac{d}{dx}(\sin x) = \lim_{\Delta x \to 0} \dfrac{\sin(x + \Delta x) - \sin x}{\Delta x}$

$\qquad = \lim_{\Delta x \to 0} \dfrac{\sin x \cos \Delta x + \cos x \sin \Delta x - \sin x}{\Delta x}$

$\qquad = \lim_{\Delta x \to 0} \left[\sin x \left(\dfrac{\cos \Delta x - 1}{\Delta x} \right) \right]$

$\qquad + \lim_{\Delta x \to 0} \left[\cos x \left(\dfrac{\sin \Delta x}{\Delta x} \right) \right]$

$\qquad = 0 + \cos x = \cos x.$

1.4 Differentiating Polynomials

1. $10x^9$ **3.** $33x^{32}$
5. $-20x^3$ **7.** $30x^9$
9. $3/2\sqrt{x}$ **11.** $-4/\sqrt{x}$

13. $f'(x) + g'(x) = 6x + 1 = (f + g)'(x)$
15. $f'(x) + g'(x) = 2x + 1 = (f + g)'(x)$
17. $f'(x) - 2g'(x)$ **19.** $5x^4 + 8$
21. $5t^4 + 12t + 8$ **23.** $4x^3 - 14x - 3$
25. $13s^{12} + 96s^7 - 21s^6/8 + 4s^3 + 4s^2$
27. $4x^3 - 9x^2 + 4x$
29. $80h^9 + 9h^8 - 113h$
31. $6x^5 + 12x^3 + 6x$
33. $24t^7 - 306t^5 + 130t^4 + 48t^2 - 18t + 17$
35. $-30r^5 + 20r^3 - 26r$
37. $7t^6 - 5t^4 + 27t^2 - 9$
39. $7u^6 + 42u^5 + 76u^3 + 15u^2 + 70u$
41. $2x - 1/2\sqrt{x}$ **43.** $3x^2 - 2 + 1/\sqrt{x}$
45. -1 **47.** 317.44
49. 5 **51.** $x^3/3 + C$
53. $x^{n+1}/(n + 1) + C$
55. $(kf)'(x) = 2akx + kb = kf'(x)$
57. $\dfrac{d}{dr}\left(\dfrac{4}{3}\pi r^3\right) = 4\pi r^2 =$ surface area
59. k is the conversion factor from miles to kilometers.

1.5 Products and Quotients

1. $3x^2 + 16x + 2$ **3.** $7x^6 - 4x^3 - 2$
5. $3x^2 + 2x - 1$ **7.** $4x^3 + 15x^2 + 16x + 6$
9. $3x^2$ **11.** $4x^3$
13. $5x^{3/2}/2$ **15.** $7x^{5/2}/2$
17. $(-x^2 + 4x + 3)/(x^2 + 3)^2$
19. $(4x^9 + 7x^6 + x^4 - 2x)/(x^3 + 1)^2$
21. $-8x/(x^2 - 2)^2$
23. $-2/x^3 + (-x^2 + 1)/(x^2 + 1)^2$
25. $(-6r^2 - 16)/r^9$
27. $2s/(1 - s)^3$
29. $(-8x^3 - 8x)/(x^4 + 2x^2)^2$
31. $-4/x^5$
33. $-2/(x + 1)^3$
35. $s^2(8s^5 + 5s^2 + 3)$
37. $8y + 6 + 1/(y + 1)^2$
39. $1/\sqrt{x}\,(\sqrt{x} + 1)^2$ **41.** $15x^{3/2}/2 + 2x$
43. $(2 + \sqrt{x})/2(1 + \sqrt{x})^2$ **45.** $-3/\sqrt{2x}\,(1 + 3\sqrt{x})^2$
47. Let $h(x) = 1/g(x)$ and differentiate $1/h(x)$ by the reciprocal rule.
49. $-1/4\sqrt{2}$
51. $32x^7 + 48x^5 - 20x^4 - 52x^3 - 52x + 13$
53. $24x^5 + 3x^2 - 26x + 2$
55. $(-8x^7 - 32x^5 + 20x^4 - 26x^3 - 13)/(4x^5 - 13x)^2$
57. $P(x) = x^2$ is an example.

1.6 The Linear Approximation and Tangent Lines

1. $y = -2x + 2$

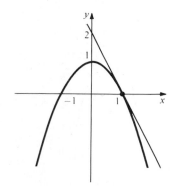

3. $y = 2x - 3$

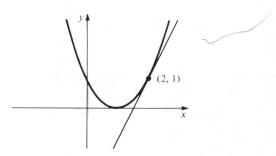

(2, 1)

5. $y = -(21/2)x$
7. $y = (-25/2)x + (17/2)$
9. $x = -1$
11. $x = 16$
13. 4.08; 4.0804 is exact.
15. 24.99; 24.990001 is exact.
17. 4.002; 4.001999 on a calculator.
19. 3.99; 3.989987 on a calculator.
21. 74.52
23. 63.52
25. 0.24π
27. 0.40π
29. 60.966
31. 0.0822
33. $y = 12x - 8$
35. 1.02
37. -4.9976
39. $1,153,433.6$
41. -14.267; -0.98; -36.00

43. $g'(3) = -16$; the graph lies below the tangent line.

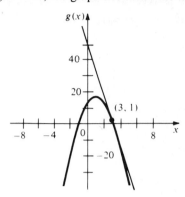

45. (a) $16 + 32(\Delta x)$
(b) Smaller
(c) $[1.730, 2.247]$

47. $1 - x$ is the linear approximation of $1/(1 + x)$ near 0.

49. $\dfrac{3600}{60 + x} \approx 60 - x + \dfrac{x^2}{60}$

Review Exercises for Chapter 1

1. $2x$
3. $3x^2$
5. 2
7. $2s + 2$
9. $-50x^4 + 24x^2$
11. $4x^3$
13. $9x^2 - 1/\sqrt{x}$
15. $50x^{49} - 1/x^2$
17. $-4x/(x^2 - 1)^2$
19. $-(5x^2 + 2)/2x^{3/2}(x^2 + 2)^2$
21. $(5s^2 + 8s^{3/2} + 6s + 8\sqrt{s} + 1)/2\sqrt{s}$
23. $(4\pi r + 3\pi r^{3/2})/2(1 + \sqrt{r})^2$
25. $-6t/(3t^2 + 2)^2$

27. $2\sqrt{2}\, p/(p^2 + 1)^2$
29. $-(2\sqrt{x} - 1)/2\sqrt{x}\,(x - \sqrt{x}\,)^2$
31. 2
33. 3
35. -192
37. 11
39. 6
41. $4x^3 + 6x$
43. $3/5$
45. 0
47. 1; 0; does not exist; does not exist; 1, 1, 2
49. 17
51. $1 - 1/2\sqrt{x}$
53. -13
55. 0
57. 0
59. $t = 2$, $1/3$
61. 1 meter/second.
63. 1.072
65. 2.000025
67. 1.0045
69. 27.00
71. $y = -6x$
73. $y = (216/361)x - (413/361)$
75. 0.16π, the exact value is calculated as $(0.1608013 \dots)\pi$ meters^3.
77. $5/2$
79. $-x$
81. $dV/dr = 9r^2$ which is $9/14$ of A.
83. (a) $dz/dy = 4y + 3$; $dy/dx = 5$.
(b) $z = 50x^2 + 35x + 5$; $dz/dx = 100x + 35$
(c) $(dz/dy) \cdot (dy/dx) = 100x + 35 = \dfrac{dz}{dx}$
(d) $x = (y/5) - (1/5)$, $dx/dy = 1/5$
(e) $dx/dy = 1/(dy/dx)$
85. $y = (2\sqrt{2} - 2)x$
87. The focal point is $(0, 1/4a)$.
89. (a) Use $x^n x^m = x^{n+m}$.
(b) Apply (a) to the numerator and denominator.
(c) Use the quotient rule. (If deg $f = 0$, then $f(x)$ might be constant, in which case, deg f' is not -1.)
91. (a) Expand $f'g - fg'$ and equate coefficients of 1, $x, x^2 \dots$ to zero.
(b) Apply (a) to $F - G$.

Chapter 2 Answers

2.1 Rates of Change and the Second Derivative

1. $y = 5x - 19$

3. $y = \frac{1}{2}x - \frac{1}{2}$

5. $\Delta P / \Delta t = 0.6$ cents/year, $P = 5$ when $t = 1987$; when $t = 1991$, $P = 7.4$

7. $v = 9.8t + 3$; $v = 150$ meters/second.

9. -60

11. $11/2$

13. πr^2

15. 69.6

17. 53

19. 400 people/day

21. $4\pi r^2$, the surface area of a sphere

23. $t = \pm\sqrt{-c/3}$; no

25. $1.00106, 0.043$

27. $12x^2 - 6$

29. $10/(x + 2)^3$

31. $168r^6 - 336r^5$

33. 2

35. $20x^3 + 84x^2$

37. $2/(x - 1)^3$

39. $4(3t^2 + 1)/(t^2 - 1)^3$

41. 3 meters/second; 0 meters/second^2

43. 0 meters/second; 16 meters/second^2

45. -2 meters/second; 0 meters/second^2

47. (a) 29.4 meters/second downward; -9.8 meters/second^2.
    (b) -20.8 meters/second.

49. 10 dollars/worker hour

51. 94 dollars/worker day

53. $[25 - 0.04x - (4 + 0.04x - 0.016x^3 + 0.00004x^4) /(1 + 0.002x^3)^2]$ dollars/boot.

55. Gas mileage (in miles/gallon).

57. Price of fuel (in dollars/gallon).

59. The first term is the rate of cost increase due to the change in price and the second term is the rate of cost increase due to the change in consumption rate.

61. $-1.6, -1.996, -1.999996$. (The derivative is -2)

63. (a) $0; 0$
    (b) $30; 30$
    (c) $0; 0$
    (d) $-12; -12$

65. $(10t^4 + 4t^3 + 12t^2 + 22t - 3)$ cm^2/second.

67. (a) $t > 6/10$
    (b) $-13/2$
    (c) $t = 1/6$

69. $a = 0$

71. (a) Accelerating when $t > 0$; decelerating when $t < 0$.
    (b) no

73. (a) $V = 4000 - 350t$
    (b) -350

75. (a)

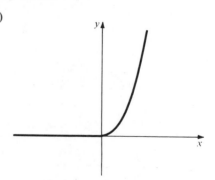

(b) $f'(x) = \begin{cases} 0 & x \leqslant 0 \\ 2x & x \geqslant 0 \end{cases}$

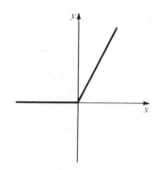

(c) $f''(x) = \begin{cases} 0 & x < 0 \\ 2 & x > 0 \end{cases}$

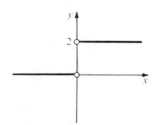

There is no second derivative at $x = 0$.

(d) The acceleration may have jumped as an engine was started.

2.2 The Chain Rule

1. $4(x + 3)^3$

3. $100(x^3 + 10x)^{99}(3x^2 + 10)$

5. $(x^2 + 8x)^2(7x^2 + 32x)$

7. $(x^9 + 8)^2(29x^{10} + 54x^8 + 16x)$

9. $(y + 1)^2(y + 2)(6y^2 + 26y + 26)$

11. $(x + 1)^2$; $x^2 + 1$

13. $(x^3 - 2)^3$; $(x - 2)^9$

15. $\frac{1}{2} - \sqrt{3} /(1 - x)$; $2/(1 + 2\sqrt{3} x)$

17. $h(x) = f(g(x))$ where $f(u) = \sqrt{u}$, $g(x) = 4x^3 + 5x + 3$.

19. $h(u) = f(g(u))$ where $f(v) = v^3$, $g(u) = (1 - u)/(1 + u)$

21. $4x(x^2 - 1)$

23. 1

25. $6(x^2 - 6x + 1)^2(x - 3)$

27. $-195x^4/(3 + 5x^5)^2$

29. $8x(x^2 + 2)[(x^2 + 2)^2 + 1]$

31. $-2x(x^2 + 3)^4(3x^2 + 4)/(x^2 + 4)^9$

33. $5x(2x^3 + 1)/\sqrt{4x^5 + 5x^2}$

35. $3x^2f(2x^2) + 4x^4f'(2x^2)$

37. (a) $f(g(h(x)))$
(b) $f'(g(h(x))) \cdot g'(h(x)) \cdot h'(x)$

39. 0.2; ≈ 0.18; as the circle grows larger, a given increase in diameter produces more area

41. 1500 gm cm^2/sec^3

43. 5¢/mile

45. -6250

47. $156(x + 1)^{11}$

49. $392(x^4 + 10x^2 + 1)^{96}[391x^6 + 3915x^4 + 53x^2 + 9700x + 5]$

51. (a) $(d/dx)f(cx) = cf'(cx)$
(b) If the x-axis is compressed by a factor 4, the slopes are multiplied by 4.

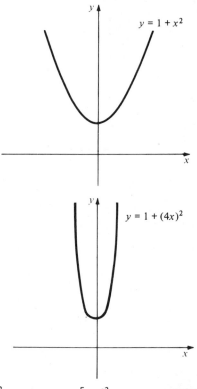

$y = 1 + x^2$

$y = 1 + (4x)^2$

53. $\dfrac{d^2}{dx^2}(u^n) = nu^{n-2}\left[u\dfrac{d^2u}{dx^2} + (n - 1)\left(\dfrac{du}{dx}\right)^2\right]$

55. $f'(x) = 1/g'(f(x))$

57. $(f \circ g)''(x) = f''(g(x))[g'(x)]^2 + f'(g(x))g''(x)$

59. (a) Use the product rule
(b) Use the quotient rule
(c) Use the power of a function rule
(d) $n_1 \dfrac{f_1'(x)}{f_1(x)} + n_2 \dfrac{f_2'(x)}{f_2(x)} + \cdots + n_k \dfrac{f_k'(x)}{f_k(x)}$
(e) $\dfrac{(x^2 + 3)(x + 4)^8(x + 7)^9}{(x^4 + 3)^{17}(x^4 + 2x + 1)^5} \times$
$\left[\dfrac{2x}{x^2 + 3} + \dfrac{8}{x + 4} + \dfrac{9}{x + 7} - \dfrac{68x^3}{x^4 + 3} - \dfrac{5(4x^3 + 2)}{x^4 + 2x + 1}\right]$

2.3 Fractional Powers and Implicit Differentiation

1. $5/(4x^{7/8})$

3. $2/x^{3/4} + 2/(3x^{5/3})$

5. $2/x^{1/3} - 5/(2\sqrt{5x})$

7. $(7/3)x^{4/3} + (10/3)x^{7/3}$

9. $35x^4/[9(x^5 + 1)^{2/9}]$

11. $-1/(2x^{3/2})$

13. $-2x/[(x^2 + 1)^{1/2}(x^2 - 1)^{3/2}]$

15. $(x + 3)/\sqrt{(x + 3)^2 - 4}$

17. $(3 - x - 5x^3)/[2\sqrt{x}(3 + x + x^3)^2]$

19. $1/[2\sqrt{x}(1 + \sqrt{x})^2]$

21. $(2 - 5x^2)/[3x^{2/3}(x^2 + 2)^2]$

23. $(18x^2 + 2x - 24x^{9/2} - 16x^{7/2} - 12x^{5/2} - 1)/$
$[4\sqrt{x}(\sqrt{x} + 2x^3)^{3/2}\sqrt{6x^2 + 2x + 1}]$

25. $7x/[4(x^2 + 5)^{1/8}]$

27. $\sqrt[3]{7}/3$

29. $(3/11)x^{-8/11} - (1/5)x^{-4/5}$

31. $-(7y + 2)/[8y^{7/8}(y - 2)^2]$

33. $-1/[2\sqrt{x}(\sqrt{x} - 1)^{3/2}(\sqrt{x} + 1)^{1/2}]$

35. 0

37. (a) $-4x^3/(2y + 1)$
(b) $-4/3$
(c) $-4x^3/(2y + 1)$

39. (a) $\sqrt{2}/4$
(b) $-\sqrt{2}/4$

41. $y = -x + 2$

43. $(-1/4)x^{-3/2} + (2/9)x^{-4/3}$

45. $y = -\sqrt{3}x + 2$

47. $x^3/(1 - x^4)^{3/4}$

49. 1.9990625

51. 8.9955

53. $[32(2 + x^{1/3})^3/x^{2/3}]$ kg/unit distance.

55. 2.4

57. $-2\sqrt{2}/27$ seconds/pound.

59. (a) $y = 50x/(x - 50)$
(b) $-2500/(x - 50)^2$
(c) $(0, 0)$ or $(100, 100)$

2.4 Related Rates and Parametric Curves

1. $dy/dt = (x/y)(dx/dt)$

3. $dy/dt = (dx/dt)/(2y + 3y^2)$

5. $dy/dt = (dx/dt)/(1 - 2y)$

7. $dy/dt = \sqrt{y/x}(dx/dt)$

9.

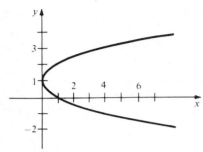

11. $y = x^3$ for $x \geq 0$

13. $y = (15x - 125)/2$

15. $y = \dfrac{10}{\sqrt{3} - 1} + \dfrac{\sqrt{159}\,(13 - 6\sqrt{3}\,)}{76(-6 + 4\sqrt{3}\,)}(x - \sqrt{159}\,)$

17. $dy/dt = -(dx/dt)/16$

19. $-41/8$

21. $-125/8\pi$ cm/sec.

23. 2.55 inches/minute.

25. (a) By the distance formula,

$$\sqrt{x^2 + y^2} = 2\sqrt{x^2 + (y - 1)^2}$$

which yields $3x^2 + 3(y - (4/3))^2 = 4/3$, a circle.

(b) 0

(c) $(\sqrt{2}\,/3, (-\sqrt{2}\, + 4)/3), (-\sqrt{2}\,/3, (\sqrt{2}\, + 4)/3)$

27. $-7/3$

29. -1

31. (a) For a horizontal line, $dy/dt = 0$, $dx/dt \neq 0$. For a vertical line, $dx/dt = 0$, $dy/dt \neq 0$.

(b) Horizontal tangents at $t = \pm\sqrt{1/3}$. Vertical tangent at $t = 0$.

33. Use $\dfrac{dh}{dt} \bigg/ \dfrac{dV}{dt} = \dfrac{dh}{dV}$

2.5 Antiderivatives

1. $x^2/2 + 2x + C$

3. $s^4/4 + s^3 + s^2 + C$

5. $-1/2t^2 + C$

7. $2x^{5/2}/5 - 2x^{3/2}/3 + C$

9. 6

11. $(1 + 4\sqrt{2}\,)/24$

13. $(3/2)x^2 + C$

15. $-x^{-1} - (1/2)x^{-2} + C$

17. $2(x + 1)^{3/2}/3 + C$

19. $2(t + 1)^{5/2}/5 + C$

21. $v = 9.8t + 1$; $x = 4.9t^2 + t + 2$

23. $v = 9.8t - 2$; $x = 4.9t^2 - 2t$

25. No.

27. Use the sum rule for derivatives.

29. $x^3/3 + 3x^2/2 + 2x + C$

31. $t^3 + t^2 + t + C$

33. $-1/[8(8t + 1)] + C$

35. $-1/[6(3b + 2)^8] + C$

37. $-1/3x^3 + x^5/5 + C$

39. $2x^{5/2}/5 + 6\sqrt{x}\, + C$

41. $(1/4)x^4 + (3/2)x^2 + C$

43. $-1/(t + 1) + C$

45. $(8x + 3)^{3/2}/12 + C$

47. $-4(8 - 3x)^{5/2}/3 + C$

49. $x + 6\sqrt{x - 1}\, + C$

51. $x^3/3 + 3x^2/2 + 9x + C$

53. 4.6 seconds

55. ≈ 3195

57. 4 seconds

59. (a) $-6x^2/(x^3 - 1)^2$

(b) $(x^3 + 1)/(x^3 - 1) + C$

61. (a) $80(x^4 + 1)^{19}x^3$

(b) $(x^4 + 1)^{20}/80 + 9x^{5/3}/5 + C$

63. $-(3/4)x^{-4} + (2/3)x^{-3} + x^{-1} + C$

65. $(1/4)x^4 + x^3 + 2x + 1$

67. (a) $\int [f(x)]^n f'(x)\,dx = [f(x)]^{n+1}/(n + 1) + C$

(b) $-1/6(x^3 + 4)^2 + C$

Review Exercises for Chapter 2

1. $18(6x + 1)^2$

3. $10(x^3 + x^2 - 1)(3x^2 + 2x)$

5. $-6/x^2$

7. $-2/(x - 1)^2$

9. $(-2x^3 - 54x)(x^2 + 1)^{12}/(x^2 - 1)^{15}$

11. $3(x^2 + x/4 - 3/8)^2(2x + 1/4)$

13. $(x^3 + 12x^2 - 6x - 8)/(x + 4)^3$

15. $24x^2 - 8x - 4$

17. $y = -1 - \sqrt[3]{2}\,(x - 1)/6$

19. $y = -5184 - 3240x$

21. $(5/3)x^{2/3}$

23. $(3\sqrt{x} + x^{5/2})/[2(1 + x^2)^{3/2}]$

25. $3\sqrt{x}\,/(1 - x^{3/2})^2$

27. $4/\sqrt{x}\,(1 + \sqrt{x}\,)^2 + 3(1 - x + \sqrt{x}\,)/(1 - \sqrt{x}\,)^2$

29. $f'(x) = (-x^2 + 2ax + c + 2ab)/(x^2 + 2bx + c)^2$
$f''(x) = [2x^3 - 6ax^2 - (6c + 12ab)x + 2ac - 8ab^2 - 4bc]/(x^2 + 2bx + c)^3$

31. $x'(t) = A/(1 - t)^2 + 2Bt/(1 - t^2)^2 + 3Ct^2/(1 - t^3)^2$
$x''(t) = 2A/(1 - t)^3 + (2B + 6Bt^2)/(1 - t^2)^3 + (6Ct + 12Ct^4)/(1 - t^3)^3$

33. $h'(r) = 13r^{12} - 4\sqrt{2}\, r^3 - (-r^2 + 3)/(r^2 + 3)^2$
$h''(r) = 156r^{11} - 12\sqrt{2}\, r^2 - (2r^3 - 18r)/(r^2 + 3)^3$

35. $f'(x) = 3(x - 1)^2 g(x) + (x - 1)^3 g'(x)$
$f''(x) = 6(x - 1)g(x) + 6(x - 1)^2 g'(x) + (x - 1)^3 g''(x)$

37. $h'(x) = 2(x - 2)^3(3x^2 - 2x + 4)$
$h''(x) = 2(x - 2)^2(15x^2 - 20x + 16)$

39. $g'(t) = (2t^7 - 3t^6 - 5t^4 + 12t^3)/(t^3 - 1)^2$
$g''(t) = (2t^9 - 4t^6 - 18t^5 + 20t^3 - 36t^2)/(t^3 - 1)^3$

41. If $V = \dfrac{4}{3}\pi r^3$, $\dfrac{dV}{dt} = 4\pi r^2 \dfrac{dr}{dt}$. You are given
$\dfrac{dV}{dt} = k \cdot 4\pi r^2$.

43. (a) 3/4 millibars per degree.
    (b) $-12°C$
45. 0
47. 32.01 miles per hour.
49. $dA/dx = 44x$; $d^2A/dx^2 = 44$.
51. $dA/dy = [120 - (25/2)\pi]y$; $\dfrac{d^2A}{dy^2} = 120 - \left(\dfrac{25}{2}\right)\pi$
53. $dA/dP = 44P/(25 + \sqrt{13})^2$; $dP/dx = 25 + \sqrt{13}$
55. $dA/dP = [120 - (25/2)\pi]P/[5\pi/2 + 32]^2$; $dP/dy = (5\pi/2) + 32$
57. (a) $[5 - (0.02)x]$ dollars per case.
    (b) \$3.32
    (c) Marginal cost is a decreasing function of x.
    (d) $x \geqslant 504$
59. $y = 0$
61. $y = \frac{1}{19} + \frac{216}{361}(x - 2)$
63. $-1/2$
65. $y = (3 + \sqrt[3]{2}) + \dfrac{(1 + 3\sqrt[3]{4})4\sqrt{2}}{(15\sqrt{2} + 2)3\sqrt[3]{4}}\left(x - \dfrac{9}{2} - \sqrt{2}\right)$
67. (a) 3.00407407
    (b) -3.979375
69. (a) $20\Delta x$
    (b) 0.42
71. $f''(x)g(x) + 2f'(x)g'(x) + f(x)g''(x)$
73. Each side gives $mn\, f(x)^{mn-1}f'(x)$.
75. $10x + C$
77. $x^4 + x^3 + x^2 + x + C$
79. $10x^{7/5}/21 + C$
81. $1/x + 1/x^2 + 1/x^3 + 1/x^4 + C$
83. $x^3/3 + 2x^{3/2}/3 + C$
85. $2x^{5/2}/5 + 2\sqrt{x} + C$

87. $2x^{9/2}/9 + 2x^{13/2}/13 + 2\sqrt{x} + C$
89. $2(x - 1)^{3/2}/3 + C$
91. $-1/(x - 1) + C$
93. $2(x - 1)^{3/2}/3 - 2(x - 2)^{7/2}/7 + C$
95. $f'(x) = \dfrac{4x^{1/2} + x - 2x^{5/6}}{12x^{7/6}(x^{1/2} - x^{1/3} + 1)^{3/2}}$;
    $\displaystyle\int f'(x)\,dx = x^{1/3}/(x^{1/2} - x^{1/3} + 1)^{1/2} + C$
97. $f'(x) = 1/2\sqrt{x} - 1/(x + 1)^{3/2}\sqrt{x - 1}$;
    $\displaystyle\int f'(x)\,dx = \sqrt{x} - \sqrt{(x - 1)/(x + 1)} + C$
99. $f'(x) = 1/2x^{3/4}(\sqrt[4]{x} + 1)^2$;
    $\displaystyle\int f'(x)\,dx = (\sqrt[4]{x} - 1)/(\sqrt[4]{x} + 1) + C$
101. $f'(x) = -2x/(x^2 - 1)^{3/2}\sqrt{x^2 + 1}$;
    $\displaystyle\int f'(x)\,dx = [(x^2 + 1)/(x^2 - 1)]^{1/2} + C$
103. (a) Replacing x by $-x$ does not change the equation, so the graph is symmetric to the y-axis. Similar argument for x-axis.
    (b) $-2/11$
    (c) $y = -2x/11 + 15/11$
105. 84 pounds per second.
107. (a) $f''(x) = n(n - 1)x^{n-2}$;
    $f'''(x) = n(n - 1)(n - 2)x^{n-3}$
    (b) $\dfrac{d^r x^n}{dx^r} = n(n - 1)(n - 2)\ldots(n - r + 1)x^{n-r}$
    (c) $[f(x)g(x)h(x)]' = f'(x)g(x)h(x) +$
    $f(x)g'(x)h(x) + f(x)g(x)h'(x)$
109. By Exercise 108, $\dfrac{d^{k-1}}{dx^{k-1}} r(x) = C_1$. Conclude
    $\dfrac{d^{k-2}}{dx^{k-2}} r(x) = C_1 x + C_2$ and repeat.

Chapter 3 Answers

3.1 Continuity and the Intermediate Value Theorem

1. (a) $(-\infty, -1)$, $(-1, 1)$, $(1, \infty)$; (b) all x;
   (c) $(-\infty, -1)$, $(-1, \infty)$
3. Use the rational function rule.
5. Use the rational function rule.
7. $(-\infty, \infty)$
9. Since $f(x)$ is defined everywhere but at $x_0 = \pm 1$, $f(x)$ is continuous by the rational function rule.
11. $\lim_{x \to 0} f(x)$ does not exist.
13. $(f + g)(x_0) = f(x_0) + g(x_0) =$
    $\lim_{x \to x_0} f(x) + \lim_{x \to x_0} g(x) = \lim_{x \to x_0} [f(x) + g(x)]$.
15. $x < -2\sqrt{2}$, $x > 2\sqrt{2}$
17. Let $f(s) = -s^5 + s^2 - 2s + 6$. Note that $f(2) = -26$, $f(-2) = 46$ and use the intermediate value theorem.
19. Consider $f(-1) = 3$, $f(0) = -1$, $f(1) = 3$ and use the intermediate value theorem.
21. Negative on $(-\infty, -\sqrt{2})$ and $(1, \sqrt{2})$, positive on $(-\sqrt{2}, 1)$ and $(\sqrt{2}, +\infty)$
23. ≈ 1.34
25. Use the second version of the intermediate value theorem.
27. No, $\lim_{x \to x_0} f(x)$ does not exist.

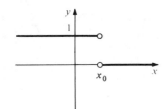

29. $f(2) = 4$
31. One possibility is $f(x) = -5x/2 + 9/2$.

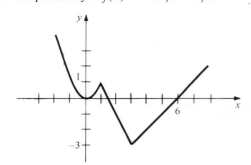

33. $f(1) = 2$
35.

37. $f(x)$ is not continuous on $[0, 2]$.
39. 2.22
41. -1.194
43. To increase accuracy by one decimal place, accuracy must increase ten times.
45. $|10f(b)/[f(b) - f(a)]|$ computed to the nearest integer marks the new division point if $[a, b]$ is divided into ten equal parts. (Other answers are possible.)
47. If f is continuous at x_0 and $f(x_0) > 0$ then $f(x) > 0$ for x in some interval about x_0.
49. In step 2, write $f(x)/x^n = 1 + a_{n-1}x^{-1} + \cdots + a_0 x^{-n}$ and $|a_{n-1}x^{-1} + \cdots + a_0 x^{-n}| \leq |a_{n-1}x^{-1}| + \cdots + |a_0 x^{-n}| \leq |a_{n-1}||x^{-1}| + \cdots + |a_0||x^{-1}| = (|a_{n-1}| + \cdots + |a_0|)|x^{-1}| < 1/2$, so $f(x)/x^n > 1/2$.

3.2 Increasing and Decreasing Functions

1. Use the interval $(2, 4)$ and the definition of increasing.
3. Verify the definition of increasing function.
5. f changes from positive to negative at $x_0 = 1/2$.
7. f changes from negative to positive at $x_0 = 0$.
9. Increasing
11. Neither
13. Right
15. January, February, March; $t = 3$ is a minimum point; inflation got worse after April 1.
17. Increasing on $(0, \infty)$; decreasing on $(-\infty, 0)$
19. Increasing on $(-\infty, 1)$, $(2, \infty)$; decreasing on $(1, 2)$.
21. (a) 3   (b) 4
    (c) 1   (d) 2
    (e) 5
23.

Question \ Function	(a)	(b)	(c)	(d)	(e)	(f)
(1) increasing	nowhere	$(-\infty, 0)$ $(0, \infty)$	$(0, \infty)$	$(0, \infty)$	$(-\infty, -1)$ $(0, 1)$	$(-\infty, \infty)$
(2) decreasing	nowhere	nowhere	$(-\infty, 0)$	$(-\infty, 0)$	$(-1, 0)$ $(1, \infty)$	nowhere
(3) local min-max points	none	none	none	0 (min)	0 (min) ± 1 (max)	none
(4) sign changes	0	0	1	nowhere	± 2	1

25. x_1, x_3, x_5 are minima; x_2, x_4 are maxima; x_6 is neither.
27. $x = 0$, a local minimum.
29. $x = 0$, a local minimum; $x = -2/3$, a local maximum.
31. $y = -1$, a local minimum; $y = 1$, a local maximum.
33. $r = -1, 0, 1$ are local minima; $r = \pm 1/\sqrt{2}$ are local maxima.
35. Increasing at -3, -36; decreasing at 1, 3/4, 25.

37. $m = -2$, positive to negative; $m = 0$, no sign change; $m = 2$, negative to positive.

39. $x_0 = -2$, positive to negative; $x_0 = -1$, negative to positive.

41. f does not change sign.

43. Positive to negative.

45. Increasing on $(-\infty, -\sqrt{5/6}\,)$, $(\sqrt{5/6}\,, \infty)$; decreasing on $(-\sqrt{5/6}\,, \sqrt{5/6}\,)$.

47. Increasing on $(-\infty, 3 - \sqrt{10}\,)$, $(3 + \sqrt{10}\,, \infty)$; decreasing on $(3 - \sqrt{10}\,, 3 + \sqrt{10}\,)$.

49. $f(x)$ is of the form $a(x^2 - 4x + 3)$ where $a > 0$.

51. (a) $\lim_{x \to 1} f(x) = 3/2$, $\lim_{x \to -1} f(x)$ does not exist.
(b) Show $f'(x) = x(x^3 - 3x + 2)/(x^2 - 1)^2 > 0$ on $(-\infty, -2)$.
(c) $f(a) < f(-2)$ if $a < -2$ or $-2 < a < -1$.
(d) If $a > -1$, then $(a^3 - 1)/(a^2 - 1) \geqslant 1$.

53. For $\quad x < x_0 < y, \quad f(x) < f(x_0) < f(y) \quad$ and $g(x) < g(x_0) < g(y)$ implies $f(x) + g(x) < f(x_0) + g(x_0) < f(y) + g(y)$.

55. If $a_1 > 0$. If $a_1 = 0 = a_2$ and $a_3 > 0$. Etc.

57. (a) $g(x)h(x)$ is increasing when $g'/g + h'/h > 0$. $g(x)h(x)$ is decreasing when $g'/g + h'/h < 0$.
(b) $g(x)/h(x)$ is increasing when $g'/g - h'/h > 0$. $g(x)/h(x)$ is decreasing when $g'/g - h'/h < 0$.

59. $f(x) = k(x^3/3 - x)$, $k < 0$

61. $2a^2 = b^2$

3.3 The Second Derivative and Concavity

1. $x = 0$ is a local minimum.

3. $x = -\sqrt[4]{1/30}$ is a local maximum; $x = \sqrt[4]{1/30}$ is a local minimum.

5. $x = 0$ is a local minimum.

7. $s = -1$ is a local minimum; $s = 1$ is a local maximum.

9. Concave up everywhere.

11. Concave up everywhere.

13. Concave up on $(1, \infty)$, concave down on $(-\infty, 1)$.

15. Concave up on $(-4/3, \infty)$, concave down on $(-\infty, -4/3)$.

17. $x = 0$

19. $x = 0$

21. None

23. $x = \pm 1/\sqrt{3}$

25. (a) Maximum   (b) Inflection point
(c) None   (d) Maximum
(e) Maximum   (f) Inflection point
(g) Inflection point   (h) Minimum

27. $x = 0$ is a local minimum, increasing on $(0, \infty)$, decreasing on $(-\infty, 0)$, concave up on $(-\infty, \infty)$.

29. $x = -2$ is a local maximum, $x = 2/3$ is a local minimum, $x = -2/3$ is an inflection point, increasing on $(-\infty, -2)$, and $(2/3, \infty)$, decreasing

on $(-2, 2/3)$, concave up on $(-2/3, \infty)$, concave down on $(-\infty, -2/3)$.

31. Inflection point at $x = 0$ for odd n, $n > 2$.

33. $f(x) = kx^4/12 - kx^3/2 + kx^2 + dx + e$ where $k \neq 0$, d and e arbitrary.

35. (a) $f''(x_0) < 0$ ($f''(x_0) > 0$) makes the linear approximation greater (less).
(b) the approximation is less for $x > 0$.

37. (a) $2(x + 1)$, $-x$, $2(x - 1)$
(b)

Δx \ x_0	-1	0	1
1	0, 2	0, -1,	6, 2
-1	-6, -2	0, 1	0, -2
0.17	0.171, 0.2	-0.099, -0.1	0.231, 0.2
-0.1	-0.231, -0.2	0.099, 0.1	-0.171, -0.2
0.01	0.019701, 0.02	-0.009999, -0.01	0.020301, 0.02
-0.01	-0.020301, -0.02	0.009999, 0.01	-0.019701, -0.02

This table shows $f(x_0 + \Delta x)$, followed by its linear approximation. $e(x) < 0$ when $f''(x_0) < 0$, $e(x) > 0$ when $f''(x_0) > 0$. $e(x)$ is the same sign as Δx when $f''(x_0) = 0$.

39. (a) $I = E/2R$
(b) $P = E^2/4R$

41. 28 feet.

43.

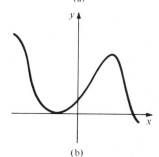

(a)

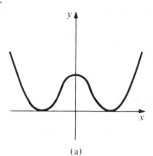

(b)

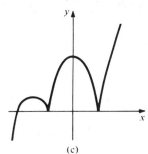

(c)

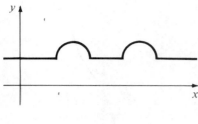

(d)

45. Not necessarily.
47. If $h'' = f''$ changes sign from negative to positive at x_0, then $h'(x)$ is decreasing to the left of x_0 and increasing to the right. Since $h'(x_0) = 0$, h' changes sign from negative to positive. Repeat to conclude the result.

3.4 Drawing Graphs

1. Odd
3. Neither
5. Near -1: $x < -1$, $f(x)$ is large and positive; $x > -1$, $f(x)$ is large and negative.
7. Near 1: $x < 1$, $f(x)$ is large and negative; $x > 1$, $f(x)$ is large and positive. ($x = -1$ is not a vertical asymptote).
9.

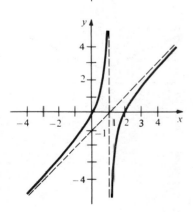

11.

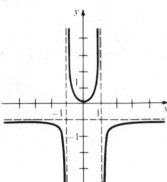

13.

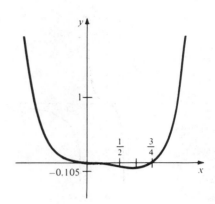

15.

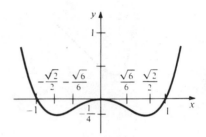

17.

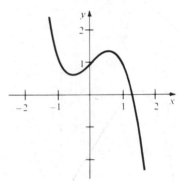

19.

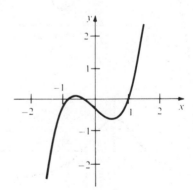

21.

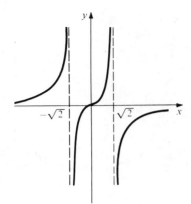

23.

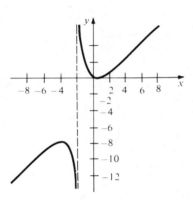

25.

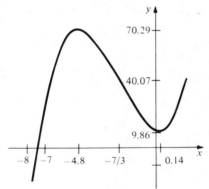

27.

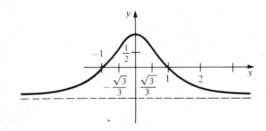

29.

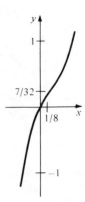

31. (a) *B*
  (b) *A*
  (c) *D*
  (d) *C*

33. (a) Increasing on $(-\sqrt{3}, 0)$ and $(\sqrt{3}, \infty)$; decreasing on $(-\infty, -\sqrt{3})$ and $(0, \sqrt{3})$.
  (b)

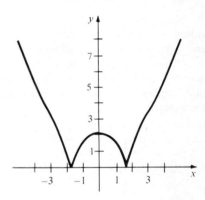

35.

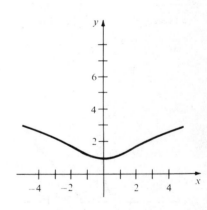

37.

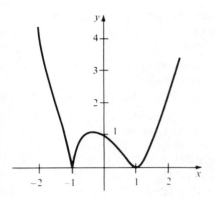

39.

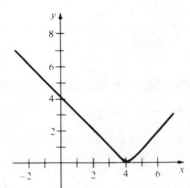

41. $f(x)$ is even if and only if $f(x) - f(-x) = 0$. When expanded, this shows that $f(x)$ is even if the even powers of x have nonzero coefficients. Use a similar argument for odd functions.

43. (a)

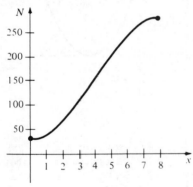

(b) $t = 4$

(c) The growth rate is zero.

45. $e(x) = \frac{1}{2}[f(x) + f(-x)]$, $o(x) = \frac{1}{2}[f(x) - f(-x)]$

47. No; for example $f(x) = 1$.

49. Locate the inflection point of g at $x = 0$ or just substitute.

51. $x^3 + (8/3)x$, type I

53. Velocity is 0, acceleration is infinite.

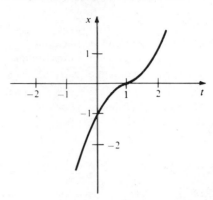

55. No critical points if $ap > 0$, 1 critical point if $p = 0$, 2 critical points if $ap < 0$.

57. Substitute

59.

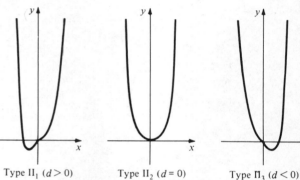

Type II$_1$ $(d > 0)$ Type II$_2$ $(d = 0)$ Type II$_3$ $(d < 0)$

61. Eliminate x from the equations $f'(x) = 0$, $f''(x) = 0$.

63. III$_1$

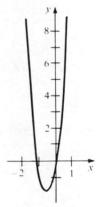

3.5 Maximum—Minimum Problems

1. $(1, -1)$ is the minimum; no maximum.
3. $(\sqrt{2}, 2\sqrt{2})$ is the minimum; no maximum.
5. No minimum. Maximum at $x = -1$.
7. $-\sqrt{5/3}$ is a minimum, $\sqrt{5/3}$ is a maximum.
9. $(1971.4, 16\%)$ is a minimum; $(1980, 34\%)$ is a maximum.

	Critical points	Endpoints	Maxima	Minima
11.	$1/2$	0	$(0,0)$	$(1/2, -1/4)$
13.	$0, \sqrt{2}, -\sqrt{2}$	$-4, 2$	$(-4, 199)$	$(\sqrt{2}, 3), (-\sqrt{2}, 3)$
15.	$1/2, -1/2$	$-10, 20$	$(1/2, 0.000005)$	$(20, 0.00000119)$
			$(-1/2, 0.000005)$	
17(a)	—	—	—	—
(b)	—	$1/2$	—	$(1/2, -1/4)$
(c)	$3/2$	2	—	$(3/2, -5/4)$
(d)	$3/2$	2	—	$(3/2, -5/4)$
(e)	$3/2$	—	—	$(3/2, -5/4)$
(f)	$3/2$	—	—	$(3/2, -5/4)$
(g)	—	$-1, 1$	$(-1, 5)$	$(1, -1)$
(h)	$3/2$	$-8, 8$	$(-8, 89)$	$(3/2, -5/4)$
19(a)	$-1/7$	$-1, 1$	$(1, 13)$	$(-1/7, 27/7)$
(b)	—	—	—	—
(c)	$-1/7$	-4	$(-4, 108)$	$(-1/7, 27/7)$
21.	1	—	—	—
23.	$\sqrt{1/3}, -\sqrt{1/3}$	$-2, 3$	$(3, 24)$	$(-2, -6)$
25.	$0, -0.6$	$-10, 10$	$(10, 999/101)$	$(-10, -1001/101)$
27.	1	$-1, 6$	$(-1, 1/2)$	—
29.	0	—	—	—

31. Minimize $f(l) = 2(l + 1/l)$

33. Maximize $f(x) = x(M - x)$
35. $l = 1$ is a minimum; the rectangle is a square of side length 1.
37. $x = M/2$; the two masses should be equal.
39. 10 cm on each side.
41. (a) $r = \sqrt[3]{500/\pi}$ cm, $h = 2\sqrt[3]{500/\pi}$ cm.
    (b) $r = \sqrt[3]{V/2\pi}$ cm, $h = 2\sqrt[3]{V/2\pi}$ cm.
    (c) $r = \sqrt{A/6\pi}$ cm, $h = \sqrt{2A/3\pi}$ cm.
43. 749 units.
45. Height and width are 12 inches, length 24 inches.
47. 32, 4
49. (a) Circle of radius $500/\pi$ feet.
    (b) Circle has radius $500/(\pi + 4)$ feet, square has side $1000/(\pi + 4)$.
51. (a) $y = \sqrt[3]{ab^2} + a$
    (b) $(a^{2/3} + b^{2/3})^{3/2}$
53. $l = (3 - \sqrt{6})c/3$
55. The square of side 1.
57. The semicircle with radius $500/\pi$ meters.
59. The right triangle with legs of length $\sqrt{2}$.

61. One possibility is $y = -|x|$.
63. One possibility is $f(x) = x$ on $(-2, 2)$, $f(2) = f(-2) = 0$.
65. $M_1 \leqslant M_2$ since M_1 is a value and M_2 is the maximum value.
67. For $p > 0$, $q > 0$, $(\sqrt{q/p}, 2\sqrt{p/q})$ is a minimum; $p < 0$, $q < 0$, $(\sqrt{q/p}, 2\sqrt{p/q})$ is a maximum. Otherwise, no solution.
69. There are none.
71. $x \approx 0.497$ or 1.503.
73. (a) A quart
    (b) If $q \geqslant g/2$, $G \leqslant Q + g/2$; if $q \leqslant g/2$, $G \leqslant Q + q$.
75. f and g must have a maximum at the same point.
77. Use the closed interval test and the definition of concavity.

3.6 The Mean Value Theorem

1. Use the mean value theorem.
3. $f(x) = 2g(x) + C$
5. Use the mean value theorem.
7. Use the mean value theorem.
9. It is between 72 and 76.
11. $x_0 = 1/2$.

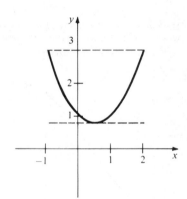

13. $F(x) = (1/x) + C_1$ for $x < 0$; $F(x) = (1/x) + C_2$ for $x > 0$; C_1 doesn't necessarily equal C_2
15. No, $f'(x)$ doesn't exist at $x = 0$.
17. $x^2/4 - 4x^3/3 + 21x + C$.
19. $-1/x + x^2 + C$.
21. $2x^5/5 + 8/5$
23. $x^5/5 + x^4/4 + x^3/3 + 13/60$
25. Use the mean value theorem twice.
27. Use the horserace theorem.
29. Show that $dN/dt \geqslant 0$ cannot hold on (t_1, t_2) by using the fact that N is nonconstant.

Review Exercises for Chapter 3

1. $(-\infty, -1)$ and $(1, \infty)$
3. $(-\infty, 1)$, $(1, 2)$ and $(2, \infty)$
5. Use the definition of continuous functions. Yes, polynomials are continuous and $h(x)$ is continuous at $x = 2$.

7.

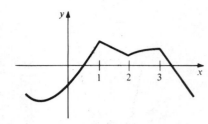

9. Use the intermediate value theorem.
11. The root is approximately 0.83.
13. Increasing on $(-\infty, 0)$ and $(1/4, \infty)$; decreasing on $(0, 1/4)$.

15. Increasing on $(-\sqrt{2}, \sqrt{2})$; decreasing on $(-\infty, -\sqrt{2})$ and $(\sqrt{2}, \infty)$.
17. Increasing for $t < -5/3$ and $t > 5$, decreasing for $-5/3 < t < 5$ for t in $[-3, 7]$. Political reaction to minimum: fulfilled promises; maximum: things are turning around now; inflection point: the rate at which things are getting worse has just turned around for the better.
19. Speeding up on $(50, 100)$; slowing down on $(0, 50)$.
21. $x_0 = 2/3$ is a local maximum; $x_0 = 1$ is a local minimum.
23. $x_0 = -3$ and $x_0 = 1$ are local minima; $x_0 = -1$ is a local maximum.

	Continuous	Differentiable	Increasing	Decreasing	Concavity Up	Concavity Down	Endpoints	Local Maximum	Local Minimum	Inflection Points
25.	$(-\infty, \infty)$	$(-\infty, \infty)$	$(0, 4/21)$	$(-\infty, 0)$ $(4/21, \infty)$	$(-\infty, 2/21)$	$(2/21, \infty)$	—	$x_0 = 4/21$	$x_0 = 0$	$x_0 = 2/21$
27.	$[-1, 2]$	$[-1, 2]$	$[-1, -\sqrt{2/3})$ $(\sqrt{2/3}, 2]$	$(-\sqrt{2/3}, \sqrt{2/3})$	$(0, 2]$	$[-1, 0)$	$-1, 2$	$x_0 = \sqrt{2/3}$ $x_0 = 2$	$x_0 = \sqrt{2/3}$ $x = -1$	$x_0 = 0$
29.	$(-\infty, 2/5)$ $(2/5, \infty)$	$(-\infty, 2/5)$ $(2/5, \infty)$	$(-\infty, 2/5)$ $(2/5, \infty)$	—	$(-\infty, 2/5)$	$(2/5, \infty)$	—	—	—	—

25.

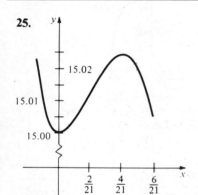

27.

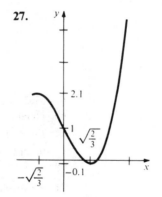

29.

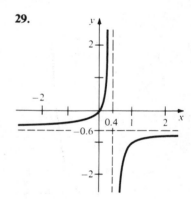

31. False.
33. False.
35. False.
37. True.
39. True.
41.

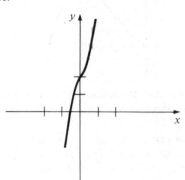

43.

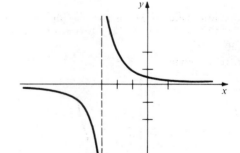

45.

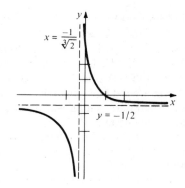

47.

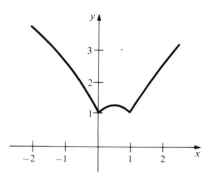

49.

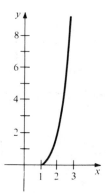

51. Maximum value is 31; minimum value is -13.
53. Maximum value is 3; minimum value is $2\sqrt{30} - 10$.
55. There are none.
57.

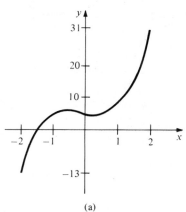

(a)

59.

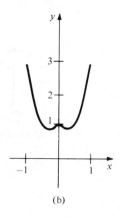

(b)

61.

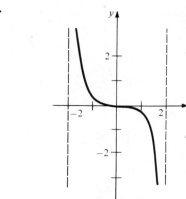

63. $(5\sqrt{2} + 2)$ inches by $(10\sqrt{2} + 4)$ inches.
65. Height is $h/3$, radius is $2r/3$.
67. $l = \sqrt[3]{5V/2}$, $h = (2\sqrt[3]{5V/2})/5$; the ratio l/h is independent of V.
69. $V = 2\pi r^3 + 40p\pi r^2$ yields radius, $h = V/\pi r^2$.
71. $x = \sqrt{1/4 + (\pi/10)^2}$, $P = 180/(5 + \sqrt{\pi^2 + 25})$
73. (a) 1
(b)

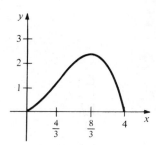

75. (a) Show that the minimum occurs at $(a + b)/2$ and use the fact $[(a + b)/2]^2 \geqslant ab$ for positive a and b.
77. (a) R = price per item $\times$ number of items. C = initial cost + (cost per item) $\times$ (number of items). P = cash received − cost.
(b) $x = 2000$ units.
(c) $x = 3000$ units.
(d) $x = 5,625,000$ units.
79. $l = 16\sqrt{5}$ feet, $w = 10\sqrt{5}$ feet.

81. (a) $x = 40/(\sqrt[3]{3} + 1)$
    (b) 16.4 miles
83. If $f'''(x) = 0$, then $f''(x) = 2A$, a constant. Then $f'(x) - 2Ax$ is a constant too.
85. (a) 119.164 lbs.
    (b) Gain is 7.204 lbs.
87. (a) Set the first derivative equal to 0.
    (b) fg has a critical point if $f'/f = -g'/g$.
89. (a)

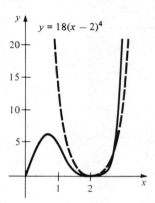

(b)

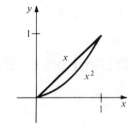

91. (a)

(b) Use the definitions.
(c) Use properties of exponentiation.
(d) Use properties of exponentiation.
(e) Consider the cases $|x - a| < 1$ or $|x - a| \geqslant 1$ and when m, n are < 1 or $\geqslant 1$.
(f)

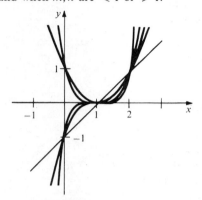

93. Assume that there is no maximum or minimum and show that this contradicts $f(0) = f(1)$.
95. The maximum and minimum points of f on $[a, b]$ cannot both be endpoints.
97. Use the mean value theorem on $[f(x)]/x$.

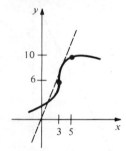

99. Consider the second derivative of $h(x) = x^2 f(x)$.

Chapter 4 Answers

4.1 Summation

1. 20.4 meters.
3. 91.0 meters.
5. 34
7. 40
9. 325
11. 1035
13. 3003
15. 9999
17. 0
19. 5865
21. $[n(n + 1) - (m - 1)m]/2$
23. Notice that $1/(1 + k^2) \leqslant 1$ for $k \geqslant 1$.
25. 100,000,000
27. 10,400
29.

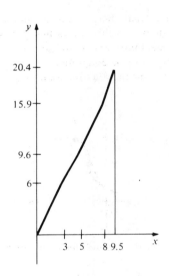

31. 14,948
33. $(n + 1)^4 - 1$
35. 30,600
37. 122
39. 124
41. (a) Apply the telescoping sum formula.
    (b) $[n(n + 1)(2n + 1)]/6 - [(m - 1)(2m - 1)m]/6$
    (c) $[n(n + 1)/2]^2$

4.2 Sums and Areas

1.

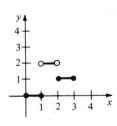

3.

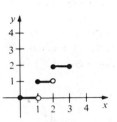

	x_0	x_1	x_2	x_3	Δx_1	Δx_2	Δx_3	k_1	k_2	k_3	Area
5.	0	1	2	3	1	1	1	0	2	1	3
7.	0	1	2	3	1	1	1	0	1	2	3

9. $L = 5$, $U = 13$

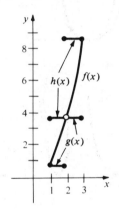

11. 91.5 meters $\leqslant d \leqslant$ 117 meters.
13. 0.01 meters $\leqslant d \leqslant$ 0.026 meters.
15. 3/2
17. $5(b^2 - a^2)/2$
19. 1/2
21. 11/6
23. 1/6

4.3 The Definition of the Integral

1. -2

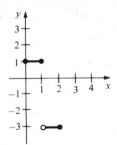

3. 0

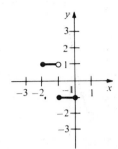

5. -2

7. 0

9. $\int_{-1}^{2} x^2 dx$

11. $\int_{-1}^{2} x\,dx$

13. $7/10$

15. 0.509

17. (a) $\int_{0}^{3} 5(t^2 - 5t + 6)\,dt$

(b) $\int_{0}^{2} 5(t^2 - 5t + 6)\,dt - \int_{2}^{3} 5(t^2 - 5t + 6)\,dt$

19. $\lim_{n \to \infty} (1/n^6) \sum_{i=1}^{n} i^5$

21. $\lim_{n \to \infty} \left(\sum_{i=1}^{n} \dfrac{2n}{5n^2 + 8in + 4i^2} \right)$

23. Choose step functions with two steps.

25. (a) $\int_{0}^{x} f(t)\,dt = \begin{cases} 2x & \text{if} \quad 0 \leqslant x < 1 \\ 2 & \text{if} \quad 1 \leqslant x < 3 \\ -x + 5 & \text{if} \quad 3 \leqslant x \leqslant 4 \end{cases}$

(b)

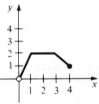

(c) F is differentiable on $(0, 4]$ except at 1, 3 and 4.

$F'(x) = \begin{cases} 2 & \text{if} \quad 0 < x < 1 \\ 0 & \text{if} \quad 1 < x < 3 \\ -1 & \text{if} \quad 3 < x < 4 \end{cases}$

27. (a)

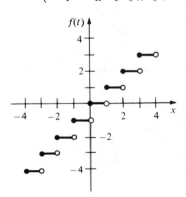

(b) 0, 15, -2, 8

(c) $(n-1)n/2$

(d)

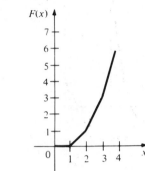

$F'(x) = \begin{cases} 0 & 0 < x < 1 \\ 1 & 1 < x < 2 \\ 2 & 2 < x < 3 \\ 3 & 3 < x < 4 \end{cases}$

29. (a) $\sum_{i=1}^{n} A_i \Delta x_i$

(b) $\int_{0}^{L} f(x)\,dx$

31. (a) Choose a partition and consider $kf(x)$ on each interval.

(b) $k \int_{a}^{b} f(x)\,dx$

33. (a) $19/2$

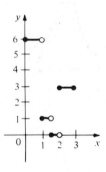

(b) $1/2$

(c) They are both 19.

(d) Calculate each explicitly.

(e) No.

35. $1/6$

4.4 The Fundamental Theorem of Calculus

1. 20

3. 30

5. $8^{1/4}$

7. 24

9. $(3/7)(b^{7/3} - a^{7/3})$

11. $(12\pi/5)(\sqrt[3]{32} - 1)$

13. $-400/3$

15. $78/5$

17. $11/1800$

19. $29/6$

21. $367/24$

23. $59/6$

25. $7/3$

27. $3/4$

29. 2

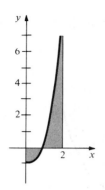

31. 4

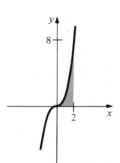

33. 25/3

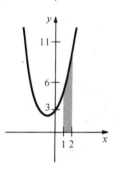

35. 22/5

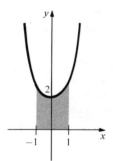

37. 93/5

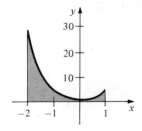

39. 82/9

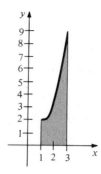

41. 604,989/5 units.

43. $-29/3$ units, 56/3 units.

45. 1600 feet.

47. (a) v_0^2/g
    (b) $5v_0^2/8g$

49. On an interval (x_i, x_{i+1}) on which $h(t) = l_i$, $f(t) \leqslant l_i$, so $F(x_{i+1}) - F(x_i) \leqslant (\Delta x_i)l_i$ by the mean value theorem. Now use a telescoping sum.

4.5 Definite and Indefinite Integrals

1. $(d/dx)(x^5) = 5x^4$

3. $(d/dx)(t^{10}/2 + t^5) = 5t^9 + 5t^4$

5. (a) $(d/dx)[t^3/(1 + t^3)] = 3t^2/(1 + t^3)^2$
    (b) 1/2

7. (a) $(3x^2 + x^4)/(x^2 + 1)^2$
    (b) 1/2

9. 370/3

11. 16512/7

13. 11/6

15. 1/6

17. 3724/3

19. (a) For any c, the area under c times $f(x)$ is c times the area under $f(x)$.

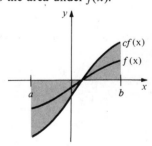

    (b) If you go c times as fast, you go c times as far.

21. 7

23. -8

25. $-5/2$

27. $-199/16200$

29. $\dfrac{d}{dx}\left(\dfrac{x^4}{4} - x - \dfrac{a^4}{4} + a\right) = x^3 - 1$

31. 1/512

33. $3/(t^4 + t^3 + 1)^6$
35. $-t^2(1 + t)^5$
37. Differentiating the distance function with respect to time gives the velocity function.
39. (a)

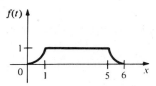

(b) $14/3$
(c) $14/3$
(d) $F(t) = \begin{cases} t^3/3 & 0 \le t \le 1 \\ t - 2/3 & 1 \le t \le 5 \\ (t - 6)^3/3 + 14/3 & 5 \le t \le 6 \end{cases}$

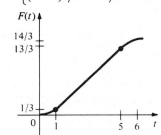

(e) $F'(t) = \begin{cases} t^2 & 0 < t \le 1 \\ 1 & 1 \le t \le 5 \\ (t - 6)^2 & 5 \le t < 6 \end{cases}$

41. (a) $\dfrac{d}{dt}[F_1(t) - F_2(t)] = f(t) - f(t) = 0$

(b) $\displaystyle\int_{a_1}^{a_2} f(s)\,ds$

43. (a) Let $u = g(t)$ and $G(u) = \displaystyle\int_a^u f(s)\,ds$. Then $\displaystyle\int_a^{g(t)} f(s)\,ds = G(g(t))$. Now use the chain rule.

(b) The rate of change of area = height times the speed of the screen.

45. $2/x$
47. $f(g(t)) \cdot g'(t) - f(h(t)) \cdot h'(t)$. The rate of change of area as both endpoints move is the sum of the rates due to the motion of each endpoint.
49. The general fact about inequalities is: if $c < a + b$ then $c = c_1 + c_2$ where $c_1 < a$ and $c_2 < b$. (Let $c_1 = a - \frac{1}{2}(a + b - c)$ and $c_2 = c - c_1$.)

51. $\displaystyle\int_a^{t+h} f(s)\,ds - \int_a^t f(s)\,ds = \int_t^{t+h} f(s)\,ds$
53. $f(c) \to f(t)$ as $h \to 0$ by continuity of f.

4.6 Applications of the Integral

1. $160/3$ **3.** \$18,856
5. $16/3$ **7.** $31/5$
9. $1/6$ **11.** $3/10$
13. $141/80$ **15.** 8
17. $207/4$ **19.** $1/4, 1/4$
21. $32/3$ **23.** $1/8$
25. 12,500 liters.   **27.** 7200
29. (a) $bh/2$
(b) $bh/2$
31. Speedometer.
33. 15 minutes, 43 seconds.
35. (a) Integrate both sides of $W'(t) = 4(t/100) - 3(t/100)^2$.
(b) $t^2/50 - t^3/10,000$
(c) 100 words
37. $y = (11 - 4\sqrt{6})/8$

Review Exercises for Chapter 4

1. 30   **3.** $21/5$
5. 379,250   **7.** 14,641
9. $223/60$

11. If x is time in seconds, $\displaystyle\int_0^1 |f(x)|\,dx$ is the distance travelled where $f(x)$ is the velocity.
13. $-718/3$ **15.** $-23/36$
17. $7/12$ **19.** $(3/7)(5^{7/3} - 3^{7/3})$
21. (a) 3.2399, 3.0399.
(b) 3.1399. [The exact integral is π.]
23. (a) $-2x/(1 + x^2)^2$
(b) $1/4$
25. $(a_2 - a_1)(ma_1 + b + ma_2 + b)/2$

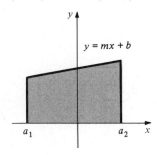

27. 6

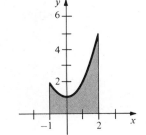

29. (a) Differentiate the right-hand side.
  (b) 2/63

31. 64/21

33. 125/6

35. The plane was at least 1700 feet above ground.

37. (a) 22 liters.
  (b) 16 liters.
  (c) (256/27) liters.

39. (a) $\int_0^a [D(x) - b]\,dx$ is the area between the curves $p = D(x)$ and $p = b$, from $x = 0$ to $x = a$.
  (b) $\int_0^a [b - S(x)]\,dx$ is the area between the curves $p = b$ and $p = S(x)$ from $x = 0$ to $x = a$.
  (c) Interpret the integral in (a) in terms of sums.
  (d) Consumer's surplus $= \frac{1}{2}$, Producer's surplus $= \frac{2}{3}$.

41. 110/3

43. (a) $\int_0^t 32x\,dx = 16t^2$ feet; $\int_0^t 1080\,dx = 1080t$ feet.
  (b) 5.2 seconds.
  (c) 432.56 feet.
  (d) 0.4 seconds.

45. (a) The fifth day.
  (b) -4000 bacteria per cubic centimeter.

47. 4

49. $x^2/(1 + x^3)$

51. 15/2

53. $3t^2/[(t^3 + 2)^2 + 1]$

55. (a) Since f is unbounded at 0, it has no upper sum and is not integrable.
  (b) $\int_\varepsilon^1 t^{-1/2}\,dt + \delta$ is a lower sum for $\varepsilon > 0$ and $\delta > 0$.
  (c) $\varepsilon f(\varepsilon) + \int_\varepsilon^1 f(t)\,dt = 2 - \sqrt{\varepsilon}$ is a lower sum for $\varepsilon > 0$.
  (d) 2

Chapter 5 Answers

5.1 Polar Coordinates and Trigonometry

1. 3.84 *m*, 19.19 *m*²

3. 5/18 rad, 45 *m*²

5. 0.5061, 0.9425, 4.4506, 2.2689, 5.5851.

7. (a) $\pi/3$, $6\pi/5$, π.
(b) 160°, 305°, 25°

9.

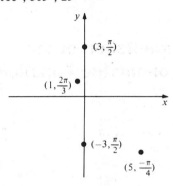

11.

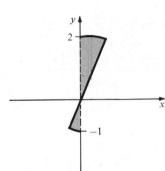

13.

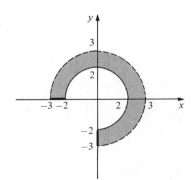

15. $(5.3852, -0.3805)$

17. (a) $(1,0)$ (b) $(5, 0.927)$
(c) $(2, \pi/6)$ (d) $(2, -\pi/6)$
(e) $(2, 5\pi/6)$

19. (a) $(\sqrt{2}, -\pi/4)$ (b) $(2, \pi/2)$
(c) $(\sqrt{197}/2, 1.5)$ (d) $(13, -2.75)$
(e) $(\sqrt{73}, 1.93)$ (f) $(3\sqrt{2}/4, \pi/4)$

21. (a) $(0,6)$ (b) $(6\sqrt{2}, -6\sqrt{2})$
(c) $(-4,0)$ (d) $(0,2)$
(e) $(-4, -4\sqrt{3})$ (f) $(0.42, -0.91)$

23. $\tan \theta = |BC|/|AC| = \cot(\pi/2 - \theta)$

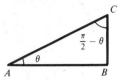

25. In the Figure, $\cos \theta = x/1 = \cos(-\theta)$.

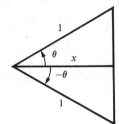

27. 84.5

29. $8\sqrt{2}$

31. 10,723 feet.

33. 917.19 meters.

35. 1.313

37. .3 radians

39. Write $\cos \theta = \cos\left(\dfrac{\theta}{2} + \dfrac{\theta}{2}\right)$ and expand.

41. Use the addition formulas on the right-hand side.

43. $\cos \theta$

45. $-\sin \theta$

47. $\sec \theta$

49. $-\sin \theta \cos \phi$

51. $(2 + \sqrt{2 + \sqrt{3}})^{1/2}/2$

53. $2/\sqrt{2 + \sqrt{3}}$

55. Use $\cos[2(\theta/2)] = 1 - 2\sin^2\left(\dfrac{\theta}{2}\right)$.

57. Take the reciprocal of both sides and use $\cos^2(\theta/2) = (1 + \cos\theta)/2$.

59. Take the reciprocal of both sides and use the product formula for $\sin\theta \sin\phi$.

61.

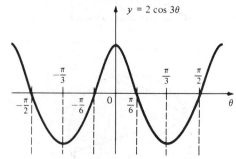
$y = 2\cos 3\theta$

63.

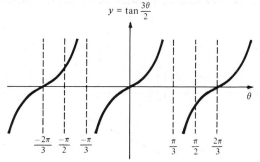

$$y = \tan\frac{3\theta}{2}$$

65.

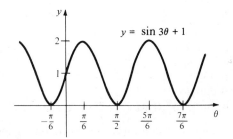

$$y = 2\sin 4x = 4\sin 2x \cos 2x$$

67.

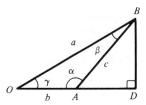

$$y = \sin 3\theta + 1$$

69. $(2n + 1)\pi/2$ where n is an integer.

71. $\sec \theta$ is differentiable for all $\theta \neq (2n + 1)\pi/2$, $\cot \theta$ is differentiable for all $\theta \neq n\pi$, n an integer.

73. (a) $\sqrt{3} \times 10^{10}$ meters per second.
(b) The angle of incidence equals the angle of refraction.
(c) 20.7° or 0.36 radians.

75. Consider this figure:

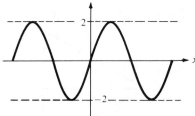

77. Consider this figure:

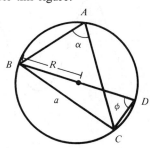

79. Expand the right-hand side using the addition formula.

81. (b) $\sin \theta = n\lambda/a$.
83. (a) Use $\sin^2(2\omega t/2) = [1 - \cos(2\omega t)]/2$
(b)

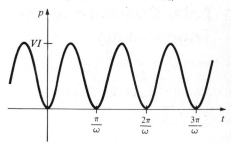

5.2 Differentiation of the Trigonometric Functions

1. $-\sin \theta + \cos \theta$
3. $-15\sin 3\theta + 20\cos 2\theta$
5. $\cos 2\theta - \theta \sin \theta + \cos \theta$
7. $-9\cos^2 3\theta \sin 3\theta$
9. $\sin \theta /(\cos \theta - 1)^2$
11. $(\cos \theta - \sin \theta - 1)/(\sin \theta + 1)^2$
13. $-3\cos^2 x \sin x$
15. $4(\sqrt{x} + \cos x)^3[(1/2\sqrt{x}) - \sin x]$
17. $(1 + 1/2\sqrt{x})\cos(x + \sqrt{x})$
19. $\dfrac{\cos x + \sin(x^2) + x \sin x - 2x^2\cos(x^2)}{\left[\cos x + \sin(x^2)\right]^2}$
21. $\sec^2 x - 2\sin x$
23. $3\tan 3x \sec 3x$
25. $1/2\sqrt{x} - 3\sin 3x$
27. $-\sin x/2\sqrt{\cos x}$
29. $12t^2\sin\sqrt{t} + [(4t^3 + 1)/2\sqrt{t}]\cos\sqrt{t}$
31.
$$\frac{-3x^2(\cos\sqrt{1 - x^3})}{2\sqrt{1 - x^3}} + \frac{(-3x^4 + 1) \cdot \sec^2[x/(x^4 + 1)]}{(x^4 + 1)^2}$$
33.
$$\frac{-3}{2(\theta^2 + 1)^{3/2}}\sqrt{\csc\left(\frac{\theta}{\sqrt{\theta^2 + 1}}\right) + 1}\cot\left(\frac{\theta}{\sqrt{\theta^2 + 1}}\right)\csc\left(\frac{\theta}{\sqrt{\theta^2 + 1}}\right)$$
35.
$$v\sec\left(\frac{1}{v^2 + 1}\right)\left[\frac{\sec^2\sqrt{v^2 + 1}}{\sqrt{v^2 + 1}} - \frac{2\tan\sqrt{v^2 + 1}}{(v^2 + 1)^2}\tan\left(\frac{1}{v^2 + 1}\right)\right]$$
37. $x^4/4 - \cos x + C$
39. $x^5/5 + (\sec 2x)/2 + C$
41. $-2\cos(u/2) + C$
43. $[\sin(\theta^2)]/2 + C$
45. $[-\cos(2\theta)]/4 + C$
47. $\theta + C$
49. $4 - 4\cos(\pi/8)$
51. 0
53. 0
55. $\pi/2$

57. (a) Prove $\sin 2\phi/(1 + \cos 2\phi) = \dfrac{\sin \phi}{\phi} \cdot \dfrac{\phi}{\cos \phi}$

(b) Manipulate $\cos \phi < (\sin \phi)/\phi < 1/\phi$.

59. a

61. Differentiate twice and substitute.

63. Differentiate once and substitute.

65. Differentiate $[-f(\cos \theta)]$ using the chain rule.

69. (a) $-\sin x \cdot \phi(3x) + [3 \cos x/\cos(3x)]$

(b) $\phi(1) - \phi(0)$

(c) $4 - 4\phi(2x)\sin(2x) + 4/\cos^2(2x)$

71. (a) $\sqrt{(dx)^2 + (dy)^2} = dt$ since speed = 1

(b) Differentiate

(c) Multiply the equation in (b) by
$\sin \theta \sin'\theta - \cos \theta \cos'\theta$ and manipulate.

(d) If $\cos \theta$ and $\sin \theta$ are positive, $\cos'\theta$ is negative
and $\sin'\theta$ is positive.

5.3 Inverse Functions

1. $(x - 5)/2$ on $[-3, 13]$

3. $\sqrt[5]{x}$ on $(-\infty, \infty)$

5. $(3t + 10)/(1 - t)$ on $[-11/2, -9/4]$

7. $(-dx + b)/(cx - a)$, domain $x \neq -a/c$. The
condition is $bc - ad \neq 0$.

9.

(a)
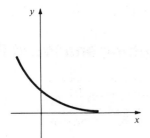

(b)
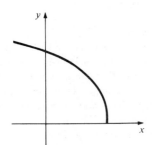

(c)
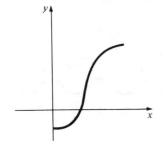

11.

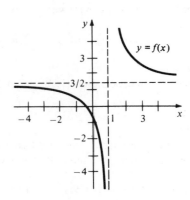

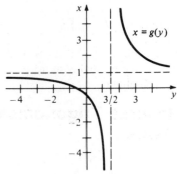

13. Largest interval is $(-\infty, -1)$ or $(1, \infty)$.

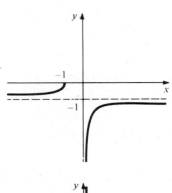

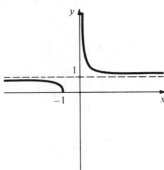

15. (a) $[0, 8/3]$

(b) $2, -1/4$

(c) $[-229/27, 1]$

(d) $1.572, -0.194$

17. Show that 1 is a local minimum point for f.

19. Show that f is increasing everywhere. The domain is $(-\infty, \infty)$.
21. Show that f is decreasing on $[-1, 2]$. The domain is $[-11, 4]$.
23. Differentiate $x^{1/3}$
25. Show that f is increasing on $(0, 2)$. $(f^{-1})'(4) = \frac{1}{5}$.
27. $1/3$
29. $1/3$, $1/3$
31. -1, $-4/3$
33. Different calculators may respond differently.
35. The inverse function gives the cost of y pounds of beans.
37. (a) $(-\infty, -5/2], (-7/3, +\infty)$
    (b) $(5 - 7y^2)/(3y^2 - 2)$
    (c) $-1/[2(3x + 7)^{3/2}\sqrt{2x + 5}]$
39. Use the definitions.
41. (a) $\Delta y/\Delta x$ is close to $f'(x_0)$ if Δx is small.
    (b) Manipulate the inequality in (a).
    (c) Use the definitions.

5.4 The Inverse Trigonometric Functions

1. 0.615
3. -0.903
5. 0
7. $\pi/3$
9. $\pi/6$
11. $8/\sqrt{1 - 64x^2}$
13. $2x\sin^{-1}x + x^2/\sqrt{1 - x^2}$
15. $2(\sin^{-1}x)/\sqrt{1 - x^2}$
17. $\dfrac{(1 - x^2)(10x^4 + 1) + 2x^2(2x^4 + 1)}{(1 - x^2)^2 + x^2(2x^4 + 1)^2}$
19. $[1 + (2/y^3)]/\sqrt{[y - (1/y)]^2[(y - (1/y^2))^2 - 1]}$
21. $[3(x^2 + 2) - (2x)(\sin^{-1}3x)\sqrt{1 - 9x^2}]/[(x^2 + 2)^2$
    $\cdot\sqrt{1 - 9x^2}]$
23. $\dfrac{2t^{11} - t^8 + 12t^7 + t^4 - 2}{(t^5 + 2t)\sqrt{(t^5 + 2t)^2 - (t^7 + t^4 + 1)^2}}$
25. $(\sin^{-1}x + \cos^{-1}x - 1)/(1 - \sin^{-1}x)^2\sqrt{1 - x^2}$
27. $\dfrac{3}{2}(x^2\cos^{-1}x + \tan x)^{1/2}$
    $\times [2x\cos^{-1}x - \dfrac{x^2}{\sqrt{1 - x^2}} + \sec^2x]$
29. $3\tan^{-1}x + x^2/2 + C$
31. $4\sin^{-1}x + C$
33. $3\tan^{-1}(2x)/2 + C$
35. $2\sec^{-1}y + C$
37. Draw a right triangle with sides $1, x, \sqrt{1 - x^2}$.
39. No.
41. $-\theta/(1 - \theta^2)^{3/2}$

43. $4\sin^{-1}(2x)/\sqrt{1 - 4x^2} + 2x$
45. 0
47. Minimum at $x = -1$, maximum at $x = 1$, point of inflection at $x = 0$.
49. Use $(d/dy)[f^{-1}(y)] = 1/[(d/dx)f(x)]$.
51. (a) $[\sqrt{2}, 2], f'(x) = -2x/\sqrt{-x^4 + 6x^2 - 8}$
    (b)

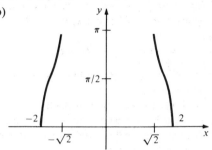

53. (a) $(\sqrt{1 - x^2y^2} - y)/(x - \sqrt{1 - x^2y^2})$
    (b) $\dfrac{[\cos(x + y) - y][1 + t^2]}{[x - \cos(x + y)][1 - t^2]^2}$
    (c) $[x - \cos(x + y)]/[\cos(x + y) - y]\sqrt{1 - t^2}$
    (d) $[\cos(x + y) - y](3t + 2)/[x - \cos(x + y)]$
55. $\tan^{-1}(2) - \pi/4$
57. $(3 - \sqrt{2})/2 + \pi/12 + \sin(\sqrt{2}/2) - \sin(1/2)$
59. $-1/\sqrt{1 - y^2}$

5.5 Graphing and Word Problems

1. -0.088 radians/second.
3. $(2.5)\pi$ meters/second.
5. 2.2 meters/second.
7. 59.2 meters before the sign.
9. 6
11. (a) $r = 1.5$, $\omega = 2\pi/3.1$
    (b) $3\pi/3.1$
13. $3\sin\theta = \sin\varphi$
15. Maxima at $x = (2n + 1)\pi/4$ for n an even integer. Minima at $x = (2n + 1)\pi/4$ for n an odd integer. Points of inflection are at $x = n\pi/2$, n an integer. Concave up on $(n\pi/2, (n + 1)\pi/2)$, n an odd integer. Concave down on $(n\pi/2, (n + 1)\pi/2)$, n an even integer.
17. $f(x)$ concave down in $(2n\pi, (2n + 1)\pi)$ for $x > 0$ and in $(-(2n + 1)\pi, -2n\pi)$ for $x < 0$, n an integer. $f(x)$ is concave up everywhere else.
19.

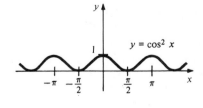

21.

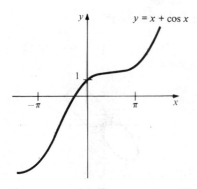

23.

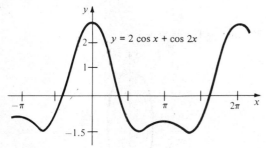

25.

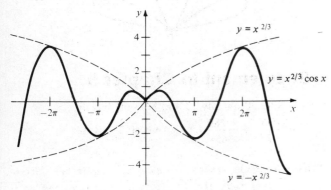

27. The sum of the distances from $(0, 1)$ to $(x, 0)$ and $(p, -q)$ to $(x, 0)$ is minimized when these points lie on a straight line.

29. $y = -3x + \pi, y = x/3 + \pi$

31.

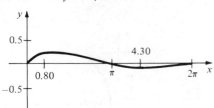

33. The local maxima and minima are about π units apart as $x \to \infty$.

5.6 Graphing in Polar Coordinates

1. $(x - \frac{1}{2})^2 + y^2 = \frac{1}{4}$

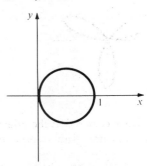

3. $x^2 + y^2 = \sqrt{x^2 + y^2} - y$

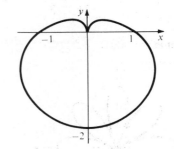

5. $x^2 + y^2 = 9$

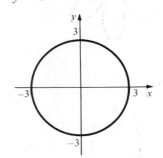

7. $x = -1, y = 0$

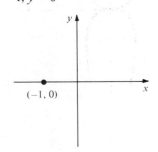

9. $(x^2 + y^2)^2 = 3x^2y - y^3$

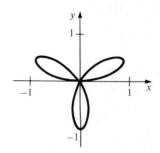

11. A circle of radius r centered at the origin.

13. The graph is symmetric with respect to the line $\theta = \pi/4$.

15. $r = 1$

17. $r^2(1 + \cos\theta\sin\theta) = 1$

19. $r\sin\theta(1 - r^2\cos^2\theta) = 1$

21. $r(\sin\theta - \cos\theta) = 1$

23. $-5\sqrt{3}$

25. 0

27. $13\sqrt{3}/9$

29. $1/3$

31. -0.08

33. max $= 1$, min $= -1$

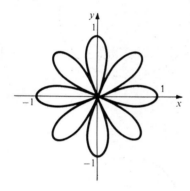

35. max $= \infty$, min $= -\infty$

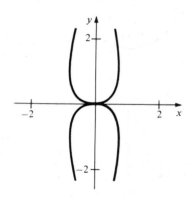

37. max $= 3$, min $= -1$

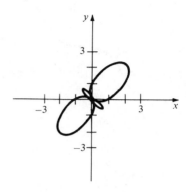

39. max $= 1$, min $= -1$

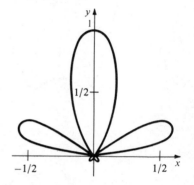

Supplement to Chapter 5

1. 8:05:30 on July 13 and 8:05 on July 14.

3. $\dfrac{d^2S}{dT^2} = \dfrac{-48\pi}{(365)^2} \tan l \sin\alpha \cos\left(\dfrac{2\pi T}{365}\right) \times$

$$\dfrac{(1 - a\cos^2x)(1 - b\cos^2x) - b\sin^2x(1 - a\cos^2x) - 2a\sin^2x(1 - b\cos^2x)}{\left[1 - \sin^2\alpha\cos^2\left(\dfrac{2\pi T}{365}\right)\right]^2\left[1 - \sin^2\alpha\sec^2l\cos^2\left(\dfrac{2\pi T}{365}\right)\right]^{3/2}}$$

5. $T = \dfrac{365}{2\pi}\cos^{-1}\sqrt{\dfrac{0.107}{\tan^2 l + 0.017}}$

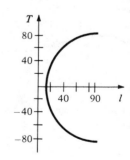

7. 5:07:47 P.M.

9. (a) 0.24 rad/hr.

    (b) 21 minutes, 49 seconds.

Review Exercises for Chapter 5

1. 1.152 radians.

3. $(0, 4)$

5. $\tan\theta \sec\theta = r$

7. $\tan(\theta + \varphi) = \dfrac{\sin(\theta + \varphi)}{\cos(\theta + \varphi)}$; expand and manipulate.

9. 1.392

11. 2.250

13. 0.38 radians.

15. $-6\cos 2x$

17. $1 + \sin 3x + 3x\cos 3x$

19. $2\theta + \csc\theta - \theta\cot\theta\csc\theta$

21. $3y^2 + 2\tan(y^3) + 6y^3\sec^2(y^3)$

23. $[-\sin(x^8 - 7x^4 - 10)](8x^7 - 28x^3)$

25. $2(1 + \cos x)/(x + \sin x)\sqrt{(x + \sin x)^4 - 1}$

27. $-36\theta\cos^2(\theta^2 + 1)\sin(\theta^2 + 1)$

29. $1/2\sqrt{x - x^2}$

31. $\cos\sqrt{x}\,\sec^2(\sin\sqrt{x})/2\sqrt{x}$

33. $(1/2\sqrt{x} - 3\sin 3x)/\sqrt{1 - \left(\sqrt{x} + \cos 3x\right)^2}$

35. $2x + 2\cos(2x + 1),\ 6t^5 + 6t^2 + 6t^2\cos(2t^3 + 3)$

37. $\sin^{-1}(x + 1) + \dfrac{x}{\sqrt{1 - (x + 1)^2}}$,

$$\left[\sin^{-1}(y - y^3 + 1) + \frac{(y - y^3)}{\sqrt{1 - (y - y^3 + 1)^2}}\right](1 - 3y^2)$$

39. $6x^2/(1 + 4x^6),\ 6b(a + bt)^2/[1 + 4(a + bt)^6]$

41. $(-\cos 3x)/3 + C$

43. $\sin 4x + \cos 4x + C$

45. $-\cos x^3 + x^2 + C$

47. $-\cos(u + 1) + C$

49. $(1/2)\tan^{-1}(y/2) + C$

51. 2

53. $\pi/3$

55. (a) Differentiate the right side.

    (b) $\pi/2$

    (c) $x\cos^{-1}x - \sqrt{1 - x^2} + C$

    (d) $x\sin^{-1}3x + \sqrt{1 - 9x^2}/3 + C$

57. (a) $[-1, 1]$

    (b) $[5, 9]$

    (c) $-1/3$

59. Use the inverse function rule.

61. 150/121 meters/second.

63. (a) $0, \pi$

    (b) 0

    (c) 2.2 meters/second.

65. Row to a point $1/\sqrt{15}$ km downstream.

67. 9.798 feet from the sign, $d\theta/dt = 0$.

69. Inflection points at $n\pi/2$, n an integer. Concave down on $[n\pi, (2n + 1)\pi/2]$, n an integer. Concave up elsewhere.

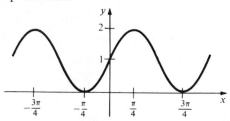

71. Show that $f'(x)$ is always positive. Deduce that $x + 1 \geqslant \cos x$ for $x \geqslant 0$.

73.

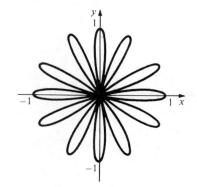

75.

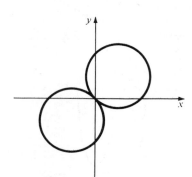

77. $y = -x - \sqrt{2}$

79. (a) On an HP-15c, $\tan^{-1}[\tan(\pi - 10^{-20})]$ and $\tan^{-1}[\tan\pi]$ both give 4.1×10^{-10}

    (b) -10^{-20} is not in the interval.

81. $g'(x) = 1/2f'(f^{-1}(\sqrt{x}))\sqrt{x}$

83. $g''(y) = -f''(x)/[f'(x)]^3$

85. To show $f''(0)$ exists, compute $[f'(\Delta x) - f'(0)]/\Delta x$ and let $\Delta x \to 0$.

Chapter 6 Answers

6.1 Exponential Functions

1. $9,\ 27,\ 27\sqrt{3},\ 3^{x/2}$
3. $1.1664,\ 2.16,\ (1.08)^x$
5. 4
7. $1 + 3^{\sqrt{3}}$
9. $5^{5\pi/2} \cdot 6^{\pi}$
11. $3^{\pi}/2^{\sqrt{3}} - 2^{\sqrt{3}}/3^{\pi}$
13. $3^{\sqrt{2}}$
15. $3^{\sqrt{2}}$

17.

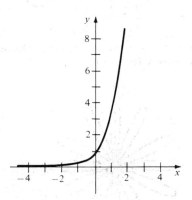

19.

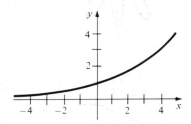

21.

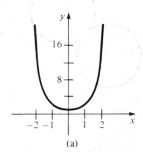

(a)

23.

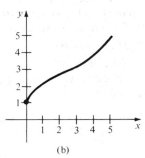

(b)

25. Reflect $y = \exp_3 x$ across the y-axis to get $y = \exp_{1/3} x$.
27. Reflect $y = \exp_3(-x)$ across the y-axis to get $y = \exp_3 x$.
29. (A)(b), (B)(d), (C)(c), (D)(a)
31. (A)(c), (B)(b), (C)(d), (D)(a)
33.

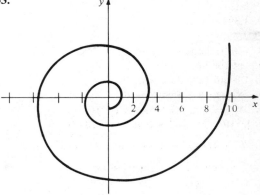

35.

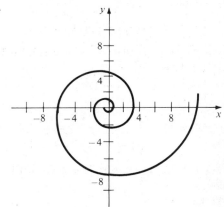

37. The shifting and stretching have the same result since $3^{x+k} = 3^k 3^x$.

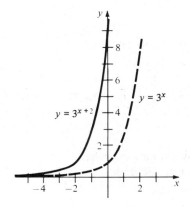

39.

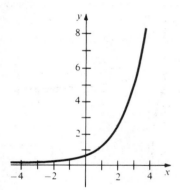

41. 1/9
43. −3
45. No solution

6.2 Logarithms

1. 2   **3.** −2
5. −3   **7.** 1
9. −1
11.

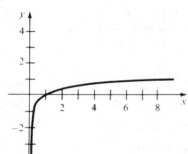

13.

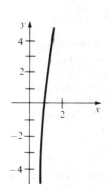

15. (A)(b), (B)(c), (C)(a), (D)(d)
17. 2   **19.** 4
21. b **23.** $(2x − 1) − \log_b 2$
25. 1.036   **27.** 0.618
29. 2.51
31. $b^{y \log_b x} = (b^{\log_b x})^y = x^y$
33. $a > 0, \neq 1$
35. $2\log_b A + \log_b B − \log_b C$
37. $3\log_b A − 2\log_b B + (1/3)\log_b C$

39. 3   **41.** 9
43. 135   **45.** $(\sqrt{5} − 1)/2$
47. Table entries 3.243, 3.486, 3.972; (a) ≈ 16,410;
     (b) $b = 1.75$, $M = 1000$
49. 0.301, 0.602, 0.903, 1, 2, 3
51. (a) 8.25, 6.7
     (b) Use laws of logarithms.
53. 27
55. (a) $x > −5$, $x \neq \pm 1$
     (b) $(−1, 1)$
     (c) $x > −5$, $x \neq \pm 1$
     (a) and (c) are the same
57. $y = 2^x + 1$; domain $(−\infty, \infty)$, range $(1, \infty)$

6.3 Differentiation of the Exponential and Logarithmic Functions

1. 8 times.   **3.** 0.707 times.
5. $x + 3$ **7.** x^3
9. $e^{4x}(4x − 2)$ **11.** $2xe^{x^2+1}$
13. $−2xe^{1−x^2} + 3x^2$ **15.** $(\ln 2)2^x + 1$
17. $3^x \ln 3 − 2^{x−1}\ln 2$ **19.** $1/x$
21. $(1 − \ln x)/x^2$ **23.** $\cot x$

25. $2/(2x + 1)$
27. $\cos x \ln x + (\sin x)/x$
29. $\dfrac{3x(1 + \ln x^2)\sec^2(3x) − 2(\tan 3x)\ln(\tan 3x)}{x \tan 3x(1 + \ln x^2)^2}$
31. $1/(\ln 5)x$
33. $(\sin x)^x[\ln(\sin x) + x \cot x]$
35. $(\sin x)^{\cos x}[\cos^2 x − \sin^2 x \ln(\sin x)]/\sin x$
37. $(x − 2)^{2/3}(4x + 3)^{8/7}[2/(3x − 6) + 32/(28x + 21)]$
39. $x^{(x^x)}[x^x(1 + \ln x)\ln x + x^{x−1}]$
41. $(\sin x + x \cos x)e^{x \sin x}$
43. $(x^6 − 5)/(x + x^7)$
45. $(2x − 8\cos x)\ln 14 \cdot 14^{x^2−8 \sin x}$
47. $(1 + 1/x)/(x + \ln x)$
49. $−\sin(x^{\sin x}) \cdot x^{\sin x}[(\cos x)\ln x + (\sin x)/x]$
51. $(2x \ln x + x)x^{(x^2)}$
53. $−x^{−\tan x^2}[\frac{1}{x} \tan x^2 + 2x(\ln x)\sec^2 x^2]$
55. $4x^3\cos(x^4 + 1) \cdot \log_8(14x − \sin x) +$
     $[(14 − \cos x)/(\ln 8)(14x − \sin x)]\sin(x^4 + 1)$
57. $3(\ln x/2\sqrt{x} + 1/\sqrt{x})x^{(\sqrt{x})}$
59. $\cos x^x \cdot (\ln x + 1)x^x$
61. $(\sin x)^{[(\cos x)^x]}(\cos x)^x[(\ln \cos x − x \tan x)\ln \sin x +$
     $\cot x]$
63. $e^{2x}/2 + C$
65. $\sin x + e^{4x}/4 + C$
67. $s^3/3 + 2 \ln|s| + C$
69. $x^2/4 + (1/2)\ln|x| + C$
71. $x + \ln|x − 1| + C$
73. $3^x/\ln 3 + C$
75. $x^2/2 + 10x + 82 \ln|x − 8| + C$
77. $3e − (8/3)$

79. $65/4 + e^4(e^2 - 1)/2$

81. $1/\ln 2$

83. $\ln(3/2)$

85. (a) $\ln x + 1$, (b) $x \ln x - x + C$

87. (a) Differentiate the right side.
(b) $e^{ax}(b \sin bx + a \cos bx)/(a^2 + b^2) + C$

89. Differentiate the right sides.

91. (a) $f'(x)e^x + f(x)e^x + g'(x)$
(b) $(f'(x) + 2x)e^{f(x) + x^2}$
(c) $f'(x)e^{g(x)} + f(x)g'(x)e^{g(x)}$
(d) $f'(e^x + g(x))(e^x + g'(x))$
(e) $f(x)^{g(x)}(g'(x)\ln f(x) + g(x)f'(x)/f(x))$

93. $e \approx 2.72$

95. (a) Differentiate using the sum and chain rules.
(b) Use the definitions.
(c) Differentiate $e^{x + x_0}$.

97. (a) 0.38629, -0.28345, -0.55765
(b) $2 \ln 2 - 2 - \lim_{\varepsilon \to 0} \varepsilon \ln \varepsilon$; $\lim_{\varepsilon \to 0} \varepsilon \ln \varepsilon = 0$
(c) $\ln x$ has no lower sums on $[0, 2]$.

99. Differentiate $\ln y = \ln f(x) + \ln g(x)$.

101. $\dfrac{dy}{dx} = y \sum_{i=1}^{k} \left[\dfrac{n_i f_i'(x)}{f_i(x)} \right]$.

6.4 Graphing and Word Problems

1.

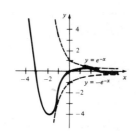

3.

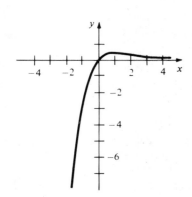

5.

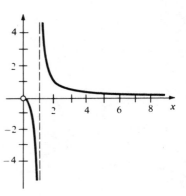

7.

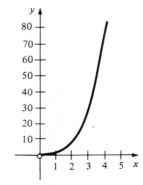

9. 29.6%

11. 0.233%

13. $\lim_{n \to \infty} (1 + \sqrt{2} \ln 3/n)^n$

15. $\lim_{h \to 0} [3(b^h - 1)/h]$

17. Use formula (8).

19. $119.72

21. $y = 3e^2 x - 2e^2$

23. $y = -\sqrt{2}\,\pi x/8 + \sqrt{2}/2$

25. $y = x + \ln 2 - 1$

27. (a) Compute directly using
$f(x) \approx f(0) + f'(0)(x - 0)$.
(b) $2^{0.01} = 1.00695555$
$1 + 0.01 \ln 2 = 1.00693147$
$2^{0.0001} = 1.000069317$
$1.0001 \ln 2 = 1.000069315$
(c) $e \approx \left(1 + \dfrac{1}{n}\right)^n$

29. $e^{-1/e}$

31. (a) Minimum perceivable sensation.
(b) $10/\ln 10$

33. $dp/dt = -0.631$.

35. Differentiate the expression.

37. (a) Show that $g''(x) \leqslant 0$ and that $\lim_{x \to \infty} g'(x) = 0$.

37a. (a) $g'(x) = f\left(1 + \dfrac{1}{x}\right) - f'\left(1 + \dfrac{1}{x}\right) \cdot \dfrac{1}{x}$
$= f\left(1 + \dfrac{1}{x}\right) - f(1) - f'\left(1 + \dfrac{1}{x}\right) \cdot \dfrac{1}{x}$

Apply the mean value theorem to $f(1 + 1/x) - f(1)$.

39. 4

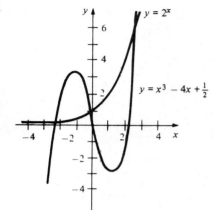

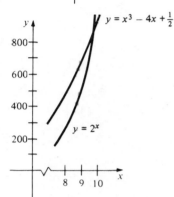

Review Exercises for Chapter 6

1. $x^{2\pi} - x^{-2\pi}$
3. 9
5. 1/2
7. -36
9. $3x^2 e^{x^3}$
11. $e^x(\cos x - \sin x)$
13. $e^{\cos 2x}(-2\sin 2x)$
15. $2xe^{10x}(1 + 5x)$
17. $6e^{6x}$
19. $\dfrac{e^{\cos x}[(-\sin x)(\cos(\sin x)) + (\cos x)(\sin(\sin x))]}{\cos^2(\sin x)}$
21. $\dfrac{[((e^x)\cos(e^x))(e^x + x^2) - (\sin(e^x))(e^x + 2x)]}{(e^x + x^2)^2}$
23. $(-e^x\sin\sqrt{1 + e^x})/2\sqrt{1 + e^x}$
25. $e^{(\cos x) + x}(-\sin x + 1)$
27. $-2xe^{-x^2}(x^2 + 2)/(1 + x^2)^2$
29. $\ln(x + 3) + x/(x + 3)$
31. $-\tan x$
33. $1/x\ln 3$
35. $(-1/\sqrt{1 - (x + e^{-x})^2})(1 - e^{-x})$
37. $(-2\ln t)/t[(\ln t)^2 + 3]^2$
39. $e^{3x}/3 + C$
41. $\sin x + (1/3)\ln|x| + C$
43. $x + \ln|x| + C$
45. $\ln 2 - 2/\pi + 1/2$
47. $3/2 + \sin 1 - \sin 2 + e - e^2$
49. $(\ln x)^x(\ln(\ln x) + 1/\ln x)$

51.
$$\left[\frac{(x + 3)^{7/2}(x + 8)^{5/3}}{(x^2 + 1)^{6/11}}\right]\left[\frac{7}{2(x + 3)} + \frac{5}{3(x + 8)} - \frac{12x}{11(x^2 + 1)}\right]$$

53.

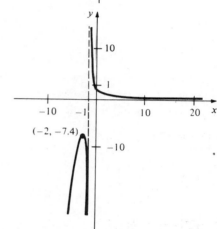

55.

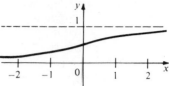

57. $(ye^{xy} - 1)/(1 - xe^{xy})$
59. $-e^{y - x}$
61. $y = 5x + 1$
63. $x\sin 6x$;
$$\int x\sin(6x)\,dx = \sin(6x)/36 - x\cos(6x)/6 + C$$
65. $xe^x/(x + 1)^2$;
$$\int[xe^x/(x + 1)^2]\,dx = e^x/(x + 1) + C$$
67. e^8
69. e^{10}
71. (a) 40% per hour, (b) 40% per hour
73. 7.70%
75. (a) 442 meters.
(b) In the first 10 seconds, the term contributes 72 meters, in the second, 35 meters.
77. (b) $11,804.41
79. $19,876
81. (b) $\lim\limits_{t\to 0} P(t) = a/b$
83. Use the mean value theorem on $(0, x)$.
85. (a) $[1 + (b - 1)]^n = 1 + n(b - 1) + \dfrac{n(n - 1)}{2}(b - 1)^2 + \cdots \geqslant 1 + n(b - 1)$
87. Use Exercise 86.
89. (a) ln is an increasing function.
(b) Use (a) to deduce
$$0 \leqslant \lim_{x\to\infty}(\ln x/x) \leqslant \lim_{n\to\infty}(1/n) = 0.$$
91. Let $z = 1/x$.
93. Let $f(x) = x - \left(\dfrac{2}{\ln 3}\right)\ln x$ and note $f'(x) > 0$ if $x \geqslant \dfrac{2}{\ln 3} \approx 1.82$. It is actually valid if $x \geqslant 5.8452\ldots$.

Index

Undergraduate Texts in Mathematics

Apostol: Introduction to Analytic Number Theory.

Armstrong: Basic Topology.

Bak/Newman: Complex Analysis.

Banchoff/Wermer: Linear Algebra Through Geometry.

Childs: A Concrete Introduction to Higher Algebra.

Chung: Elementary Probability Theory with Stochastic Processes.

Croom: Basic Concepts of Algebraic Topology.

Curtis: Linear Algebra: An Introductory Approach.

Dixmier: General Topology.

Driver: Why Math?

Ebbinghaus/Flum/Thomas Mathematical Logic.

Fischer: Intermediate Real Analysis.

Fleming: Functions of Several Variables. Second edition.

Foulds: Optimization Techniques: An Introduction.

Foulds: Combinatorial Optimization for Undergraduates.

Franklin: Methods of Mathematical Economics. Linear and Nonlinear Programming. Fixed-Point Theorems.

Halmos: Finite-Dimensional Vector Spaces. Second edition.

Halmos: Naive Set Theory.

Iooss/Joseph: Elementary Stability and Bifurcation Theory.

Jänich: Topology.

Kemeny/Snell: Finite Markov Chains.

Lang: Undergraduate Analysis.

Lax/Burstein/Lax: Calculus with Applications and Computing, Volume 1. Corrected Second Printing.

LeCuyer: College Mathematics with A Programming Language.

Lidl/Pilz: Applied Abstract Algebra.

Macki/Strauss: Introduction to Optimal Control Theory.

Malitz: Introduction to Mathematical Logic: Set Theory - Computable Functions - Model Theory.

Marsden/Weinstein: Calculus I, II, III. Second edition.

Martin: The Foundations of Geometry and the Non-Euclidean Plane.

Martin: Transformation Geometry: An Introduction to Symmetry.

Millman/Parker: Geometry: A Metric Approach with Models.

Owen: A First Course in the Mathematical Foundations of Thermodynamics

Undergraduate Texts in Mathematics

continued from ii

29. $\int \operatorname{csch} x \, dx = \ln\left|\tanh\dfrac{x}{2}\right| = -\dfrac{1}{2}\ln\dfrac{\cosh x + 1}{\cosh x - 1}$

30. $\int \sinh^2 x \, dx = \dfrac{1}{4}\sinh 2x - \dfrac{1}{2}x$

31. $\int \cosh^2 x \, dx = \dfrac{1}{4}\sinh 2x + \dfrac{1}{2}x$

32. $\int \operatorname{sech}^2 x \, dx = \tanh x$

33. $\int \sinh^{-1}\dfrac{x}{a} \, dx = x\sinh^{-1}\dfrac{x}{a} - \sqrt{x^2 + a^2} \qquad (a > 0)$

34. $\int \cosh^{-1}\dfrac{x}{a} \, dx = \begin{cases} x\cosh^{-1}\dfrac{x}{a} - \sqrt{x^2 - a^2} & \left[\cosh^{-1}\left(\dfrac{x}{a}\right) > 0, a > 0\right] \\ x\cosh^{-1}\dfrac{x}{a} + \sqrt{x^2 - a^2} & \left[\cosh^{-1}\left(\dfrac{x}{a}\right) < 0, a > 0\right] \end{cases}$

35. $\int \tanh^{-1}\dfrac{x}{a} \, dx = x\tanh^{-1}\dfrac{x}{a} + \dfrac{a}{2}\ln|a^2 - x^2|$

36. $\int \dfrac{1}{\sqrt{a^2 + x^2}} \, dx = \ln(x + \sqrt{a^2 + x^2}\,) = \sinh^{-1}\dfrac{x}{a} \qquad (a > 0)$

37. $\int \dfrac{1}{a^2 + x^2} \, dx + \dfrac{1}{a}\tan^{-1}\dfrac{x}{a} \qquad (a > 0)$

38. $\int \sqrt{a^2 - x^2} \, dx = \dfrac{x}{2}\sqrt{a^2 - x^2} + \dfrac{a^2}{2}\sin^{-1}\dfrac{x}{a} \qquad (a > 0)$

39. $\int (a^2 - x^2)^{3/2} \, dx = \dfrac{x}{8}(5a^2 - 2x^2)\sqrt{a^2 - x^2} + \dfrac{3a^4}{8}\sin^{-1}\dfrac{x}{a} \qquad (a > 0)$

40. $\int \dfrac{1}{\sqrt{a^2 - x^2}} \, dx = \sin^{-1}\dfrac{x}{a} \qquad (a > 0)$

41. $\int \dfrac{1}{a^2 - x^2} \, dx = \dfrac{1}{2a}\ln\left|\dfrac{a + x}{a - x}\right|$

42. $\int \dfrac{1}{(a^2 - x^2)^{3/2}} \, dx = \dfrac{x}{a^2\sqrt{a^2 - x^2}}$

43. $\int \sqrt{x^2 \pm a^2} \, dx = \dfrac{x}{2}\sqrt{x^2 \pm a^2} \pm \dfrac{a^2}{2}\ln\left|x + \sqrt{x^2 \pm a^2}\,\right|$

44. $\int \dfrac{1}{\sqrt{x^2 - a^2}} \, dx = \ln\left|x + \sqrt{x^2 - a^2}\,\right| = \cosh^{-1}\dfrac{x}{a} \qquad (a > 0)$

45. $\int \dfrac{1}{x(a + bx)} \, dx = \dfrac{1}{a}\ln\left|\dfrac{x}{a + bx}\right|$

46. $\int x\sqrt{a + bx} \, dx = \dfrac{2(3bx - 2a)(a + bx)^{3/2}}{15b^2}$

47. $\int \dfrac{\sqrt{a + bx}}{x} \, dx = 2\sqrt{a + bx} + a\int \dfrac{1}{x\sqrt{a + bx}} \, dx$

48. $\int \dfrac{x}{\sqrt{a + bx}} \, dx = \dfrac{2(bx - 2a)\sqrt{a + bx}}{3b^2}$

49. $\int \dfrac{1}{x\sqrt{a + bx}} \, dx = \dfrac{1}{\sqrt{a}}\ln\left|\dfrac{\sqrt{a + bx} - \sqrt{a}}{\sqrt{a + bx} + \sqrt{a}}\right| \qquad (a > 0)$

$\qquad\qquad\qquad = \dfrac{2}{\sqrt{-a}}\tan^{-1}\sqrt{\dfrac{a + bx}{-a}} \qquad (a < 0)$

50. $\int \dfrac{\sqrt{a^2 - x^2}}{x} \, dx = \sqrt{a^2 - x^2} - a\ln\left|\dfrac{a + \sqrt{a^2 - x^2}}{x}\right|$

51. $\int x\sqrt{a^2 - x^2} \, dx = -\dfrac{1}{3}(a^2 - x^2)^{3/2}$

52. $\int x^2\sqrt{a^2 - x^2} \, dx = \dfrac{x}{8}(2x^2 - a^2)\sqrt{a^2 - x^2} + \dfrac{a^4}{8}\sin^{-1}\dfrac{x}{a} \qquad (a > 0)$

Continued on overleaf.

53. $\displaystyle \int \frac{1}{x\sqrt{a^2 - x^2}}\, dx = -\frac{1}{a}\ln\left|\frac{a + \sqrt{a^2 - x^2}}{x}\right|$

54. $\displaystyle \int \frac{x}{\sqrt{a^2 - x^2}}\, dx = -\sqrt{a^2 - x^2}$

55. $\displaystyle \int \frac{x^2}{\sqrt{a^2 - x^2}}\, dx = -\frac{x}{2}\sqrt{a^2 - x^2} + \frac{a^2}{2}\sin^{-1}\frac{x}{a} \qquad (a > 0)$

56. $\displaystyle \int \frac{\sqrt{x^2 + a^2}}{x}\, dx = \sqrt{x^2 + a^2} - a\ln\left|\frac{a + \sqrt{x^2 + a^2}}{x}\right|$

57. $\displaystyle \int \frac{\sqrt{x^2 - a^2}}{x}\, dx = \sqrt{x^2 - a^2} - a\cos^{-1}\frac{a}{|x|}$

$\displaystyle \qquad\qquad = \sqrt{x^2 - a^2} - a\sec^{-1}\left(\frac{x}{a}\right) \qquad (a > 0)$

58. $\displaystyle \int x\sqrt{x^2 \pm a^2}\, dx = \frac{1}{3}(x^2 \pm a^2)^{3/2}$

59. $\displaystyle \int \frac{1}{x\sqrt{x^2 + a^2}}\, dx = \frac{1}{a}\ln\left|\frac{x}{a + \sqrt{x^2 + a^2}}\right|$

60. $\displaystyle \int \frac{1}{x\sqrt{x^2 - a^2}}\, dx = \frac{1}{a}\cos^{-1}\frac{a}{|x|} \qquad (a > 0)$

61. $\displaystyle \int \frac{1}{x^2\sqrt{x^2 \pm a^2}}\, dx = \mp \frac{\sqrt{x^2 \pm a^2}}{a^2 x}$

62. $\displaystyle \int \frac{x}{\sqrt{x^2 \pm a^2}}\, dx = \sqrt{x^2 \pm a^2}$

63. $\displaystyle \int \frac{1}{ax^2 + bx + c}\, dx = \frac{1}{\sqrt{b^2 - 4ac}}\ln\left|\frac{2ax + b - \sqrt{b^2 - 4ac}}{2ax + b + \sqrt{b^2 - 4ac}}\right| \qquad (b^2 > 4ac)$

$\displaystyle \qquad\qquad = \frac{2}{\sqrt{4ac - b^2}}\tan^{-1}\frac{2ax + b}{\sqrt{4ac - b^2}} \qquad (b^2 < 4ac)$

64. $\displaystyle \int \frac{x}{ax^2 + bx + c}\, dx = \frac{1}{2a}\ln|ax^2 + bx + c| - \frac{b}{2a}\int \frac{1}{ax^2 + bx + c}\, dx$

65. $\displaystyle \int \frac{1}{\sqrt{ax^2 + bx + c}}\, dx = \frac{1}{\sqrt{a}}\ln|2ax + b + 2\sqrt{a}\sqrt{ax^2 + bx + c}| \qquad (a > 0)$

$\displaystyle \qquad\qquad = \frac{1}{\sqrt{-a}}\sin^{-1}\frac{-2ax - b}{\sqrt{b^2 - 4ac}} \qquad (a < 0)$

66. $\displaystyle \int \sqrt{ax^2 + bx + c}\, dx = \frac{2ax + b}{4a}\sqrt{ax^2 + bx + c} + \frac{4ac - b^2}{8a}\int \frac{1}{\sqrt{ax^2 + b + c}}\, dx$

67. $\displaystyle \int \frac{x}{\sqrt{ax^2 + bx + c}}\, dx = \frac{\sqrt{ax^2 + bx + c}}{a} - \frac{b}{2a}\int \frac{1}{\sqrt{ax^2 + bx + c}}\, dx$

68. $\displaystyle \int \frac{1}{x\sqrt{ax^2 + bx + c}}\, dx = \frac{-1}{\sqrt{c}}\ln\left|\frac{2\sqrt{c}\sqrt{ax^2 + bx + c} + bx + 2c}{x}\right| \qquad (c > 0)$

$\displaystyle \qquad\qquad = \frac{1}{\sqrt{-c}}\sin^{-1}\frac{bx + 2c}{|x|\sqrt{b^2 - 4ac}} \qquad (c < 0)$

69. $\displaystyle \int x^3\sqrt{x^2 + a^2}\, dx = \left(\frac{1}{5}x^2 - \frac{2}{15}a^2\right)\sqrt{(a^2 + x^2)^3}$

70. $\displaystyle \int \frac{\sqrt{x^2 \pm a^2}}{x^4}\, dx = \mp \frac{\sqrt{(x^2 \pm a^2)^3}}{3a^2 x^3}$

71. $\displaystyle \int \sin ax \sin bx\, dx = \frac{\sin(a - b)x}{2(a - b)} - \frac{\sin(a + b)x}{2(a + b)} \qquad (a^2 \neq b^2)$

Continued on inside back cover.